AF905292

Springer Proceedings in Business and Economics

More information about this series at http://www.springer.com/series/11960

P.K. Kapur • Uday Kumar • Ajit Kumar Verma
Editors

Quality, IT and Business Operations

Modeling and Optimization

Springer

Editors
P.K. Kapur
Amity Center for Interdisciplinary Research
Amity University
Noida, UP, India

Uday Kumar
Division of Operation & Maintenance Engineering
Luleå University of Technology
Luleå, Sweden

Ajit Kumar Verma
Western Norway University of Applied Sciences
Haugesund, Norway

ISSN 2198-7246 ISSN 2198-7254 (electronic)
Springer Proceedings in Business and Economics
ISBN 978-981-10-5576-8 ISBN 978-981-10-5577-5 (eBook)
DOI 10.1007/978-981-10-5577-5

Library of Congress Control Number: 2017951856

© Springer Nature Singapore Pte Ltd. 2018
This work is subject to copyright. All rights are reserved by the Publisher, whether the whole or part of the material is concerned, specifically the rights of translation, reprinting, reuse of illustrations, recitation, broadcasting, reproduction on microfilms or in any other physical way, and transmission or information storage and retrieval, electronic adaptation, computer software, or by similar or dissimilar methodology now known or hereafter developed.
The use of general descriptive names, registered names, trademarks, service marks, etc. in this publication does not imply, even in the absence of a specific statement, that such names are exempt from the relevant protective laws and regulations and therefore free for general use.
The publisher, the authors and the editors are safe to assume that the advice and information in this book are believed to be true and accurate at the date of publication. Neither the publisher nor the authors or the editors give a warranty, express or implied, with respect to the material contained herein or for any errors or omissions that may have been made. The publisher remains neutral with regard to jurisdictional claims in published maps and institutional affiliations.

Printed on acid-free paper

This Springer imprint is published by Springer Nature
The registered company is Springer Nature Singapore Pte Ltd.
The registered company address is: 152 Beach Road, #21-01/04 Gateway East, Singapore 189721, Singapore

Contents

Contributors

Vernika Agarwal Department of Operational Research, University of Delhi, New Delhi, India

Deepti Aggrawal University School of Management and Entrepreneurship, Delhi Technical University, East Delhi Campus, Delhi, India

Ram Bhushan Agnihotri Amity Institute of Information Technology, Amity University, Noida, India

Akash Agrawal Quality Council of India, Delhi, India

Anand Mohan Agrawal GLA University, Mathura, Uttar Pradesh, India

Vivek Agrawal IBM, GLA University, Mathura, Uttar Pradesh, India

Dmitriy Aidarov Russian Presidential Academy of National Economy and Public Administration, Togliatti, Russia

Bashir Alam Department of Computer Engineering, Jamia Millia Islamia, New Delhi, India

Adarsh Anand Department of Operational Research, University of Delhi, Delhi, India

Pooja Chaudhary Department of Computer Engineering, National Institute of Technology Kurukshetra, Kurukshetra, Haryana, India

Prasad Chitimalla Global Consulting Practice, Tata Consultancy Services, Pune, India

Chetna Choudhary Amity School of Engineering & Technology, Amity University, Noida, UP, India

Jyoti Dhingra Darbari Department of Operational Research, University of Delhi, New Delhi, India

Subhrata Das Department of Operational Research, University of Delhi, Delhi, India

M. N. Doja Department of Computer Engineering, Jamia Millia Islamia, New Delhi, India

Nitika Doohan MediCapsIndore, Indore, India

Frolova Elena Department of General Engineering, Orenburg State University, Orenburg, Russian Federation

Elena Eskina Department “Production of Aircraft and Quality Management”, Samara University, Samara, Russia

Diego Galar Luleå University of Technology, Luleå, Sweden

Kanika Gandhi Bhavan’s Usha & Lakshmi Mittal Institute of Management, New Delhi, India

Ashish Garg Department of Electronics & Communication Engineering, Amity School of Engineering & Technology, New Delhi, India

Albina Gazizulina Monitoring Centre for Science and Education, Peter the Great St. Petersburg Polytechnic University, St. Petersburg, Russia

Prerna Gautam Department of Operational Research, Faculty of Mathematical Sciences, University of Delhi, Delhi, India

Amit Gupta Maharaja Agarsen Institute of Technology (MAIT), G.G.S. Indraprastha University, Delhi, India

B. B. Gupta Department of Computer Engineering, National Institute of Technology Kurukshetra, Kurukshetra, Haryana, India

Ritu Gupta AIAS, Amity University, Noida, UP, India

Sanket Gupta ATC, Indore, India

Seema Gupta School of Mathematics, Devi Ahilya University Indore, Indore, India

Shashank Gupta Department of Computer Engineering, National Institute of Technology Kurukshetra, Kurukshetra, Haryana, India

Chandra K. Jaggi Department of Operational Research, Faculty of Mathematical Sciences, University of Delhi, Delhi, India

Anuradha Jain Institute of Basic Science, Khandari, Agra, UP, India

Madhu Jain Department of Mathematics, Indian Institute of Technology Roorkee, Roorkee, Uttarakhand, India

Sumit Jain SCSIT-DAVV, Indore, India

Shailesh Jaloree Department of Applied Mathematics and Computer Science, S.A.T.I., Vidisha, Madhya Pradesh, India

P. C. Jha Departmentof Operational Research, University of Delhi, New Delhi, India

Sachin Kamley Department of Computer Applications, S.A.T.I., Vidisha, Madhya Pradesh, India

Yogita Kansal Amity Institute of Information Technology, Amity University, Noida, India

P. K. Kapur Amity Center for Interdisciplinary Research, Amity University, Noida, UP, India

Aditi Khanna Department of Operational Research, Faculty of Mathematical Sciences, University of Delhi, Delhi, India

Sunil K. Khatri Amity Institute of Information Technology, Amity University, Noida, UP, India

Arjen Kraaijeveld Stord/Haugesund University College, Haugesund, Norway

Deepak Kumar Amity Institute of Information Technology, Amity University, Noida, India

Gaurav Kumar Department of Electronics & Communication Engineering, Amity School of Engineering & Technology, New Delhi, India

Sachin Kumar Department of Computer Science and CIC, University of Delhi, Delhi, India

Uday Kumar Luleå University of Technology, Luleå, Sweden

Vijay Kumar Department of Mathematics, Amity School of Engineering & Technology, New Delhi, India

Mahesh Kuruba Global Consulting Practice, Tata Consultancy Services, Pune, India

Torgrim Log Stord/Haugesund University College, Haugesund, Norway

Nisha Malhotra Division of Computer Engineering, NSIT, University of Delhi, New Delhi, India

Rajesh Mehta Department of Computer Science, Amity School of Engineering and Technology, New Delhi, India

Gaurav Mishra Department of Applied Sciences, Amity University, Greater Noida, India

Prabhanjan Mishra Amity Institute of Information Technology, Amity University, Noida, UP, India

Rekha Mishra University School of Management Studies (USMS), Guru Gobind Singh Indraprastha University, New Delhi, India

Sushama Nagpal Division of Computer Engineering, NSIT, University of Delhi, New Delhi, India

Reena Nupur Department of Applied Mathematics, School of Vocational Studies and Applied Sciences, Gautam Buddha University, Noida, Uttar Pradesh, India

Maria Ostapenko Department "Machines and Tools", Tyumen Industrial University, Tyumen, Russia

Saibal K. Pal Scientist 'G', SAG, DRDO, Delhi, India

Nitin Pandey Amity Institute of Information Technology, Amity University, Noida, India

Saurabh Panwar Department of Operational Research, University of Delhi, Delhi, India

K. N. Rajeshwari School of Mathematics, Devi Ahilya University Indore, Indore, India

Shubhangi Rastogi Ajay Kumar Garg Engineering College, Ghaziabad, India

Sangeeta Sabharwal Division of Computer Engineering, NSIT, University of Delhi, New Delhi, India

Nitin Sachdeva Department of Operational Research, University of Delhi, Delhi, India

Ramita Sahni Department of Applied Mathematics, Amity Institute of Applied Sciences, Amity University, Noida, India

A. K. Saini University School of Management Studies (USMS), Guru Gobind Singh Indraprastha University, New Delhi, India

Keshav Seth Division of Computer Engineering, NSIT, University of Delhi, New Delhi, India

Nand Kishore Sharma ATC, Indore, India

Rashi Sharma Department of Operational Research, University of Delhi, New Delhi, India

Rucche Sharrma SOCIS, IGNOU, Delhi, India

Ruchi Sharma Indira Gandhi Delhi Technical University for Women, Kashmere Gate, New Delhi, India

Vivek Sharma Department of Computer Engineering, Jamia Millia Islamia, New Delhi, India

A. K. Shrivastava Amity Center for Interdisciplinary Research, Amity University, Noida, UP, India

Ompal Singh Department of Operational Research, University of Delhi, Delhi, India

Pratyush Singh Division of Computer Engineering, NSIT, University of Delhi, New Delhi, India

R. K. Singh Indira Gandhi Delhi Technical University for Women, Kashmere Gate, New Delhi, India

Ram Pal Singh Department of Computer Science, DDUC, University of Delhi, Delhi, India

V. B. Singh Delhi College of Arts & Commerce, University of Delhi, Delhi, India

Shakshi Singhal Department of Operational Research, University of Delhi, Delhi, India

Anjana Solanki Department of Applied Mathematics, School of Vocational Studies and Applied Sciences, Gautam Buddha University, Noida, Uttar Pradesh, India

Ajit Srividya Stord/Haugesund University College, Haugesund, Norway

Adithya Thaduri Luleå University of Technology, Luleå, Sweden

R. S. Thakur Department of Computer Applications, M.A.N.I.T., Bhopal, Madhya Pradesh, India

Anshul Tickoo Amity School of Engineering, Amity University, Noida, UP, India

Ajit Kumar Verma Western Norway University of Applied Sciences, Haugesund, Norway

Shekhar Verma Indian Institute of Information Technology, Allahabad, India

Virendra P. Vishwakarma University School of Information and Communication Technology, USICT, Guru Gobind Singh Indraprastha University, New Delhi, India

Said Eid Younes Department of Information Systems, College of Economics, Management and Information Systems, University of Nizwa, Nizwa, Sultanate of Oman

Hira Zaheer Indian Institute of Technology, Roorkee, India

About the Editors

P. K. Kapur
Professor P.K. Kapur is director of the Amity Centre for Interdisciplinary Research, Amity University, Noida, India; former dean of the Faculty of Mathematical Sciences (FMS), New Delhi; and former head of the Department of Operational Research, University of Delhi. He is the author of books such as: *Software Reliability Assessment with OR Applications* (Springer UK, 2011) and *Contributions to Hardware and Software Reliability* (World Scientific, Singapore, 1999). He has executed various research projects funded by the University Grants Commission (UGC) and the Defence Research and Development Organisation (DRDO) in the field of mathematical modeling in marketing and software reliability. He has been the president of SREQOM since 2000 and is a former president of the Operational Research Society of India. He is the editor-in-chief of *IJSAEM* (Springer), has edited five volumes of conference proceedings, and has been the guest editor for special issues of *IJRQSE*, *CDQM*, *OPSEARCH*, etc.

Uday Kumar
Professor Uday Kumar holds a B.Tech. from IIT BHU (1979) and a Ph.D. in mining equipment engineering (reliability and maintenance) from Luleå University of Technology, Sweden (1990). Currently, Dr. Kumar is the chaired professor of operation and maintenance engineering and director of the Strategic Area of Research and Innovation—Sustainable Transport at Luleå University of Technology. His primary research interests are in reliability and maintainability engineering, maintenance modeling, condition monitoring, LCC and risk analysis, etc. Currently he is engaged in many EU Framework Projects as a steering board member, work package leader, or principal investigator. He has published over 250 papers in international journals and made contributions to many edited books. He is one of the chief editors of *IJSAEM*, is area editor (Europe) for the *Journal of Quality in Maintenance Engineering*, and is on the editorial advisory board of many international journals.

Ajit Kumar Verma
Ajit K. Verma is a Professor (Technical Safety) since March 2012 and is working with ATØM, Western Norway University of Applied Sciences, Haugesund, Norway. Prior to this, he was affiliated to the Reliability Engineering, Department of Electrical Engineering at IIT Bombay for around 15 years with a research focus in reliability, risk and safety engineering and soft computing applications in various domains. He is also a Guest Professor at Lulea University of Technology, Sweden, and was an adjunct at the University of Stavanger. He is the Series Editor of the Springer series *Asset Analytics: Performance and Safety Management*, as well as *Reliable and Sustainable Electric Power and Energy Systems Management*, and has jointly edited and authored many books published by Springer. He has over 250 publications in various journals and conferences and has been a supervisor for 37 PhD theses. He is a senior member of IEEE and a life fellow of IETE. He has been the Editor-in-Chief of *OPSEARCH* published by Springer (January 2008–January 2011) as well as the Founder Editor-in-Chief of *International Journal of Systems Assurance Engineering and Management* (IJSAEM) published by Springer and an Editor-in-Chief of *Journal of Life Cycle Reliability and Safety Engineering* (Springer). He is on the editorial board of various international journals. He has served as a Guest Editor of Special Issues of journals including *IEEE Transactions on Reliability*.

A Conceptual Architectural Design for Intelligent Health Information System: Case Study on India

Sachin Kumar, Saibal K. Pal, and Ram Pal Singh

1 Introduction

In India there is a famous saying that 'Health is wealth,' yet the situation in India in terms of health performance is not good. India is seeking to play a global power role in the world because of its demographic dividend and emerging market. But to accomplish that, the health of the people will play a vital role. This can be safeguarded through an operative, competent, and comprehensive intelligent public health system [5].

1.1 State of Indian Health System

India's current health care situation for rural areas brings deeply problematic situations to the surface that are different from other social sectors. People living in rural and remote areas contribute approximately 86% of all medical visits in our country where the majority of people still travel long distances to receive medical help and medical cost 70–80% together from their own hard-earned money lands them into poverty [8, 12]. In urban areas the government has been able to create

S. Kumar (✉)
Department of Computer Science and CIC, University of Delhi, Delhi, India
e-mail: sachin.blessed@gmail.com

S.K. Pal
Scientist 'G', SAG, DRDO, Delhi, India
e-mail: skptech@yahoo.com

R.P. Singh
Department of Computer Science, DDUC, University of Delhi, Delhi, India
e-mail: rprana@gmail.com

© Springer Nature Singapore Pte Ltd. 2018

P.K. Kapur et al. (eds.), *Quality, IT and Business Operations*, Springer Proceedings in Business and Economics, DOI 10.1007/978-981-10-5577-5_1

infrastructure but could not do so effectively in rural or semi-urban areas which sustain 70% of the population. This forces most of the people to access private health facilities which are generally unregistered and unauthorized but provide health care at affordable cost in their villages [8, 12]. Primary public health care systems' ineffectiveness and inefficiency have induced a broken recommendation system which serves as a starting point for people [10]. Proper use of facilities and utilisation of services and products have displayed dependence on residence and educational level. For instance, 70% of the illiterate population do not avail themselves of absolute neutrophil count (ANC) care [8, 12]. If literates are taken into account, 15% of them and 43% of rural women do not avail themselves of ANC services. Whereas in the urban sphere 74% avail themselves of ANC services [10].

1.2 ICT and Big Data Analytics

Information and communication technology (ICT) has changed the landscape in India and the world. It has created solutions to real- life problems in various sectors in the form of products and services which elucidate real-world problems in an effective and efficient manner, and provided infrastructure and other essentials [10, 14]. There has been two-way development: developing technologies for solving problems and then applying them to the real world and seeking their application and suitability. Solutions such as Web portals, devices, network solutions, access devices, data management applications, and analytics solutions can be efficient and effective for challenges due to volume, veracity, variety, and velocity, known as the 4Vs [14]. To extract knowledge insights from data having such features involves advanced techniques which support the capture, storage, dissemination, management, and information analysis with good data visualisation techniques [1, 14]. Big data mean data are big as well as having a high level of complexity and variety. Data obtained in many forms are considered as raw with or without proper formatting, mix of different types of data, with or without filters, and so on [1, 14]. Data need transformation to information by applying some logic or rule in a particular context. In fundamental thinking, big data are associated with a volume of data having quantities on the order of petabytes, exabytes, and zettabytes. In terms of velocity big data become more complex with data streaming in at an unprecedented speed and analysis of the data in the dynamic situation in a timely manner becomes difficult. Variety in big data describes various types of data formats [1, 14]. For instances, data can be structured numeric data as in traditional databases, data from e- commerce and business applications, mismanaged and unstructured text and image documents, sets of email, or sets of video, audio, or financial transactions [1, 14]. In such a mix of data, managing, merging, and governing diversity of datatypes are very complex and difficult and beyond the control of traditional data analysis systems. The veracity concept in big data is the sudden increase in varieties and velocities of data, in other words, inconsistent flow and volume of data with periodic peaks [1, 14]. For instance, if something is trending on social media such as Twitter and Facebook for a time period then it is eligible for veracity of big data. In real-time, daily, periodically seasonal, and event-triggered peak data loads become challenges which test the robustness of systems of data collection and analysis, and the information retrieved from them. Today collected data are generated in distributed locations, from many fonts and still must be connected, matched, cleaned, and then transformed across the system. For analysis and solution of such problems, it is essential to make connections and correlate relationships, hierarchies, and multiple data linkages in a timely and efficient manner and as data can quickly curved out of computational machine or analysing systems control.

1.3 Big Data in Public Health in India

Big data in public and private healthcare systems are a mix of both electronic data coming from urban areas in different formats manually or on standalone automated systems, or from semi-rural and rural areas but the scope of population of these data is very large and with even higher volumes in the future it would be very complex [1]. Traditional software and hardware ICT systems cannot handle or manage such data. They require new technology in collection, processing, and analysis with the capability of collection and processing data in distributed locations. Due to e-governance systems, integration of ICT, modernisation of systems, and processing of information, Indian hospitals are using electronic health records (EHR) and hospital information systems (HIS) for maintaining data and processing for daily tasks [7]. This has made their organisation productive and profitable at the local level or in a decentralised way. But macro-level integration of technology and the concept of data-driven health decision making and policy are not being implemented. The Indian healthcare industry with its current potential is generating zettabytes of data equal to 1021 GB of data per day by taking records for patient care and other details, medicines, diagnostic tests, insurance claims, and data generated with the help of equipment for monitoring vital indications and symptoms [9, 11]. In the future in Indian health care systems, due to the digital Indian mission and push for e-governance systems, digital data growth will be exponential and explosive creating an even more conducive environment for big data and technology to gain insights from data collected in public health systems [11].

2 Theoretical Prospects and Big Data Analytics

In this section, some key elements about big data which are crucial for public health care are discussed.

2.1 Data Collection in Public Health

In Indian circumstances, data collection from decentralisation locations, merging, and processing are challenging due to volume, connectivity, veracity, diversity, and dynamic changes taking place in real-time. The data-collection groups are heterogeneous and make extraction and integration a real challenge. Stakeholders including health care service providers, patients, payers, employers and employees, disease monitoring and management companies, social media, and the patients themselves are part of the data collection process.

2.2 Integrating Data

Many real-life case studies on business and political movements have linked contrasting sources of data to gain knowledge for choices and decision making for the nation, customers, and users. After this forward-thinking analysis and study is applied for transforming existing schemes or to establish a new fashion based on the collected data, computations are performed. Integration of data from unrelated sources and formats which are heterogeneous datasets affords real insights, ways of changing policy direction, and new products and services formation. This understanding of smart linking can possibly lead to improvements in public health care with the help of correct classification of the exact and feasible cure for the patient's diseases or problems.

2.3 Standardisation of Data

This is a real challenge and there is no common set of rules which solves problems of health care data standardisation deficiency. The immense volume of collected data was generated with the help of stakeholders in health care in many formats including text, audio, video, and other multimedia formats. These formats are distributed for several purposes: physician records such as notes and patient records in image format or other diagnostic reports of patient scans, protection claims such as insurance and medical claims, health discussions on social media, and information from patients' wearable and other monitoring devices.

2.4 Generating New Knowledge

After standardisation and data collection from heterogeneous locations and formats are complete, the real challenge lies in obtaining or generating new knowledge from the data. Here new technological improvements and algorithmic approaches are required. With the usual medical and administrative data available, addition of other integrated patient and environmental data most probably would give more correlated and better classification, regression, and clustering with predictions. This would help in targeting interventions to the right set of patients. Such data-driven predictions may assist in identification of areas for progress at both the quality and efficiency levels in public health care [2]. This would assist in areas including the readmissions process, adverse events taking place because of environmental change or some other means, treatment optimisation with cost reduction, and early identification of worsening health problems due to outbreaks. Lack of good analytics, prediction, and lack of governance important for the analysis of health care data puts this sector behind.

Many other domains of society and governance such as merchandising have been utilising the new techniques, for example, graph and data analytics and machine learning for gleaning better knowledge from the same data sources and collection. Things need to be changed and changing. For instance, hospitals in urban areas with data available to them have started customising graph and data analytics for calculating the relation through several complex and dependent variables including patients' lab results, nurses' records, patients' family history, diagnostic details and history, medications, and, for identification purposes, surveys of patients and preventing their exposure to disease or risk of an adverse outcome. Better knowledge management with relationships, data visualisations, and efficient assessment of disparate facts about patients at risk could be helpful with timely intervention. Natural language processing (NLP) and other artificial intelligence methods have also become more mainstream in data analysis, prediction, and automation.

2.5 *Translating Knowledge into Practice*

Some new analytical approaches and collection of standardised data are precarious for big data and data visualisation for gaining insight. Developing usable practical applications will be a key to its success. Traditional mind-sets and cultural rituals pose challenges for data collection and knowledge extraction groups. Stakeholders and users such as physicians, patients, research staff, employees, and employers in hospitals and policy makers need from the start to have an indication of the procedure for converting knowledge into practice. The knowledge gained from big data is able to simplify many different phases of health care. From various case studies [2] within and outside India, there is evidence of improved effectiveness and efficiency of different types of treatments. With the help of diverse delivery models many outcomes have been achieved which can be used for comparison purposes, and for diagnoses and treatment, predictive models have been used. Such data- based approaches may improve understanding regarding the effectiveness of patient analysis and disease analysis with consumer behaviour which would provide a solid basis for the government to design its benefits packages. For example, as fresh insight is added from data for comparing advantages of secondary representatives for diabetes treatment and correlation with human behaviour and symptoms, policy-making experts could utilise the information for developing plans/recommendations or for guiding upcoming random trials.

3 Solution for Health System

Big data give various opportunities in health care systems. This section deals with what big data can offer in terms of process improvisation in public health systems in India.

3.1 Big Data Offering in Health Care Operations

Table 1 discusses the solutions with products and services which big data can offer in the health domain. In this table, activities and subactivities are discussed.

3.2 Motivation for Indian Context

We have examples throughout the world of how big data analytics is benefiting the area of health care stakeholders immensely in better services, satisfaction to people, and revenue models. Some examples of medical data in combination with cost data to deduce a key characteristic dataset are cited by The Institute for Health Technology Transformation of US [6]. When this analytics proceeding was applied it led to the detection of adversarial drug effects in the case of one drug, Vioxx, which was subsequently withdrawn from the market in the United States [6]. In addition to this, one IBM report explained the North York General Hospital examples, 450-bed community hospitals, and one teaching hospital in Canada which use real-time analytics for improving patient outcomes and maintenance focusing operations of health care delivery. Apart from this, scalable real-time analytics has been applied by North York for providing different viewpoints, on the same types of data collected during patients' stays including clinical, administrative, and financial insights [3]. In Toronto, big data analytics is used by The Hospital for Sick Children based on advanced data visualisation and machine-learning components for improving the results of nosocomial infections threatening the life of newborns and insights gained were applied as advanced analytics as vigorous symbolic data collected with the help of easily observing devices for categorising probable signs of contamination. In one study at Apollo Hospital, an infection resistor with the help of data analytics has played a significant role in performing slightly capable outbreak state, together with early acknowledgement of clusters and outbreaks [3]. Dengue is a real threat in different parts of the nation repeatedly affecting numerous residents

Table 1 Medical domain and activities

Activities	Description of activities
Research & development	Prediction and model for lowering erosion and producing a fast, extra-targeted research and development tube for medicines and tools. Devices based on advanced algorithms and statistics for improving the patient admission process and clinical trial design. Analysis of records of patient and clinical trials needs to recognise the consequent signs as well as determine adverse signals before services and products reach the marketplace
Public health	Patterns of Disease and disease itself with the process of tracing sickness eruptions and broadcasting for improving health and reconnaissance need speedy reaction. This depends on the fast and accurate analysis of data. Faster development and production of targeted vaccines and medicines are required, for example, selecting an annual bug strain in public health. Transforming a huge amount of data into usable facts is a must and may help in identification of needs and services to predict and prevent crises
Evidence-based	Analyse variety of unformatted, structured and unstructured data in various forms, financial and operational data for hospitals and patients, medical data and genomic facts and figures for matching the treatment process so that patients at risk for readmission and similar diseases can be recognised and provided extra care
Genomic analytics	this type of activity is execution of gene sequencing with more refined search, efficient cost-effective correlation of algorithms, and development of systems which adopt genomic analysis as one part of regular medicinal care decision-making policy and growing patient health records
Patient profile	Capturing, analysing in real-time the huge amount of responsive data available from hospitals, other medical colleges, devices which can be used at home to deliver safety and management of various resources, with the ability for adverse event prediction can be done by a big data domain. It also helps in advanced analytics for patient profiling for group segmentation, clustering, classification, and predictive modelling. This identifies persons taking advantage of a proactive care system and suggested living style fluctuations, for instance, those patients in danger of a particular disease such as diabetes disorder
Clinical data	Medical data includes doctors' records, notes, prescriptions for patients, data generated by machines, large format images from devices, cine sequences, and scanned documents, mostly not analysed as part of normal text data analysis in traditional systems. Data collected from numerous equipment such as sensors, in-home devices, and tele-health are also included in this part which must be analysed by specific technology for meaning and insights
Web and social media	Data regarding health issues collected from tweets, publications, and posts on social media are also good sources of data analysis. In health care big data analytics leads to better results. With the application of advanced analytics in patients' data to segment and identify individuals or build fresh revenue streams with aggregation and synthesis of patient medical data for providing information and guidance to a third person

along with their families with financial as well as human losses [13]. Many features in the development of the disease can be easily recognised and alerts can be sent proactively to handle an outbreak situation with the help of big data analytics. For maintaining standards and in hope of great revolutions on the Indian subcontinent, ICT was adopted by health care industries to determine diagnostic accuracy and then generate reports by matching the datasets varying from diagnostic reports and images, cultures, and other clinical reports and notes, and diagnostic identifications. The present time requirement is spreading these enterprises by constructing an open digital platform that can be connected through different types of appliances and devices, which helps doctors in feeding data of patients, diseases, and behavior, helping patients and their relatives as well as doctors to stay connected with one another and to perform large-scale analytics. The idea is that to diagnose or prescribe effectively and efficiently the doctor present at any location can access the whole history of a patient. To figure out what actions or decisions should be taken for which patient, nurses and guardians can frequently receive information and they can collect data which can be used in making algorithms to improve diagnoses.

3.3 Challenges and Opportunities in India

Lack of ICT introduction in the governance process and less penetration of the Internet and e-governance systems are problems faced by many governance products and services in India. One of them is health care. Health care data are not consistent and mostly fragmented, generated in traditional IT systems with formats that are not compatible. In India for data analysis this incompatibility is also a challenge. Due to the digital India mission, India is also building up access to information processing in rural and remote areas [4]. An astounding volume of health care data is distributed amongst medical institutions, hospitals, centres providing primary care, scientists and research scholars, health guarantors, and state and central government. The veracity of data would be the next challenge with the many kinds of data received from diverse systems in different formats, semantic observance to keep a format standard, interoperability questions, and homogeneity. Private and public health data both are subjects of complexity for the success of big data which does not imply that private information can be easily available to the public. In the proposed framework one of the major challenges of concern is security during processing to leverage the information for delivering quality care to sufferers. Policies related to privacy, security, intellectual assets, and even liabilities would be mentioned in the big data domain. To optimise the big data concept use, along with putting the right talent and technology at the right place, the organisations also need to structure incentives and workflows. Machine learning is very useful in obtaining knowledge from data and prediction. It can help in transforming India's health care system. The present need is to better manage these data.

In addition to this, there is big data analytics' potential utilisation by the mobile care space. Either with the help of third-party or independently in the health care field, value-added services have been launched by mobile telecom service providers. However, after all of these efforts there has been no real big data winner in health care. In India the opportunity of working with big data is dissimilar and comprises many stakeholders including service providers, insurance related to health, research and development, research scholars and scientists, pharmaceuticals and medical devices, experimental trials, clients and customers, salespersons, and lastly the government.

4 Proposed Intelligent Health Information System

The proposed system and its design are based on big data analytics and advanced analytics policy in the field of health care in India on the basis of its political and executive structures. The key stakeholders are research centres, medical institutes, hospitals, doctors, researchers, surgeons, quality teams, citizens, government officials, and policy-formulating people and agencies. The Indian political system is hierarchical as shown in Fig. 1 with central government and decentralisation of power in states and local administrations. Health is a state subject, therefore involvement, access, and modification by state and local administration are important for the platform for health systems using ICT.

In the development of a model for the intelligent health system, major goals are dependent on the type of information required from the systems. Figure 2 describes

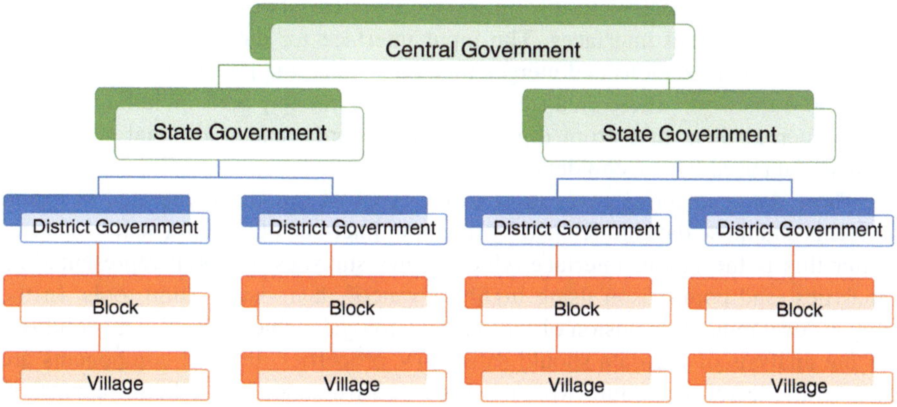

Fig. 1 Political system in India

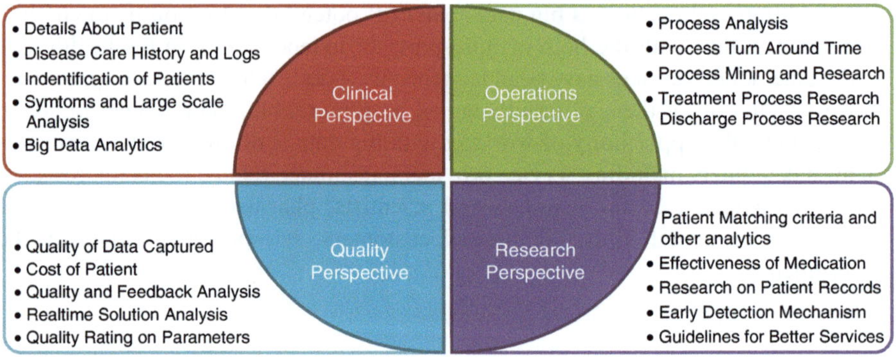

Fig. 2 Domain of activities

the four important perspectives listing certain activities which this system would be accomplishing: clinical, operational, quality, and research for better execution and implementation. In the execution of the platform, the requirement is to give rights of data collection and modification at each stakeholder level and also show some analytics at each level which can be customised. For that reason, one centralised interface, state interface, and local interface with each stakeholder having editing rights is required for the system to work. This is described in Fig. 3. Regional nodes are established as per the requirements in district hospitals and zonewise data insights can be generated with prediction and analytics. Similar things at the central level will also help a possible solution. The three figures mentioned above set the solution in model form. All of these set the stakeholders' role at each place. Figure 4 describes the internal component and structure of the software and Web-integrated solution for an intelligent health information system. Two main components are the input and output interfaces. The input interface for data insertion takes data from distributed locations and merges them, analysing the information through the Hadoop framework. These data can be categorised in many parts including external data, internal data, data from digitisation of records, and traditional standalone system data used by stakeholders.

These data categories cover most of the types of raw data generated in health systems and those data are fetched by the input interface of the information system. After this is the output interface which shows statistics, pictorial representation, analytics, and comparison using highly advanced data visualisation tools. In the major development, inclusion of various technologies which are efficient solutions of the analysis of big data should satisfy 4V properties. Internal components are a machine-learning module which helps in classification of outliers, regression, and prediction with implemented algorithms and clustering to learn the associated reason for the disease and other common attributes in patients and diseases. Hadoop integration helps in processing the large amount of diverse data obtained from

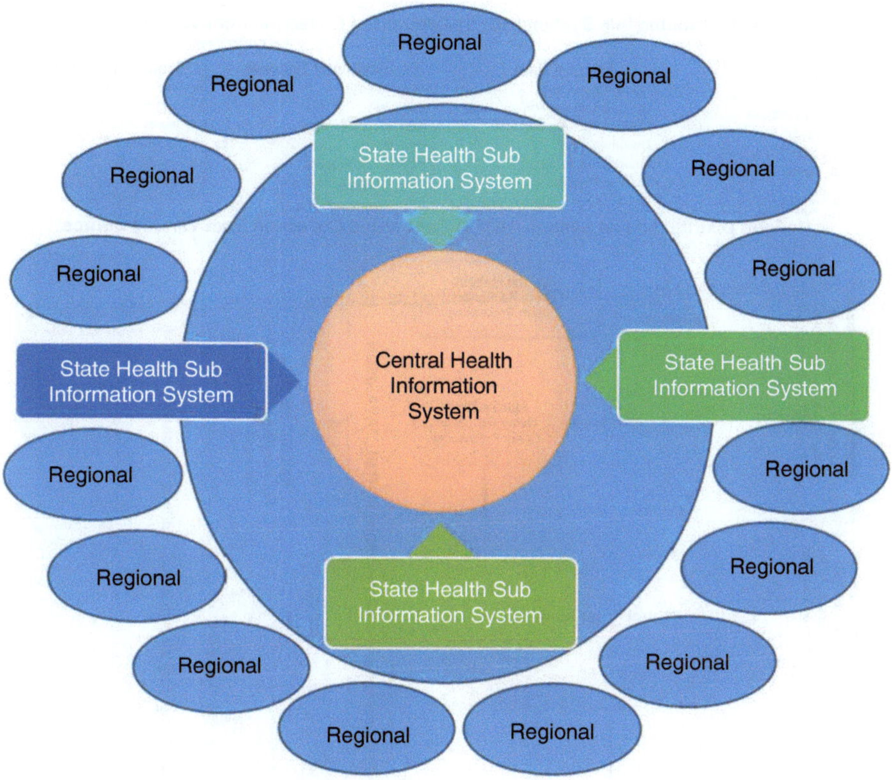

Fig. 3 Set of data collection and visualisation

distributed systems and gaining insight from the data in static and dynamic form. The data visualisation module shows the information in a very easy and simple way with the latest graphics so that ordinary citizens and other research people can understand the direction. It helps in policy feedback and decision making to the executive agency. All this is being implemented on the concept of data-driven decision-making policy and D-governance. The D-governance concept means that the decision is based on the knowledge obtained from the data collected from e-governance systems and other traditional systems dynamically and on a real-time basis.

4.1 Algorithmic Approach

The following are some of the typical algorithms that can be used for knowledge extraction and to gain insights from the platform for health care data. Explanations given below are for reference for some of the use cases:

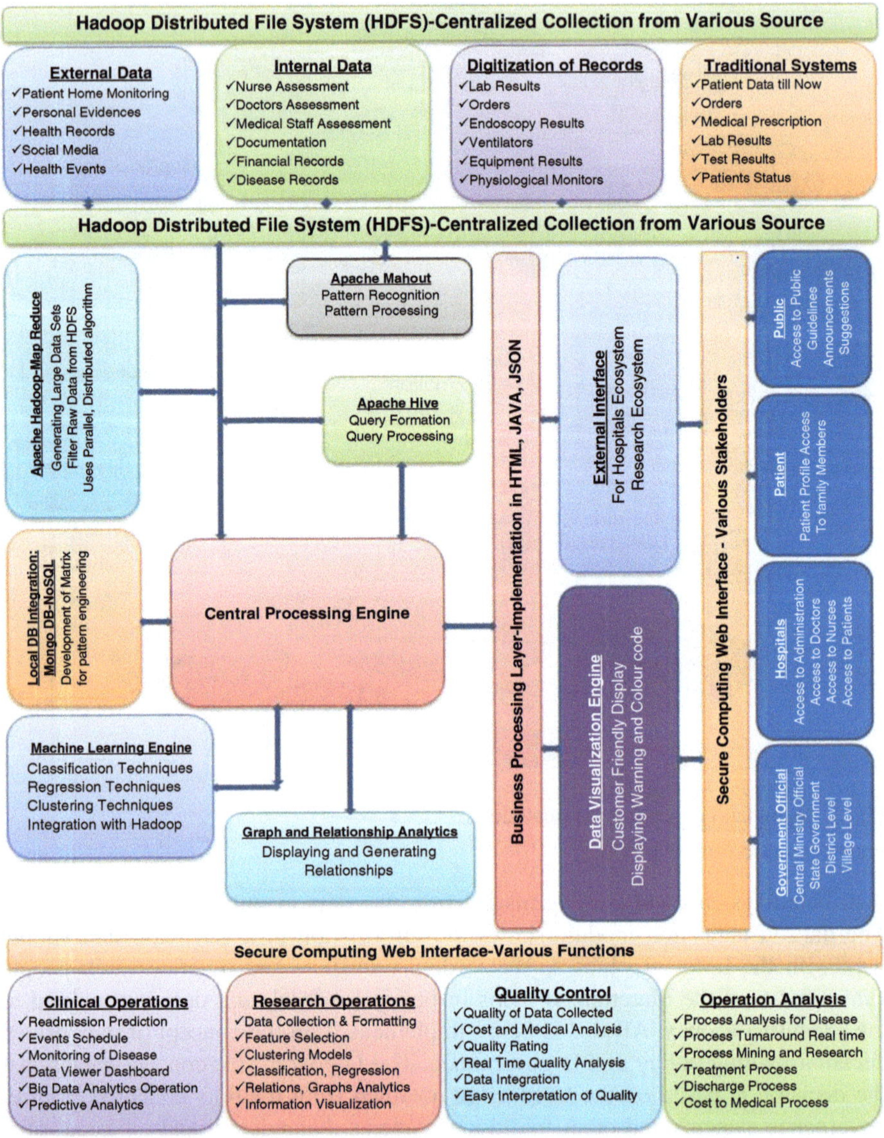

Fig. 4 Conceptual architectural diagram

- Algorithms for reduction of patient readmission rates and visualisation
- Process for chronic diseases and approaches for patient detection
- Using machine-learning and data-mining techniques a patient's condition irregularities in real-time can be detected.

- Analysis and prediction for a patient's length of stay (LOS) for disease in hospitals.
- Prediction and clustering of infection rates, mortality rates, and other trends in disease and area.
- Analysis of tests, their efficiency, their effectiveness in the cure of numerous patient disorders suffering from different diseases.
- Help and assistance to doctors to record quality data from patients with the help of supervised machine learning and its visualisation.

Some of the main functional abilities of the implemented system will be the following.

4.2 *Discussion on Outcome and Functionality*

There are various modules to identify and predict several types of insights from data related to disease and patients. System-generated modules for detection of patients at high risk of some disease and planning of attentional directions for delivering health care to avoid readmissions is part of the framework. Good estimation of readmission guarantees better care for legitimate patients at the hospital/medical research centre with reduced financial cost and proper management of hospital resources. Models based on previous year and seasonwise data of previous patient care help in implementing predictive algorithms and hence help in decision making. Another component of disease management and care would also work on real-time analysis of patient data: its mapping on top of process- driven predictive models for syndromes and patient disorders. Data visualisation with motives of health monitoring is important for the government. In order to prevent mortalities without causing any delay, some relevant interventions are necessary which can be done only after getting processed data in the correct format such as patient temperature, weight, blood pressure, respiration, heart rate, glucose level, and other parameters and the processing step includes receiving, storing, examining, and pushing the data onto real-time metrics and attributes generating a patient scorecard and profile.

4.3 *Technological Assessment of Framework*

The developed architectural concept uses proven leading open source technologies in the area of big data and advanced analytics. This big data architectural system empowers the government to incubate such software developers within a poor expenses barrier during delivering very powerful abilities at great scale. Acceptance of open source tools and techniques also enables the team continuously to refine, embrace, and upgrade several superior upcoming generation products into the policy for delivering the scale, enhancement, and proficiencies for a predictive model and decision taking. Technologies of choice are Apache Hadoop 2.x suite with HDFS, HBase, Map Reduce, Sqoop, Hive, Pig, Oozie, Zookeeper, R Studio for Data Mining Pattern definitions, MongoDB NOSQL, Apache Solr and Apache Lucene, Tableau and D3JS for Advanced Visualizations, HTML5/JSON constructs for UI Rendering, and Neo4J for graph capabilities in the system.

5 Expected Results and Conclusion

This prototype model would show a decrease in patient readmissions for a particular disease or disorder and an increase in the percentage of detection of occurrences of disease areas patientwise. This would also show many successful cases of advance recognition of patient behavior pathways for the disease or disorder of interest. The operational module implemented with the program will reduce patients waiting at the diagnostic centres and hospitals through the health process chain. Modelling in real-time and graph-based relationships with financial and other resources will lead to successful and proactive involvements for management. It basically promotes an idea for getting data through a system and making recommendations to people who require understanding and information from data. Data visualisation encourages an effective customer navigation of a large amount of datasets as a measure of supervised machine-learning implementation. It can be constructed on established open source technologies with a relatively lower cost. The framework should leverage natural language processing-driven innovative text mining and analytics assimilates formless data with partly or fully organised datasets to extract quantifiable business value. Big data analytics and machine learning together with data visualisation technology produce actionable insights that can be used to predict disease outcomes, their patterns, plan treatment protocols, and for strategic organisational planning. By digitising, combining, and effectively using big data, health care organisations and government ranging from single doctors' practices to small and large hospitals to national hospital networks stand to benefit. Data analytics with algorithmic application consenting to necessities and with a high data processing approach can help hospitals in financial planning, supply chain management, human resource management, and quality care delivery which ultimately results in good public health. Reduction in readmission rates, predictive algorithms for diagnostics, and real-time monitoring of ICU vacancies will be some of the practical applications of big data in hospitals and e-governance systems. In the public health domain, big data help to analyse disease patterns to improve public health surveillance and speedy response. Big data analysis with the discussed framework can help public health departments understand disease trends, help control sudden breakouts, and help in building awareness with facts and data. The platform leverages some of the finest open source big data technologies including data mining and machine learning.

References

1. Burghard B (2012) Big data and analytics key to accountable care success. IDC Health Insights
2. Chaulagai CN, Moyo CM, Koot J, Moyo HB, Sambakunsi TC, Khunga FM, Naphini PD (2005) Design and implementation of a health management information system in Malawi: issues, innovations and results. Health Policy Plan 375–384. doi:10.1093/heapol/czi044
3. Connected Health: The drive to integrated healthcare delivery. Accenture. http://www.himss.eu/sites/default/files/Accenture-Connected-Health-Global-Report-Final-Web.pdf
4. Digital India a programme to transform India into digital empowered society and knowledge economy. http://pib.nic.in/newsite/erelease.aspx?relid=108926. Accessed on 26 Jan 2015
5. Fan V, Luthra S (2012) Healthcare in India: a call for innovative reform. The National Bureau of Asian Research for the Senate India Caucus. http://www.nbr.org/downloads/pdfs/Outreach/
6. Greener M (2005) Drug safety on trial. EMBO reports
7. Health Care in India – Vision 2020 Issues and Prospects (2012) Planning Commission. http://www.planningcommission.nic.in/reports
8. Kumar R (2012) Academic institutionalization of community health services: way ahead in medical education reforms. J Family Med Prim Care 1019. doi:10.4103/2249-4863.94442
9. Mozambique through Malaria Incidence Prediction (2013) IIMC International Information Management Corporation. http://www.planningcommission.nic.in/reports/genrep/bkpap2020/
10. Raghupati W KS (2010) Data mining in health care. In: Health informatics: improving efficiency and productivity. Taylor & Francis, USA, pp 211–223
11. Raghupathi W, Raghupathi V (2014) Big data analytics in healthcare: promise and potential. Health Inf Sci Syst 2:1–10
12. Singh S, Badaya S (2014) Health care in rural India: a lack between need and feed. South Asian J Cancer 143144. doi:10.4103/2278-330X.130483
13. Wootton R, Nivritti GP, Scott RE, Telehealth in the developing world. http://www.idrc.ca/EN/Resources/Publications/openebooks/396-6/index.html
14. Yoo C, Ramirez L, Liuzzi J (2014) Big data analysis using modern statistical and machine learning methods in medicine. Int Neurourol J 18(2):50–57

A General Framework for Modeling of Multiple-Version Software with Change-Point

Gaurav Mishra, P. K. Kapur, and A. K. Shrivastava

1 Introduction

Promotion of information technology and software world has changed the pace of life and society and the development of software. Due to this, the software developers face a tough time satisfying and pleasing their customers. The developers are always in a hunt for new ways and paths to customer satisfaction and remarkable market presence. One possible way to this goal is multi-upgradations of the software. Most software engineers have always devoted much time and exertion trying to puzzle out how to design true and authentic software with minimal faults. During the past 30 years, several non-homogeneous Poisson process (NHPP) software reliability growth models (SRGMs) have been presented and embraced for the evaluation and appraisal of the maturation of the reliability of the software [1–6]. Normally, three critical factors, fault sensing or detection, troubleshooting, and operating profile, are responsible for evaluation and appraisal of the reliability of software. In wide-eyed, many SRGMs assume that the troubleshooting is done with a fixed precalculated speed.

In wide-eyed, at the start of the testing phase, it is possible to detect many defects in the visual scanning, whereas the sensing of errors or bugs based on the efficiency of resistance measurement, speed of scanning, defect detection, etc. In the middle of the testing phase, the anomaly detection frequency is normally based on parameters

G. Mishra (✉)
Department of Applied Sciences, Amity University, Greater Noida, India
e-mail: gmishra@gn.amity.edu

P.K. Kapur • A.K. Shrivastava
Amity Center for Interdisciplinary Research, Amity University, Noida, UP, India
e-mail: pkkapur1@gmail.com; kavinash1987@gmail.com

© Springer Nature Singapore Pte Ltd. 2018

P.K. Kapur et al. (eds.), *Quality, IT and Business Operations*, Springer Proceedings in Business and Economics, DOI 10.1007/978-981-10-5577-5_2

such as the speed of execution of the processor instructions, the failure rate, the spreading code, and the coefficient of the normal duration of the CPU calendar day. But in fact, in the period of debugging, the frequency of errors depends largely on the debugger's ability, the environment of debugging, tools and methods, etc. Thus, the speed of bug detection and removal is probably not fixed and unperturbed, called change-point [3, 5]. The deficiency of developers and testers, ability of bug detection and removal can be changed with increasing time. In addition, you can also make new employees so you do not waste resources or waste time. Also, it is always advisable to offer and implement new debugging techniques and tools in advanced software development which are significantly different from those used before [4]. These techniques and tools are used to ensure uninterrupted progress in testing and execution of the software. Thus, in a point of time, adding new techniques and tools can be viewed as a change-point.

In recent years, many SRGMs are projected with the conceptions of change-point in reliability of the software [3, 5]. In software and hardware reliability, modeling and analysis of change-point were innovated by M. Zhao [7]. Various testing effort controls and testing effort functions with change-point and modeling of SRGMs with different severities incorporating single and multiple change-points for software have been introduced by Kapur et al. [8–11]. An SRGM with environmental function in terms of time and change-point is presented by Zhao et al. [12]. Also, the testing time and testing effort factors depend on framework with change-point in the two-dimensional model proposed by Singh et al. [13]. Many models with hazard rate and change-points for reliability analysis are introduced by Inoue et al. [14–18]. Chiu [19] proposed SRGMs with staff improvements and change points. Also, few reliability change-point models have been proposed and used in the recent years, such as the Littlewood CPM, the Jelinski-Moranda de-eutrophication and the Weibull distribution with change-point [1, 3, 5, 20–23].

Conventional software failure curves used for modeling SRGMs cannot reduce the number of errors due to regular updates of the software [24]. However, when the software is in the operation, the company updates the software by deleting errors of previous versions and the introduction of new features to improve performance. These upgrades are at risk that future version may contain new bugs that cause software error. The company cannot ignore the failure of previous versions during the operational phase in the last published version. The companies cannot ignore the defects of the previous version during the operational phase of the latest published version. This can lead to an increase in the error rate during the update process. This failure rate is gradually reduced as the testing team detects and corrects the error before the next version is released. The adding of new software features causes to increase failure rate and also increase the complexity of the software as the software functions improved [25]. The purpose of the company's software has a software update to improve the reliability, which is possible only through consolidation and the reintroduction of some modules, using the best engineering techniques to reduce the intensity of software errors. Various researchers have worked in this direction to model multiple versions of software. Singh et al. [26] have developed a model of detection and correction in the two phases under the influence of two types of imperfect debugging for multiple versions of the software on the bases of time and effort. Kapur et al. [25, 27] presented the growth model of reliability of software based on detection and correction in two phases, with imperfect debugging and testing effort in multiple-version software. They suggested that the exponential distribution is a process of discovery and logistic distribution a correction process. Kapur et al. [24] stretched their work to formulate multiple-version models in the imperfect debugging environment. They also proposed a two-dimensional model for modeling multiple versions of software [28]. In the existing literature, no research has been done for incorporating change-point phenomenon in modeling multiple versions of the software. We have taken the first step to fill this gap by proposing a general framework for multiple versions of software considering single change-point in each version. Furthermore, we have developed a new change-point model by using a generalized modified Weibull (GMW) distribution presented by Carrasco [29]. The comparison criteria result shows that our model is best fitted to the given data set as compared to the existing SRGMs. The rest of the article is organized as follows: Sect. 2 describes the basic notations, model assumptions, and the generalized framework for modeling multiple releases with single change-point. Section 3 describes the process of a GMW distribution with single change-point in each version. In Sect. 4, we validated the formulated model on the given data set and compared it with existing SRGMs. Finally, the article concludes in Sect. 5.

2 General Framework for Multiple Versions of Software with Change-Point

2.1 Basic Notations

$F_{ij}(t)$	Fault distribution function for i^{th} release before and after change-point, $i = 1, 2, 3, 4; j = 1, 2$.
$f_{ij}(t)$	Probability density function (j is representing the change-point).
$m_i(t)$	Expected number of defects erased in time $(0, t]$ for i^{th} release, $i = 1, 2, 3, 4$.
a_i	Initial fault content, $i = 1, 2, 3, 4$.
τ_i	Change-point in successive release, $t_{i-1} < \tau_i < t_i$.
t_i	Release time of ith version.
$\alpha_{ij}, \gamma_{ij}, \lambda_{ij}, \beta_{ij}$	Parameters of the flexible GMW distribution before and after change-point, $i = 1, 2, 3, 4$ and $j = 1, 2$.

2.2 *Basic Assumptions*

The following hypotheses are made for modeling where the number of detectable defects is fixed and finite. The hypotheses are:

(A1) Once there is a software error, the error which led to it will be detected and corrected in real time, and no new errors are introduced in the course correction.
(A2) Error detection and removal rate can be changed at some point of time called change-point.
(A3) Fixing of issue before and after the change-point follows the same distribution with modified parameter values.
(A4) There are very few possibilities of survival of the remaining number of defects of ith version to $(i+2)^{th}$ version.

In practice, a software debugging process is a counting process $\{N(t), t \geq 0\}$, is said to be an NHPP, follows Poisson distribution with mean function $m(t)$, and is formulated as

$$\Pr\{N(t) = r\} = \frac{\{m(t)\}^r e^{-m(t)}}{r!}, r = 0, 1, 2, \ldots\ldots \quad (1)$$

Now we will propose the generalized framework with change-point for multiple versions of the software. For the validation of the same framework, we propose a new change-point model in Sect. 4 with a flexible generalized modified Weibull distribution [19] with four parameters.

2.3 *Modeling Multi-upgradation for Release 1*

Depending upon the above hypothesis,suppose $f(t)$ and $F(t)$ are intensity and cumulative distribution functions of fault and $1 - F(t)$ represent the fraction of undetected defects at time t, and then the defect removal process of SRGM for the very first release with single change-point may be outlined as (Fig. 1)

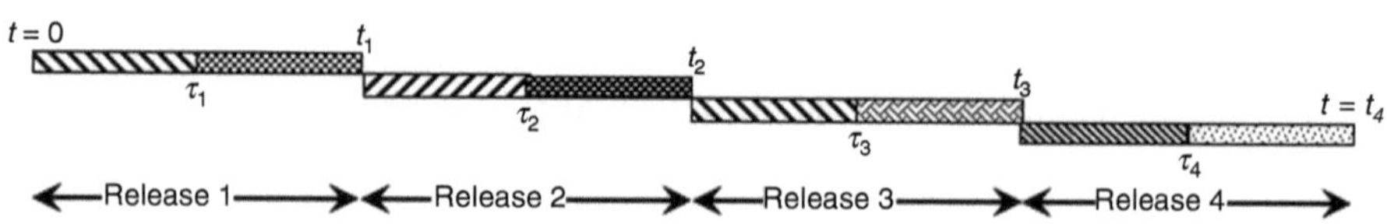

Fig. 1 Release time and change-point distribution on timeline

$$\frac{dm_1(t)}{dt} = \begin{cases} \frac{f_{11}(t)}{1-F_{11}(t)}\{a_1 - m(t)\} & 0 \le t \le \tau_1 \\ \frac{f_{12}(t)}{1-F_{12}(t)}\{a'_1 - m(t)\} & t > \tau_1 \end{cases} \tag{2}$$

where $\frac{f_{1j}(t)}{1-F_{1j}(t)}, j = 1, 2$ is outlined as the instantaneous failure intensity and $a'_1 = a_1 - m_1(\tau_1) = a_1(1 - F_{11}(\tau_1))$ is the remaining defects after the first change-point τ_1. Solving Eq. (2) under the boundary situations $m(t=0) = 0$ and $m(t=\tau) = 0$, we have

$$m_1(t) = \begin{cases} a_1 F_{11}(t) & 0 \le t \le \tau_1 \\ a_1(1 - F_{11}(\tau_1))\left(1 - \frac{(1-F_{12}(t))}{(1-F_{12}(\tau_1))}\right) & \tau_1 < t \le t_1 \end{cases}$$
$$= \begin{cases} a_1 F_{11}(t) & 0 \le t \le \tau_1 \\ a_1(1 - F_{11}(\tau_1))F_{12}(t - \tau_1) & \tau_1 < t \le t_1 \end{cases} \tag{3}$$

So, the aggregate defects corrected in time $[0, t_1]$ are given by

$$m_1(t) = a_1 F_{11}(\tau_1) + a_1(1 - F_{11}(\tau_1))\left(1 - \frac{(1 - F_{12}(t))}{(1 - F_{12}(\tau_1))}\right)$$
$$= a_1\left(1 - \frac{\left(1 - F_{11}(\tau_1)(1 - F_{12}(t))\right)}{(1 - F_{12}(\tau_1))}\right) \tag{4}$$

2.4 *Modeling of Successive Releases (i = 2, 3, 4)*

After the initial release, the remaining number of failures, that is, $a_1(1 - F_{11}(\tau_1))(1 - F_{12}(t_1 - \tau_1))$, which cannot be discovered in time $[0, t_1]$, is moved to the immediate version. If $F_{ij}(t)$ is representing the defect distribution function before and after $(j = 1, 2)$ the change-point at τ_i in i^{th} release, where $t_{i-1} < \tau_i < t_i$, then the failure removal process for upcoming versions is defined as

$$\frac{dm_i(t)}{dt} = \begin{cases} \frac{f_{i1}(t)}{1-F_{i1}(t)}\left(a'_i - m_i(t)\right) & t_{i-1} \le t \le \tau_i \\ \frac{f_{i2}(t)}{1-F_{i2}(t)}\left(a''_i - m_i(t)\right) & \tau_i < t \le t_i \end{cases} \tag{5}$$

where $a'_i = a_i + a_{i-1}\left(1 - F_{(i-1)1}(\tau_{i-1} - t_{i-2})\right)\left(1 - F_{(i-1)2}(t_{i-1} - \tau_{i-1})\right)$ is the aggregate number of defects of the present version and previous version only and $a''_i = \left[a_i + a_{i-1}\left(1 - F_{(i-1)1}(\tau_{i-1} - t_{i-2})\right)\left(1 - F_{(i-1)2}(t_{i-1} - \tau_{i-1})\right)\right](1 - F_{i1}(\tau_i - t_{i-1}))$is the remaining number of faults after τ_i in the present version.

On solving (5) with boundary conditions $m(t = t_i) = 0$and $m(t = \tau_i) = 0$, we get

$$m_i(t)=\begin{cases} \left[a_i+a_{i-1}\left(1-F_{(i-1)1}\left(\tau_{i-1}-t_{i-2}\right)\right)\left(1-F_{(i-1)2}\left(t_{i-1}-\tau_{i-1}\right)\right)\right] \\ F_{i1}\left(t-t_{i-1}\right) & t_{i-1}\leq t\leq\tau_i \\ \left[a_i+a_{i-1}\left(1-F_{(i-1)1}\left(\tau_{i-1}-t_{i-2}\right)\right)\left(1-F_{(i-1)2}\left(t_{i-1}-\tau_{i-1}\right)\right)\right] \\ \left(1-F_{i1}\left(\tau_i-t_{i-1}\right)\right)F_{i2}\left(t-\tau_i\right) & \tau_i<t\leq t_i \end{cases} \tag{6}$$

Hence, aggregate defects corrected in $[t_{i-1}, t_i]$ are

$$m_i(t) = a_i' F_{i1}\left(\tau_i - t_{i-1}\right) + a_i'' F_{i2}\left(t_i - \tau_2\right) \tag{7}$$

3 Derivation of SRGM with GMW

The bathtub-shaped failure curves are very utile in survival analysis of software. The four-parameter flexible GMW is capable of modeling bathtub-shaped distributions. Also it is capable to model monotone and non-monotone failure curves, which is very common in reliability problems.

The density function of flexible $GMW(\alpha, \gamma, \lambda, \beta)$ distribution is given by

$$f(t) = \frac{\alpha\beta t^{\gamma-1}\left(\gamma+\lambda t\right)\exp\left\{\lambda t-\alpha t^{\gamma}\exp\left(\lambda t\right)\right\}}{\left[1-\exp\left\{-\alpha t^{\gamma}\exp\left(\lambda t\right)\right\}\right]^{1-\beta}}, t > 0 \tag{8}$$

and cumulative distribution function (cdf) is given by

$$F(t) = \left[1-\exp\left\{-\alpha t^{\gamma}\exp\left(\lambda t\right)\right\}\right]^{\beta} \tag{9}$$

where $\alpha > 0, \beta > 0, \gamma \geq 0$ and $\lambda \geq 0$.

Here, α controls the scale of distributions, and γ, βcontrols its shape. The parameter λ is a kind of accelerating factor in the imperfection time. The GMW distribution has flexibility with various forms of known distributions. It is a generalization for many existing SRGMs. Some of them are listed below.

All the described SRGMs in Table 1 reflect the adoptability of GMW distribution for modeling of multiple versions of the software product. Now we will find the mean value function for each version using GMW distribution.

Table 1 Different SRGMs

Distribution	Parameters	Expression $F(t)$
Weibull distribution	$\lambda=0, \beta=1$	$(1-\exp(-\alpha t^{\gamma}))$
Exponentiated Weibull distribution [6, 8]	$\lambda=0$	$(1-\exp(-\alpha t^{\gamma}))^{\beta}$
Modified Weibull distribution [2]	$\beta=1$	$(1-\exp\{-\alpha t^{\gamma}\exp(\lambda t)\})$
Exponential distribution	$\lambda=0, \beta=1, \gamma=1$	$(1-\exp(-\alpha t))$
Generalized exponential distribution [28]	$\lambda=0, \gamma=1$	$(1-\exp(-\alpha t))^{\beta}$
Rayleigh distribution	$\lambda=0, \beta=1, \gamma=2$	$(1-\exp(-\alpha t^{2}))$
Log-gamma distribution	$\beta=1, \gamma=0$	$(1-\exp\{-\alpha \exp(\lambda t)\})$
Generalized Rayleigh distribution [5]	$\lambda=0, \gamma=2$	$(1-\exp(-\alpha t^{2}))^{\beta}$

3.1 For Release 1

Using Eqs. (8) and (9) in (3), the mean value function of the first version can be obtained as follows:

$$m_1(t) = \begin{cases} a_1\left[1-e^{\{-\alpha_{11}t^{\gamma_{11}}\exp(\lambda_{11}t)\}}\right]^{\beta_{11}} & 0 \le t \le \tau_1 \\ a_1\left\{1-\left[1-e^{\{-\alpha_{11}\tau_1{}^{\gamma_{11}}\exp(\lambda_{11}\tau_1)\}}\right]^{\beta_{11}}\right\} \\ \left(1-\dfrac{\left\{1-\left[1-e^{\{-\alpha_{12}t^{\gamma_{12}}\exp(\lambda_{12}t)\}}\right]^{\beta_{12}}\right\}}{\left\{1-\left[1-e^{\{-\alpha_{12}\tau_1{}^{\gamma_{12}}\exp(\lambda_{12}\tau_1)\}}\right]^{\beta_{12}}\right\}}\right) & \tau_1 < t \le t_1 \end{cases} \quad (10)$$

So, the aggregate defects corrected in time $[0, t_1]$ are

$$m_1(t)=a_1\left(1-\frac{\left\{1-\left[1-e^{\{-\alpha_{11}\tau_1{}^{\gamma_{11}}\exp(\lambda_{11}\tau_1)\}}\right]^{\beta_{11}}\right\}\left\{1-\left[1-e^{\{-\alpha_{12}t^{\gamma_{12}}\exp(\lambda_{12}t)\}}\right]^{\beta_{12}}\right\}}{\left\{1-\left[1-e^{\{-\alpha_{12}\tau_1{}^{\gamma_{12}}\exp(\lambda_{12}\tau_1)\}}\right]^{\beta_{12}}\right\}}\right) \quad (11)$$

3.2 For Release 2, 3, and 4 (i = 2, i = 3 and i = 4)

$$m_i(t) = \begin{cases} a'_i\left[1-e^{\{-\alpha_{i1}t^{\gamma_{i1}}\exp(\lambda_{i1}t)\}}\right]^{\beta_{i1}} & t_{i-1} \le t \le \tau_i \\ a'_i\left\{1-\left[1-e^{\{-\alpha_{i1}\tau_i{}^{\gamma_{i1}}\exp(\lambda_{i1}\tau_i)\}}\right]^{\beta_{i1}}\right\} \\ \left(1-\dfrac{\left\{1-\left[1-e^{\{-\alpha_{i2}t^{\gamma_{i2}}\exp(\lambda_{i2}t)\}}\right]^{\beta_{i2}}\right\}}{\left\{1-\left[1-e^{\{-\alpha_{i2}\tau_i{}^{\gamma_{i2}}\exp(\lambda_{i2}\tau_i)\}}\right]^{\beta_{i2}}\right\}}\right) & \tau_i < t \le t_i \end{cases} \quad (12)$$

where $a_i' = a_i + a_{i-1}\left(1 - \left[1 - e^{\left\{-\alpha_{(i-1)1}(\tau_{i-1}-t_{i-2})^{\gamma_{(i-1)1}}\exp\left(\lambda_{(i-1)1}(\tau_{i-1}-t_{i-2})\right)\right\}}\right]^{\beta_{(i-1)1}}\right)$

$$\left(1 - \left[1 - e^{\left\{-\alpha_{(i-1)2}(t_{i-1}-\tau_{i-1})^{\gamma_{(i-1)2}}\exp\left(\lambda_{(i-1)2}(t_{i-1}-\tau_{i-1})\right)\right\}}\right]^{\beta_{(i-1)2}}\right).$$

So, the aggregate defects corrected in time $[t_{i-1}, t_i]$ are

$$m_i(t) = a_i'\left(1 - \frac{\left\{1 - \left[1 - e^{\{-\alpha_{i1}\tau_i^{\gamma_{i1}}\exp(\lambda_{i1}\tau_i)\}}\right]^{\beta_{i1}}\right\}\left\{1 - \left[1 - e^{\{-\alpha_{i2}t^{\gamma_{i2}}\exp(\lambda_{i2}t)\}}\right]^{\beta_{i2}}\right\}}{\left\{1 - \left[1 - e^{\{-\alpha_{i2}\tau_i^{\gamma_{i2}}\exp(\lambda_{i2}\tau_i)\}}\right]^{\beta_{i2}}\right\}}\right) \tag{13}$$

4 Numerical Illustration

The data set used for analysis and validation is from the four versions of the software product. The initial version is counted 100 bugs in 20 weeks, 120 bugs in 19 weeks, 61 bugs in 12 weeks and 42 bugs in 19 weeks of the testing period [1]. The eighth week, the sixth week, the third week, and the fifth week are marked as a change-point in the successive release of the software.

The model proposed in this work is not linear and provides additional problems in the estimation of the parameters. A software package such as SPSS, which was used here to estimate the parameters, helps to overcome this problem. The parameter estimation is done by SPSS and comparison results for all the releases given in Tables 2, 3 and 4, respectively. The goodness of fit curves is given in Figs. 2, 3, 4, and 5.

Table 2 Estimated parameters of the proposed model

Release (i)	a_i	α_1	γ_1	λ_1
1	100.587	0.002	1.342	0.230
2	119.999	0.004	1.024	0.282
3	61.000	0.049	0.038	0.448
4	39.978	0.037	1.274	0.002

Table 3 Estimated parameters of the proposed model

β_1	α_2	γ_2	λ_2	β_2
0.310	0.048	0.874	0.107	0.001
0.438	0.074	0.908	0.070	0.090
0.899	0.189	1.394	0.001	5.646
0.994	0.065	0.815	0.067	0.003

Table 4 Comparison results of the model

Release (i)	MSE	Bias	Variance	RMSPE	TS	R^2
1	0.882	−0.006	0.964	0.964	1.251%	0.999
2	1.467	−0.104	1.258	1.262	1.411%	0.999
3	0.296	0.117	0.741	0.750	1.260%	0.999
4	0.669	0.014	0.840	0.841	2.806%	0.996

Fig. 2 Goodness of fit curve for release 1

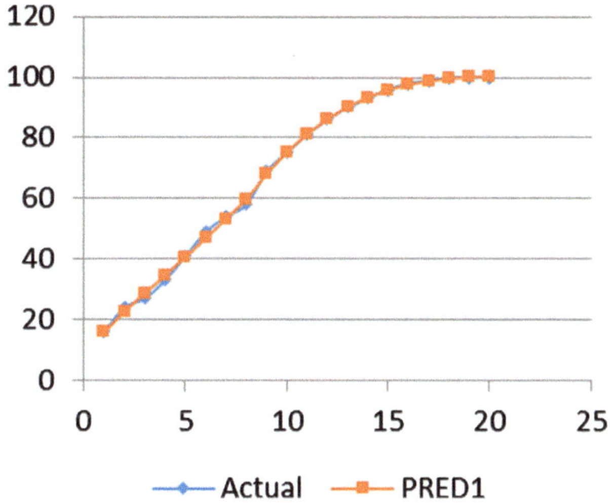

Fig. 3 Goodness of fit curve for release 2

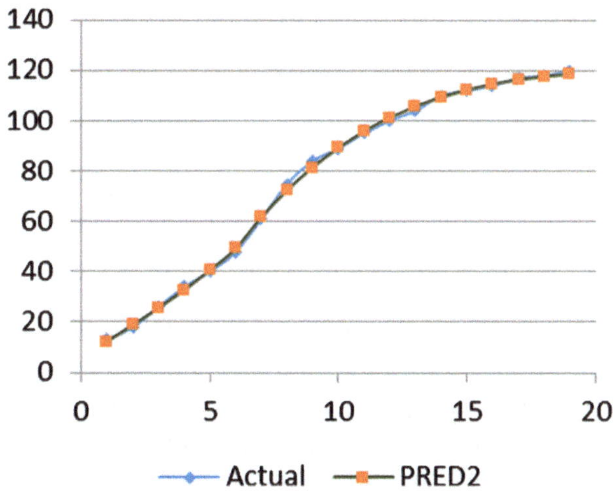

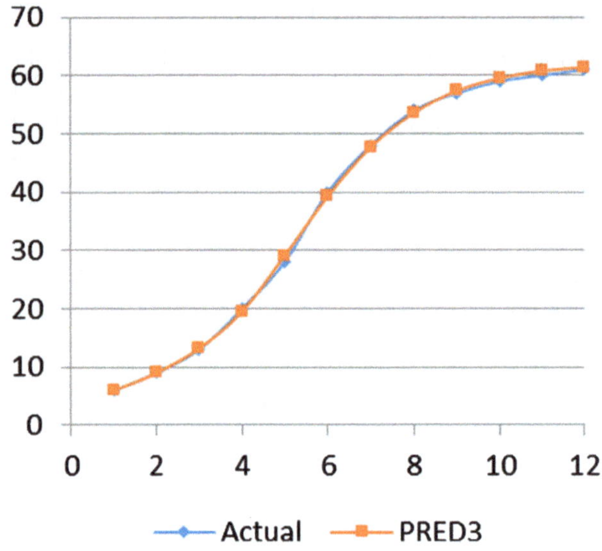

Fig. 4 Goodness of fit curve for release 3

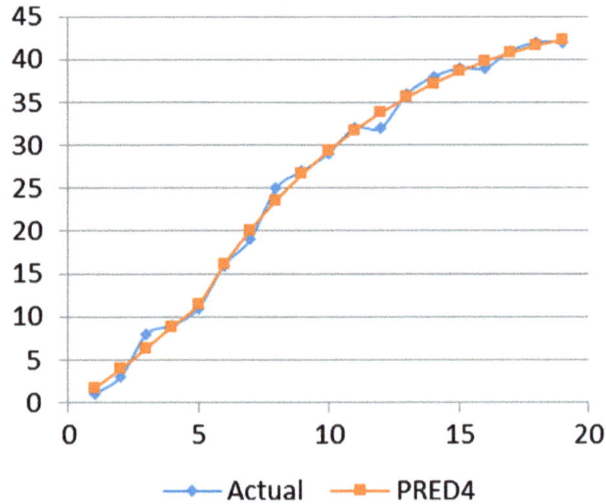

Fig. 5 Goodness of fit curve for release 4

4.1 Performance Comparison Criteria

The data we are using for analysis is for failure counts in the successive releases, so we are using the following criteria to compare various models based on their prognostic and retrodictive power:

1. The mean square of fitting error (MSE) [3, 5]
2. The prediction bias [3, 5]
3. The variance [3, 5]
4. The root mean square prediction error (RMSPE) [3, 5]

Table 5 Distribution function of SRGMs [7, 19]

$i=1,2,3,4$	$F_{i1}(t), t<\tau$	$F_{i2}(t), t>\tau$
SRGM 1 (exponential)	$\left(1-e^{-b_{i1}t}\right)$	$\left(1-e^{-b_{i2}t}\right)$
SRGM 2 (delayed S-shaped)	$\left(1-(1+b_{i1}t)\,e^{-b_{i1}t}\right)$	$\left(1-(1+b_{i2}t)\,e^{-b_{i2}t}\right)$
SRGM 3 (logistic)	$\left(1-\frac{(1+\beta_{i1})}{(1+\beta_{i1}e^{-b_{i1}t})}e^{-b_{i1}t}\right)$	$\left(1-\frac{(1+\beta_{i2})}{(1+\beta_{i2}e^{-b_{i2}t})}e^{-b_{i2}t}\right)$
SRGM 4 (Weibull)	$\left(1-e^{-b_{i1}t^k}\right)$	$\left(1-e^{-b_{i2}t^k}\right)$
SRGM 5 (Rayleigh)	$\left(1-e^{-b_{i1}t^2}\right)$	$\left(1-e^{-b_{i2}t^2}\right)$
SRGM 6 (normal) (μ, σ)—mean and standard deviation	$\Phi(t;\mu_{i1},\sigma_{i1})$	$\Phi(t;\mu_{i2},\sigma_{i2})$
SRGM 7 (gamma) (α, β)—shape and scale parameters	$\Gamma(t;\alpha_{i1},\beta_{i1})$	$\Gamma(t;\alpha_{i2},\beta_{i2})$
SRGM 8 (modified Weibull)	$\left(1-e^{-\alpha_{i1}t-\beta_{i1}t^{k_{i1}}}\right)$	$\left(1-e^{-\alpha_{i2}t-\beta_{i2}t^{k_{i2}}}\right)$

Table 6 Estimated parameters of SRGM 1 (exponential)

Release (i)	a_i	b_{i1}	b_{i2}
1	103.599	0.105	0.240
2	123.128	0.080	0.194
3	66.392	0.057	0.210
4	38.877	0.052	0.142

Table 7 Estimated parameters of SRGM 2 (delayed S-shaped)

Release (i)	a_i	b_{i1}	b_{i2}
1	102.445	0.267	0.293
2	119.461	0.241	0.265
3	58.422	0.256	0.366
4	37.949	0.202	0.229

Table 8 Estimated parameters of SRGM 3 (logistic)

Release (i)	a_i	b_{i1}	b_{i2}	β_{i1}	β_{i2}
1	102.013	0.108	0.321	0.002	6.200
2	118.341	0.081	0.219	0.003	0.565
3	54.720	0.088	0.729	0.098	49.905
4	41.856	0.198	0.228	3.933	3.322

5. The Theil statistic [3, 5]
6. The coefficient of multiple determination (R^2)

We have compared our proposed model with the well-known (eight) existing SRGMs given in Table 5. The parameter estimation and comparison criteria results of the SRGMs are given below in Tables 6, 7, 8, 9, 10, 11, 12 and 13 and 14, 15, 16, 17, 18, 19, 20, and 21, respectively. Looking at the comparison criteria results, we see that our proposed model is best fitted on the given data set. Some of the distributions which give nearby comparison criteria results are logistic, normal, gamma, and modified Weibull.

Table 9 Estimated parameters of SRGM 4 (Weibull)

Release (i)	a_i	b_{i1}	b_{i2}	k
1	103.806	0.128	0.372	0.879
2	123.355	0.086	0.230	0.945
3	55.621	0.022	0.020	2.171
4	44.887	0.039	0.071	1.255

Table 10 Estimated parameters of SRGM 5 (Rayleigh)

Release (i)	a_i	b_{i1}	b_{i2}
1	100.832	0.017	0.010
2	118.777	0.017	0.011
3	60.345	0.025	0.028
4	41.446	0.015	0.011

Table 11 Estimated parameters of SRGM 6 (normal)

Release (i)	a_i	μ_{i1}	σ_{i1}	μ_{i2}	σ_{i2}
1	100.938	6.527	5.919	5.888	5.337
2	120.256	7.097	4.878	2.123	7.894
3	54.331	6.230	4.071	5.377	2.169
4	41.563	7.319	3.931	2.168	8.042

Table 12 Estimated parameters of SRGM 7 (gamma)

Release (i)	a_i	α_{i1}	β_{i1}	α_{i2}	β_{i2}
1	103.364	0.832	0.078	0.917	0.246
2	120.442	0.958	0.075	1.537	0.248
3	61.289	0.766	0.041	7.627	1.297
4	47.095	1.355	0.099	1.142	0.150

Table 13 Estimated parameters of SRGM 8 (modified Weibull)

Release (i)	a	α_{i1}	β_{i1}	α_{i2}	β_{i2}	k_{i1}	k_{i2}
1	103.630	0.093	0.118	0.056	0.188	0.001	0.890
2	129.026	0.070	0.081	0.019	2.115	0.001	0.224
3	48.978	0.063	0.074	0.036	0.002	0.006	3.229
4	49.136	0.050	0.104	0.000	1.055	0.443	0.148

Table 14 Comparison results of SRGM 1 (exponential)

Release (i)	MSE	Bias	Variance	RMSPE	TS	R^2
1	3.345	0.291	1.946	1.968	2.436%	0.996
2	1.601	0.095	1.311	1.315	1.474%	0.999
3	4.796	0.134	2.339	2.343	5.075%	0.989
4	0.781	0.084	0.920	0.924	3.033	0.996

Table 15 Comparison results of SRGM 2 (Yamada)

Release (i)	MSE	Bias	Variance	RMSPE	TS	R^2
1	25.036	−1.451	5.745	5.925	6.665%	0.969
2	12.607	−0.963	4.030	4.143	4.136%	0.990
3	5.287	−0.376	2.532	2.560	5.329%	0.988
4	0.846	−0.074	0.954	0.957	3.157%	0.995

Table 16 Comparison results of SRGM 3 (K-G logistic)

Release (i)	MSE	Bias	Variance	RMSPE	TS	R^2
1	3.426	−0.272	1.910	1.930	2.400%	0.996
2	1.793	−0.087	1.385	1.387	1.560%	0.999
3	0.324	−0.005	0.731	0.731	1.318%	0.999
4	0.774	0.039	0.907	0.907	3.020%	0.996

Table 17 Comparison results of SRGM 4 (Weibull)

Release (i)	MSE	Bias	Variance	RMSPE	TS	R^2
1	2.463	0.078	1.616	1.618	2.090%	0.997
2	1.506	0.031	1.262	1.262	1.429%	0.999
3	3.898	0.491	2.285	2.337	4.575%	0.991
4	0.683	0.034	0.851	0.852	2.836%	0.996

Table 18 Comparison results of SRGM 5 (Rayleigh)

Release (i)	MSE	Bias	Variance	RMSPE	TS	R^2
1	45.319	2.246	7.977	8.287	8.967%	0.944
2	21.745	1.384	5.387	5.562	5.431%	0.984
3	3.923	0.446	2.261	2.304	4.591%	0.991
4	1.247	0.127	1.169	1.176	3.833%	0.993

Table 19 Comparison results of SRGM 6 (normal)

Release (i)	MSE	Bias	Variance	RMSPE	TS	R^2
1	0.962	0.015	1.007	1.007	1.307%	0.999
2	3.560	0.267	1.996	2.014	2.198%	0.997
3	0.276	0.000	0.695	0.695	1.219%	0.999
4	0.930	0.029	0.992	0.993	3.310%	0.995

Table 20 Comparison results of SRGM 7 (gamma)

Release (i)	MSE	Bias	Variance	RMSPE	TS	R^2
1	2.328	−0.061	1.569	1.570	2.033%	0.997
2	1.766	−0.045	1.368	1.368	1.548%	0.999
3	0.316	0.090	0.744	0.749	1.303%	0.999
4	0.665	0.022	0.839	0.839	2.799%	0.996

Table 21 Comparison results of SRGM 8 (modified Weibull)

Release (i)	MSE	Bias	Variance	RMSPE	TS	R^2
1	1.700	−0.003	1.338	1.338	1.737%	0.998
2	1.058	−0.023	1.058	1.058	1.198%	0.999
3	0.449	−0.094	0.837	0.842	1.553%	0.999
4	0.773	0.079	0.914	0.918	3.018%	0.996

5 Conclusion

Under the present market conditions, software engineers are working hard to fulfil the rapid increasing demand of reliable software by releasing new versions of software from time to time. For the market support, companies are changing their strategies from time to time during the entire time of the software development life cycle; therefore, existing multi-upgradation framework may not serve the purpose of estimating the number of faults in every version. In this paper we presented a generalized framework for software reliability growth modeling of multi-release accommodating the concept of change-point. In the proposed framework, generalized modified Weibull distribution is considered before and after the change-point for the fault removal process. This comprehensive modeling framework may help to derive various models by considering different probability density and distribution function without making any more assumptions. The formulated model is validated on real data sets and compared with several existing SRGMs. Numerical illustration shows that our proposed model is best fitted on the given data set as compared to the existing SRGMs. In the future, the possibility of including multiple change-points in each release, fault severity, or imperfect debugging can be worked out.

References

1. Huang CY (2005) Performance analysis of software reliability growth models with testing-effort and change-point. J Syst Softw 76(2):181–194
2. Huang CY, Lin CT (2010) Analysis of software reliability modeling considering testing compression factor and failure-to-fault relationship. IEEE Trans Comput 59(2):283–288
3. Pham H (2006) System software reliability. Springer, London
4. Musa JD, Iannino A, Okumoto K (1987) Software reliability, measurement, prediction and application. McGraw-Hill, New York

5. Kapur PK, Pham H, Gupta A, Jha PC (2011) Software reliability assessment with OR applications. Springer, London
6. Kapur PK, Pham H, Anand S, Yadav K (2011) A unified approach for developing reliability growth models in the presence of imperfect debugging and error generation. IEEE Trans Reliab 60(1):331–340
7. Zhao M (1993) Change-point problems in software and reliability. Commun Stat Theory Methods 22(3):757–768
8. Kapur PK, Gupta A, Shatnawi O, Yadavalli VSS (2006) Testing effort control using flexible software reliability growth model with change point. Int J Performability Eng 2(3):245–263
9. Kapur PK, Kumar A, Yadav K, Khatri SK (2007) Software reliability growth modelling for error of different severity using change point. Int J Qual Reliab Saf Eng 14(4):311–326
10. Kapur PK, Singh VB, Anand S (2007) Software reliability growth model of fielded software based on multiple change point concept using a power function of testing time. In: Quality reliability and Infocom Technology. MacMillan India Ltd, New Delhi, pp 171–178
11. Kapur PK, Singh VB, Anand S, Yadavalli VSS (2008) Software reliability growth model with change point and effort control using a power function of the testing time. Int J Prod Res 46(3):771–787
12. Zhao J, Liu H, Cui G, Yang X (2006) Software reliability growth model with change point and environmental function. J Syst Softw 79(11):1578–1587
13. Singh O, Garmabaki AHS, Kapur PK (2011) Unified framework for developing two dimensional software reliability growth models with change points. In: IEEE International Conference on Quality and Reliability (ICQR), held during 14–17 Sep 2011, Bangkok, pp. 570–574
14. Hayashida S, Inoue S, Yamada S (2014) Software reliability assessment using exponential type change point hazard rate models. Int J Reliab Qual Saf Eng 21(4). doi:10.1142/S0218539314500193
15. Inoue S, Hayashida S, Yamada S (2013) Extended hazard rate models for software reliability assessment with effect at change point. Int J Reliab Qual Saf Eng 20(2). doi:10.1142/S0218539313500095
16. Inoue S, Taniguchi S, Yamada S (2015) An all-stage truncated multiple change point model for software reliability assessment. Int J Reliab Qual Saf Eng 22(4). doi:10.1142/S0218539315500175
17. Inoue S, Yamada S (2010) Change point modeling for software reliability assessment depending on two-types of reliability growth factors. Ind Eng Eng Manag:616–620. doi:10.1109/IEEM.2010.5674522
18. Inoue S, Yamada S (2008) Optimal software release policy with change-point. In: Proceedings of 2008 IEEE international conference on industrial engineering and engineering management, pp 531–535
19. Chiu K-C (2015) An exploration on debugging performance for software reliability growth models with learning effects and change-points. J Ind Prod Eng 32(6):369–386
20. Zhao M (2003) Statistical reliability change point estimation models. In: Handbook of reliability engineering. Springer, London/New York, pp 157–163
21. Singh O, Singh VB, Kumar Jyotish J, Kapur PK (2009) Generalized framework for fault detection – correction process incorporating change-point. Commun Depend Qual Manag Int J 12(1):35–46
22. Kapur PK, Kumar J, Kumar R (2008) A unified modeling framework incorporating change-point for measuring reliability growth during software testing. OPSEARCH, Oper Res Soc India J 45(4):317–334
23. Chang YP (2001) Estimation of parameters for non homogeneous poisson process software reliability with change point model. Commun Stat Simul Comput 30(3):623–635
24. Kapur PK, Tandon A, Kaur G (2010) Multi up-gradation software reliability model, 2nd international conference on reliability, safety and hazard, held during 14–16 Dec. 2010, (ICRESH-2010), Mumbai, pp 468–474

25. Kapur PK, Garmabaki AHS, Singh J (2010) Multi up-gradation software reliability model with imperfect debugging. Int J Syst Assur Eng Manag 1(4):299–306
26. Singh O, Kapur PK, Shrivastava AK, Das L (2014) A unified approach for successive release of a software under two types of imperfect debugging. In: IEEE Xplore conference Proceedings of 3rd International Conference on Reliability, Infocom Technologies and Optimization held during October 8–10, 2014 at Amity University, Noida, Uttar Pradesh, pp. 275–280
27. Kapur PK, Singh O, Shrivastava AK, Singh JNP (2015) A software up-gradation model with testing effort and two types of imperfect debugging. IEEE Xplore proceedings of International conference on Futuristic Trends in Computational Analysis and Knowledge Management, pp 613–618
28. Kapur PK, Pham H, Aggarwal AG, Kaur G (2012) Two dimensional multi-release software reliability modeling and optimal release planning. IEEE Trans Reliab 61(3):758–768
29. Carrasco JMF, Ortega EMM, Corderio GM (2008) A generalized modified Weibull distribution for lifetime modeling. Comput Stat Data Anal, Elsevier 53:450–462. doi:10.1016/j.csda.2008.08.023
30. Mudholkar GS, Srivastava DK, Kollia GD (1996) A generalization of the Weibull distribution with application to the analysis of survival data. J Am Stat Assoc 91:1575–1583
31. Lai CD, Xie M, Murthy DNP (2003) A modified Weibull distribution. IEEE Trans Reliab 52:33–37
32. Gupta RD, Kundu D (1999) Generalized exponential distributions. Aust N Z J Stat 41:173–188
33. Kundu D, Rakab MZ (2005) Generalized Rayleigh distribution: different methods of estimation. Compu Stat Data Anal 49:187–200
34. Wood (1996) Predicting software reliability. IEEE Comput 9:69–77

An Improved Scoring System for Software Vulnerability Prioritization

Ruchi Sharma and R.K. Singh

1 Introduction

IT vulnerabilities cannot be eradicated to reach a state of zero, but spotting the most critical ones is essential. It is practically infeasible for an organization to patch or remediate all software vulnerabilities as soon as they are discovered. Delay in fixing the vulnerabilities might be there due to lack of administrative, financial, or human resources. In some cases, compliance may dictate no changes or patches to the system. Sometimes the impacted system cannot be allowed to go out of service during the remediation process. In addition, meeting certain hard deadlines to survive in the market or scarcity of available resources causes undue delay. So, there could be varied reasons which lead to an organization's inability to fix all the detected vulnerabilities [1, 2, 4, 9].

If there are n vulnerabilities detected in a software but the organization has the resources and time to fix only k vulnerabilities where

$$k < n,$$

then deciding *k vulnerabilities* out of n is challenging and difficult, and vulnerability prioritization can address this difficulty. Vulnerability prioritization is the process of assigning a priority to vulnerabilities so as to schedule their remediation process while minimizing the risk.

Various organizations, companies, and researchers have their own rating systems to rank and prioritize vulnerabilities based on qualitative and quantitative rating systems. Qualitative rating systems constitute an intuitive approach to describe the severity of vulnerabilities, while quantitative scoring systems associate a score

R. Sharma (✉) • R.K. Singh
Indira Gandhi Delhi Technical University for Women, Kashmere Gate, New Delhi, India
e-mail: rs.sharma184@gmail.com; rksingh@igdtuw.ac.in

© Springer Nature Singapore Pte Ltd. 2018

P.K. Kapur et al. (eds.), *Quality, IT and Business Operations*, Springer Proceedings in Business and Economics, DOI 10.1007/978-981-10-5577-5_3

with each vulnerability. One of the widely used quantitative scoring systems is the US-CERT's vulnerability scoring system, the Common Vulnerability Scoring System Version 2 (CVSS V2). National Vulnerability Database uses CVSS V2 for vulnerability scoring. Vulnerability rating and scoring system (VRSS) is a hybrid methodology which gives a rating to vulnerabilities while associating a score with each of them. Both CVSS V2 and VRSS give a static rating to vulnerabilities which remain associated with the vulnerability throughout and do not change with time.

1.1 Common Vulnerability Scoring System (CVSS)

Common Vulnerability Scoring System Version 2 (CVSS V2) is an open framework to assess the severity of security vulnerabilities in a software. It establishes a measure of impact and threat posed by a vulnerability so as to prioritize the efforts during vulnerability fixation [1, 3, 7, 8].

CVSS analyzes a vulnerability based on the base metric which uses vulnerability impact and exploitability levels to generate a score between 0 and 10. The higher the score, the more is the priority. Based on this quantitative score, qualitative rating of high, medium, or low severity is assigned to vulnerability.

- Vulnerabilities having a base score value in the range 7.0–10.0 have high severity.
- Vulnerabilities having a base score value in the range 4.0–6.9 have medium severity.
- Vulnerabilities having a base score value in the range 0–3.9 have low severity level.

The base metric group considers those attributes of a vulnerability that do not change with time. Access vector, access complexity, and authentication metrics are the three metrics used in this group [1]. It measures how a vulnerability will directly affect an IT asset if it is successfully exploited. The metrics of this group are shown in Table 1.

The impact is defined independently as the degree of loss of confidentiality, integrity, and availability. The possible values for components of impact metric along with quantitative score are as shown in Table 2.

Table 1 Exploitability metrics [1]

Exploitability metric	[a]Metric value	Quantitative score
Access vector	L/A/Nw	0.395/0.646/1.0
Access complexity	H/Md/L	0.35/0.61/0.71
Authentication	N/S/M	0.704/0.56/0.45

[a]Legend
L local, *A* adjacent network, *Nw* network
H high, *Md* medium, *L* low
N none, *S* single, *M* multiple

Table 2 Impact metrics [1]

Impact metric	Metric value	Quantitative score
Confidentiality	C, P, N	0.66/0.275/0
Integrity	C, P, N	0.66/0.275/0
Availability	C, P, N	0.66/0.275/0

C complete, *P* partial, *N* none

The base equation is the foundation of CVSS V2 scoring and is used to generate the final severity score of a vulnerability [1, 3]. It is evaluated by the following equations:

$$\text{Base score} = \left(\left(\begin{matrix}(0.6^{*}\text{Impact}) + \\ (0.4^{*}\text{Exploitability})\ 1.5\end{matrix}\right)^{*} f\,(\text{Impact})\right) \quad (1)$$

$$\text{Impact} = 10.41^{*}\left(\begin{matrix}1 - (\text{1-ConfImpact}) \\ {}^{*}(\text{1-IntegImpact})^{*}\ (\text{1-AvailImpact})\end{matrix}\right) \quad (2)$$

$$\begin{aligned}\text{Exploitability} &= 20^{*}\ \text{AccessVector}^{*}\text{AccessComplexity} \\ &\quad {}^{*}\text{Authentication}\end{aligned} \quad (3)$$

This score is used for vulnerability prioritization in National Vulnerability Database (NVD). The higher the value of base score, the higher is the severity associated with that vulnerability [1].

1.2 Vulnerability Rating and Scoring System (VRSS)

Vulnerability rating and scoring system was given by Qixu Liu and Yuqing Zhang in 2010 after analyzing vulnerability datasets from three famous vulnerability databases (IBM ISS X-Force, Vupen Security, and National Vulnerability Database) [2]. VRSS is a hybrid approach that gives a qualitative rating and quantitative score to vulnerabilities by combining respective advantages of both qualitative and quantitative systems. VRSS analyzes a vulnerability based on its impact and assigns a qualitative rating to it. It then uses the impact score and exploitability levels to generate a score between 0 and 10. The higher the score, the more is the priority [2].

Quantitative score in VRSS method is taken as the sum of the impact of a vulnerability and its exploitability. The impact is evaluated using three subcomponents, namely, confidentiality impact (C), integrity impact (I), and availability impact (A). Impact score is evaluated using some predefined rules as shown in Table 3. A qualitative rating is given to the vulnerability, and a quantitative score is defined

Table 3 Rules for qualitative rating and impact evaluation [2]

ID	Description	Severity	Impact
1.	Each of confidentiality, integrity, and availability properties has a "complete" loss	High	9
2.	One of confidentiality, integrity, and availability properties has a "partial" loss. The other two have a "complete" loss	High	8
3.	One of confidentiality, integrity, and availability properties has a "none" loss. The other two have a "complete" loss	High	7
4.	One of confidentiality, integrity, and availability properties has a "complete" loss. The other two have a "partial" loss	High	6
5.	One of confidentiality, integrity, and availability properties has a "complete" loss. One of them has a "partial" loss, and one has a "none" loss	Medium	5
6.	One of confidentiality, integrity, and availability properties has a "complete" loss. The other two have a "none" loss	Medium	4
7.	Each of confidentiality, integrity, and availability properties has a "partial" loss	Medium	3
8.	One of confidentiality, integrity, and availability properties has a "none" loss. The other two have a "partial" loss	Medium	2
9.	One of confidentiality, integrity, and availability properties has a "partial" loss. The other two have a "none" loss	Low	1
10.	Each of confidentiality, integrity, and availability properties has a "none" loss	Low	0

which suggests its impact score. The qualitative rating is given on the basis of impact of the vulnerability and is governed by certain rules that are summarized in Table 3. Exploitability is calculated by taking twice the product of access vector, access complexity, and authentication given by Eq. 4. These three components of exploitability are evaluated in the same manner as given in Table 1 for CVSS 2.

Overall score is calculated using the following formula:

$$\text{Exploitability Score} = 2^{*}\text{Access Vector}^{*}\text{Access Complexity}^{*}\text{Authentication} \tag{4}$$

$$\text{Quantitative Score} = \text{Impact Score} + \text{Exploitability Score} \tag{5}$$

The score generated by Eqs. (4) and (5) is used to prioritize vulnerabilities in VRSS. The higher the value of quantitative score, the higher is the severity of associated vulnerability, and therefore priority of that vulnerability is higher than all the vulnerabilities having a lower score.

1.3 Limitations of the Existing Prioritization Methodologies

The prioritization methods currently used for vulnerabilities suffer from certain limitations. Severity of a vulnerability is a dynamic attribute that changes with changing levels of temporal factors associated with it. This may have significant impact on the intrinsic characteristics of a vulnerability.

In CVSS V2-based prioritization, vulnerabilities are prioritized based on the score generated by base metric group [1, 6]. But the base score does not take into account the attributes which are time based due to which the score associated with a particular vulnerability is same with time and might not suit the industrial scenario where temporal attributes significantly affect the severity of a vulnerability [1, 9].

VRSS also does not consider any temporal attribute while generating the severity score [1, 2, 9]. Time-based or temporal attributes consider the changing severity which is a more practical scenario in the software world.

For example, if a vulnerability Y has been detected in year 2014 and that time there was no official fix available for it. Say the score associated with Y in 2014 is S_1. Now, the organization worked upon it to release a stable official fix. Now, if Y is again encountered in year 2015 and the score is S_2, then $S1 > S_2$, if all other attributes of vulnerability are kept the same (due to an available patch for the vulnerability developed on initial encounter).

Therefore, including temporal attributes while generating the quantitative score for prioritization leads to a dynamic prioritization scheme which efficiently handles the changing levels of vulnerability severity and gives a more realistic and practical approach toward vulnerability prioritization.

2 Proposed Approach

The proposed approach is a hybrid approach derived by combining both VRSS and CVSS. It gives both qualitative rating and quantitative score for prioritization. We took into consideration the remediation level and the rate at which a particular type of vulnerability is growing with time termed as vulnerability index (VI). The first attribute, i.e., vulnerability index, raises the static score to take into consideration the changing rate of vulnerability with time. On the other hand, remediation level adjusts the severity score downward leading to declining urgency as remediation becomes final.

2.1 Vulnerability Index (VI)

The vulnerability index (VI) is a temporal metric to capture the effect of rate at which a particular vulnerability is increasing or decreasing with time. VI is evaluated using the historical data [6] to find the rate of change of a particular type of vulnerability. If the number of vulnerabilities of type Z in year of observation is Z_{NVO} and the number of vulnerabilities of type Z in its previous year is Z_{NVP}, then

$$\text{vulnerability index } (VI) = \frac{Z_{NVO}}{Z_{NVO} + Z_{NVP}} \tag{6}$$

The formula obtained in Eq. (6) takes in consideration the fact that rate of change cannot exceed 1 and minimum value can be equal to 0. These values are obtained in case of extremes. In all the other cases, value of VI lies between 0 and 1, i.e., $0 \leq VI \leq 1$. Following is the impact of vulnerability index [VI]:

- A value equal to one suggests that no vulnerability of type Z was detected in the previous year.
- While a value equal to zero suggests that no vulnerability of type Z is detected in the year of observation.
- A value equal to 0.5 suggests that the number of vulnerabilities in the year of observation and its previous year is same.
- A value between 0 and 0.5 suggests that the number of vulnerabilities has decreased from the previous year.
- A value between 0.5 and 1 suggests that the number of vulnerabilities has increased from the previous year.

The quantity obtained by Eq. (6) gives the value of the rate at which a vulnerability is increasing or decreasing with time. Vulnerability index is a dynamic attribute and captures the changing rate of vulnerability detection for a particular vulnerability over the years. The reason behind introducing this metric is that the rate at which a vulnerability is encountered also affects the severity associated with it and, therefore, would be a helpful attribute while performing vulnerability prioritization. If the rate is increasing with time, i.e., the number of vulnerabilities is showing an upward trend with time, then it means that there is a need to give more attention. While if the rate is decreasing, then the severity associated with it also decreases. The vulnerability index is applied after all the static attributes are considered and evaluated. It has an impact of raising the quantitative score that is used for vulnerability prioritization. The score obtained after considering the effect of VI will either increase the severity or it will show no effect. But, it can never decrease the severity score obtained after considering static attributes only.

2.2 *Remediation Level (RI)*

Remediation level takes into account the extent to which an organization is ready to deal with a vulnerability. This is a time-based attribute, i.e., it can change over time. The remediation or preparation level of an organization against a vulnerability can significantly affect the severity associated with it and therefore forms an important tool for vulnerability prioritization methodology. Remediation level is used in temporal metrics of CVSS but is not used for prioritization [1, 6].

Among all the factors, the remediation is an important attribute for vulnerability prioritization. When published initially, a vulnerability is generally unpatched. A

Table 4 Remediation levels [1]

Remediation level (RL)	Weightage
Official fix	0.87
Temporary fix	0.90
Workaround	0.95
Unavailable	1.00
Not defined	1.00

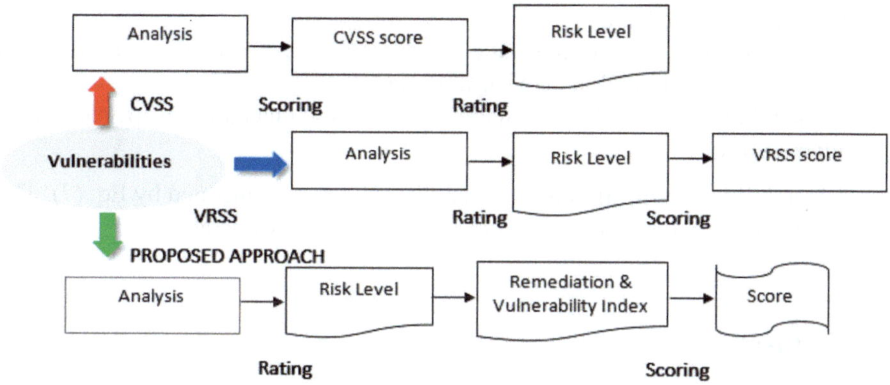

Fig. 1 Proposed prioritization methodology

short-term remediation may be offered by workarounds or hotfixes before the issue of an official patch or upgrade. With all these stages, the severity score of a vulnerability adjusts downward suggesting a decreasing level urgency as remediation becomes final as shown in Table 4. The less official a fix, the higher the vulnerability score. Once the fix is officially available, vulnerability score decreases. The weightage gives a numeric equivalent of each level which is shown in Table 4 [1, 3, 7, 8].

The diagrammatic representation of existing and proposed prioritization methodology is as shown in Fig. 1.

The base of our approach is the concept used in VRSS [2]. The impact is evaluated using the same approach as followed in VRSS and given in Table 3. The exploitability score associated with a vulnerability is evaluated using the values given in Table 1.

The qualitative rating is given on the basis of impact of the vulnerability and is governed by rules summarized in Table 3. The vulnerability index and remediation level are used to generate the final quantitative score termed as Q score. The Q score is generated using the attributes described above and is given by Eq. (7). The value of Q score lies between 0 and 10.

$$\text{Q score} = \frac{(\text{Impact} + \text{Exploitability})^{*}\ \text{RL}^{*}\ (1 + \text{VI})}{2} \tag{7}$$

The Eq. (7) formulated to generate Q score is characterized by the following points:

- The value obtained by evaluating (Impact + Exploitability) is the static score and is same as the one obtained in VRSS approach.
- The value of static score lies between 0 and 10.
- The factor of RL accounts for decreasing the static score generated after taking the sum of impact and exploitability. It will lead to no impact if the fix is not available or not defined.
- VI leads to an increase in the static score or no impact if there are no reported vulnerabilities of type Z in the year of observation.
- By including RL and VI in the static score leads to changing severity levels for vulnerabilities.

Vulnerability prioritization is done based on the score generated by Eq. (7). The higher the Q score, the more is the priority of associated vulnerability.

3 Datasets

The datasets used in this work are collected from National Vulnerability Database (NVD) [5, 6]. NVD contains the dataset for 23 types of vulnerabilities. Dataset comprises of vulnerabilities of different types, namely, buffer overflow, memory corruption, privilege escalation, denial of service, cross-site scripting (XSS), etc. The dataset contain the number of vulnerabilities of each type over the years. We used datasets for two types of vulnerabilities, i.e., (i) buffer overflow and (ii) memory corruption.

The data for various security attributes of a vulnerability is collected from CVE available at National Vulnerability Database [6].

4 Illustrative Example

- *CVE-2003-0062: Buffer Overflow in NOD32 Antivirus*

Discovered in February 2003, CVE-2003-0062 is a buffer overflow vulnerability in Linux and UNIX versions and could allow local users to execute arbitrary code with the privileges of the user executing NOD32. The attacker must wait for another user to scan a directory path of excessive length to trigger this vulnerability [5, 6].

- According to CVSS, this vulnerability was given a score of 6.2 and thus medium severity.
- According to VRSS, the qualitative rating is medium and quantitative score is 5.00 [32].
- According to our proposed approach:

- Impact vector: [C:C/I:C/A:C].
- Exploitability vector: [AV:N/AC:L/Au:N].
- The qualitative rating of this vulnerability is high according to Table 3 and impact score is 9 [how not clear].
- Remediation level = official patch available. So, RL = 0.87 [reference table].
- Vulnerability type = buffer overflow.
- No. of buffer overflow vulnerabilities in 2003=$Z_{NVO} = 370$.
- No. of buffer overflow vulnerabilities in 2002=$Z_{NVP} = 435$.
- So, vulnerability index=$\frac{370}{370+435} = 0.46$

$$\text{Q score}=\frac{(9+2*0.395*0.35*0.704)^{*}0.87(1+0.46)}{2} = 5.8$$

- *CVE-2002-0392: Apache Chunked-Encoding Memory Corruption Vulnerability*

The value of Q score which gives the severity rating to a vulnerability in our prioritization methodology is between the values of existing approaches of CVSS and VRSS. This is due to the fact that the organization already had a patch available for the vulnerability in this example which is not considered while generating the score in CVSS and VRSS. This leads to a lower severity level for CVE-2003-0062 as compared to CVSS score. The value of VI considered in our approach further takes into account the rate at which the vulnerability is changing and adds to the severity of CVE-2003-0062.

In June 2002, a vulnerability was discovered in the Apache web server which handles requests using chunked encoding. According to the Apache Foundation, a successful exploit can lead to the execution of arbitrary code with the privileges of the web server in some cases and denial of service in others.

- According to CVSS, this vulnerability was given a score of 7.5 and thus high severity [31].
- According to VRSS, the qualitative rating is medium and quantitative score is 5.00 [32].
- According to our proposed approach:
  - Impact vector: [C:N/I:N/A:C].
  - Exploitability vector: [AV:N/AC:L/Au:N].
  - The qualitative rating of this vulnerability is medium according to Table 13 and impact score is 4.
  - Remediation level = official patch available. So, RL = 0.87.
  - Vulnerability type = memory corruption.
  - No. of memory corruption vulnerabilities in 2002=$Z_{NVO} = 2$.
  - No. of memory corruption vulnerabilities in 2001=$Z_{NVP} = 0$.
  - So, vulnerability index=$\frac{2}{2+0} = 1$

$$\text{Q score} = \frac{(4+2*1*0.71*0.704)^{*}0.87(1+1)}{2} = 4.3$$

The Q score which gives the severity rating to a vulnerability in our prioritization methodology comes out to be lower than both the existing approaches of CVSS and

VRSS. This is due to the fact that the organization already had a patch available for the vulnerability in this example which is not considered while generating the score in CVSS and VRSS. This leads to a lower severity level for CVE-2002-0392 in our proposed approach.

5 Conclusion

The new prioritization approach proposed in this study takes into account two temporal metrics, namely, remediation level and vulnerability index. They change with time and therefore lead to changing severity levels of the same vulnerability with time which is a more practical scenario in the current software world. The existing approaches for vulnerability prioritization, i.e., CVSS and VRSS, do not consider any time-based metrics for prioritization and therefore cannot take into account the changing levels of severity with time.

In a practical scenario, if the same vulnerability is encountered again and again by the organization, then its remediation level (RL), i.e., the organization's level of preparation to counter that vulnerability, changes. Generally, the urgency decreases as the remediation becomes final.

Also the rate at which a vulnerability is encountered determines the severity associated with it. If the rate is increasing with time, i.e., the number of vulnerabilities is showing an upward trend with time, then it means that there is a need to give more attention. While if the rate is decreasing, then the severity associated with it also decreases. We have termed this metric as vulnerability index (VI). Therefore, both the metrics introduced in this study, namely, RL and VI, effectively perform the prioritization of vulnerabilities while introducing the concept of changing levels of score associated with a vulnerability.

References

1. Mell, P, Scarfone K, Romanosky S (2007, June) A complete guide to the common vulnerability scoring system version 2.0
2. Liu Q, Zhang Y (2011) VRSS: a new system for rating and scoring vulnerabilities. Comput Commun 34(3):264–273
3. Fruhwirth C, Mannisto T (2009) Improving CVSS-based vulnerability prioritization and response with context information. In: Proceedings of the 2009 3rd international symposium on empirical software engineering and measurement. IEEE Computer Society, Washington, DC
4. Information Week Dark Reading. Connecting the Information Security Community. www.darkreading.com/vulnerabilities. [Online], June 1, 2015
5. CVE Details. The Ultimate Security Vulnerability Data source. www.cvedetails.com. [Online], May 12, 2015
6. National Vulnerability Database, nvd.nist.gov/, [online], March 10, 2015
7. Scarfone K, Mell P (2009) An analysis of CVSS version 2 vulnerability scoring. In: Proceedings of the 2009 3rd international symposium on empirical software engineering and measurement. IEEE Computer Society, Washington, DC

8. Ibidapo, Ayodele Oluwaseun, et al (2011) An analysis of cvss v2 environmental scoring. In: Privacy, security, risk and trust (PASSAT) and 2011 IEEE third international conference on Social Computing (SocialCom), 2011 IEEE Third International Conference on. IEEE
9. Tripathi A, Singh UK (2011) On prioritization of vulnerability categories based on CVSS scores. In: Computer Sciences and Convergence Information Technology (ICCIT), 2011 6th International Conference on. IEEE

Analysis of Feature Ranking Techniques for Defect Prediction in Software Systems

Sangeeta Sabharwal, Sushama Nagpal, Nisha Malhotra, Pratyush Singh, and Keshav Seth

1 Introduction

Software quality is the main concern in software development process. Some of the software quality attributes are correctness, maintainability, robustness, and fault proneness. Out of the above attributes, fault proneness is one of the most important attributes. To identify fault proneness, we need to develop software defect prediction model, which predicts whether a particular module is faulty or not. The metrics (features) play an important role in developing an efficient fault prediction models. Therefore the selection of relevant features is an important step in building the model. The feature selection algorithm extracts the relevant features. The feature selection algorithms include feature ranking and feature subset selection. Our paper focuses on ranking the software metrics to develop the defect prediction model. Our approach consists a part of the preprocessing step which enables one to analyze the software metrics in the dataset and find the relevant set of metrics, in order to build a successful defect prediction model. In feature ranking, the feature which fulfills the criterion is selected. Feature subset selection, on the other hand, selects the subset of relevant features because sometimes combination of features performs well as compared to a single feature. Evaluating the feature subset selection method is impractical if the search space is too large. Our paper uses a hybrid feature selection approach which helps us to depict the use of feature ranking along with feature subset selection. Feature ranking ranks the features and selects only those useful features which are used to minimize the dataset. It is followed by feature subset selection to find the exact feature subset. In our paper, we have analytically explored five filter-based feature ranking techniques: chi-square (CS) [1], information gain

S. Sabharwal (✉) • S. Nagpal • N. Malhotra • P. Singh • K. Seth
Division of Computer Engineering, NSIT, University of Delhi, New Delhi, 110078, India
e-mail: ssab63@gmail.com; sushmapriyadarshi@yahoo.com; nisha.jan89@gmail.com; webhead561@gmail.com; setkeshav@gmail.com

© Springer Nature Singapore Pte Ltd. 2018
P.K. Kapur et al. (eds.), *Quality, IT and Business Operations*, Springer Proceedings in Business and Economics, DOI 10.1007/978-981-10-5577-5_4

(IG) [2], gain ratio (GR) [2], Kolmogorov–Smirnov statistic (KS) [3], and relief algorithm (RA) [4], and three different feature subset selection methods: exhaustive search (ES) [5], best first search (BFS) [6], and automatic hybrid search (AHS) [7]. For training the defect predictors, we have considered three different learners: Naïve Bayes (NB) [8], multilayer perceptron (MLP) [9], and support vector machine (SVM) [10].

In our work, we are using public dataset KC1, which contains class-level and method-level metrics [11]. There are only four method-level metrics. Koru et al. [19] converted these KC1 method-level metrics into class-level ones using minimum, maximum, average, and sum operations. For example, the metric LOCBLANK is converted into minLOC_BLANK, maxLOC_BLANK, avgLOC_BLANK, and maxLOC_BLANK. The final dataset contains 94 metrics with 145 instances and one class attribute. Defect level is the class attribute which is true if the module contains one or more defects.

The remaining sections are: Sect. 2 describes the related work done by various researchers. Sect. 3 briefly explains the feature ranking and feature subset selection methods used. Sect. 4 explains the classifiers used. Sect. 5 discusses the experimental design. Sect. 6 provides the experimental results, and Sect. 7 contains the conclusion and future work.

2 Related Work

This section includes the work done by previous researchers on feature ranking.

Gao et al. [7] proposed hybrid attribute selection approach for attribute selection to build software defect prediction model. In the hybrid approach, they have firstly applied feature ranking to reduce the search space and then applied feature subset selection. They considered filter feature ranking techniques and also used learners to build the models. They have concluded that automatic hybrid search performs well, and also there is no major difference in the performance of the models if we remove most of the metrics. Khoshgoftaar et al. used neural network (NN) on large telecommunication dataset to predict whether a module is faulty or not [12]. They also compared neural network model with a nonparametric discriminant model and found that NN model performed well in fault prediction. Weston et al. [15] used SVM feature selection which is based upon selecting only those attributes which minimize bounds on the leave-one-out error. The feature selection for SVM was outperformed on the datasets tested. Guyon and Elisseeff [16] introduced variable and feature selection. Their study focused on selecting subsets of features that are necessary to build a good predictor. They introduced filter methods, wrapper methods, and some of the embedded methods. They recommended to use linear SVM and selected variables using mutual information and nested subset selection methods. Rodriguez et al. [17] applied feature selection to various databases. They used filter as well as wrapper models and tested them on five software engineering datasets. They also used different data mining algorithms for classification of faulty modules. They have also found that the reduced search space gives better accuracy of fault prediction than the original dataset. Gayatri et al. [18] proposed decision tree induction method which is based on feature selection. The features which construct rules for the learners are considered as the relevant features. And these relevant features are known as decision tree induction rule-based (DTIRB) feature set. They used 18 learners for training. They also compared support vector machines (SVM) and relief feature selection techniques with the proposed method (DTIRB) and observed that the proposed method performed well.

3 Filter-Based Feature Ranking Techniques and Feature Subset Selection Methods

In our approach we have used five filter-based feature ranking techniques and three feature subset selection methods. The filter feature ranking technique does not use any learning algorithm to determine the relevant features. Next we briefly describe each of these techniques.

3.1 Feature Ranking Techniques Used

1. *Chi-square method* – This method evaluates the relevant feature by calculating the CS statistic ($\chi 2$) value with reference to the class [1], and it is a nonparametric statistical technique that finds out the difference between the observed and the expected frequencies. This test uses frequencies because CS statistic uses nominal data. The CS statistic $\chi 2$ is calculated as

$$\chi^2 = \sum_{i=1}^{n} \frac{(O_i - E_i)^2}{E_i}$$

CS statistic ($\chi 2$) asymptotically reaches a $\chi 2$ distribution, *Oi* is the observed frequency, *Ei* is the expected frequency, and *n* is the count of the outcomes of each event.

2. *Information gain* – This method evaluates the worth of a feature by calculating the information gain with respect to the class.

   IG is the difference in information entropy from a previous state to a state that takes some information [2]. The gain is given by

$$\text{IG}\,(\text{Class}, \text{Feature}) = H\,(\text{Class}) - H\,(\text{Class}|\,\text{feature})$$

where H denotes the entropy. IG has the tendency to favor features with more values which sometimes treat irrelevant feature as a relevant feature. Therefore, IG is not recommended if a feature has more distinct values.

3. *Gain ratio* – It overcomes the disadvantage of IG by modifying the IG by involving the number of outputs produced by the feature [2]. GR chooses a feature by taking the number and size of branches, and then IG corrects by taking the intrinsic information. Intrinsic information is entropy of distribution of instances into branches. Value of attribute decreases as intrinsic information gets larger. GR is calculated as

$$\mathrm{GR}\,(\mathrm{Class}, a) = IG\ (\mathrm{Class}, a)\,/\mathrm{Intrinsic\_info}\,(\mathrm{Attribute})$$

4. *Kolmogorov–Smirnov statistic* – This method is based on the empirical distribution function. The KS method [3] makes use of the KS statistic to find out the maximal change among the empirical distribution function of the feature values of every class' instances. The ability to classify and differentiate between two classes is directly proportional to the distance between the two functions. Based on the ranking, the features having high KS score are selected. The KS score for i^{th} independent variable is

$$\mathrm{KS} = \max \mid {S_{Xi}}^{\mathrm{true}}\,(\mathrm{Xi}) - {S_{Xi}}^{\mathrm{false}}\,(\mathrm{X}_i) \mid$$

5. *Relief method* – This method is based on the weights of the features [4]. In a dataset, it randomly selects a row, and for an instance R, relief searches nearest neighbor from the same class called as "nearest hit H" and nearest neighbor from the different class called as "nearest miss M." Depending on the values of H, R, and M, relief updates the weight vector W[A] for all features. This step is repeated n number of times, where n is specified by a user. The W[A] is calculated as:

$$\mathrm{W}\,[\mathrm{A}] \quad = \mathrm{W}\,[\mathrm{A}] - \mathrm{diff}\,(\mathrm{A}, \mathrm{R}, \mathrm{H})\,/n + \mathrm{diff}\,(\mathrm{A}, \mathrm{R}, \mathrm{M})\,/n$$

Function diff (feature, instance 1, instance 2) is defined according to different types of features. Features can have either nominal or numerical values. For nominal features, if *value* $(F, I_1) = Value(F, I_2)$, then the value of this function is zero; otherwise it is one. For numerical features, it is defined as

$$\mathrm{diff}\,(F, I_1, I_2) = \frac{|\mathrm{value}\,(A, I_1) - \mathrm{value}\,(A, I_2)|}{\max(A) - \min(A)}$$

The relevant feature is the one which can differentiate between the instances that belong to distinct classes and show no change when the instances belong to the same class.

In the following subsection, we explain the techniques used to select subset of metrics from the top ranked features obtained as a result of feature ranking. The feature subset selection methods we have used are exhaustive search, best first search, and automatic hybrid search.

3.2 Feature Subset Selection Methods

1. *Exhaustive search* – This method [5] constructs all the combinations of feature subsets and calculate the consistency rate (CR) of every subset. Initially, the first CR for the subset is considered as the threshold, and that subset is also considered as the perfect subset. When the algorithm proceeds, the subset with higher CR value is replaced by the current one. ES constructs subsets by either starting with full features or then discarding features from the set or by a single feature set and then starts inserting features into the set. The disadvantage of using ES is that it is time-consuming as well as computationally expensive because it constructs all the combination of subsets even if some of them are redundant or irrelevant. This method is impractical to use if the search space is too large.
2. *Best first search* – This method is one of the heuristic search algorithms [6]. It overcomes the disadvantage of ES method by not constructing all the combination of subsets. BFS searches the feature subsets by greedy hill climbing build up by backtracking. Best first constructs subsets by either starting with full features and searching in backward direction or starting with empty set and then searching in forward direction. The best subset is the one which has higher CR value.
3. *Automatic hybrid search* – This method also employs the evaluation of the consistency rate (CR). First of all, the CR of the whole dataset is computed [7]. By repeatedly selecting the subsets with the locally highest CR, supersets are generated until the attribute subsets with the equal CR as the original set are extracted.

4 Classifiers Used

This section briefly describes the three classifiers used for training the defect prediction models.

1. *Naïve Bayes* – It assumes conditional independence of the features and therefore is termed as Naïve [8]. It is based on Bayes' rule of conditional probability.
2. *Multilayer perceptron* – It is an artificial neural network model that maps sets of inputs into desired outputs [9]. Multiple nodes are connected in a directed graph fashion. MLP consists of an input, output, and one or more hidden layers. The weighted sum of the inputs from the prior layer is the output of each node.

Multiple hidden layers of a neural network can proximate any function. In our work, the hidden layer is set to "four" which means a network with one hidden layer having four nodes, and the size of the validation is set to "ten" to let the classifier to rule out 10% of the training data away so that it can be used as a validation set to decide when to stop the iterative training process.

3. *Support vector machine* – This is a generalized linear classifier that can map input vectors into a large dimensional space, to construct a maximum margin hyperplane [10]. The instances that are located on the margins of the maximum margin hyperplane are called support vectors. In our paper, the SVM classifier gives the worst result.

5 Experimental Design

In this section, we will discuss the experimental design of the above-mentioned techniques.

Firstly, we individually apply the five filter-based feature ranking techniques on the full dataset and generate ranking from the individual ranking techniques. We fetch only the top [$\log_2$ n] features, where n is the total number of features in the dataset. We have considered $\log_2$n features (i.e., seven metrics) as prior work done by K. Gao et al. [7] has shown that it was apt to use $\log_2$ n features when working on Weka for imbalanced datasets and to construct random forest learners for binary classification. We also applied feature subset select methods, but as the number of software metrics in the original dataset is large, it is not possible to directly apply these methods. So, for this we used a hybrid approach, in which we first individually applied all the five feature ranking techniques on the original dataset, and then we have selected the top 14 features. Now we have applied the feature subset selection methods only on these 14 features. Finally, we have used three learners to build the defect prediction models. We have also applied the three learners to the original dataset for the comparisons.

6 Experimental Results

This section presents the results in Tables 1 and 2.

For training the models, the experiments include tenfold cross validation with ten runs. One layer is the test data, and the rest layers are treated as the training data. We have considered the Weka tool [14] for performing the five feature rankings (except KS) along with the subset selection methods (except AHS). The KS and AHS methods are implemented in R tool [13]. We used area under curve (AUC) for comparing the classifiers. The AUC values of learners are presented in Table 1. In the last row of the Table 1, the results on the original datasets are shown in which no ranking techniques and search methods have been used.

Table 1 The AUC values of learners

Ranking technique	Search algorithm	Learner	Dataset (AUC Value)
CS		NB	0.851 0.800
		MLP	0.719 0.811
		SVM	0.796 0.705
		NB	0.815 0.795
	ES	MLP	0.699 0.824
		SVM	0.816
		NB	0.746
	BFS	MLP	
		SVM	
		NB	
	AHS	MLP	
		SVM	
IG		NB	0.841 0.799
		MLP	0.719 0.821
		SVM	0.797 0.687
		NB	0.810 0.810
	ES	MLP	0.727 0.825
		SVM	0.815
		NB	0.723
		MLP	
	BFS	SVM	
		NB	
		MLP	
	AHS	SVM	
GR	ES	NB	0.847 0.793
		MLP	0.713 0.842
		SVM	0.802 0.724
		NB	0.853 0.810
		MLP	0.741 0.840
	BFS	SVM	0.841
		NB	0.655
		MLP	
		SVM	
	AHS	NB	
		MLP	
		SVM	
RA		NB	0.820 0.772
		MLP	0.712 0.821
		SVM	0.797 0.687
		NB	0.807 0.796
	ES	MLP	0.710 0.811
		SVM	0.794
		NB	0.722

(continued)

Table 1 (continued)

Ranking technique	Search algorithm	Learner	Dataset (AUC Value)
	BFS	MLP	
		SVM	
		NB	
		MLP	
	AHS	SVM	
KS		NB	0.819 0.794
		MLP	0.688 0.836
		SVM	0.850 0.713
		NB	0.836 0.850
	ES	MLP	0.713 0.837
		SVM	0.829
		NB	0.713
		MLP	
	BFS	SVM	
		NB	
		MLP	
		SVM	
	AHS		
No ranking technique	No search algorithm	NB	0.809
		MLP	0.808
		SVM	0.721

From Table 1, we observe that from the three learners, SVM performs the worst, and NB gives the best result. AHS when combined with KS, CS, IG ranking techniques, BFS when combined with GR ranking technique, and ES when combined with relief method give better results. AHS showed better performance and, therefore, is recommended. From Table 2, the top seven ($\log_2 94$) metrics are maxLOC_TOTAL, maxNUM_UNIQUE_ OPERANDS, maxHALSTEAD_ CONTENT, COUPLING_BETWEEN_ OBJECTS, avgNUM_UNIQUE_ OPERANDS, avgLOC_TOTAL, and avgHALSTEAD_LENGTH. The results are more or less similar to the results obtained by K. GAO et al. [7]. In [7] it is concluded that information gain, gain ratio, and Kolmogorov–Smirnov statistic if applied separately perform well as compared to other ranking techniques, and the automatic hybrid search method performs the best. Also, the search algorithms outperformed when the search space was less [7]. Gayatri et al. [18] used different ranking techniques from those which we have used and concluded that metric maxLOC_BLANK is the most important.

Table 2 Frequency of selected metrics

Metrices	Frequency	Metrics	Frequency
maxLOC_TOTAL	22	PERCENT_PUB_ DATA	0
maxNUM_UNIQUE_ OPERANDS	16	ACCESS_TO_PU B_DATA	0
maxHALSTEAD_ CONTENT	15	DEPTH	0
COUPLING_BETWE EN_ OBJECTS	14	NUM_OF_CHIL DREN	0
avgNUM_UNIQUE_ OPERANDS	12	FAN_IN	0
avgLOC_TOTAL	11	minLOC_CODE_ AND_ COMMENT	0
avgHALSTEAD_LEN GTH	4	minLOC_COMM ENTS	0
avgNUM_OPERATO RS	4	minCYCLOMATIC_ COMPLEXITY	0
sumHALSTEAD_EFF ORT	3	minDESIGN_ COMPLEXITY	0
sumNUM_UNIQUE_ OPERATORS	3	minESSENTIAL_ COMPLEXITY	0
sumLOC_TOTAL	3	minLOC_EXECU TABLE	0
minHALSTEAD_LE VEL	2	minHALSTEAD_ CONTENT	0
maxHALSTEAD_ DIFFICULTY	2	minHALSTEAD_ DIFFICULTY	0
maxNUM_OPERANDS	2	minHALSTEAD_ EFFORT	0
maxNUM_UNIQUE_ OPERATORS	2	minHALSTEAD_ ERROR_EST	0
avgHALSTEAD_LEV EL	2	minHALSTEAD_ LENGTH	0
sumHALSTEAD_ DIFFICULTY	2	minHALSTEAD_ PROG_ TIME	0
LACK_OF_COHESI ON_ OF_METHODS	1	minHALSTEAD_ VOLUME	0
DEP_ON_CHILD	1	minNUM_OPER ANDS	0
maxLOC_EXECUTA BLE	1	minNUM_OPER ATORS	0

(continued)

Table 2 (continued)

Metrices	Frequency	Metrics	Frequency
maxHALSTEAD_EFF ORT	1	minNUM_UNIQU E_ OPERANDS	0
maxHALSTEAD_LE NGTH	1	minNUM_UNIQU E_ OPERATORS	0
maxHALSTEAD_VO LUME	1	minLOC_TOTAL	0
avgLOC_EXECUTA BLE	1	maxLOC_BLANK	0
avgHALSTEAD_ DIFFICULTY	1	maxBRANCH_C OUNT	0
avgNUM_OPERANDS	1	maxLOC_CODE_ AND_ COMMENT	0
avgNUM_UNIQUE_ OPERATORS	1	maxLOC_COMM ENTS	0
sumLOC_EXECUTA BLE	1	maxCYCLOMATIC_ COMPLEXITY	0
sumHALSTEAD_PR OG_ TIME	1	maxDESIGN_ COMPLEXITY	0
sumNUM_OPERANDS	1	maxESSENTIAL_ COMPLEXITY	0
sumNUM_UNIQUE_ OPERANDS	1	maxHALSTEAD_ ERROR_EST	0
avgLOC_BLANK	0	maxHALSTEAD_ LEVEL	0
avgBRANCH_COUNT	0	maxHALSTEAD_ PROG_ TIME	0
avgLOC_CODE_AN D_ COMMENT	0	maxNUM_OPER ATORS	0
avgLOC_COMMENTS	0	avgHALSTEAD_ CONTENT	0
avgCYCLOMATIC_ COMPLEXITY	0	avgHALSTEAD_ EFFORT	0
avgDESIGN_ COMPLEXITY	0	avgHALSTEAD_ ERROR_EST	0
avgESSENTIAL_ COMPLEXITY	0	avgHALSTEAD_ PROG_ TIME	0
sumLOC_BLANK	0	avgHALSTEAD_ VOLUME	0
sumBRANCH_COUNT	0	sumESSENTIAL_ COMPLEXITY	0
sumLOC_CODE_AN D_COMMENT	0	sumHALSTEAD_ CONTENT	0
sum_LOC _COMMENTS	0	sumHALSTEAD_ ERROR_EST	0
sumCYCLOMATIC_ COMPLEXITY	0	sumHALSTEAD_ LENGTH	0
sumDESIGN_ COMPLEXITY	0	sumHALSTEAD_ LEVEL	0
sumNUM_OPERATORS	0	sumHALSTEAD_ VOLUME	0

7 Conclusion and Future Work

In our paper we have applied five filter-based feature ranking techniques followed by three feature subset selection methods to build defect prediction models. It is observed that the search space is reduced by applying the proposed hybrid approach. For each feature selection process, the software defect prediction models are built. It was found that among the 94 metrics in the KC1 dataset, maxLOC_TOTAL is the most important metric. The results of our study concluded that on elimination of the most of the irrelevant metrics, the classification of the defect prediction model stays same or gives better results.

In the future, we will use wrapper-based techniques for feature selection by adding a different aspect on filter-based techniques, and we will also try to model the cost of feature ranking techniques.

References

1. Meesad P, Boonrawd P, Nuipian V (2011) A Chi-Square-Test for Word Importance Differentiation in Text Classification. 2011 International conference on information and electronics engineering, Singapore, IPCSIT vol 6
2. Patil LH, Atique M. A novel feature selection based on information gain using WordNet. Sant Gadge Baba Amravati University, Amravati, India
3. Gao K, Khoshgoftaar TM, Wang H (2009) An empirical investigation of filter attribute selection techniques for software quality classification. Eastern Connecticut State Univ, Willimantic
4. Durgabai RPL (2014) Feature selection using relief algorithm. Int J Adv Res Comput Commun Eng 3(10)
5. Furlanello C, Serafini M, Merler S, Jurman G (2003) Entropy-based gene ranking without selection bias for the predictive classification of microarray data. BMC Bioinforma 4:54
6. Thangaraju P, Mala N (2015) Effectiveness of searching techniques in feature subset selection. IRACST - Int J Comput Sci Inf Technol Secur (IJCSITS), 5(2). ISSN: 2249-9555
7. Gao K, Khoshgoftaar TM, Wang H, Seliya N (2011) Choosing software metrics for defect prediction: an investigation on feature selection techniques. Softw Pract Exper 41:579–606. doi:10.1002/spe.1043. Published online in Wiley Online Library (wileyonlinelibrary.com)
8. Klawonn F, Angelov P (2006) Evolving extending Naïve Bayes classifiers. Data Mining Workshops, 2006. ICDM workshops 2006. Sixth IEEE international conference
9. Panchal G, Ganatra A, Kosta YP, Devyani P (2011) Behaviour Analysis of Multilayer Perceptrons with Multiple Hidden Neurons and Hidden Layers. Int J Comput Theory Eng, 3(2), ISSN: 1793-8201
10. Srivastava DK, Bhambhu L. Data classification using support vector machine. J Theor Appl Inf Technol. JATIT 2005–2009
11. http://promise.site.uottawa.ca/SERepository/datasets/kc1-class-level-numericdefect.arff
12. Khoshgoftaar TM, Nguyen L, Gao K, Rajeevalochanam J (2003) Application of an attribute selection method to CBR based software quality classification. In: Fifteenth IEEE Conference. Sacramento, CA
13. R tool: http://www.rdatamining.com/resources/tools
14. weka:http://www.cs.waikato.ac.nz/ml/weka
15. Weston J et al (2001) Feature selection for SVMs. In: Todd KL, Thomas GD, Volker T (eds) Advances in neural information processing systems 13. The MIT Press, Cambridge, MA, pp 668–674

16. Guyon I, Elisseeff A (2003) An introduction to variable and feature selection. J Mach Learn Res 3:1157–1182
17. Rodr'ıguez D, Ruiz R, Cuadrado-Gallego J, Aguilar-Ruiz J, Garre M (2007) Attribute selection in software engineering datasets for detecting fault modules. 33rd EUROMICRO Conference on Software Engineering and Advanced Applications. Germany
18. Gayatri N, Nickolas S, Reddy AV (2010) Feature selection using decision tree induction in class level metrics dataset for software defect predictions. Proceedings of the world congress on engineering and computer science 2010 vol I WCECS 2010, October 20–22, 2010, San Francisco, USA
19. Koru AG, Liu H (2005) —An investigation of the effect of module size on defect prediction using static measures‖, In: Workshop on predictor models in software engineering, St. Louis, Missouri pp. 1–5

Analysis of Impediments to Sustainability in the Food Supply Chain: An Interpretive Structural Modeling Approach

Jyoti Dhingra Darbari, Vernika Agarwal, Rashi Sharma, and P. C. Jha

1 Introduction

Over the last few years, sustainable supply chain management has become one of the most prominent research fields [1–3]. According to the Department of Economic and Social Affairs, United Nations, the world population is projected to reach 8.5 billion by 2030. Thus, global production will also have to increase in order to match the needs of the increasing population, which will further lead to depletion of the already scarce natural resources. Hence, need of the hour demands a change in the production methods and redesigning of the supply chains in order to minimize any adverse impacts on the environment, society, and economy [4]. Manufacturing organizations also need to focus on regional and local economic development rather than only on their individual growth and performance. Incorporating sustainability into their supply chains must take precedence over other targets and goals to be achieved. Absence of the same can even harm their corporate social responsibility (CSR) image as well as the society. For instance, Mattel toys had to recall its 9 million toys in 2007 due to toxic coating of lead paint in the toy [5]. Another company, Amalgamated Bean Coffee (Tata group subsidiary in India), was reprimanded by the World Bank for not adhering to the social concerns in their plantations. These recent incidents have had a negative impact on the environment and the society. Issues such as these can taint the image of the company which may prove to be economically disastrous for the company. Moreover, the consumers have also now become more vigilant toward the product consumed and its manufacturing conditions. This constant scrutiny by consumers along with the mounting pressure from the government, nongovernmental organizations (NGOs),

J.D. Darbari (✉) • V. Agarwal • R. Sharma • P.C. Jha
Department of Operational Research, University of Delhi, New Delhi, India
e-mail: jydbr@hotmail.com; vernika.agarwal@gmail.com; rashilakhanpal9@gmail.com; jhapc@yahoo.com

© Springer Nature Singapore Pte Ltd. 2018

P.K. Kapur et al. (eds.), *Quality, IT and Business Operations*, Springer Proceedings in Business and Economics, DOI 10.1007/978-981-10-5577-5_5

employees, and shareholders for more environment-friendly products and services poses serious threats to the manufacturing firms [6]. Therefore, there is a severe rise in the number of firms adhering to sustainability standards and practices in order to improve their sustainable image.

The food manufacturing industry is one of the basic examples for such an active business environment in which the consumers demand food products which are sustainably produced [7]. Food supply chains (FSCs) are different from other supply chains because of the perishable nature of inventory, which has a direct influence on the logistics of each part of the supply chain [8]. A typical FSC includes manufacturing enterprises that convert agricultural products and livestock into products used for intermediate or final consumption [9]. FSC is quite complex in nature, often running across boundaries, both internal and external, with many players in action – farmers, fertilizer and pesticide factories, finance providers, component suppliers, transport and logistics services, government, retail shops, and other interconnected businesses. Just like any other supply chain, it also has a negative environmental impact due to greenhouse gas emissions, usage of energy and water resources, soil degradation, biodiversity destruction, etc. In addition to these, some of the other challenges faced by this industry are increasing public attention toward high food standards, price fluctuations [10], incorporation of social practices [11], government laws and regulations, environmental policies, and health and safety guidelines. Thus, in order to address these constraints of FSC, adoption of sustainable practices at each level can provide the much needed impetus for the sector.

In India, which is predominantly an agro-based economy, both food production and food processing sectors play vital roles. As per estimates of the Central Statistics Office, agriculture and allied sectors together contributed 15.35 per cent toward the gross value added during 2015–2016. India is the world's largest milk producer, second largest producer of fruit and vegetables [12], and third largest fish producer. Despite the rise in the agriculture production, there is a lot of wastage due to inadequate infrastructural support like lack of availability of sustainable supply [13, 14], lack of financial resources [15], lack of knowledge and expertise [16], conflict among stakeholders, etc. On the financial and social forefront, farmers are often found to be at the losing end due to involvement of multiple stakeholders in FSC. They have no or little contribution in decisions relating to standards, metrics, best practices, professional negotiations with food buyers, etc. Though there are government policies and acts in place to support farmers like the APMC Act and the minimum support price act, they require intensive monitoring efforts, in terms of enforcement and control. Also, many a times, insufficient access to right information at the right time [17] becomes a deterrent in the path of development.

India has the potential to become a leading food supplier for the whole world with its key competitive advantages – favorable climatic regions, huge percentage of cultivable land, different soil types, large livestock base, nearness to major export terminals, domestic water resources, extended coastline, and economies of scale. In order to thrive on these promising avenues and transforming them into capital gains, adoption of global sustainable practices in FSC is of paramount importance. For a deeper understanding of FSC and to uncover opportunities for development in terms of sustainability, it has become imperative to identify and comprehend barriers along the various stages of the supply chain [4, 18]. Impediments can be of any type – regulatory, social, or environmental. Complete understanding of all of them and their influence on one another can help the manufacturers in improving their triple bottom line performance which relies on the three dimensions of sustainability, namely, economic, environmental, and social. In many studies, importance of sustainability in FSC has been recognized and has been analyzed upon [19–21]. Research shows that incorporating high ecological and societal standards can result in competitive advantage and help in strengthening the overall position of all the parties involved in an FSC [13]. Many initiatives have also been taken in this direction like intensive agriculture [22], food safety [23], food traceability [24], reverse logistics, etc. However, in India, there is a lack of substantial academic research on the feasible ways to effectively incorporate sustainability in the food industry. There is gap between understanding the importance of sustainability and its actual application at different levels in FSC. Among the existing research on FSC, the study conducted by [15] provides the taxonomy of barriers for inclusion of sustainability by food retailers in Sweden. Out of these impediments, ten have been shortlisted based on the empirical evidence and interactions with few of the experts in the food industry. The paper reflects the barriers that are being faced in India obstructing the adoption of sustainability initiatives.

Researchers and practitioners have used many techniques, methods, and processes in the past, to facilitate this understanding. Interpretive structural modeling (ISM) is one such technique which has been used widely in different areas of supply chain management like vendor management [25], identification of enablers of the green supply chain [26], barriers in third-party logistics implementation [27], classification of dependent and independent barriers in implementation of CSR in supply chain management [28], etc. It is a multi-criteria decision-making approach which helps in structuring a set of attributes based on their interactions and interrelationships. The idea of ISM is to use decision-makers' opinions and experience to decompose a set of criteria into subsystems, thereby forming a multilevel hierarchy system. [29] suggest that the implicit and explicit relations between factors describe a situation far more accurate than any factor taken individually. In this study, ISM has been utilized to identify the prominent barriers and understand relationships among them. Taking into account the opinions of industrial experts and academicians as well as an extensive literature survey, ten impediments/barriers have been considered. The core research criteria addressed in the study are the analysis of these impediments of sustainability, and the work is based on the following key objectives:

- To classify the impediments of sustainable FSC in India
- To establish the interrelationships among these identified inhibitors using ISM and MICMAC analysis

The rest of the paper discusses the list of identified impediments in Sect. 2, broadly describes the ISM methodology adopted in Sect. 3, discusses results and conclusion in Sect. 4.

2 Impediments for Adoption of Sustainability Initiatives in FSC

Various impediments for adoption of sustainability initiatives in Indian FSC are identified based on literature review and interactions with the decision-makers, and ten barriers are finalized as listed in Table 1.

3 ISM

Investigation of problems involving multiple complexities requires deliberation of specialists as well as the stakeholders. Structural modelers play an important role in understanding the group view, breaking the complexity down by structuring the problem, and presenting a simplified pictorial representation of the problem for rational decision-making. ISM is a useful method which aids in capturing group perceptions in complex decision-making situations using elementary notions of graph theory [41]. It is a powerful mathematical tool for structuring issues or problems which can be defined in terms of sets of elements and relations [42]. In the present study, ISM methodology is used to derive a structured framework for sustainable FSC to accomplish the following goals:

1. To investigate the interdependence among the barriers (elements) which impact sustainable FSC
2. To classify these barriers in accordance with their degree of reachability
3. To develop a digraph model portraying the complexity of the interdependent relationships in the form of a hierarchical network

3.1 ISM Methodology

A stepwise summary of the ISM methodology as applied in the present work is as follows [43]:

Step 1. Establish pairwise relationships among barriers (elements) using the symbols V, A, X, or O where $i \rightarrow j$ symbolizes that barrier i "influences" barrier j. Mathematically it is interpreted as: element i is reachable from element j. The pairwise matrix so obtained is called the structural self-interaction matrix (SSIM).

Table 1 Barriers to sustainability initiatives in FSC

	Barriers	Description	Source
B1	Conflict of interests between product sustainability policy and free trade	Lack of integration of sustainability objective in the free trade policy	[15]
B2	Lack of governmental leadership in outlining the vision for sustainability and responsibilities of food retailers	Absence of government incentives for innovative, efficient, and environmentally effective products. Also, there is leniency in enforcement of existing laws pertaining to sustainability	[30]
B3	Lack of financial resources	Insufficient funds for sustainability projects or the consideration that after implementing sustainable initiatives, the return on investment (ROI) period will be very long	[31]
B4	Lack of knowledge and expertise	Inadequate awareness to understand the benefits of sustainability and little or no skills to integrate the same at different stages of FSC	[32]
B5	Lack of power over suppliers	With lack of proper guidelines and policies, there is no accountability of the suppliers as far as the quality, quantity, and timing of the supply	[15, 33]
B6	Lack of availability of sustainable supply	Reliable and sustainable supply is not easily available which hinders the manufacturing of a sustainable product	[34]
B7	Higher prices for sustainable products	Initial investment for the incorporation of sustainability is high, thereby leading to higher prices of the final product	[35]
B8	Complexity of supply chains' configuration	FSC is complex in nature due to involvement of multiple players and perishable nature of inventory. Also, it is effected by the risks associated with regularity, natural, political, and economic instability	[36, 37]
B9	Lack of sufficient levels of consumer demand	Higher prices for sustainable products and lack of consumer awareness result in low demand	[38, 39]
B10	Lack of scientific framework to identify the most profound sustainability impacts	There is no structured analytical framework developed due to lack of research and experience for a course of action to be followed for adopting sustainability practices	[40]

Table 2 Contextual relation

Symbol	$i \to j$	$j \to i$	*(i,j)*th entry	*(j,i)*th entry
V	√	×	1	0
A	×	√	0	1
X	√	√	1	1
O	×	×	0	0

Table 3 SSIM

	B10	B9	B8	B7	B6	B5	B4	B3	B2
B1	O	O	V	V	V	V	O	A	A
B2	O	O	O	O	V	O	V	O	
B3	V	V	V	O	O	O	V		
B4	V	O	O	O	O	O			
B5	A	0	A	O	A				
B6	O	O	A	O					
B7	O	A	O						
B8	O	V							
B9	O								

Step 2. Derive the adjacency matrix from SSIM by replacing "1" or "0" for the symbols (as given in Table 2). It is also called the initial reachability matrix.

Step 3. Develop the final reachability matrix from the adjacency matrix by incorporating all the transitivities. This ensures that if barrier *i* influences barrier *j* and barrier *j* influences barrier *k*, then barrier *i* influences barrier *k*.

Step 4. Partition the reachability matrix into various levels by checking the reachability and antecedent sets at each level.

Step 5. Derive the digraph from the reachability matrix and convert into an ISM model after deleting all the transitivities.

Step 6. Review the ISM model for any conceptual inconsistency and modify if necessary.

Step 7. MICMAC analysis can be carried out for analyzing the driving and dependence power of the barriers.

The pairwise matrix for the barriers is given in Table 3. The *reachability matrix* is now derived from the *adjacency matrix* by ensuring the transitivity of the binary relation (Table 4). Value "1" at the *(i,j)*th entry of the reachability matrix indicates that the element *j* is reachable from element *i* or else the value is "0." This matrix forms the basis for the graphical representation of the system.

3.2 Level Partitioning

The hierarchical structure is developed by deriving the reachability sets and the antecedent sets at each level. For each level *k*, the *reachability set* of element *i* denoted as R_i consists of all those elements which are reachable from element *i*,

Table 4 Adjacency matrix

	B1	B2	B3	B4	B5	B6	B7	B8	B9	B10
B1	1	0	0	0	1	1	1	1	0	0
B2	1	1	0	1	0	1	0	0	0	0
B3	1	0	1	1	0	0	0	1	1	1
B4	0	0	0	1	0	0	0	0	0	1
B5	0	0	0	0	1	0	0	0	0	0
B6	0	0	0	0	1	1	0	0	0	0
B7	0	0	0	0	0	0	1	0	0	0
B8	0	0	0	0	1	1	0	1	1	0
B9	0	0	0	0	0	0	1	0	1	0
B10	0	0	0	0	1	0	0	0	0	1

Table 5 Final reachability matrix

	B1	B2	B3	B4	B5	B6	B7	B8	B9	B10	Dr. P
B1	1	0	0	0	1	1	1	1	0	0	5
B2	1	1	0	1	1	1	1	1	0	1	8
B3	1	0	1	1	1	0	1	1	1	1	8
B4	0	0	0	1	0	0	0	0	0	1	2
B5	0	0	0	0	1	0	0	0	0	0	1
B6	0	0	0	0	1	1	0	0	0	0	2
B7	0	0	0	0	0	0	1	0	0	0	1
B8	0	0	0	0	1	1	1	1	1	0	5
B9	0	0	0	0	0	0	1	0	1	0	2
B10	0	0	0	0	1	0	0	0	0	1	2
Dp. P	3	1	1	3	7	4	6	4	3	4	

Dr. P – driving power, Dp. P – dependence power

whereas the *antecedent set* of element *i* denoted as A_i is set of all those elements from which element *i* is reachable. If the intersection of the two sets is same as R_i, then the element can be denoted as the *k*th-level element and its corresponding row and column are deleted from the matrix for the next level. The process is continued till each element is assigned a level. Table 5 demonstrates the iterations and partitioning of the reachability matrix into six levels.

The *directed graph* representing the hierarchical network is drawn with the elements represented by the nodes and the directed line connecting element *i* and element *j* denoting the nature of the relation between them. The elements in the last level appearing at the bottommost level of the diagraph and the upward-directed links are drawn using the contextual relation defined in the adjacency matrix. To simplify the above process, a *conical matrix* can be derived by rearranging the rows and columns of the reachability matrix in the decreasing order in which the levels are obtained (Table 6). A graphical representation of this matrix (after removing its transitivities) generates the desired ISM model as demonstrated by Fig. 1 (Table 7).

Table 6 Level partitions

	RS	AS	RS∩ AS	Level
B1	1,5,6,7,8	1,2,3	1	IV
B2	1,2,4,5,6,7,8,10	2	2	VI
B3	1,3,4,5,6,7,8,9,10	3	3	VI
B4	4,10	2,3,4	4	V
B5	5	1,2,3,5,6,8,10	5	I
B6	5,6	1,2,3,6,8,10	6	II
B7	7	1,2,3,7,8,9	7	I
B8	5,6,7,8,9	2,8	8	III
B9	7,9	1,2,3,8,9	9	II
B10	5,6,10	2,3,4,10	10	IV

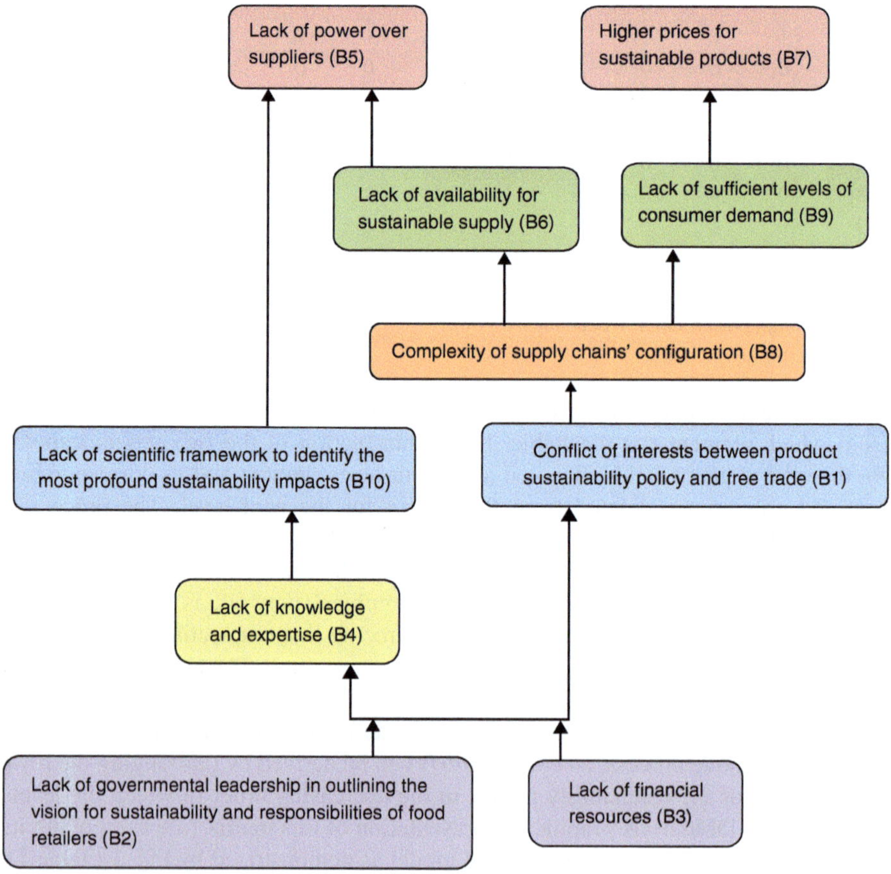

Fig. 1 ISM model

Table 7 Conical matrix

	B1	B2	B3	B4	B5	B6	B7	B8	B9	B10	Level
B2	1	1	0	1	~~1~~	1	~~1~~	~~1~~	0	~~1~~	VI
B3	1	0	1	1	~~1~~	0	~~1~~	1	1	1	VI
B4	0	0	0	1	0	0	0	0	0	1	V
B1	1	0	0	0	1	1	1	1	0	0	IV
B10	0	0	0	0	1	0	0	0	0	1	IV
B8	0	0	0	0	1	1	~~1~~	1	1	0	III
B6	0	0	0	0	1	1	0	0	0	0	II
B9	0	0	0	0	0	0	1	0	1	0	II
B5	0	0	0	0	1	0	0	0	0	0	I
B7	0	0	0	0	0	0	1	0	0	0	I

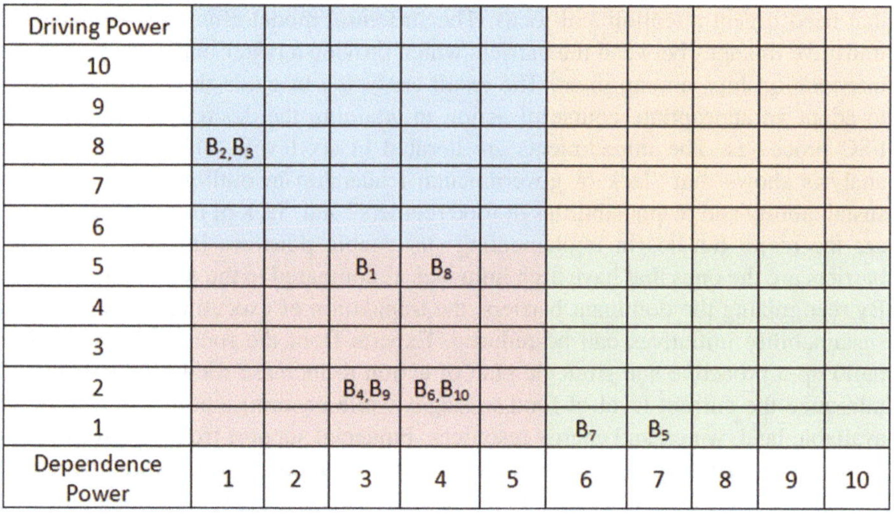

Fig. 2 Classification of enablers

3.3 *MICMAC Analysis*

The driving power of an element *i* is the total number of elements which are reachable from element *i*, and its dependence power is the total number of elements from which element *i* is reachable. For each element *i*, Table 5 indicates values of both by summing the "1s" in the *i*th row and the *j*th column, respectively. Using these values, the elements are classified into four categories [44]: autonomous (quadrant I), dependent (quadrant II), linkage (quadrant III), and independent (quadrant IV) as shown in Fig. 2.

4 Conclusion

The notion of sustainability is developing into the de facto baseline and common code of practice in global food supply chain. Increasing awareness among customers and strict imposition of international norms have channelized the food sector's initiatives toward redesigning their business strategies. Emphasis is now being laid on inclusion of sustainable practices in the form of animal welfare issues, farmer equity, and socio-ecological performance. However, in India, it is still a challenge to emulate these sustainable practices, and there is a pressing need to first identify the barriers which hinder the implementation of sustainability practices in FSC. The present study delves upon some of the influential barriers that thwart the growth of sustainability in FSC and create a substantial challenge for the decision-makers. ISM methodology is utilized for identifying the most crucial barriers that need urgent attention and focus. The structural model efficiently displays the transitive linkages between the barriers which provide a better understanding of the interrelationships among them. The result analysis can guide the decision makers to adopt an appropriate course of action in attaining the desired sustainability in FSC processes. The impediments are iterated in six levels. The outcome of the analysis shows that "lack of governmental leadership in outlining the vision for sustainability and responsibilities of food retailers" and "lack of financial resources" are the major hurdles in implementing sustainable practices in FSC. These two barriers are the ones that have high influence as compared to the remaining barriers. By recognizing the dominant barriers, the trepidation of executing the adoption of sustainability initiatives can be reduced. Experts from the food industry need to build up a proactive and strategic plan of action to increase food productivity and minimize the current level of food wastages while ensuring optimal utilization of available land, water, and energy resources. Financial support from the government in the form of subsidy, rewards, and awards and enforcement of laws would provide the much needed encouragement to the food industry in successful execution of the course of action. It would eventually lead to a wider market for sustainably produced food products, thereby creating favorable opportunities for socioeconomic growth for all the stakeholders.

References

1. Seuring S, Martin M (2008) From a literature review to a conceptual framework for sustainable supply chain management. J Clean Prod 16(15):1699–1710
2. Carter CR, Liane Easton P (2011) Sustainable supply chain management: evolution and future directions. Int J Phys Distrib Logist Manag 41(1):46–62
3. Ahi P, Searcy C (2013) A comparative literature analysis of definitions for green and sustainable supply chain management. J Clean Prod 52:329–341
4. Kaplinsky R, Morris M (2013) A handbook for value chain research. International Development Research Center, Ottawa

5. Hora M, Bapuji H, Roth AV (2011) Safety hazard and time to recall: the role of recall strategy, product defect type, and supply chain player in the US toy industry. J Oper Manag 29(7):766–777
6. Byrne PJ, Ryan P, Heavey C (2013) Sustainable logistics: a literature review and exploratory study of Irish based manufacturing organizations. Int J Eng Technol Innov 3(3):200–213
7. Beske P, Land A, Seuring S (2014) Sustainable supply chain management practices and dynamic capabilities in the food industry: a critical analysis of the literature. Int J Prod Econ 152:131–143
8. Yu M, Nagurney A (2013) Competitive food supply chain networks with application to fresh produce. Eur J Oper Res 224(2):273–282
9. Christopher M (2010) Logistics and supply chain management: creating value adding networks, 4th edn. Financial Times Prentice Hall, Harlow
10. Van der Vorst JGAG, Beulens AJM, van Beek P (2000) Modelling and simulating multi-echelon food systems. Eur J Oper Res 122(2):354–366
11. Hassini E, Surti C, Searcy C (2012) A literature review and a case study of sustainable supply chains with a focus on metrics. Int J Prod Econ 140(1):69–82
12. Saxena M (2015) Indian horticulture database-2014. National Horticulture Board, Gurgaon
13. Smith BG (2008) Developing sustainable food supply chains. Philos Trans R Soc B: Biol Sci 363(1492):849–861
14. Johnson M (2004) Marks & Spencer implements an ethical sourcing program for its global supply chain. J Organ Excell 23(2):3–16
15. Chkanikova O, Mont O (2015) Corporate supply chain responsibility: drivers and barriers for sustainable food retailing. Corp Soc Responsib Environ Manag 22(2):65–82
16. Hall J (2006) Environmental supply chain innovation. In: Greening the supply chain. Springer, London, pp 233–249
17. Joshi R, Banwet DK, Shankar R (2009) Indian cold chain: modeling the inhibitors. Br Food J 111(11):1260–1283
18. Grandori A (2015) Improving organization forms in the agri-food industry. Br Food J 117(10):2418–2434
19. Gold S, Hahn R, Seuring S (2013) Sustainable supply chain management in "base of the pyramid" food projects – a path to triple bottom line approaches for multinationals? Int Bus Rev 22(5):784–799
20. Egilmez G, Kucukvar M, Tatari O, Bhutta MKS (2014) Supply chain sustainability assessment of the US food manufacturing sectors: a life cycle-based frontier approach. Resour Conserv Recycl 82:8–20
21. Del Borghi A, Gallo M, Strazza C, Del Borghi M (2014) An evaluation of environmental sustainability in the food industry through life cycle assessment: the case study of tomato products supply chain. J Clean Prod 78:121–130
22. Brussaard L, Caron P, Campbell B, Lipper L, Mainka S, Rabbinge R, Babin D, Pulleman M (2010) Reconciling biodiversity conservation and food security: scientific challenges for a new agriculture. Curr Opin Environ Sustain 2(1):34–42
23. Trienekens J, Zuurbier P (2008) Quality and safety standards in the food industry, developments and challenges. Int J Prod Econ 113(1):107–122
24. Aung MM, Chang YS (2014) Traceability in a food supply chain: safety and quality perspectives. Food Control 39:172–184
25. Mandal A, Deshmukh SG (1994) Vendor selection using interpretive structural modelling (ISM). Int J Oper Prod Manag 14(6):52–59
26. Diabat A, Govindan K (2011) An analysis of the drivers affecting the implementation of green supply chain management. Resour Conserv Recycl 55(6):659–667
27. Diabat A, Khreishah A, Kannan G, Panikar V, Gunasekaran A (2013) Benchmarking the interactions among barriers in third-party logistics implementation: an ISM approach. Benchmarking: An Int J 20(6):805–824
28. Faisal MN (2010) Sustainable supply chains: a study of interaction among the enablers. Bus Process Manag J 16(3):508–529

29. Attri R, Dev N, Sharma V (2013) Interpretive structural modelling (ISM) approach: an overview. Res J Manag Sci 2(2):3–8
30. Jones P, Comfort D, Hillier D (2008) Moving towards sustainable food retailing? Int J Retail Distrib Manag 36(12):995–1001
31. Ki-Hoon L (2009) Why and how to adopt green management into business organizations? Manag Decis 47(7):1101
32. Wong KY, Aspinwall E (2004) Characterizing knowledge management in the small business environment. J Knowl Manag 8(3):44–61
33. Grimm JH, Hofstetter JS, Sarkis J (2014) Critical factors for sub-supplier management: a sustainable food supply chains perspective. Int J Prod Econ 152:159–173
34. Dincer I, Rosen MA (1999) Energy, environment and sustainable development. Appl Energy 64(1):427–440
35. Tukker A, Tischner U (2006) Product-services as a research field: past, present and future. Reflections from a decade of research. J Clean Prod 14(17):1552–1556
36. Lambert DM, Cooper MC, Pagh JD (1998) Supply chain management: implementation issues and research opportunities. Int J Logist Manag 9(2):1–20
37. Vurro C, Russo A, Perrini F (2009) Shaping sustainable value chains: network determinants of supply chain governance models. J Bus Ethics 90(4):607–621
38. Jørgensen TH (2008) Towards more sustainable management systems: through life cycle management and integration. J Clean Prod 16(10):1071–1080
39. Priem RL, Li S, Carr JC (2012) Insights and new directions from demand-side approaches to technology innovation, entrepreneurship, and strategic management research. J Manag 38(1):346–374
40. Kajikawa Y (2008) Research core and framework of sustainability science. Sustain Sci 3(2):215–239
41. Malone DW (1975) An introduction to the application of interpretive structural modeling. Proc IEEE 63(3):397–404
42. Janes FR (1988) Interpretive structural modelling: a methodology for structuring complex issues. Trans Inst Meas Control 10(3):145–154
43. Kannan G, Pokharel S, Sasi Kumar P (2009) A hybrid approach using ISM and fuzzy TOPSIS for the selection of reverse logistics provider. Resour Conserv Recycl 54(1):28–36
44. Kannan G, Noorul Haq A (2007) Analysis of interactions of criteria and sub-criteria for the selection of supplier in the built-in-order supply chain environment. Int J Prod Res 45(17):3831–3852

Analysis of Module-Based Software Reliability Growth Model Incorporating Imperfect Debugging and Fault Reduction Factor

Madhu Jain, Anuradha Jain, and Ritu Gupta

1 Introduction

In the context of software development life cycle, testing has a vital role for the assessment of the quality metrics which may be helpful in fault detection and removal process. The prime objective of any level of testing is to detect software failures so that the defects may be identified and corrected or removed from the software. The software testing mainly involves module testing, integration testing, and system and acceptance testing. Initially software modules are tested independently, and then after the interfaces among modules are tested during the integration testing which is performed by the developers as well as test engineers. System testing is performed by professional testing engineers on the completed software product including all the external interfaces before it is introduced to the market. The acceptance testing is mainly done during the operational phase. The prediction of the system reliability of the complex software system has been turned out as a most important quality metric in present scenario, and thus reliability of the software modules or the system software as a whole needs to be quantified under certain reliability requirements subject to testing schedule and resource constraints. In recent past, the performance prediction of the software system

M. Jain
Department of Mathematics, Indian Institute of Technology Roorkee, Roorkee 247667, Uttarakhand, India
e-mail: madhufma@iitr.ernet.in

A. Jain
Institute of Basic Science, Khandari, Agra 282002, UP, India
e-mail: anujain_ibs@yahoo.com

R. Gupta (✉)
AIAS, Amity University, 125, Noida 201303, UP, India
e-mail: rgupta@amity.edu

© Springer Nature Singapore Pte Ltd. 2018

P.K. Kapur et al. (eds.), *Quality, IT and Business Operations*, Springer Proceedings in Business and Economics, DOI 10.1007/978-981-10-5577-5_6

as far as software reliability is concerned was done by Kumar et al. [1]. They considered fault detection phenomenon in the testing phase when modification can be made to the software during the testing phase. To allocate the available testing resources, software developer divides the software into modules to allocate the testing resources. Reference [2] has described the optimal allocation problem in modular software systems during the testing phase. They elaborated the problem of how to allocate and optimize the specified testing resources among all the modules. In this direction, reference [3] developed a model for large software systems by integrating a number of relatively small and independent modules that are tested independently during module testing phase and discussed optimal testing resource allocation problem too. Quasi renewal process which can be used to estimate the time delay period and to check the priority of the observed faults has been employed in order to integrate the information about the software testing and fault debugging activities in the software testing environment [4]. Reference [5] has discussed the software modularity and investigated how metrics for high level abstractions of complex and large module-based system can help to inform on the composition of high level modules.

In the software reliability modeling and analysis, fault detection and correction activities reflect the failure occurrence rate directly. Some faults which are not detected during development phase may produce software unreliability and create unexpected problems for the users. To overcome such types of problems, reference [6] proposed pre-compiled fault detection technique to detect and correct faults before a source code is compiled. Generally it is assumed that the fault removal time becomes a constant, but fault removal time usually changes with the moment when the fault occurs. A time-dependent delay function has also been proposed that is used for the delay of fault removal process [7]. In the literature, the effect of imperfect debugging is prevalent in favor of removal of the faults during software reliability modeling. In this context, reference [8] presented discrete and continuous time models under software failure occurrence phenomenon incorporating imperfect debugging and fault generation with learning practice. The software reliability growth models with imperfect debugging have also been studied subject to some other failure factors like fault reduction factor and multiple change point [9]. The consumption of testing resources in the testing is important for the developers and to be measured from cost view point. Keeping this concern in attention, reference [10] discussed the testing effort-dependent model which includes fault detections and corrections under imperfect debugging. A recent application on mobile software has been presented to study the reliability analysis and to examine the effect of the debugging process on mobile software for the network traffic in mobile clouds [11]. Moreover, for estimating the number of remaining faults and measuring the reliability of the software, an imperfect software debugging model has been proposed which incorporates a log-logistic distribution-based fault content function [12]; it can be helpful to capture the increasing and decreasing nature or characteristics of the fault introduction rate per fault.

For the software reliability assessment, some factors besides the imperfect debugging which are directly related to the environmental factors are debugging time lag and human learning process. Any changes on these factors show the influence on the fault reduction factor. Fault reduction factor (FRF) is defined by the ratio of net fault reduction to failure experienced (in terms of ratio of faults and failures) or net number of faults removed in the proportion to the failure experienced. Reference [13] analyzed the effect of various parameters which are needed for the modeling of the software reliability. They studied some parameters like fault reduction factor which may be used to control the growth of the software reliability. The fault reduction factor (FRF) plays significant role to handle the situation of failure occurrence while studying the growth of software reliability. In this direction, reference [14] has presented an inflection S-shaped software reliability model that incorporates fault reduction function.

Optimization problem related to software release with the reliability and absolute/optimal cost of the system is a major concern from the developer as well as customer view point. The problem of determining the optimal time at which the system testing is stopped and then it is ready for the use in the operational phase has attracted many researchers who have developed software reliability models to propose the software release policies in different contexts. References [15–19] and many others have investigated software release policies in variant contexts to minimize the development cost while satisfying a reliability objective. Reference [20] developed the software cost model by incorporating the logistic-exponential testing coverage function and suggested the optimal release policy based on the number of remaining faults. As per the requirement in global competition, reference [21] explored multiple optimal release policies to confine the effect of high levels of severe faults generated in the software.

In this paper, we develop a software reliability model which consists of N-independent modules. The modules are different from each other and have different kinds of faults. The faults are considered as simple, hard, and complex faults in each module and are tested individually for removing the faults that remain in the modules of the software. A flexible software reliability growth model is proposed by considering the imperfect debugging and fault reduction factor during the software testing in the module phase of the software. We derive explicit expression for the mean value function for N-independent module-based system which is categorized from the different kinds of faults. The rest of the paper is organized as follows. Section 2 describes the software reliability growth model by stating the requisite assumptions and notations. Section 3 contains governing equations and reliability analysis. For the prediction of reliability indices, some formulae are derived in Sect. 4. To validate the analytical results, numerical results are provided in Sect. 5. The concluding remarks and significance of the work done are stated in the final Sect. 6.

2 Model Description

Consider a software system consisting of N modules; each module has different characteristics from the other module and has different types of faults contents. To develop a module-based software reliability growth model, it is assumed that the software has K-types of faults. The three-stage time-dependent processes (fault observation, fault isolation, and fault removal) are considered for each module to develop module-based software reliability growth model. To make more versatile model, we incorporate more realistic features by incorporating the concepts of fault reduction factor and imperfect debugging.

To develop the module-based software reliability growth model (SRGM) which incorporates the imperfect debugging and fault reduction factor (FRF), we consider the nonhomogeneous Poisson process (NHPP). For mathematical formulation, the following assumptions are made:

- The software system is composed of N-independent modules that are tested individually. The software faults in each module are categorized as simple-, hard-, and complex-type faults which can be estimated by SRGM by including the concepts of the imperfect debugging and fault reduction factor.
- Each time when the failure occurs, the corresponding faults are removed immediately, but besides it, some faults are introduced during the testing process as such imperfect debugging of faults is taken into account.
- Each module has different types of faults as simple-, hard-, and complex-type faults; the different type faults have the different failure characteristics.
- When detected, faults are removed at time t; it is possible to introduce some new faults with introduction rate $\eta_{k_n}^{(n)}(t)$ in the nth module.

Notations

$m(t)$	Expected number of total faults removed in the given time interval (0, t] from the module-based software.
$m_{\text{obs},k_n}^{(n)}(t)$	Expected number of k_n types (simple, hard, and complex) faults observed in the nth module at time t.
$m_{\text{iso},k_n}^{(n)}(t)$	Expected number of k_n (simple, hard, and complex) types of faults isolated from the nth module.
$m_{\text{rem},kn}^{(n)}(t)$	Expected number of k_n-types of faults removed from nth module at time t.
$a(t)$	Sum of the number of faults to be eventually detected and the number of new faults introduced by time t.
$a_{k_n}^{(n)}$	Initial number of k_n-types of faults (as simple, hard, and complex faults) in the nth module.
$\eta_{k_n}^{(n)}(t)$	Imperfect debugging factor for k_n-types of faults in the nth module at time t.
$\mu_{k_n}^{(n)}(t)$	Time-variable fault reduction factor for k_n-types of faults for nth module at time t.
$r_{k_n}^{(n)}(t)$	Fault observation (detection)/isolation/removal rate function for k_n-types of faults for nth module at time t.
η	Fault introduction rate.

x	Mission time of the software.
P	Additional fraction of faults observed during testing.
C_0	Cost of setup for the software.
C_1	Cost of removing a fault during software testing before releasing.
C_2	Cost of removing a fault after releasing the software.
C_3	Loss due to software failure or risk cost.
$C_4(T)$	Opportunity cost function.
T	Testing time of the software.
T_c	Software life cycle length.
α_0	The scale coefficient.
α	The intercept value.
α_2	The degree of opportunity loss in time.

3 Governing Equations and Analysis

All faults which are presented in the module-based software system have been categorized by simple-, hard-, and complex-type faults of the software. The reliability of each module will be different due to the different failure characteristics of fault contents. These faults may require different fault reduction factor and different error detection rate for each module. Software system consists of the N modules which are different from each other from the fault content and testing requirement point of view. Each module has different initial number of faults and dissimilar type of faults of different severity. There are total K-types of faults in the software whereas nth ($n = 1,2,\ldots,N$) module has k_n type of faults ($k_n = 1, 2, \ldots, K_n$) which may be simple, hard, and complex. Each nth module consists of p_n, simple type faults; q_n, hard type faults; and r_n, complex type faults, i.e.:

$$K = \sum_{n=1}^{N} K_n = \sum_{n=1}^{N} \left[\sum_{i=1}^{p_n} k_i^{(n)} + \sum_{j=p_n+1}^{p_n+q_n} k_j^{(n)} + \sum_{k=p_n+q_n+1}^{p_n+q_n+r_n} k_k^{(n)} \right] \tag{1}$$

where $k_n^{(n)} = \sum_{i=1}^{p_n} k_i^{(n)} + \sum_{j=p_n+1}^{p_n+q_n} k_j^{(n)} + \sum_{k=p_n+q_n+1}^{p_n+q_n+r_n} k_k^{(n)}$ and $K_n = \sum_{n=1}^{N} k_n^{(n)}$.

Mean Value Function

Mean value function for the software reliability growth model (SRGM) with fault reduction factor under imperfect debugging environment for observation, isolation, and removal processes of nth ($n = 1, 2, \ldots, N$) module can be expressed in terms of the following differential equations:

$$\frac{\partial}{\partial t} m_{\text{obs},k_n}^{(n)}(t) = r_{k_n}^{(n)}(t)\left[a_{k_n}^{(n)}(t) - m_{\text{obs},k_n}^{(n)}(t)\right] \tag{2}$$

$$\frac{\partial}{\partial t} m_{\text{iso},k_n}^{(n)}(t) = r_{k_n}^{(n)}(t)\left[m_{\text{obs},k_n}^{(n)}(t) - m_{\text{iso},k_n}^{(n)}(t)\right] \tag{3}$$

$$\frac{\partial}{\partial t} m_{\text{rem},k_n}^{(n)}(t) = r_{k_n}^{(n)}(t)\left[m_{\text{iso},k_n}^{(n)}(t) - m_{\text{rem},k_n}^{(n)}(t)\right] \tag{4}$$

$$\text{where } r_{k_n}^{(n)}(t) = r_{k_n}^{(n)} \mu_{k_n}^{(n)}(t) \tag{5}$$

It is assumed that the fault reduction factor is a constant, so that:

$$\mu_{k_n}^{(n)}(t) = \mu_{k_n}^{(n)}, 0 < \mu_{k_n}^{(n)} \leq 1.$$

When the detected faults are removed, then there is a possibility that the new faults are introduced with rate $\eta_{k_n}^{(n)}(t)$; the imperfect debugging fault analysis for nth module is taken as:

$$\frac{\partial a_{k_n}^{(n)}(t)}{\partial t} = \eta_{k_n}^{(n)}(t) \frac{\partial m_{\text{obs},k_n}^{(n)}(t)}{\partial t} \tag{6}$$

where $\eta_{k_n}^{(n)}(t) = \eta_{k_n}^{(n)}$ is considered to be constant. The initial condition is imposed as $t = 0, m_{\text{obs},k_n}^{(n)}(t = 0) = 0$ and $a_{k_n}^{(n)}(0) = a_{k_n}^{(n)}$. Now the number of faults to be eventually detected is determined by:

$$a_{k_n}^{(n)}(t) = a_{k_n}^{(n)} + \eta_{k_n}^{(n)} m_{\text{obs},k_n}^{(n)}(t) \tag{7}$$

By substituting the values of imperfect debugging parameter from (7) to (2), we obtain: $t = 0, m_{\text{obs}}^{(n)}(0) = m_{\text{iso}}^{(n)}(0) = m_{\text{corr}}^{(n)}(0) = 0$

$$\begin{aligned} m_{\text{obs},k_n}^{(n)}(t) &= \frac{a_{k_n}^{(n)}}{\left(1-\eta_{k_n}^{(n)}\right)}\left[1 - e^{-\mu_{k_n}^{(n)} r_{k_n}^{(n)} t\left(1-\eta_{k_n}^{(n)}\right)}\right] \\ &= \frac{a_{k_n}^{n}}{u^{(n)}}\left[1 - e^{-v^{(n)} u^{(n)} t}\right] \end{aligned} \tag{8}$$

$$m_{\text{iso},k_n}^{(n)}(t) = \frac{a_{k_n}^{(n)}}{\left(1-\eta_{k_n}^{(n)}\right)}\left[\left\{1-e^{-\mu_{k_n}^{(n)} r_{k_n}^{(n)} t}\right\} - \frac{e^{-\mu_{k_n}^{(n)} r_{k_n}^{(n)} t}}{\eta_{k_n}^{(n)}}\left\{e^{-\left(\eta_{k_n}^{(n)}-1\right)} - 1\right\}\right]$$
$$= \frac{a_{k_n}^{(n)}}{u^{(n)}}\left[\left\{1-e^{-v^{(n)}t}\right\} - \frac{e^{-v^{(n)}t}}{\eta_{k_n}^{(n)}}\left\{e^{-u^{(n)}} - 1\right\}\right] \tag{9}$$

$$m_{\text{rem},k_n}^{(n)}(t) = \frac{a_{k_n}^{(n)}}{\left(1-\eta_{k_n}^{(n)}\right)}\left[\left\{1-e^{-\mu_{k_n}^{(n)} r_{k_n}^{(n)} t}(1+t)\right\}\right.$$
$$\left.+\frac{r_{k_n}^{(n)}\mu_{k_n}^{(n)}}{\eta_{k_n}^{(n)}}\left\{\left(1-e^{-\left(\eta_{k_n}^{(n)}-1\right)}\right) te^{-\mu_{k_n}^{(n)} r_{k_n}^{(n)} t}\right\}\right] \tag{10}$$
$$= \frac{a_{k_n}^{(n)}}{u^{(n)}}\left[\left\{1-e^{-v^{(n)}t}(1+t)\right\} + \frac{v^{(n)}}{\eta_{k_n}^{(n)}}\left\{\left(1-e^{u^{(n)}}\right) te^{-v^{(n)}t}\right\}\right]$$

where $1-\eta_{k_n}^{(n)} = u^{(n)}$and $\mu_{k_n}^{(n)} r_{k_n}^{(n)} = v^{(n)}$.

The mean value function for our proposed model that incorporates the fault reduction factor under imperfect debugging environment is obtained by using:

$$m^*(t) = \sum_{n=1}^{N} m_{\text{obs},k_n}^{(n)}(t) + \sum_{n=1}^{N}\sum_{k_n=1}^{K_n}\left[m_{\text{iso},k_n}^{(n)}(t) + m_{\text{rem},k_n}^{(n)}(t)\right] \tag{11}$$

Here, our main interest is to evaluate the total expected number of faults removed from each of the modules. The mean value function for the software reliability growth model based on module-based system consisting of the fault reduction factor and operating under imperfect debugging environment is:

$$m(t) = \sum_{n=1}^{N}\sum_{k_n=1}^{K_n}\left[m_{\text{rem},k_n}^{(n)}(t)\right]$$
$$= \sum_{n=1}^{N}\sum_{k_n=1}^{K_n}\frac{a_{k_n}^{(n)}}{u^{(n)}}\left[\left\{1-e^{-v^{(n)}t}(1+t)\right\} + \frac{v^{(n)}}{\eta_{k_n}^{(n)}}\left\{\left(1-e^{u^{(n)}}\right) te^{-v^{(n)}t}\right\}\right] \tag{12}$$

where $\mu_{k_n}^{(n)} > 0, \eta_{k_n}^{(n)} > 0, r_{k_n}^{(n)} > 0, a_{k_n}^{(n)} > 0$.

4 Performance Measures

We assume the modules of the software system are independent from the software testing point of view. Software testing is carried out at the modular level. Now we obtain the software reliability and expected maintenance cost for modular-based SRGM as follows:

4.1 Software Reliability

Environment factors are directly related to the fault reduction factor. Imperfect debugging and debugging time lag between fault detection and correction process affect the fault reduction factor. The impact of changes in environment factors is directly related to fault reduction factor that affects the software testing time.

The software reliability function is given by:

$$\begin{aligned} R(x/T) &= \exp\left[-\{m(T+x) - m(T)\}\right] \\ &= \exp\left[\sum_{n=1}^{N} \sum_{k_n=1}^{K_n} \frac{a_{k_n}^{(n)}}{u^{(n)}} \left[\left\{\left(e^{-v^{(n)}T}\left\{e^{-v^{(n)}x}(1+T+x) - (1+T)\right\}\right)\right\}\right.\right. \\ &\quad \left.\left. + \left\{\frac{v^{(n)}}{\eta_{k_n}^{(n)}}\left(Te^{-v^{(n)}T}\left\{1 - e^{u^{(n)}}\right\} - (T+x)\,e^{-v^{(n)}(T+x)}\left\{1 - e^{u^{(n)}}\right\}\right)\right\}\right]\right] \end{aligned} \tag{13}$$

4.2 Expected Maintenance Cost

The objective of our model is to find the reliability of each module so that the reliability of the system will be maximized within a given budget. For evaluating the expected maintenance cost of the software, various cost factors, viz., risk cost and opportunity cost, are important from the customer's point of view. The total expected maintenance cost includes the setup cost, cost of removing a fault before and after releasing a software, risk cost, and opportunity cost.

The total expected maintenance cost of the module-based software is given by:

$$\begin{aligned} E\{C(T)\} = C_0 + C_1(1+P)\,m(T) + C_2\,[m(T_c) \\ - (1+P)\,m(T)] + C_3(T) + C_4(T) \end{aligned} \tag{14}$$

where the risk cost due to software failure after releasing the software is computed using:

$$C_3(T) = C_3\{1 - R(x/T)\} \tag{15}$$

The opportunity cost of the software is computed in the form of power law as:

$$C_4(T) = \alpha_0(\alpha_1 + T)^{\alpha_2} \tag{16}$$

By using (12), (13), (15), and (16), (14) gives:

$$\begin{aligned}
E\{C(T)\} = {} & C_0 + (1+P)(C_1 - C_2)\sum_{n=1}^{N}\sum_{k_n=1}^{K_n}\frac{a_{k_n}^{(n)}}{u^{(n)}}\left[\left\{1 - e^{-v^{(n)}T}(1+T)\right\}\right. \\
& \left. + \frac{v^{(n)}}{\eta_{k_n}^{(n)}}\left\{\left(1 - e^{u^{(n)}}\right)Te^{-v^{(n)}T}\right\}\right] \\
& + C_2\sum_{n=1}^{N}\sum_{k_n=1}^{K_n}\frac{a_{k_n}^{(n)}}{u^{(n)}}\left[\left\{1 - e^{-v^{(n)}T_c}(1+T_c)\right\} + \frac{v^{(n)}}{\eta_{k_n}^{(n)}}\left\{\left(1 - e^{u^{(n)}}\right)T_c e^{-v^{(n)}T_c}\right\}\right] \\
& + C_3\left\{1 - \exp\left[\sum_{n=1}^{N}\sum_{k_n=1}^{K_n}\frac{a_{k_n}^{(n)}}{u^{(n)}}\left[\left\{\left(e^{-v^{(n)}T}\left\{e^{-v^{(n)}x}(1+T+x) - (1+T)\right\}\right)\right\}\right.\right.\right. \\
& \left.\left.\left. + \left\{\frac{v^{(n)}}{\eta_{k_n}^{(n)}}\left(Te^{-v^{(n)}T}\left\{1 - e^{u^{(n)}}\right\} - (T+x)e^{-v^{(n)}(T+x)}\left\{1 - e^{u^{(n)}}\right\}\right)\right\}\right]\right]\right\} \\
& + C_4\alpha_0(\alpha_1 + T)^{\alpha_2}
\end{aligned} \tag{17}$$

The prime goal of present investigation is to determine the optimal software release time which minimizes the total expected maintenance cost of the software to achieve the desired level of reliability. The prespecified minimum achieved reliability objective is R_0. Therefore the optimization problem to determine the optimal release time for the proposed model can be formulated as follows:

Minimize $E\{C(T)\}$

$$\text{Subject to } R(x/T) \geq R_0\,, \quad 0 < R_0 < 1 \tag{18}$$

5 Numerical Illustration

In this section, we perform the numerical experiments to examine the effect of different parameters on the reliability of the module-based software system by varying testing time T. We set the default parameters as $a^{(1)}$=50, $a^{(2)}$=200, $a^{(3)}$=300, $\eta^{(1)}$=0.4, $K_1 = 2$, $K_2 = 3$, $\eta^{(2)}$=0.6, $\eta^{(3)}$=0.7, $\mu^{(1)}$=0.7, $\mu^{(2)}$=0.8, $\mu^{(3)}$=0.9, $r^{(1)}$=0.5, $r^{(2)}$=0.6, $r^{(3)}$=0.7, and $x = 0.005$.

Figures 1, 2, 3, and 4 illustrate the software reliability R by varying testing time T for different values of initial number of faults ($a^{(1)}$), fault removal rate ($r^{(1)}$), fault reduction factor ($\mu^{(1)}$) in module 1, and fault introduction rate ($\eta^{(2)}$) in module 2, respectively. It is noted that initially reliability increases sharply up to $T = 15$ then it becomes almost constant. In Fig. 1, as we increase the initial number of faults ($a^{(1)}$), a decreasing pattern of software reliability is observed. We see in Fig. 2 that the reliability of the software increases on increasing the fault removal rate ($r^{(1)}$). Figure 3 depicts that there is no significant effect of fault reduction factor $\mu^{(1)}$ on the reliability with the variation in testing time. The effect of fault introduction rate $\eta^{(2)}$ on the reliability by varying T is shown in Fig. 4. It is found that on increasing the testing time T, reliability decreases significantly for higher values of fault introduction rate.

Fig. 1 Software reliability by varying $a^{(1)}$

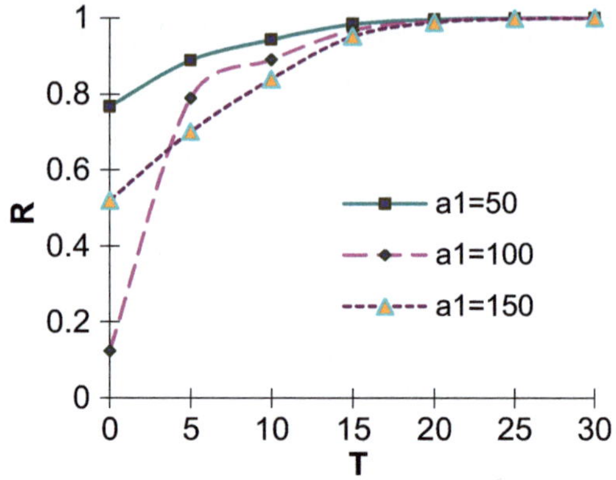

Fig. 2 Software reliability by varying $r^{(1)}$

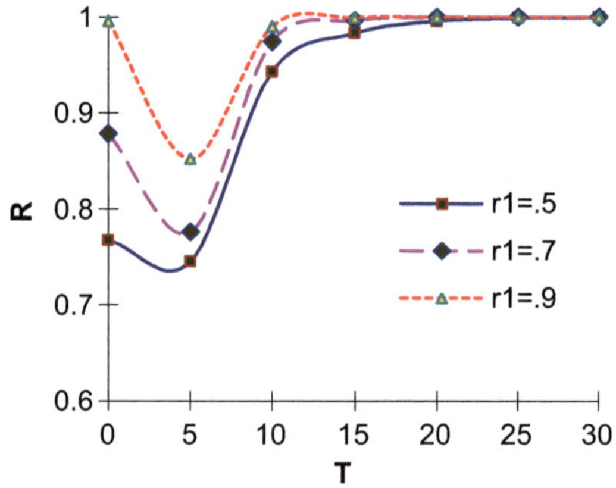

Fig. 3 Software reliability by varying $\mu^{(1)}$

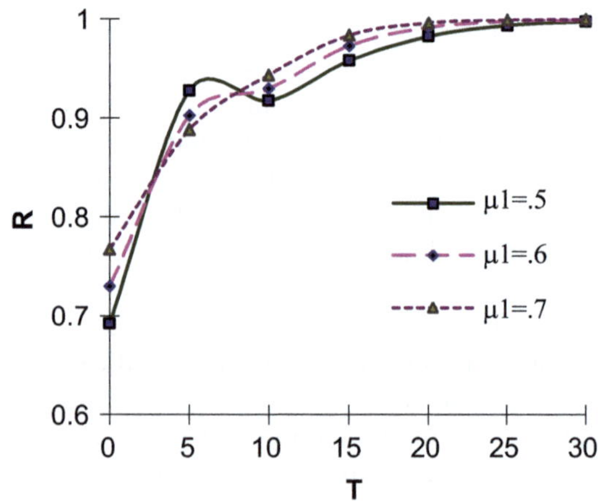

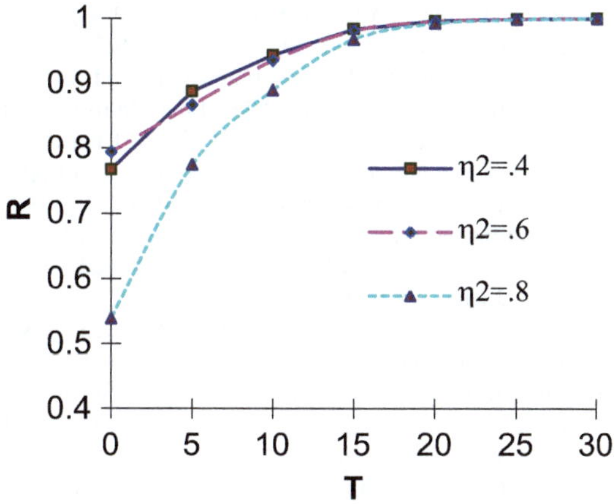

Fig. 4 Software reliability by varying $\eta^{(2)}$

Based on the numerical results, we conclude that the reliability of module-based software system increases on increasing the fault removal rate ($r^{(1)}$) but decreases on increasing the initial number of faults ($a^{(1)}$) and fault reduction factor ($\eta^{(2)}$).

6 Conclusion

To analyze the realistic scenarios of testing process for complex software system, the module-based software reliability growth model is developed by incorporating the key factors, namely, fault reduction factor and imperfect debugging. From the software testing point of view, by using these factors together, our investigation portrays the more versatile and practical issues of the software testing and debugging processes. We have formulated the expressions for the mean value function and software reliability. The optimal release time is established by minimizing the total expected maintenance cost of the software subject to reliability constraints of a modular software system. Our model provides valuable insight to improve the reliability of modular system wherein failure is an unavoidable phenomenon due to imperfect debugging. Numerical illustration provided exhibits the tractability of testing issues of the complex modular-based software system.

References

1. Kumar S, Zeephongsekul P, Balasubramanian S (1995) Modelling and analysis of a software system with improvements in the testing phase. Math Comput Model 22:183–191
2. Huang CY, Lo JH (2006) Optimal resource allocation for cost and reliability of modular software systems in the testing phase. J Syst Softw 79:653–664

3. Jha PC, Gupta D, Yang B, Kapur PK (2009) Optimal testing resource allocation during module test considering cost, testing effort and reliability. Comput Ind Eng 57:1122–1130
4. Jain M, Jain A (2013) Quasi renewal analysis of software reliability growth model incorporating testing efforts and time delay. Int J Maths Oper Res 5:721–742
5. English M, Buckley J, Collins JJ (2016) Chapter 8- Investigating software modularity using class and module level metrics. Soft Qual Assur pp 177–200
6. Deeprasertkul P, Bhattarakosol P, O'Brien F (2005) Automatic detection and correction of programming faults for software applications. J Syst Softw 78:101–110
7. Yang X, Sang N, Lei H (2008) An improved NHPP model with time- varying fault removal delay. J Elec Sci Tech China 6:270–273
8. Satnawi O (2009) Discrete time NHPP models for software reliability growth phenomenon. Int Arab J Info Tech 6:124–131
9. Jain M, Manjula T, Gulati TR (2012) Software reliability growth model (SRGM) with imperfect debugging, fault reduction factor and multiple change point concepts. Adv Intel Soft Comput 131:1027–1037
10. Peng R, Li YF, Zhang WJ, Hu QP (2014) Testing effort dependent software reliability model for imperfect debugging process considering both detection and correction. Reliab Eng Syst Saf 126:37–43
11. Tamura Y, Yamada S (2015) Air application for reliability analysis considering debugging process and network traffic in mobile clouds. Simul Model Pract Theo 50:165–175
12. Wang J, Wu Z, Shu Y, Zhang Z (2015) An imperfect software debugging model considering log-logistic distribution fault content function. J Syst Softw 100:167–181
13. Hsu CJ, Huang CY, Chang JR (2011) Enhancing software reliability modeling and prediction through the introduction of time-variable fault reduction factor. Appl Math Model 35:506–521
14. Pachauri B, Dhar J, Kumar A (2015) Incorporating inflection S-shaped fault reduction factor to enhance software reliability growth. Appl Math Model 39:1463–1469
15. Jain M, Priya K (2002) Optimal policies for software testing time. J Comp Soc Ind 32:25–30
16. Jha PC, Gupta A, Kapur PK, Gupta D (2009) Release time decision policy of software employed for the safety of critical system under uncertainty. Opsearch 45:209–224
17. Kapur PK, Singh O, Bardhan A (2005) A software reliability growth model for operational use with testing coverage. Qual Reliab Inform Tech, pp 61–73
18. Pham H, Zhang X (1999) Software release policies with gain in reliability justifying the cost. Ann Softw Eng 8:147–166
19. Singh O, Kapur PK, Anand A (2012) A multi-attribute approach for release time and reliability trend analysis of a software. Int J Syst Assur Eng Manag 3(3):246–254
20. Chatterjee S, Singh JB (2014) A NHPP based software reliability model and optimal release policy with logistic-exponential test coverage under imperfect debugging. Int J Syst Assur Eng Manag 5(3):399–406
21. Singh O, Aggrawal D, Anand A, Kapur PK (2015) Fault severity based multi-release SRGM with testing resources. Int J Syst Assur Eng Manag 6(1):36–43

Analytics for Maintenance of Transportation in Smart Cities

Adithya Thaduri, Ajit Kumar Verma, and Uday Kumar

1 Introduction

The population of cities and the percentage of migration of people toward cities are increasing day by day. As from the report by the UN [1], 54% of the population of the world lives in cities, and the forecasted population will be 66% in 2050 or higher. These cities might face tremendous challenges because of the problems arising due to increased congestion of the people. To maintain the assets of the cities, some part of the problems can be overcome by the use of digitization using the present technologies available to cater the needs. As per the definition of Smart Cities Council, the usage of digital technology embedded across different functions of the city is termed as "smart city." The main objectives of the smart city include, but are not limited to, the following:

- Environmental sustainability and efficiency
- Sustainable homes and buildings
- Efficient use of resources
- Efficient and sustainable transportation
- Better urban planning – livable cities

The necessity of the smart cities is quite multi-faceted because cities themselves act as a platform for all existing technologies and future technologies that are governed by the policies for the betterment of the society. Management of assets

A. Thaduri (✉) • U. Kumar
Luleå University of Technology, Luleå, Sweden
e-mail: adithya.thaduri@ltu.se; Uday.Kumar@ltu.se

A.K. Verma
Western Norway University of Applied Sciences, Haugesund, Norway
e-mail: AjitKumar.Verma@hvl.no

© Springer Nature Singapore Pte Ltd. 2018
P.K. Kapur et al. (eds.), *Quality, IT and Business Operations*, Springer Proceedings in Business and Economics, DOI 10.1007/978-981-10-5577-5_7

is also a difficult task to consider in the case of smart cities. The main reasons for the purpose of smart cities are:

- Growing population
- Traffic congestion
- Space – homes and public space
- Resource management
- Global warming
- Aging infrastructure
- Tighter city budgets

Smart cities are booming around the world though the definitions are more or less similar to the one that is defined above. In a typical smart city, one of the important vertical solutions is transportation and mobility. Transportation is one of the driving areas for growth, mobility, services, and support. For effective and efficient operation of the transportation field, there is a requisite to maintain these assets in a proper condition. Maintenance of these assets is crucial in reducing cost, traffic congestion, inconvenience to the society, etc.

One of the interesting aspects in transportation is that the operation not only depends on itself but also depends on the surrounding assets and the people that use these assets. Hence, usage of advanced technologies such as Internet of things (IoTs) and data analytics can be applied for envision of the smart cities.

This paper provides an overview of the smart cities and challenges faced by the maintenance of the transportation in smart cities. This paper also focuses on the possibilities of application of data analytics for transportation in smart cities.

2 Definitions of Smart Cities

There are different ways of defining and conceptualizing smart cities that are still in progress [20, 21]. There are several definitions that exist in the literature [20, 30]:

- A city well performing in a forward-looking way in economy, people, governance, mobility, environment, and living, built on the smart combination of endowments and activities of self-decisive, independent and aware citizens [22].
- A city that monitors and integrates conditions of all of its critical infrastructures, including roads, bridges, tunnels, rails, subways, airports, seaports, communications, water, power, even major buildings, can better optimize its resources, plan its preventive maintenance activities, and monitor security aspects while maximizing services to its citizens [23].
- A city "connecting the physical infrastructure, the IT infrastructure, the social infrastructure, and the business infrastructure to leverage the collective intelligence of the city" [24].
- A city striving to make itself "smarter" (more efficient, sustainable, equitable, and livable) [25].

- A city "combining ICT and Web 2.0 technology with other organizational, design and planning efforts to dematerialize and speed up bureaucratic processes and help to identify new, innovative solutions to city management complexity, in order to improve sustainability and livability" [26].
- "The use of Smart Computing technologies to make the critical infrastructure components and services of a city–which include city administration, education, healthcare, public safety, real estate, transportation, and utilities more intelligent, interconnected, and efficient" [27].
- "A smart city brings together technology, government and society to enable the following characteristics: smart cities, a smart economy, smart mobility, a smart environment, smart people, smart living, smart governance" – IEEE Smart Cities [28].
- "The concept is not static, there is no absolute definition of a smart city, no end, but rather a process, or series of steps, by which cities become more livable and resilient and, hence, able to respond quicker to new challenges" – DBIS, UK [29].

Out of all these definitions, the one that was defined by DBIS, UK, seems realistic to the context relevant to transportation in a broad perspective. It also provided that there is no static definition but depends on the objectives of the smart city.

3 Architectures

It is interesting to note that the generic architecture of the smart city is similar to the standards of cyber-physical systems. This architecture includes basic architectural levels as shown in Fig. 1 [31]. The adoption of this information architecture will enable the developed smart assets to be integrated into other governing systems enabling new business models and innovations.

Smart connectivity level: Acquisition of reliable information from components and systems is a prerequisite for higher-order functions. Information can be acquired directly from embedded sensor systems in smart components and from higher-level services of enterprise data acquisition and control systems using the service-oriented architecture.

Data-to-information conversion level: Inference of useful information from data is important in condition monitoring systems due to the high complexity and data rate of signals (e.g., vibration sampled at several kHz). Efficient dimension reduction and feature extraction from raw sensor signals are critical for reliable results and predictions at higher levels of the architecture.

Cyber level: At this level, information about every connected component and system is in principle available, including virtual models of components and simulation models that can be used in combination with historical condition monitoring data stored in data centers to evaluate different maintenance and operational

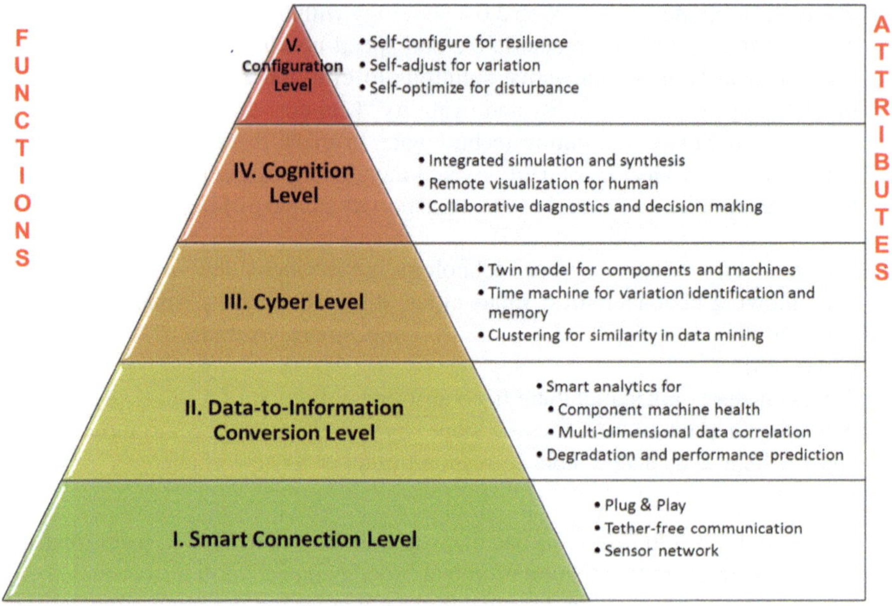

Fig. 1 Cyber-physical systems

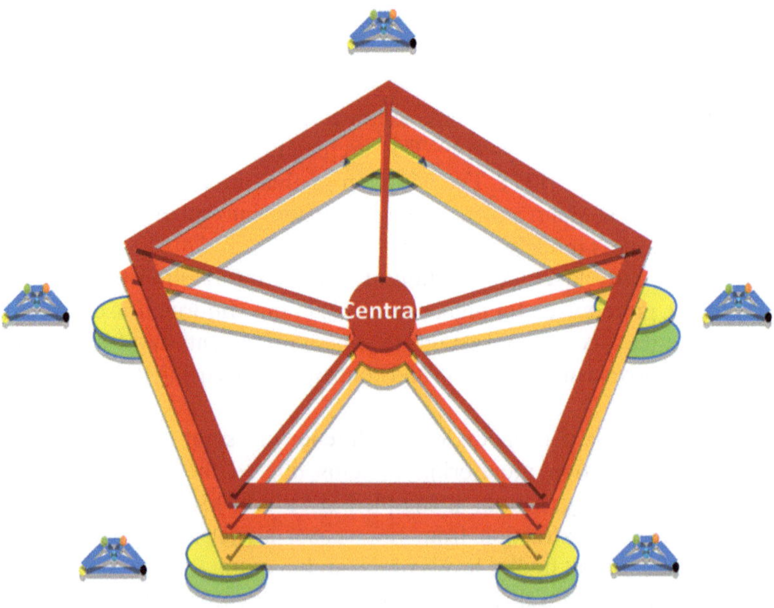

Fig. 2 Internet of cities

scenarios. Data analytics queries can be executed to extract information providing new insights.

Cognition level: Making of correct maintenance and operational decisions requires interactions with expert users through efficient human-machine interfaces and the use of artificial cognitive systems capable to interact with entities at the cyber level for evaluating possible strategies and action plans.

Configuration level: At the highest level, the virtual architecture is orchestrated by meaningful queries and operational goals configured by humans. The purpose of the lower levels of the architecture is to realize as far as possible the aims configured at this level (e.g., using optimization methods and historically guided prediction heuristics) leading to a self-configuring and self-optimizing system of systems.

There are several architectures defined for the smart city in the literature, and more or less most of them follow the similar structure defined in the Fig. 1.

4 Big Data Analytics and IoT

The implementation of big data analytics were discussed on several authors [10, 33]. A comprehensive illustration of the usage and implementation of the smart analytics is provided by Pendse [9]. For the need of accessibility and storage, there must be a mechanism for implementing cloud-based approach for the big data analytics for smart cities [4, 5]. The urban planning of the cities can be embraced with the Internet of things for the better usage of facilities for the big data analytics [6]. An architecture was developed by Puiu et al. [7], for the large-scale implementation of the data analytics framework for smart cities. For the data management, there are several models and topological models to be followed that are available in the literature [8]. The utilization of the data analytics [19] along with the way the analytics was perceived by the user with the visualization using interactive phenomenon was developed [11].

A conceptual model for the smart cities is proposed as in Fig. 2. But, the actual operation and execution have not lived up to the standards because of several reasons. The incorporation of new technologies also faces governance, security, and cost issues. The possibilities of these technologies are given by [2]:

- The amount of traffic can be measured by attaching the vehicles with radio frequency identification (RFID) tags. The concentration of vehicles can be observed from the central facility through this RFID. The similar information through big data cloud-based approach where users can access through mobile apps.
- By using different types of tags for different types of vehicles, the amount of breakdowns by each of the vehicle can be provided to the customer for better selection of the vehicle
- IoTs are installed on maintenance equipments and other facilities to make sure that maintenance actions are carried out in the smart city and these

information are fed to the central facility. The commuters are thus making of these information can reroute their path to reach the destination.
- The amount of load on the different transport systems can be predicted by gathering information and processing through the advanced tools. By this way, the addition of vehicles can be incorporated to the system to reduce the traffic congestion.
- Big data and IoT can also help to reduce carbon emissions for the sustainability of the cities. The amount of traffic is gathered by installing sensors on the road/adjacent to the road for monitoring. These information are fed to the traffic police for diversion and to reduce the congestion. By this way, the amount of energy wasted at the traffic signals can be controlled to some extent.
- Parking has been a nightmare in the cities because availability of space is a more major concern. This problem can be dealt by providing real-time information from the cloud-based analytics to reduce fuel usage.

The operators of transportation are presently utilizing these technologies to make sure that the operation of fleet runs smoothly. It allowed them to input data from events to predict the numbers of who would be traveling and make sure that transport was running effectively to make sure that spectators and athletes could be effectively transported to and from the stadiums.

To create an efficient, effective, and flexible public transport in the future smart cities, it is essential to apply the above techniques. The data collected from different sources can be used not only to monitor the present condition of the assets but also to predict the future behavior. This may decrease the transport delays with minimal maintenance activities by predicting future failures thus by increasing the transport availability [3].

5 Maintenance Analytics for Transportation

Improving transportation infrastructure and services is a high priority for many cities.

The maintenance of the transport is the main driver for the society. It is also the main platform for different types of systems that are interdependent to each other such as industries, companies, education, and other societal needs. Hence, it is important to have better maintenance capabilities for all means of transport and develop a smart transportation infrastructure for future needs and have the capability for better efficiency of the city itself.

Beyond cost considerations, various modes of transport all require power and communications to function properly within a smart transportation environment – especially true as cities move to electrified light rail or buses and set up recharging infrastructure for electric vehicles [32].

A role of the transport in the smart cities was discussed by [19]. A methodological framework for the purpose of transport was developed by the Debnath et al. [12]. For the purpose of information gathering and assessment, sensor modeling and sensor architecture were developed [13]. Also, a crowd-based tool was developed where the users are the actual customers for the transportation [14]. The visualization of the transportation can be carried out by combining the information from the users and the geographical layout of the city [15]. By keeping the sustainability in the mind, a case study was conducted with development of models precisely for the energy efficiency for freight transport [16]. For future purposes, an artificial intelligence-based methodology was developed for intelligent transportation in smart cities [17]. Similarly, a framework was developed using the autonomous transportation system [18].

The several prospects and possibilities of the maintenance analytics for the transportation are briefed below [32].

5.1 *Reducing Traffic Congestion*

- The application of novel means of data integration and analytics to visualize the necessary information to minimize congestion.
- By monitoring the historical data related to the number of vehicles and amount of time spent in the traffic lights and bottlenecks, maintenance analytics can be able to predict the amount of time/vehicles required to cross a junction. By this means, the time of traffic lights is synchronized and adjusted according to the traffic needs. This can also be utilized to reduce traffic-related accidents to prevent congestion and other factors.
- Data obtained from different sensors, especially cameras, are useful to monitor and detect particular vehicles or zones which are more prone to incidents and accidents. Thus, maintenance analytics can provide alternative routes, alerting drivers, and better notification system to pass information to public.
- Parking has been a bigger problem because of increase in the number of vehicles and less place inside the city. It also leads to the increase in pollution that might affect directly and indirectly to the working behavior of people in the city. Hence, maintenance analytics can facilitate parking areas using mobile apps with real-time information and also predict future behavior. One case that is related to future parking congestion is special events that are planned in the future.

5.2 *Reducing Trip Time*

- By applying the advanced analytics, according to the traffic congestion, the traffic planning system can suggest the drivers to provide the best traffic route based on time, place, and distance.
- The travelers are being suggested by the system that is integrated with ICT and can facilitate multimodal transport with optimal distance, cost, and time.
- By introducing the future weather data into the system, one can predict the scenarios of drainage and congestion to reduce the trip time.

5.3 Choice and Control

- The recommender system with specialized agents can provide multiple alternatives to the user to suit the behavior. This choice is further given as a feedback into the system to train the alternatives and rank them accordingly.
- Users are also welcome to suggest the best alternatives into the system for being smart in the future.
- The ridesharing apps such as Uber, Ola, and other national cab services can facilitate the users to select them of their choice.

5.4 Public Safety

- Smart transportation has a strong link to public safety.
- First responders require mobility to perform their lifesaving work.
- Information and Communication Technology (ICT) can make their jobs easier by optimizing traffic lights when necessary and empowering them to see potential traffic snarls in real time, so they can select the most efficient travel routes.

5.5 Situational Awareness

- Use of smart devices that are already installed or newly constructed system to be deployed at various places in the cities to get better context awareness. This will be useful to obtain better recommendation systems according to the location.
- Sustainability and resilience of the cities are the major concerns of the smart cities. These issues can be better monitored using the smart transport system to reduce wastages, increase robustness, and decrease accidents to maintain the city in the best possible condition.
- Context-awareness provided by the system also enables not only location-specific but also can provide classify the different clusters in cities as zones to plan the best traffic solutions, especially in occurrence of major catastrophic events, disasters and recovery plans.

5.6 Asset Optimization

- The application of these technologies can also enhance the optimization of the assets in terms of usage and resource allocation to reduce the energy consumption.
- The maintenance analytics can provide answers related to the different parts of the system such as when to carry out maintenance, what resources are required, and how to carry out for maximum efficiency at reduced costs.

5.6.1 Internet of Smart Cities

In the future, there will be interconnection of smart cities for the data and knowledge sharing of ideas for the better execution of the technology for the betterment of the society.

The architecture defined in Fig. 1 is implemented at several smart cities. Instead of all layers connected to each other, only the top three layers are required for knowledge sharing and advanced data analytics with respect to real-life problems. This also helps to provide interesting insights through visualization to compare the performance of each of the assets in different transportations.

6 Challenges and Future Areas

There are different challenges to be taken care of by smart cities for transportation, and some of them are crash fatalities/injuries, energy shifts, congestion, transportation infrastructure costs, unexpected delays of agency/governmental budgets, urbanization and land use patterns, and parking option

Apart from the abovementioned challenges, there are other concerns:

- For developing countries, due to the deficit of funds, faced with providing basic amenities, it is difficult to allocate funds for smart cities, and the implementation of this technology can be backlashed with huge economical and ideological issues.
- As similar to the issues of digitization, privacy and security are also going to be big issues. It is a tough task to convince the government, especially in this case, because the loopholes in this issue can be a major problem for their future prospects.
- Each city is different, and the methodologies or models can be developed and applied based on the geographical condition, economical condition, demographics, and feasibility.

7 Conclusion

The vision of smart city is difficult to deploy at this time, but it is faced by different challenges from different perspectives. It has to be an inclusive concept. The idea of smart cities is in the nascent stage, and also the implementation of technologies is also not utilized to the major extent. The points to consider are:

- Enormous possibilities of the analytics in the upcoming smart cities especially for transportation.
- The benefits can be directly visible to the public to increase efficiency in the transportation.

- Analytics are not only limited to maintaining the assets in good condition but also provide interesting insights about governance.
- Interconnection and knowledge transfer of Internet of cities can be helpful in the future for better decision-making.

References

1. UN (2014) World's population increasingly urban with more than half living in urban areas. [Online]: http://www.un.org/en/development/desa/news/population/world-urbanization-prospects-2014.html
2. Pal K. How Big data helps build smart cities. http://www.kdnuggets.com/2015/10/big-data-smart-cities.html 10th October, 2015
3. Barton D (2016) 7 Uses For Analytics In Smart Cities. https://channels.theinnovationenterprise.com/articles/158-7-uses-for-analytics-in-smart-cities,
4. Khan Z, Anjum A, Kiani SL (2013) Cloud based big data analytics for smart future cities. In: Proceedings of the 2013 IEEE/ACM 6th international conference on utility and cloud computing, IEEE Computer Society, pp 381–386
5. Khan Z, Anjum A, Soomro K, Tahir MA (2015) Towards cloud based big data analytics for smart future cities. J Cloud Comput 4(1):1
6. Rathore MM, Ahmad A, Paul A, Rho S (2016) Urban planning and building smart cities based on the internet of things using Big Data analytics. Comput Netw 101:63–80
7. Puiu D, Barnaghi P, Tönjes R, Kümper D, Ali MI, Mileo A, Parreira JX et al (2016) CityPulse: large scale data analytics framework for smart cities. IEEE Access 4:1086–1108
8. Bradley PE (2015) Supporting data analytics for smart cities: an overview of data models and topology. In: International symposium on statistical learning and data sciences. Springer, Cham, pp 406–413
9. Pendse P (2015) Smart cities and analytics
10. Ismail A (2016) Utilizing big data analytics as a solution for smart cities. In: 2016 3rd MEC International Conference on Big Data and Smart City (ICBDSC). IEEE, Piscataway, pp 1–5
11. Behl M, Mangharam R (2016) Interactive analytics for smart cities infrastructures. In: 2016 1st International Workshop on Science of Smart City Operations and Platforms Engineering (SCOPE) in partnership with Global City Teams Challenge (GCTC)(SCOPE-GCTC). IEEE, Piscataway, pp 1–6
12. Debnath AK, Chin HC, Haque MM, Yuen B (2014) A methodological framework for benchmarking smart transport cities. Cities 37:47–56
13. Farkas K, Feher G, Benczur A, Sidlo C (2015) Crowdsending based public transport information service in smart cities. IEEE Commun Mag 53(8):158–165
14. Bakillah M, Liang SHL, Zipf A (2012) Toward coupling sensor data and volunteered geographic information (VGI) with agent-based transport simulation in the context of smart cities. In: Proceedings of the first ACM SIGSPATIAL workshop on sensor web enablement. ACM, New York, pp 17–23
15. Navarro C, Roca-Riu M, Furió S, Estrada M (2016) Designing new models for energy efficiency in urban freight transport for smart cities and its application to the spanish case. Transp Res Procedia 12:314–324
16. Agarwal PK, Gurjar J, Agarwal AK, Birla R (2015) Application of artificial intelligence for development of intelligent transport system in smart cities. J Int J Trans Eng Traffic Syst 1(1):20–30
17. Schlingensiepen J, Mehmood R, Nemtanu FC (2015) Framework for an autonomic transport system in smart cities. Cybern Inf Technol 15(5):50–62

18. Simonis I (2015) Sensing models and sensor network architectures for transport infrastructure monitoring in smart cities. EGU Gen Assembly Conference Abstr 17:9183
19. Santana Beneyto, Marta Irene (2015) The role of transport in the smart cities
20. Boulton A, Brunn SD, Devriendt L (2011) Cyberinfrastructures and "smart" world cities: physical, human, and soft infrastructures. In: Taylor P, Derudder B, Hoyler M, Witlox F (eds) International handbook of globalization and world cities. Edward Elgar, Cheltenham. Available from http://www.neogeographies.com/documents/cyberinfra structure_smart_world_cities.pdf
21. Hollands RG (2008) Will the real smart city please stand up? City 12(3):303–320
22. Giffinger R, Fertner C, Kramar H, Kalasek R, Pichler-Milanović N, Meijers E (2007) Smart cities: ranking of European medium-sized cities. Centre of Regional Science (SRF), Vienna University of Technology, Vienna
23. Hall RE (2000) The vision of a smart city. In: Proceedings of the 2nd international life extension technology workshop, Paris, France, September 28. Available from http://www.osti.gov/bridge/servlets/purl/773961-oyxp82/webviewable/773961.pdf
24. Harrison C, Eckman B, Hamilton R, Hartswick P, Kalagnanam J, Paraszczak J, Williams P (2010) Foundations for smarter cities. IBM J Res Dev 54(4)
25. Natural Resources Defense Council. What are smarter cities?, Available from http://smartercities.nrdc.org/about
26. Toppeta D (2010) The Smart City vision: how Innovation and ICT can build smart, "Livable", sustainable cities. The Innovation Knowledge Foundation. Available from phttp://www.thinkinnovation.org/file/research/23/en/Top peta_Report_005_2010.pdf
27. Washburn D, Sindhu U, Balaouras S, Dines RA, Hayes NM, Nelson LE (2010) Helping CIOs understand "Smart City" initiatives: defining the Smart City, its drivers, and the role of the CIO. Cambridge, MA: Forrester Research, Inc. Available from http://public.dhe.ibm.com/partnerworld/pub/smb/smarterplanet/forr_help_cios_und_smart_city_initiatives.pdf
28. IEEE, http://smartcities.ieee.org/about
29. Department of Business Innovation & Skills, Smart cities: background Paper. October 2013
30. Chourabi H, Nam T, Walker S, Ramon Gil-Garcia J, Mellouli S, Nahon K, Pardo TA, Scholl HJ (2012) Understanding smart cities: an integrative framework. In: System science (HICSS), 2012 45th Hawaii international conference on. IEEE, Maui, pp 2289–2297
31. Lee J, Bagheri B, Kao H-A (2015) A cyber-physical systems architecture for industry 4.0-based manufacturing systems. Manuf Lett 3:18–23
32. Smart cities Council, "Transportation", http://smartcitiescouncil.com/smart-cities-information-center/transportation
33. Kumar S, Prakash A (2016) Role of big data and analytics in smart cities. Intl J Sci Res 5(2):31–33. https://www.ijsr.net/archive/v5i2/NOV161007.pdf

Business Strategy Prediction System for Market Basket Analysis

Sumit Jain, Nand Kishore Sharma, Sanket Gupta, and Nitika Doohan

1 Introduction

In current scenario, computer is more popular with development of database technology; their large numbers of data are stored in huge amount of database. Obviously, this very useful information without using efficient methods is impossible to obtain. Data mining is one of the technologies that seemed to be as a reflection of the demand. Association rule mining is one of the important researches, and its main aim is to find out the requirement between various fields based on support and reliability. So many examples are available such as Market Basket Analysis.

It has a two-phase activity, namely, data analysis (Phase 1) and data acquisition (Phase 2), that performed sequentially.

Phase 1: In this phase we analyze all the items (i.e., common between various customers) on the basis of popularity, best sellers, recently used, and their frequency with the help of their shopping log record. By the log record, we can easily find out the frequent items.

S. Jain (✉)
SCSIT-DAVV, Indore, India
e-mail: sumitjain1679@gmail.com

N.K. Sharma • S. Gupta
ATC, Indore, India
e-mail: er.nksharma.mtechcs@gmail.com; sanket.jec@gmail.com

N. Doohan
MediCapsIndore, Indore, India
e-mail: nitika.doohan@gmail.com

© Springer Nature Singapore Pte Ltd. 2018

P.K. Kapur et al. (eds.), *Quality, IT and Business Operations*, Springer Proceedings in Business and Economics, DOI 10.1007/978-981-10-5577-5_8

Phase 2:-In order of 1st phase, it is to produce the instructions from the frequent item sets (Output of 1st Phase) and often is an important step in terms of acquisition of set items Mining Association Rules of Procedure.

These two-phase methods can help prepare the strategies for marketing, and by these strategies (Market Basket Analysis), a manager can easily find the optimal and feasible layout between various layouts [1]. For example, text copy is strongly connected to pen so they are purchased together (mostly seen during starting session of schools/colleges), that's why they are located nearby to each other or at the same place.

Currently, the association rule mining problems are excessively valued by examiners in the various fields such as database, artificial intelligence, information retrieval, visualization, informatics, and other fields. So many unbelievable results have been found in this process. Mining association rules have problems from large database that has the most mature, important, and active research contents.

Problems with large databases in the mining association rules have become the most mature, important, and active research material.

The rest of the chapter is organized as follows. In Sect. 2, we describe related tasks in market basket analysis. In Sect. 3, the market basket analysis approach is described. In Sect. 4, the system architecture of the business strategy prediction system is described. In Sect. 5, the experiment analysis is described. In Sect. 6, the performance results are described. In Sects. 7 and 8, the limitations and future scope are described. In Sect. 9 conclusions and future work are given.

2 Related Work

Association rules are set of statements which is in the form of $\{X1, X2 \ldots Xn\} \rightarrow Y$, meaning that when we find all of statement X1, X2 ... Xn in the market basket, after calculating all the set of X, then we can find Y. We typically were finding merely the rules which are confident on given threshold value. If items are placed into baskets in random order, then it results in higher confidence.

The market research is the framework activity that will work as an umbrella activity that links the consumers and end users directly to the market; it is done only through information which is used to recognize and explore marketing prospect and their problems. The aim of research is to balance the producer-consumer problem, i.e., sale, relationship, demand, availability, and facility to find the useful item in short time duration in today's fast life, and, finally, this balance proves beneficial for all, i.e., end user and producer and retailer are the key part of market. So marketing research specifies the required information to deal with these issues, designs the method for collecting information, manages and implements the data collection to understand the market, and overcomes all deficiencies for proper tuning between the end user and market [2, 3].

This framework activity is the systematic gathering, recording, and analyzing of qualitative and quantitative data about issues that relate to running products in the market and services provided by them. The marketing research is mainly focused to identify and assess how and what changing elements of the marketing mix impact consumer's behavior. The term is commonly interchanged with market research. Marketing research is concerned specifically about marketing processes that drastically improve the sells with consumer satisfaction [4].

To maintain the proper balancing between all key factors which are related directly and indirectly with market, we need to find out the requirement and think like consumers, and on the basis of that result, we have to find out the data and arrange them in fruitful manner. So here Apriori algorithm is used for generating frequent item sets based on some rules from the large database. These association rules are one of the major techniques of data mining. In this fast world due to heavy demand, the volume of data is increasing day by day [5–7]. Hence we need to sort out that data; therefore, the mining association rules work on the databases and collect the frequent item sets from massive amount of database. That's why many marketers interested in that part help in many decision-making processes, such as basket data analysis, retail sectors deciding store layout, and making promotional strategies.

This helps in finding the relationship between a large number of database items. The customer's purchasing behavior is dependent on the product's order, their grouping, their arrangement and layout of store. Based on the norms of sales transaction, we get some effective rules, like the sale of other items are dependent on the purchase of certain items and it reflects the purchasing behaviour of the customer [6]. These generated rules can be used in many fields, such as analysis of consumer's need and their shopping, additional sales, goods shelf design, storage planning, and classifying the users according to the buying patterns or vice versa. Apriori algorithm has sound theory base, but the focus is its efficient issue [8, 9].

3 Market Basket Analysis Approach

Market Basket Analysis is one of the frequent and valuable kinds of data analysis in the marketing and retailing sector, in that it concerns our main motto which is to find out and sort out the products which are mutually purchased by the customers and retailers during marketing and retailing (in the same order). In that way an "Apriori algorithm" plays a vital role in Market Basket Analysis.

Through survey it is observed that, during the marketing, all the items purchased by the customers are strongly or loosely coupled to each and other; it means they have some certain relations that must be based on their nature and type of use. On the other hand, we can also find out that all the retail shops, stores, and wholesale shops are also arranged in the same way as we discussed in previous survey, so it means the way of arrangement and choosing of items during purchasing are the same. To improve the sale or to help the customers during purchasing (in directly giving hints of item whether they are useful or not), the layout of retail shops must be the same as the customer's need and their approach to select the items for their use and vice versa.

Our research is working in the same field, and our approach is to try to find out that relation and convert it in the form of method, so definitely with help of this method, we will give paradigm in the marketing and retailing sector.

Here we have some dependencies shown below:

No. of customers (visitors/purchaser) → Popularity of store
Collection of items and layout → No of customers (visitors/purchaser)
Customer's need and useful items → Marketing

4 Market Basket Analysis Approach System Architecture of Business Strategy Prediction System

It describes the architecture for a system to satisfy specified requirements in Fig. 1.

Transaction File
In this case, first, all transaction files reside in the same system in which we run the software. Its database is maintained by MySQL. This type of transaction file should be or needs to be loaded for further processing. The file is loaded and shown by searching it through name.

Select Transaction File
User needs to enter the name of transaction file required from database. In that case user will first select the file as per the requirement. The selected file will be used by the algorithm when user clicks "run algorithm" button.

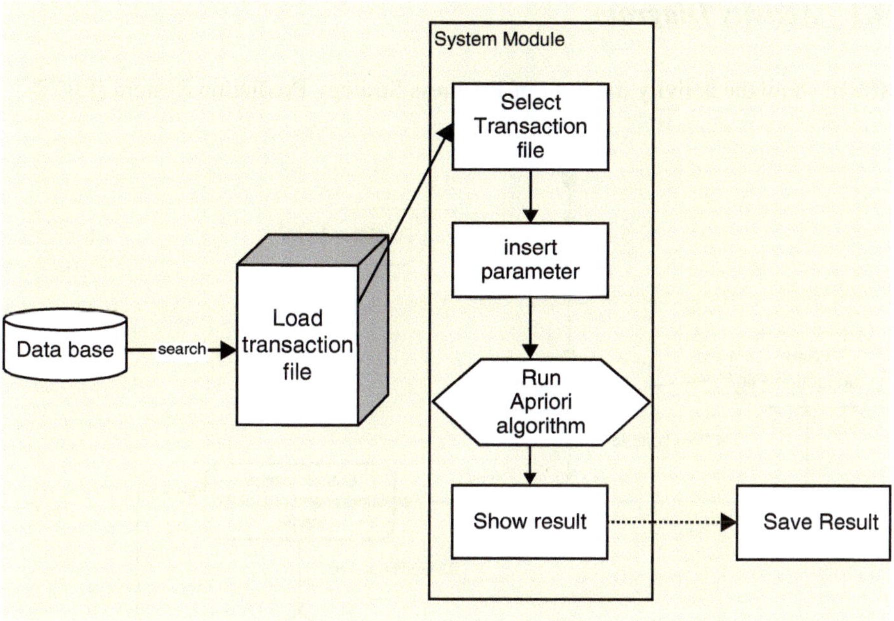

Fig. 1 System architecture of business strategy prediction system

Insert Parameter

For running the algorithm, we should write the name of the selected transaction file in the field from transaction file name. Then users also need to enter parameter value, i.e., minimum support. When this parameter is set, the user will run algorithm over the selected file from database by clicking "run algorithm" button. Then the user will give the name of the output file in which you want to save the result. In case the result file will not be displayed, it means you need to increase the value of minimum support.

Show and Save Result

This module will show final frequent item sets as the result of Market Basket Analysis. These rules are then used by sales executive or any business analyst to predict business strategy for earning profit.

4.1 Activity Diagram

It will show the activity diagram of Business Strategy Prediction System (Fig. 2).

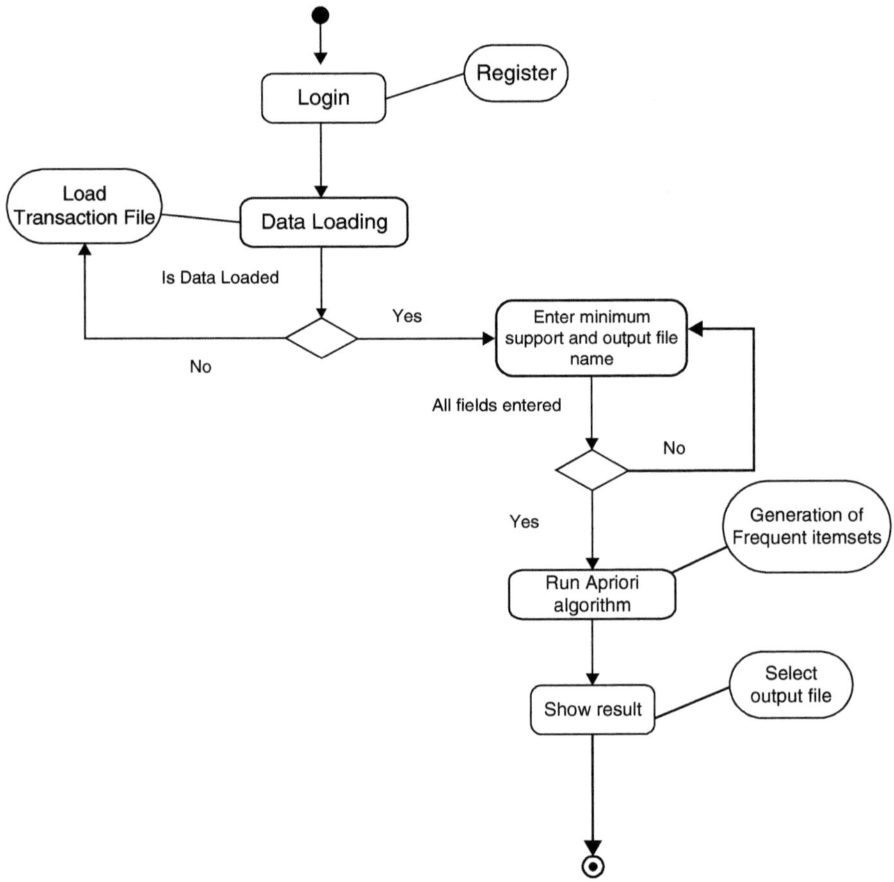

Fig. 2 Activity diagram of business strategy prediction system

4.2 Algorithms for Business Strategy Prediction System

Algorithm design	
(a)	Combining step: Ck is generated by joining
(b)	Lk-1with itself
(c)	Prune step any (k-1): item set that is not frequent cannot be a subset of a frequent k-item set
(d)	Code: Ck, candidate item set of size k
(e)	Lk: frequent item set of size k
(f)	L1 = {frequent items}
(g)	For (K = 1; Lk! = ∅; k++) do begin
(h)	Ck + 1 = candidates generated from Lk
(i)	For each transaction t in database do
(j)	Increase the count of all candidates in Ck + 1
(k)	That are contained in t
(l)	Lk + 1 = candidates in
(m)	Ck + 1with min_support
(n)	End return ∪k Lk

5 Experimental Analysis

In the proposed method, it takes less time to find more and more frequent item sets. This screenshot consists of 13 transactions of 9 items as input, and this item set provides outcome for maximum frequent item set. The output results are given in the form of screenshots which are as follows:

1. It is the first page of Business Strategy Prediction System where we click on start button and the system will get started.

2. Click on view help button to proceed. You can go back to login page after reading the help manual.

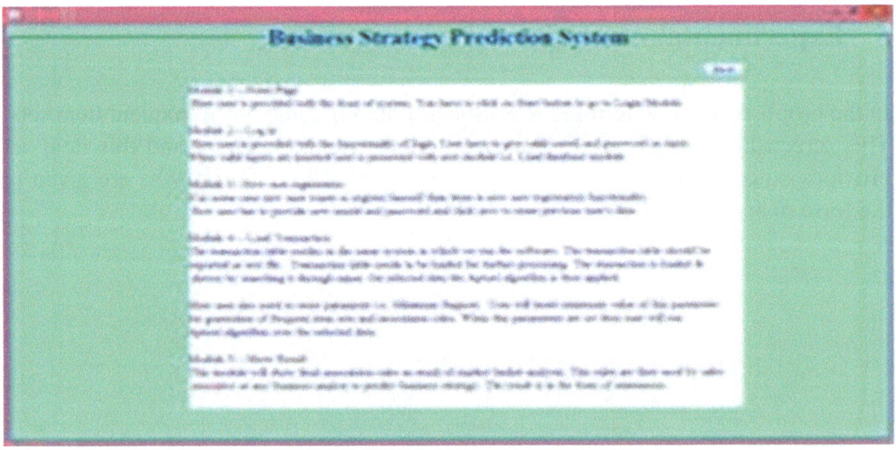

3. To Enter a Valid Username and Password

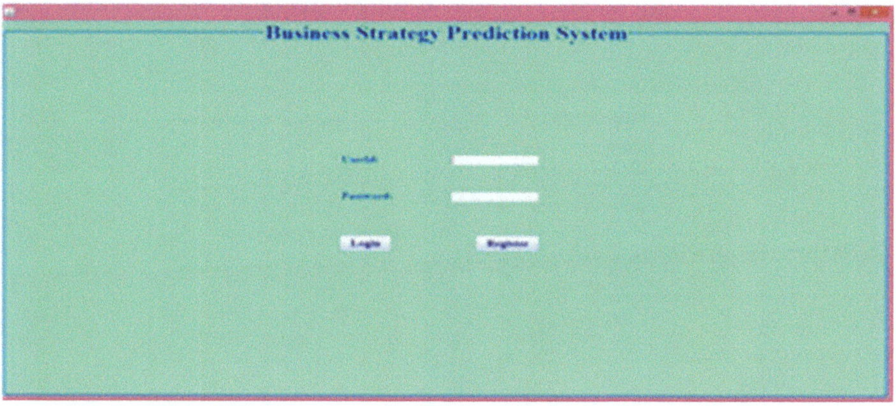

4. View Load Transaction file module. Just click on load button and you will be redirected to the selection menu for the file; just go to the path where the file is saved on your computer, select it, and click on open. **The selected file will get loaded like this**

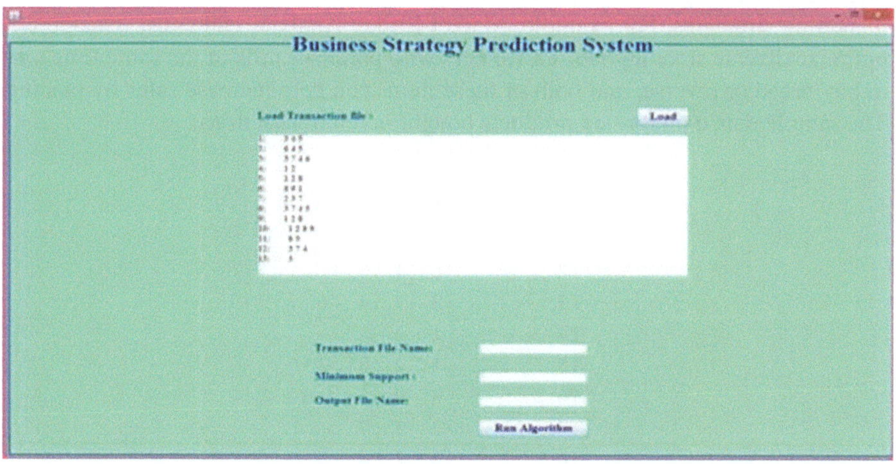

5. For running the algorithm, please write the name of the selected transaction file in the field transaction file name. After this enter the minimum support value desirable for your strategy. Then give the name of output file in which you want to save the result. In case the result file will not be displayed, it means you need to increase the value of minimum support.
6. Clicking on show button will make you select the output file; you just need to select it and open, and the result will be displayed in the frequent rule frame, and the corresponding references will be displayed in another frame.

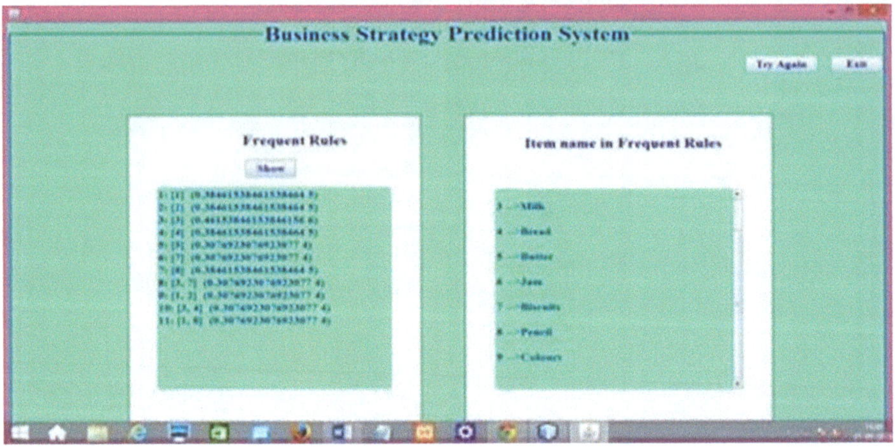

Example

An Apriori algorithm is used in Market Basket Analysis. In the process, customers place in their shopping basket by finding relationships between different items that analyzes customer buying habits. Such retailers have discovered items which are often purchased together by customers to get information that can help in developing marketing strategies. For example, Market Basket Analysis can assist managers to optimize diverse store layouts. Customers who purchase milk at the same time tend to buy bread or reverse, and both of these items can help increase sales by placing. The sample data contains six products bought in ten transactions:

Transactions	Items			
T1	3	4	5	
T2	6	4	5	
T3	3	7	4	6
T4	1	2		
T5	1	2	8	
T6	8	9	1	
T7	2	3	7	
T8	3	7	4	5
T9	1	2	8	
T10	1	2	8	9
T11	8	5		
T13	3	7	4	3

The Business Strategy Prediction System scans the transaction to see which items are bought together frequently.

Item name	Pen	Copy	Milk	Bread	Butter	Jam	Biscuits	Pencil	Colors
Items Id	1	2	3	4	5	6	7	8	9

To do this the transactions are analyzed to calculate:

N: Total number of transactions
Xij: Total number of transactions in which i and j are bought together

From this measures support is calculated using Apriori Algorithm.

$$\text{Sij} = (\text{Xij}/\text{N}) \times 100\%$$

The table shows the different items and their support; these items having support with 20% or above are bought frequently if the minimum support given is 20%.

Support	Pen	Copy	Milk	Bread	Butter	Jam	Biscuits	Pencil	Colors
Pen		30%						30%	
Copy	30%						7%	23%	
Milk				30%		7%	30%		
Bread			30%		23%	15%	23%		
Butter				23%				7%	
Jam			7%	15%					
Biscuits		7%	30%	23%					
Pencil									15%
Colors								7%	

6 Performance Result

These experiments were performed with the help of Market Basket Analysis. In this section system will perform with the different parameters that are used for estimating the model, and their obtained results are shown in the form of time complexity and space complexity graph. Our method gives efficient result on all frequent item sets generated by Market Basket Analysis approach.

Time Complexity
The quantity of time essential for execution of algorithm with the input data is termed as time complexity of the system. Both time slices show that as in the graph they are based on the number of transaction and number of itemsets. Based on the obtained results there is increase in the samples of the database with time.

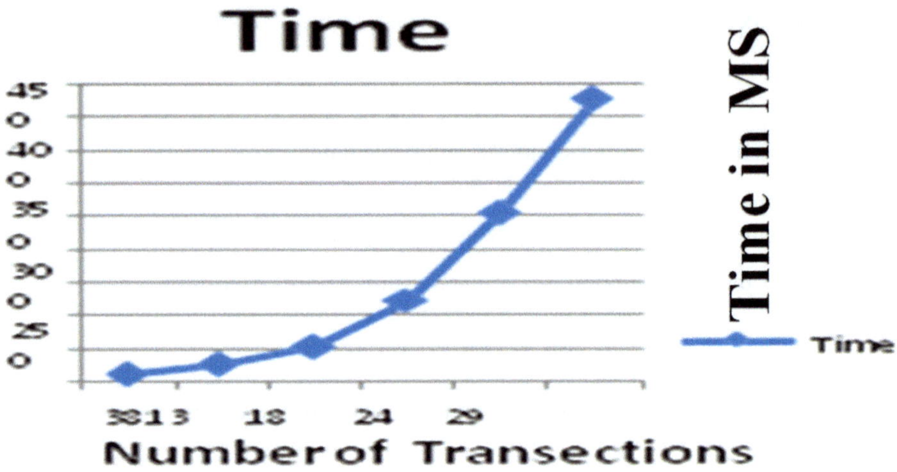

Fig. 3 Time in terms of millisecond vs number of transaction

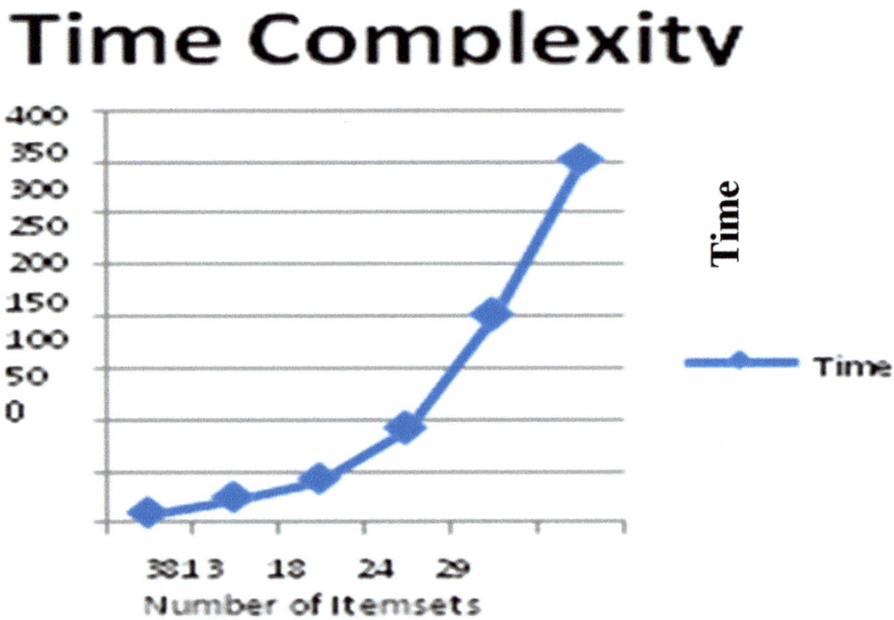

Fig. 4 Time in terms of millisecond vs number of item sets

The following Fig. 3 shows the time vs number of transaction, where X-axis shows the number of transaction and Y-axis shows the amount of time in millisecond.

The following Fig. 4 shows the time vs number of item sets, where X-axis shows the number of item sets and Y-axis shows the amount of time in millisecond.

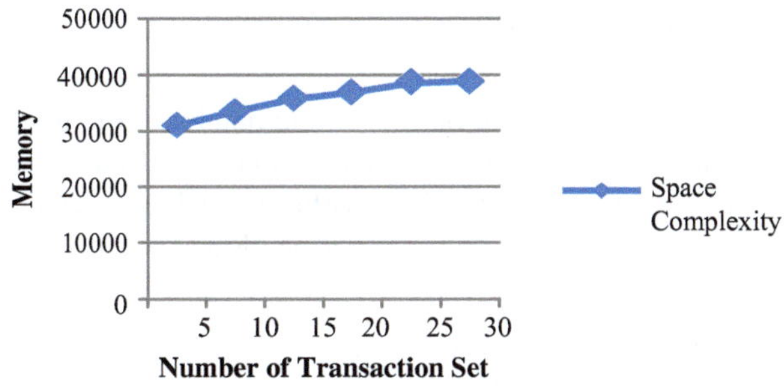

Fig. 5 Memory vs number of transaction

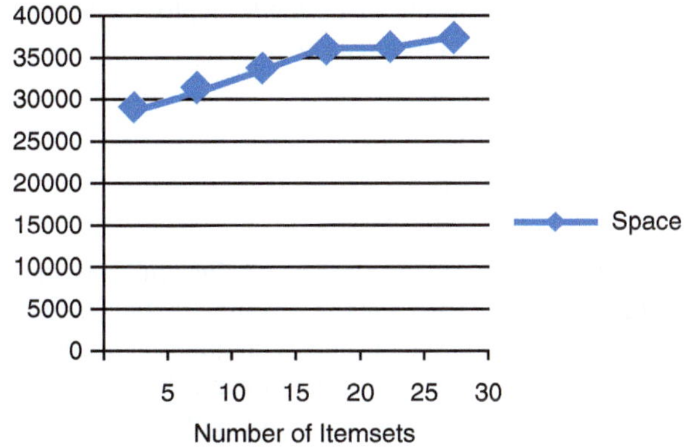

Fig. 6 Memory vs number of item sets

Space Complexity

In this case memory is misused by an algorithm at many places. It includes both extra space or temporary space and space used by input. Both time slices show that, as in the graph, they are based on the number of transaction and number of item sets. According to the given results, the memory consumption of the algorithm increases as the number of samples in database increases.

The following Fig. 5 shows the time vs number of transaction, where X-axis shows the number of transaction and Y-axis shows the amount of memory.

The following Fig. 6 shows the time vs number of transaction, where X-axis shows the number of item sets and Y-axis shows the amount of memory.

7 Limitation

This system is only in desktop application and is used by only single user at a time. This is specifically designed for business strategy. Transactions of any business should be in a proper format, i.e., it should be in transactional database; if it is non-transactional, it needed to be converted into transactional database first.

8 Future Scope

Here we are focusing on this single database; the nodes of a distributed system for the database partitions in a large centralized database to mining association rules can be applied. Proposed algorithm gives better result when data set is too large. It can be enhanced by deploying it on cloud so that it can work on different data sets. It will provide:

- A better access to the product.
- A better way to find the business strategies to help the businesses.
- It will be accessible to all due to deployment on cloud.
- It can work for large data sets.

9 Conclusion

The "Business Strategy Prediction System" provides a better way of analyzing the behavior of the customer for making a better business strategy. It makes use of the frequent item sets which makes frequent data sets which are helpful for analyzing the purchasing behavior of customers. We have generated Market Basket Analysis and computed the results in the form of time complexity and space complexity.

Further we can elaborate this approach and remove all the complexities which are coming in the performance.

References

1. Sun W, Pan M, Qiang Y (2011) Improved association rule mining method based on t statistical. Appl Res Comput 28(6):2073–2076
2. Yong-qing W, Ren-hua Y, Pei-yu L (2009) An improved apriori algorithm for association rules of mining. IEEE
3. Chen Z, Shibang C, Song Q, Zhu C (2011) An improved apriori algorithm based on pruning optimization and transaction reduction. IEEE
4. Chang R, Liu Z (2011) An improved apriori algorithm. In: 2011 international conference on electronics and optoelectronics (ICEOE)
5. Wang H, Liu X (2011) The research of improved association rules mining apriori algorithm. In: 2011 eighth international conference on fuzzy systems and knowledge discovery (FSKD)
6. Han J, Kamber M (2007) Conception and technology of data mining. China Machine Press, Beijing
7. JN Wong (trans) (2003) Tutorials of data mining. Tsinghua University Press, Beijing
8. Yuan Y, Yang C, Huang Y, Mining D (2007) And the optimization technology and its application. Science Press, Beijing
9. Kon YS, Rounteren N (2010) Rare association rule mining and knowledge discovery: technologies for frequent and critical event detection. H ERSHEY. Information Science Reference, PA

Defending the OSN-Based Web Applications from XSS Attacks Using Dynamic JavaScript Code and Content Isolation

Pooja Chaudhary, B. B. Gupta, and Shashank Gupta

1 Introduction

Worms that exploit cross-site scripting (XSS) vulnerability are present in almost 80% of the OSN-based web applications. Sharing of personal as well as professional information on OSN platform attracts the attackers to launch the attacks like XSS. XSS attack [1–4, 8, 10, 11] is an assault against the web application wherein an attacker introduces the malicious JavaScript code into the source code of the OSN-based web applications. This is done in an attempt to steal the user's login credentials and any other sensitive information like session tokens or financial account information. This attack may lead to severe consequences like cookie stealing, account hijacking, misinformation, denial-of-service (DOS) attack, and many more [5, 6, 9, 12].

Blueprint [7] provides robust protection from the XSS attacks in spite of having irregular browser parsing behavior. However, it requires modifications at both the sides, i.e., the client and server side. It does not protect against the non-scripting attacks. It is difficult to apply as it changes the way application generates the HTML content for browser. PathCutter [13] is a server-side approach that attempts to reduce the impact of the XSS worm by constraining its dissemination path in OSN. However, it is not able to protect each and every view from being infected from the XSS attack vector. Li et al. [19] uses machine learning algorithm to classify malicious Web pages from benign Web pages in OSN-based web applications. However, its effectiveness depends on the training phase to build up the feature database. Saner [14] makes use of both static and dynamic analysis for detecting the presence of malicious scripting code in the source code of web applications.

P. Chaudhary (✉) • B.B. Gupta • S. Gupta
Department of Computer Engineering, National Institute of Technology Kurukshetra, Kurukshetra, Haryana, India
e-mail: pooja.ch04@gmail.com; gupta.brij@gmail.com; shashank.csit@gmail.com

© Springer Nature Singapore Pte Ltd. 2018

P.K. Kapur et al. (eds.), *Quality, IT and Business Operations*, Springer Proceedings in Business and Economics, DOI 10.1007/978-981-10-5577-5_9

However, its deployment cost is high. Nandji et al. [16] proposed a client-server-based architecture to enforce document structure integrity. It combines runtime tracking and randomization to thwart XSS attack. However, it is not effective to detect the DOM-based XSS attack.

Usually, an untrusted JavaScript variable may be in different context along the same path or in the different path along the program, called path-sensitive analysis. It arises due to the presence of the rich language constructs like string operations, IF-ELSE, loops, etc. Hence, it is problematic to accurately determine the context. Besides this, if the web application programming language does not include the rich language constructs like looping, control-flow, and string operations, then sanitization is a simple task. Usually, sanitization is a three-step procedure: (1) parse the document, (2) determine the context, and (3) sanitize the XSS vulnerable untrusted variable. However, the presence of these sophisticated operations makes the auto-sanitization a challenging task. Based on this, this paper presents a robust technique that is based on the determination of the context of untrusted JavaScript malicious code and also performs sanitization on them, to provide the protection from the XSS attack. Our approach accomplishes context identification at the server side instead of the client side during the runtime. It is divided into two phases: first, CASG phase is responsible for the generation of the typed HTML documents by assigning type qualifier to untrusted JavaScript variable. In addition to this, it also injects the sanitizer into the documents on the basis of context determined. Second, CADP phase identifies the context of the dynamically qualified variable/expression through runtime parsing. Finally, it processes the sanitizer routines on the HTML document to isolate the malicious content from benign content and displays the resultant document to the user.

2 Proposed Design

The proposed system comprises of two phases: CASG phase and CADP phase. The first phase performs the static analysis of HTML document to determine the context in which the XSS vulnerable untrusted JavaScript code can be securely rendered. In addition to this, it also injects sanitizers in the HTML documents extracted from the externally available sanitizers' library. The second phase initially performs the dynamic parsing of HTML document produced as a result of the first phase. Secondly, it performs the sanitization on the basis of the context determined to achieve malicious content isolation from benign content. Basically, it is a client-server-based framework wherein sanitizers are injected in the Web page at the server side and applied on the untrusted variables at the client side (only for dynamically determined context). In the next subsection, we have described the abstract view of our proposed model. Detailed description is provided in further subsections.

2.1 Abstract View of Proposed Framework

Our approach integrates static and dynamic type analysis. Statically unambiguous context are determined at the server side for the untrusted JavaScript variables. In addition to this, it also inserts sanitizers' routines in the document. However, statically ambiguous contexts of the untrusted variable/expression are determined at the client side dynamically. In this context-type-based method, a type qualifier is attached with each untrusted variable/expression. Type qualifier describes the context for the untrusted JavaScript variable/expression, in which it can be safely rendered. On the basis of these qualifiers, we represent the context of each variable and guarantee that the document is context sensitively sanitized. Figure 1 describes the abstract view of our proposed framework. Our technique works under two phases: Context-Aware Sanitization Generator (CASG) and Context-Aware Dynamic Parsing (CADP).

CASG phase: This phase carries out the static analysis of HTML document produced by the OSN server, in response to the HTTP request generated by the user. Each untrusted JavaScript variable/expression is accompanied a type qualifier. On the basis of the context indicated by this qualifier, sanitizers are injected into the HTML document (typed document) and returned it to the user.

CADP phase: The input to this phase is the HTML typed document from the OSN server. For all those untrusted JavaScript variables/expressions, whose contexts cannot be determined statically due to the presence of the rich language constructs, undergo parsing mechanisms to determine the context at the runtime. In compliance with the identified context, sanitizers are implemented on the untrusted inputs to perform malicious content separation. Lastly, the sanitized document is displayed to the user in the browser window.

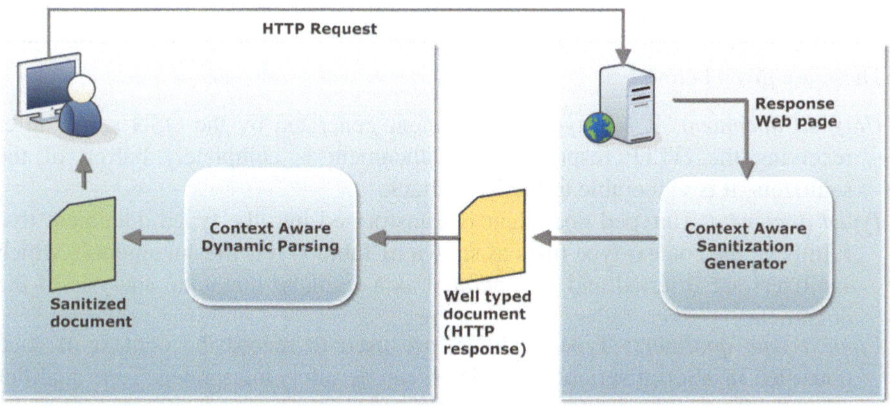

Fig. 1 Design of abstract view of proposed framework

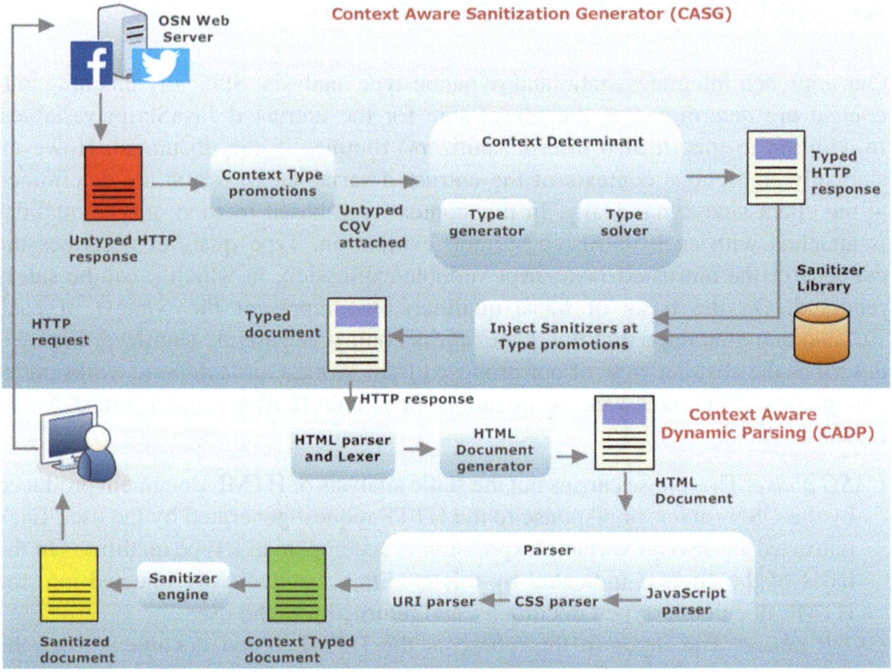

Fig. 2 Detailed design of proposed framework

2.2 Detailed Overview

Here, we outline our context-type qualifier-based method to defend against the XSS worms. Each component of the system is elaborated to have a deep understanding of the planned mechanisms. Figure 2 elaborates the detailed framework of our proposed system. The framework uses some of the key terms and components. These are given below:

Untyped document: It is a type of document generated by the OSN server after receiving the HTTP response. This document is completely barren of the sanitizers. It is vulnerable to the XSS attack.

Typed document: Untyped document is transformed into the typed document that fulfills all the context type rules as shown in Table 1. It is the document in which sanitizers are inserted and is generated as a result of the static analysis of the document.

Context type qualifiers: Type qualifiers are used to record the context of each untrusted JavaScript variable in which it can be safely interpreted. Type qualifier is a type of variable attached to each untrusted input as v = (Q)e, attached to the expression e, stores the type of the context to be identified by the system, where Q represents context-type qualifier. For constant and string (static), it captures

Table 1 Content type rules

Type of expression	Type rules
Constants	**If** v$\neq$**String** **then** $\Gamma\mapsto$v: $STA_{C}\rightarrow_{C}$ where C $\in$**Context**
Constant string	**If** s = **String** **then** $\Gamma\mapsto$s: $STA_{C'}\rightarrow_{C''}$ where C′, C″$\in$**Context**, if s parses validly from C′ to C″
Variables	**If** x = **Variable** **then** $\Gamma\mapsto$x: Q where Q = Type qualifier indicating context
Integers	**If** p: $STA_{C}\rightarrow_{C}$ and q = $STA_{C}\rightarrow_{C}$ such that p, q $\in$**Integer** **Then** $\Gamma\mapsto$p$\oplus$q: STA $_{C}\rightarrow_{C}$ where C $\in$**Context**
Static string concatenation	**If** S_1: $STA_{C1}\rightarrow_{C2}$ and S_2: $STA_{C2}\rightarrow_{C3}$**such that** S_1, S_2 $\in$**String** **Then** $\Gamma\mapsto$ Con_cat(S_1 S_2): STA $_{C1}\rightarrow_{C3}$ where C1, C2, C3 $\in$**Context**
Dynamic string concatenation	**If** S_1: $DYN_{S'}$ and S_2: $DYN_{S''}$**such that** S_1, S_2 $\in$**String** **Then** $\Gamma\mapsto$ Con_cat(S_1 S_2): $DYN_{S'}\infty_{S''}$ where $\infty$is inner- join and S′, S″ = Set of contexts

the transition between the context produced when it is parsed. For example, static string <img src=" is qualified as STAPCDATA URL-Start when parsed. It shows up the transition from PCDATA to URL-Start context. Initially, all untrusted input is qualified as UNSAFE. Our system qualifies them as $STA_{C}\rightarrow_{C'}$ (C and C′ indicate contexts) only when variable is filtered by the sanitizer primitives with corresponding context C.

Handling ambiguous context typing: As a result of the path-flow sensitivity, we address static type ambiguity. For example, at the joint point of IF-ELSE, context cannot be statically determined because in one branch of IF-ELSE, it prints some string and in other it prints some other result. To handle such issues, we have divided type qualifiers into two types: static type qualifiers (STAC) and dynamic type qualifiers (DYNS). In case of dynamic qualified input, S represents the set of the context transition that the input can make. Selection of the sanitizers is done at the server side statically and inserted in the document. For dynamically qualified untrusted input, sanitizers' identification is accomplished by performing the parsing on the document at the client side. Our method incorporates two phases: Context-Aware Sanitization Generator (CASG) and Context-Aware Dynamic Parsing (CADP).

Context-Aware Sanitization Generator (CASG): In this phase, HTML document is generated corresponding to the HTTP request. The goal of this phase is to convert the untyped document into the typed document. To accomplish this goal, static analysis of the HTML document is done to define the context of the untrusted JavaScript code in which it can be carefully handled. This phase consists of multiple components. The detailed description of each of these components is described below:

Context type promotions: These are the operations used to identify places where untrusted input initially qualified as UNSAFE is converted to the STAC and DYNS. Type promotions are written as v = (Q)e. Q denotes the context-type qualifier variable and records the context to be recognized. Input to this component is the untyped document, and it aims to convert this document into an internal representation that includes the type qualifier for the untrusted JavaScript variables. It aids in the identification of the locations, where sanitizers are inserted lastly by the CASG phase. Finally, the resultant document is passed for the generation of the type constraints and to resolve these constraints.

Context determinant: The goal of this component is to identify the context for each type qualifier (Q) corresponding to each untrusted input. It will represent the context in which untrusted JavaScript input is embedded and sanitizers are selected accordingly. It splits into two small components: type generator and type solver. Type generator produces the rules for the variables and expressions to be satisfied as shown in Table 1. Type solver, on the basis of these type rules, determines the context for each context type qualifier. This step receives the internal representation of the document embedded with the type qualifier and converts it into the typed document. Finally, the resultant HTML document is typed document.

Insert sanitizers at type promotions: The key purpose of this component is to inject the sanitizers at the type promotions operations. Selection of the sanitizers is done on the basis of the context determined for the untrusted JavaScript input. Sanitizers' library is externally available, and the function SMAP(f, a) which maps the sanitizers f to their matching context on the untrusted JavaScript variable a is also available externally. The generated HTML document is the typed document with sanitizers inserted into it and, finally, it is returned to the client. Figure 3 describes the algorithm implemented at the server side to accomplish the functionalities of all the components defined above.

The working process of the algorithm is explained below:

1. QV_LOG contains the candidate type promotions qualifier variable. These are utilized to inject the sanitizers. S_LIB is the externally available sanitizers' library. It contains the sanitizers that are to be applied on the untrusted JavaScript input to provide protection against the XSS worms.
2. On each untrusted JavaScript variable *e*, in the HTML document, we attach a type qualifier Q in the form as $v = (Q)e$. The resultant Web page is the internal representation of the document embedded with the type qualifier Q which corresponds to each untrusted JavaScript input. After this, it is merged with the QV_LOG as $QV\_LOG \leftarrow VI \cup QV\_LOG$.
3. For each $V_I \in$QV_LOG, context determinant generates and solves the type constraints according to the rules in the Table 1. Here, Γ represents the type environment that performs the mapping of the program variable/expression to the type qualifier Q. In the path-sensitive system, variables' context changes from one point in the program to other point. Thus, to handle this issue, untrusted expressions are represented through the typing judgments as $\Gamma \mapsto e : Q$.

Algorithm: Context Aware Sanitization Generator

Input: Set of HTTP request ($H_1, H_2, \ldots H_N$)
Output: Sanitizers injected document for each request ($D_1, D_2, \ldots D_N$)
Context type qualifier: $Q_1 | Q_2 | \ldots | Q_N$
Qualifiers (Q):= UNSAFE
$| \; STA_{C'} \rightarrow_{C''} \quad c', c'' \in C$
$| \; DYN_S \quad S \in 2^{C*C}$
Context (C):= RCDATA| PCDATA| TAGNAME.....
Start
QV_LOG ← Set of type qualifier variables
S_LIB ← Sanitizers Library
For Each HTTP request H_I
  For Each untrusted variable/ expression e
    Do $V_I \leftarrow (Q)e$
      QV_LOG ← $V_I \cup$ QV_LOG
    For Each $V_I \in$ QV_LOG
      $\Gamma \mapsto V_I : STA_{C'} \rightarrow_{C''} \; | \; \Gamma \mapsto V_I : DYN_S$
      Inject SMAP(f, V_I) in D_I f ∈ S_LIB ↦ Q
    End For Each
  End Eor Each
Return document D_I
End For Each
End

Fig. 3 Detailed algorithm of CASG phase

It indicates that at any program location e has context type qualifier Q in the type environment Γ.

4. Each type constraint is solved by inferring the value of the type qualifier Q as static or dynamic. This is done by $\Gamma \mapsto V_I$: $STA_{C'} \rightarrow_{C''} \mid \Gamma \mapsto V_I$: DYN_S. First expression states that the V_I is attached a type qualifier now qualified as static under the type environment Γ. Second expression indicates that the V_I has type qualifier qualified as the dynamic with S as the approximation set of the context under the type environment Γ. This step will generate the typed document.
5. SMAP(f, V_I) is the function which maps the sanitizers from library to the program variables according to the matching context indicated by the type qualifier attached to the V_I. Lastly, the typed document is returned to the client as the HTTP response.
6. *Context-Aware Dynamic Parsing (CADP)*: Its key goal is to perform the runtime parsing of this typed document to determine the context of the variable that cannot be statically determined by the CASG phase. It includes many small components as described below:

HTML parser and lexer: Its main goal is to construct a parse tree, i.e., Document Object Model (DOM). During parsing, executable script nodes are determined and nodes are created for them in the parse tree.

HTML document generator: It stores and processes the Web content represented by the parse tree. It also performs the separation of the content and gives it to other parts of the browser for rendering. For example, scripting code is supplied to the JavaScript parser for processing.

Parser: Document generator separates the visual, linking elements and scripting elements and passed them to the CSS parser, URI parser, and JavaScript parser, respectively. These parsers process the content received. In addition to this, they also dynamically determine the context of the untrusted input to which a dynamic type qualifier is attached.

Sanitizer engine: This component receives the context type document with the injected sanitizers' primitives. Therefore, its aim is to execute these sanitizers' primitives on to the untrusted variable corresponding to them, to achieve the malicious content isolation from the benign content.

Figure 4 describes the algorithm used to implement the CADP phase. It provides the stepwise working process of the phase.

1. HTML parser generates document D which corresponds to each HTTP response HR_I. Then, for each static type qualifier Q, applies the sanitizers on the variables injected by the CASG phase and displays the sanitized document to the user.
2. For all type promotions operations as ($\Gamma \mapsto V_I$: DYN_S), we parse the document to dynamically identify the context of the untrusted input in which it can be safely interpreted. D_S stores the result of the JS_Parser. D_C holds the result of CSS_Parser and D_U stores the result of URI_Parser.
3. After applying parsers, we execute the sanitizer routines on the resultant typed document. San (f, V_I) function will executes sanitizers' primitive f on V_I according to the matching context indicated by the attached type qualifier.
4. Finally, the sanitized HTML document is displayed to the user at the browser window with malicious content quarantined.

3 Implementation and Experimental Evaluation

We have implemented the proposed framework as an extension of Google Chrome for mitigating the effect of XSS worms from OSN-based web applications. This implementation of our proposed framework has been performed in the form of a JavaScript function. This function of JavaScript comprises of few lines of source code of JavaScript. Websites can invoke this function on an HTML form. In addition to this, we have also tested the XSS worm detection capability of our framework on the tested suite of two real-world OSN web applications (i.e., HumHub [18] and Elgg [21]). We have also referred the XSS cheat sheet [17] for retrieving the XSS attack vectors. In order to include the capabilities of our proposed framework in these web applications, developers of both these websites have to do less quantity of effort. These web applications have already built-in validation mechanisms deployed at the server side. The motivation to select these web applications is that we simply need to mark an argument that web sites can utilize

```
Algorithm: Context Aware Dynamic Parsing
Input: Set of sanitizers injected HTTP responses (HR1, HR2, ... HRN)
Output: Sanitized HTTP response (HR1', HR2', ... HRN')
Start
D ← ϕ
For Each HTTP response HRI
        D ← HTML_Parser(HRI)
        For Each VI
        If (VI : STA C' → C'' ) then
                Apply sanitizers at VI
                Return sanitized document HRI'
            End If
        End For Each
        For Each type qualifier variable (VI)    VI ∈ (HTML element/attribute ∩ Γ ↦ VI : DYNS)
                San(VI, f)   f∈ S (S∈ 2^(C*C))
        End For Each
        DS ← JS_Parser(D)
        For Each type qualifier variable (VI)    VI ∈ (JavaScript context ∩ Γ ↦ VI : DYNS)
                San(VI, f)   f∈ S (S∈ 2^(C*C))
        End For Each
        DC ← CSS_Parser(DS)
        For Each type qualifier variable (VI)    VI ∈ (CSS context ∩ Γ ↦ VI : DYNS)
                San(VI, f)   f∈ S (S∈ 2^(C*C))
        End For Each
        DU ← URI_Parser(DC)
        For Each type qualifier variable (VI)    VI ∈ (URL context ∩ Γ ↦ VI : DYNS)
                San(VI, f)   f∈ S (S∈ 2^(C*C))
        End For Each
    Apply sanitizers at each VI
    Return sanitized document HRI'
End For Each
End
```

Fig. 4 Detailed algorithm of CADP phase

the capabilities of our proposed framework irrespective of the input testing. This will aid in alleviating XSS vulnerability concerns and would include supplementary security layer. We deploy these web applications on an XAMPP web server with MySQL server as the backend database. We have selected different categories of XSS attack vectors. These categories include event handlers, character encoding scripts, URL obfuscation, HTML quote encapsulation, HTML entity elements, and embedded character tags. The observed results of proposed design on two OSN web applications corresponding to the chosen categories of XSS attack vectors have been shown in Figs. 5 and 6.

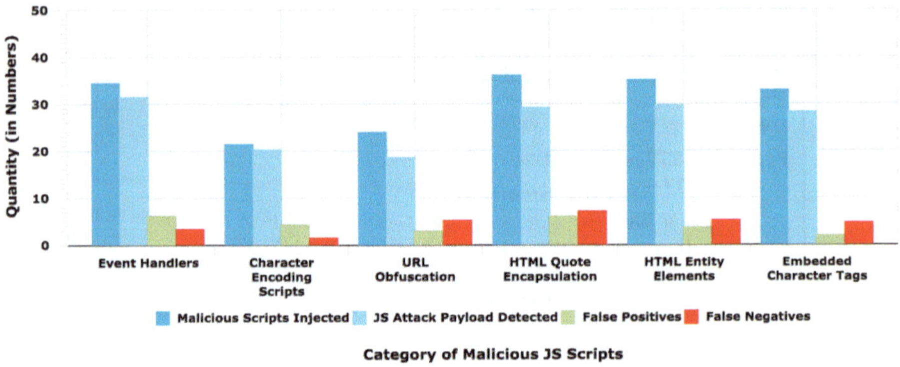

Fig. 5 Observed results of HumHub

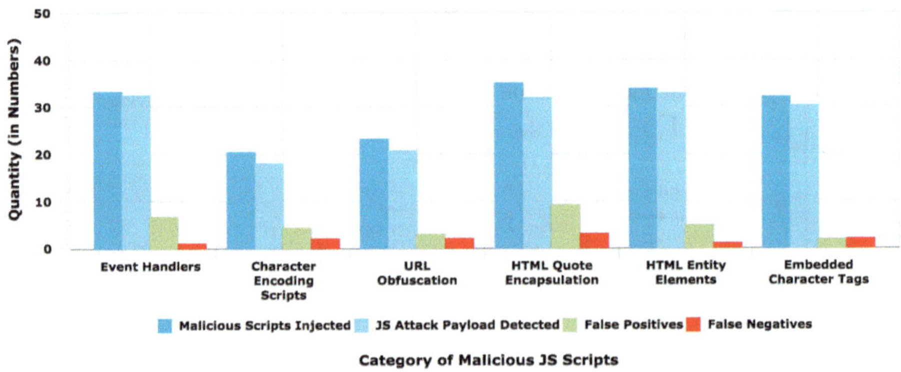

Fig. 6 Observed results of Elgg

3.1 Performance Analysis

We have presented a detailed performance analysis of our proposed design by conducting a statistical analysis method (i.e., F-Measure). The analysis conducted reveals that the proposed framework produces better results as compared to existing state-of-art techniques. To perform binary classification, precision and recall are the parameters used for evaluations. And their harmonic mean is F-measure. Here we calculate the precision, recall, and, finally, F-Measure of observed experimental results of proposed framework. The analysis conducted reveals that the framework exhibits high performance as the observed value of F-Measures in all the platforms of OSN web applications is almost close to 0.9. Therefore, the proposed framework exhibits 90% success rate in the two OSN-based web applications. Table 2 highlights the detailed performance analysis of proposed framework on two web applications.

Table 2 Statistics of performance analysis

Web application	TP	FP	FN	Precision	Recall	F-Measure
HumHub	166	30	11	0.846	0.937	0.890
Elgg	157	25	26	0.862	0.857	0.860

$$\text{Precision} = \frac{\text{True Positives (TP)}}{\text{True Positives (TP)} + \text{False Positives (FP)}}.$$

$$\text{Recall} = \frac{\text{True Positives (TP)}}{\text{True Positives (TP)} + \text{False Negatives (FN)}}.$$

$$\text{F-Measure} = \frac{2\ \text{(True Positives)}}{2\ \text{(True Positives)} + \text{False Negatives} + \text{False Positives}}.$$

3.2 Comparison

Table 3 illustrates the comparison of existing XSS defensive solutions with our proposed XSS defensive framework. We have utilized nine different metrics, i.e., CAP, Context-Aware Parsing; CAS, Context-Aware Sanitization; SA, static analysis; DA, dynamic analysis; P-XSS, persistent cross-site scripting; R-XSS, reflected cross-site scripting; PXD, polymorphic XSS worm detection; CSM, client-side modifications; and SSM, server-side modifications. It is clearly reflected from the Table 3 that except blueprint [7] and our work, all other recent techniques has not performed context-aware parsing to achieve content isolation. And it also reflects that except Saner [14] and our work, none of other XSS defensive techniques has performed context-aware sanitization. In addition to this, most of the existing techniques require several alterations at the server as well as browser side. However, our proposed XSS defensive solution is entirely based on the context-aware sanitization and dynamic parsing. These two unique techniques adequately validate the effectiveness of our XSS defensive technique on different platforms of OSN-based web applications.

4 Conclusion

This paper presents a context-aware sanitization and dynamic parsing-based XSS defensive framework. It initially determines the context of untrusted JavaScript variable and injects the context-aware sanitizers in the locations of untrusted

Table 3 Summary of comparison of related XSS defensive techniques with our work

Parameters	Techniques Blueprint [7]	PathCutter [13]	Saner [14]	DSI [16]	Wang, Li et al. [19]	Wang et al. [15]	Sun et al. [20]	Our work
CAP	Yes	No	No	No	No	No	No	Yes
CAS	No	No	Yes	No	No	No	No	Yes
SA	No	Yes	Yes	No	Yes	Yes	No	Yes
DA	Yes	No	Yes	Yes	Yes	Yes	Yes	Yes
P-XSS	Yes	Yes	Yes	Yes	Yes	Yes	Yes	Yes
R-XSS	Yes	Yes	Yes	Yes	Yes	Yes	Yes	Yes
PXD	Yes	Yes	No	Yes	No	No	No	Yes
CSM	Yes	No	No	Yes	No	No	Yes	Yes
SSM	Yes	Yes	No	Yes	No	No	No	Yes

* *CAP* context-aware parsing, *CAS* context-aware sanitization, *SA* static analysis, *DA* dynamic analysis, *P-XSS* persistent cross-site scripting, *R-XSS* reflected cross-site scripting, *PXD* polymorphic XSS worm detection, *CSM* client-side modifications, *SSM* server-side modifications

JavaScript variable. Finally, it performs the runtime parsing on the generated HTML document to determine the context of the untrusted JavaScript variable that is statically ambiguous to identify initially. The evaluation and testing of proposed framework was done on real-world OSN-based web applications. Evaluation results revealed that our framework correctly determines the context of untrusted variables at runtime with acceptable false-positive and false-negative rate.

References

1. Gupta S, Gupta BB (2014) "BDS: browser dependent XSS sanitizer" book on cloud-based databases with biometric applications IGI-Global's advances in information security, privacy, and ethics (AISPE) series. IGI-Global, Hershey, pp 174–191
2. Gupta BB, Gupta S, Gangwar S, Kumar M, Meena PK (2015) Cross-site scripting (XSS) abuse and defense: exploitation on several testing bed environments and its defense. J Inf Privacy Secur 11(2):118–136
3. Gupta S, Gupta BB (2016) JS-SAN: defense mechanism for HTML5-based web applications against javascript code injection vulnerabilities. Security and Communication Networks
4. Gupta S, Gupta BB (2015) Cross-site scripting (XSS) attacks and defense mechanisms: classification and state-of-Art. Int J Syst Assur Eng Manage, Springer
5. Gupta S, Gupta BB (2015) PHP-sensor: a prototype method to discover workflow violation and XSS vulnerabilities in PHP web applications. In: Proceedings of the 12th ACM international conference on computing frontiers. ACM
6. Gupta S, Gupta BB (2015) XSS-SAFE: a server-side approach to detect and mitigate cross-site scripting (XSS) attacks in JavaScript code. Arab J Sci Eng, Springer, pp 1–24
7. Louw MT, Venkatakrishnan VN (2009) Blueprint: robust prevention of cross-site scripting attacks for existing browsers. In: Security and Privacy, 2009 30th IEEE symposium on IEEE, pp 331–346

8. Gupta S, Gupta BB (2016) An infrastructure-based framework for the alleviation of JavaScript worms from OSN in mobile cloud platforms. In: International conference on network and system security. Springer International Publishing
9. Gupta S, Gupta BB (2016) XSS-secure as a service for the platforms of online social network-based multimedia web applications in cloud. Multimedia Tools and Applications, pp. 1–33
10. Gupta S, Gupta BB (2016) Automated discovery of JavaScript code injection attacks in PHP web applications. Proc Comput Sci 78:82–87
11. Gupta S, Gupta BB (2016) XSS-immune: a Google chrome extension-based XSS defensive framework for contemporary platforms of web applications. Security and Communication Networks
12. Chaudhary P, Gupta S, Gupta BB (2016) Auditing defense against XSS worms in online social network-based web applications. In: Handbook of research on modern cryptographic solutions for computer and cyber security. IGI Global, pp 216–245
13. Cao Y, Yegneswaran V, Porras PA, Chen Y (2012) PathCutter: severing the self-propagation path of XSS JavaScript worms in social web networks. In: NDSS
14. Balzarotti D, Cova M, Felmetsger V, Jovanovic N, Kirda E, Kruegel C, Vigna G (2008) Saner: composing static and dynamic analysis to validate sanitization in web applications. In: Security and privacy, 2008. SP 2008. IEEE Symposium on IEEE, pp 387–401
15. Wang W-H, et al. (2013) A static malicious Javascript detection using SVM. In: Proceedings of the international conference on computer science and electronics engineering, vol 40
16. Nadji Y, Saxena P, Song D (2009) Document structure integrity: a robust basis for cross-site scripting defense. In: NDSS
17. OWASP, XSS. Prevention Cheat-sheet., https://www.owasp.org/index.php/XSS_%28Cross_Site_Scripting%29_Prevention_Cheat_Sheet
18. Humhub social networking site, https://www.humhub.org/en
19. Wang R et al. (2014) Machine learning based cross-site scripting detection in online social network. High performance computing and communications, 2014 IEEE 6th international symposium on cyberspace safety and security, 2014 IEEE 11th international conference on embedded software and system (HPCC, CSS, ICESS), 2014 IEEE international conference on. IEEE
20. Sun F, Liang X, Zhendong S (2009) Client-side detection of XSS worms by monitoring payload propagation. Computer security–ESORICS 2009. Springer, Berlin/Heidelberg, pp 539–554
21. Elgg social networking engine, https://elgg.org

Development of QFD Methodology

Frolova Elena, Albina Gazizulina, Elena Eskina, Maria Ostapenko, and Dmitriy Aidarov

1 Introduction

The present paper discusses anticreep locking mechanism Garant Consul (Fig. 1), which is mounted into gearboxes of the Nissan cars in the Nissan service centres in the Russian Federation. It prevents the gear change, and therefore it does not allow the illegal use of a car.

Gearbox locking mechanism "Garant Consul" uses a latch secret mechanism Abloy Protec (1.97 billion combinations of the secret mechanism) housed in a thick-walled case which guarantees a high level of protection against the illegal use. The absence of removable elements guarantees easy and simple use, and the concealed installation and a very small keyhole do not spoil the design of the car interior.

F. Elena (✉)
Department of General Engineering, Orenburg State University, Orenburg, Russian Federation
e-mail: fev_2004@list.ru

A. Gazizulina
Monitoring Centre for Science and Education, Peter the Great St. Petersburg Polytechnic University, St. Petersburg, Russia
e-mail: albinagazizulina@gmail.com

E. Eskina
Department "Production of Aircraft and Quality Management", Samara University, Samara, Russia
e-mail: elena2002.83@mail.ru

M. Ostapenko
Department "Machines and Tools", Tyumen Industrial University, Tyumen, Russia
e-mail: ms_ostapenko@mail.ru

D. Aidarov
Russian Presidential Academy of National Economy and Public Administration, Togliatti, Russia
e-mail: adv_tol@mail.ru

© Springer Nature Singapore Pte Ltd. 2018

P.K. Kapur et al. (eds.), *Quality, IT and Business Operations*, Springer Proceedings in Business and Economics, DOI 10.1007/978-981-10-5577-5_10

Fig. 1 Gearbox locking mechanism for "Nissan" cars

The brackets of the locking mechanism "Garant Consul" are designed as a welded structure of a sheet steel (5 mm thick). A coat of black matt powder paint is applied to the surface of the brackets that provides a high level of corrosion resistance and fits in the car interior. A coat of black matt powder paint is applied to the surface of the brackets that provides a high level of corrosion resistance and fits in the car interior and consequently an improved usability of the locking mechanism. The size of the exit section of the secret mechanism is below 14 mm that is 2.5–3 times less than that of the locking mechanisms of other manufacturers. That enables installation on surfaces of complex geometry with minimal harm to the car interior. Placing of the lock exit section on the upper part of the centre console provides the maximal comfort for a driver.

Locking and unlocking of the gearbox selecting mechanism is performed with a half turn of the Abloy key clockwise and anticlockwise. By spring force, the locking mechanism locks the gearbox using the tension element attached to it. The mechanism is locked by setting the speed-change lever in a particular position indicated in the user's manual.

2 Customers' Requirements

For the analysis of the installation process of the gearbox locking mechanism, we carried out a written customers' survey in order to find out customers' requirements and requests to the process. The survey took place in the Nissan service centre in 2014. The target group consisted of 100 users of Nissan cars who ordered the installation of the gearbox locking mechanism "Garant Consul."

Firstly, we made a list of the customers' requirements and determined which expectations are the most important. Table 1 shows the customers' requirements to the installation process of the gearbox locking mechanism [1–6].

Table 1 Customers' requirements and the significance of the planned quality parameters

Customers' requirements	Significance
Simple use, less actions	10
Functionality, the lock is easy to reach	10
To improve the reliability of the mechanism	10
To improve the reliability of the change lever lock	7
Compatibility with autostart in case of manual gearbox	10
Compatibility	5
Possibility of gearbox locking without a key (using a chip instead)	7
Reduce the price of the device and its installation	10
Provide a possibility to instal the device not in the Nissan service centre with retaining warranty	10
Reduce the changes of the standard car interior	10
Installation time	10

Table 2 Symbols of weight

Dependence	Weight
● – Strong dependence	9
○ – Medium	3
▲ – Weak	1

Further, we shall transform the customers' requirements into the installation process stages of the gearbox locking mechanism. To do that, we assign the values to the symbols describing the dependences according to their weight (Table 2).

Assigning the values "9–3–1" to the levels of dependence allows to separate the significant and not very significant elements of the discussed dependencies.

Table 3 shows the obtained survey data. The rows show the customers' requirements to the process of the gearbox locking mechanism installation, and the columns present data on the installation process stages [7–9].

3 Calculate the Absolute Value of the Requirement to the Installation Process

Next, we calculate the absolute value of the requirement to the installation process as the sum of the products of customers' assessment times the weight corresponding to the dependence degree [10–14]. The result is given at the end of the column demonstrating the importance of a requirement, i.e., the priority quality characteristics for a customer.

$$B = \sum (I_x \bullet R_x),$$

Table 3 QFD analysis of the installation process of the gearbox locking mechanism “Garant Consul”

		Process stages							Competitors		
Customers’ requirements	Significance	Order acceptance	Car disassembly (partly)	Fit test	Installation	Adjustment	Reassembly	Delivery and acceptance	Nissan service centre	Service centre Bluesmo-bile	Ya7auto
Simple use, less actions	10			●				●	5	4	5
Functionality, the lock is easy to reach	10						●	○	4	5	3
To improve the reliability of the mechanism	10			●		●		○	5	5	5
To improve the reliability of the change lever lock	7			○	●	○			4	5	4
Compatibility with autostart in case of manual gearbox	10	●	●			▲	▲	▲	4	4	5
Compatibility	5	○	○	▲			▲	▲	5	5	4

Possibility of gearbox locking without a key (using a chip instead)	7				●	○	●		3	5	4
Reduce the price of the device and its installation	10	●		▲		○		●	4	4	5
Provide a possibility to instal the device not in the Nissan service centre with retaining warranty	10	●							5	4	5
Reduce the changes of the standard car interior	10		○		●		●		4	4	4
Installation time	10	▲	○	●	●	●	○	▲	5	5	5
Absolute weight	1807	295	165	306	306	262	288	185			
Relative weight, %		16.4	9.2	16.9	16.9	14.5	15.9	10.2			

Table 4 Data about the satisfaction of the customers who ordered the gearbox locking mechanism installation to the competitors

Requirement	Nissan service centre	Service centre Bluesmobile	Ya7auto
Simple use, less actions	5	4	5
Functionality, the lock is easy to reach	4	5	3
To improve the reliability of the mechanism	5	5	5
To improve the reliability of the change lever lock	4	5	4
Compatibility with autostart in case of manual gearbox	4	4	5
Compatibility	5	5	4
Possibility of gearbox locking without a key (using a chip instead)	3	5	4
Reduce the price of the device and its installation	4	4	5
Provide a possibility to instal the device not in the Nissan service centre with retaining warranty	5	4	5
Reduce the changes of the standard car interior	4	4	4
Installation time	5	5	5

where:

B – significance weight of the process stage
I_x – dependence between the requirement and the process stage (▲ – 1; ○ – 3; ● – 9)
R_x – absolute weight of the requirement

To determine the competitiveness of the Nissan service centre under discussion, we surveyed the customers in order to find out the satisfaction degree of competitors' customers. To determine the competitiveness of the Nissan service centre under discussion, we surveyed the customers in order to find out the satisfaction degree of competitors' customers of service centres "Blues mobile" and "Ya7auto". Table 4 shows the data of the survey.

We shall calculate the significance values for the process stages of the gearbox locking mechanism installation with account of the market situation according to the dependence since the gap with the competitors is minimal, i.e., the level of the competition is high:

$$\mathrm{W}_2\text{of the process with account of the market situation} = \sum (I_{\mathrm{x}} \cdot R_{\mathrm{x}})\,(\Pi_{OTC_x} + 1)\,,$$

Table 5 calculation results for the process stage significance with account of the market situation and the Weber-Fechner law

	a	b	c	d	e	f	g
$B_{1\text{процесса}} = \sum (I_x \bullet R_x)$	**295**	165	**306**	**306**	262	288	185
$W_{2\text{ of the process}} = \sum (I_x \cdot R_x)(\Pi_{OTC_x} + 1)$	**475**	285	337	**585**	365	**604**	235
$W_{3\text{ of the process}} = \sum (I_x \cdot R_x)(\Pi_{OTC_x}{}^k + 1)$	**475**	285	337	**711**	407	**730**	235

where

I_x – relation between the requirement and the process stage (▲– 1; ○– 3; ● – 9)
R_x – absolute weight of the requirement
$\Pi_{OTC_x} + 1$ – difference between the target and the obtained values of the customers' satisfaction level (or the difference between the values at the competitor's organisation and the target organisation) (considered only if the calculation results are not negative)

Let us calculate the significance values for the process stages of the gearbox locking mechanism installation with account of the market situation and the Weber-Fechner law (coefficient $k = 2$, which considers the degree of perception intensity) according to the dependence:

$$B_3 \text{ of the process with account of the Weber–Fechner law} = \sum (I_x \cdot R_x)\left(\Pi_{OTC_x}{}^k + 1\right)$$

Table 5 shows the calculation results for the process stage significance with account of the competition and the Weber-Fechner law.

4 Conclusion

Comparing the satisfaction degree of the customers of the Nissan service centre and that of the competitors' customers, we can determine the stages which require improvements. Considering the competition and the Weber-Fechner law (Table 5), we can draw the conclusion that Nissan service centre under discussion should improve the following stages of the gearbox locking mechanism installation:

1. Car interior assembly, interior panel follow-up revision
2. Installation
3. Order acceptance

References

1. Glushkov SV, Skvortsov YV, Perov SN (2014) Comparison of the results of solving the problem of fracture mechanics for pipe with non-through crack. PNRPU Mech Bull 2014(3):36–49
2. Akao Y (1997) QFD: past, present, and future, Asahi University, International Symposium on QFD '97 – Linköping
3. Papic L (2007) Deploying customer requirements via four-stage team approach in business planning. Int J Reliab Qual Saf Eng 14(03):263–274
4. Cohen L (1995) Quality function deployment: how to make QFD work for you. Addison-Wesley Publishing Company, Reading
5. Stamatis DH (1995) Failure mode and effect analysis: FMEA from theory to execution. ASQC Quality Press, Milwaukee
6. Grunske L, Colvin R Winter K. Probabilistic model-checking support for FMEA from: http://staff.itee.uq.edu.au/kirsten/publications/qest.pdf
7. Stamatis D (2003) Failure mode and efects analysis, FMEA from theory to execution. ASQ Quality Press, Milwaukee, p 487
8. Horvth A, Varr D, Schoofs T (2010) Model-driven development of ARINC 653 configuration tables. IEEE/AIAA 29th Digital Avionics Systems Conference (DASC)
9. Dubois H, Ibanez V, Lopez C, Machrouh J, Meledo N, Mouy P, Silva A (2012) The product line engineering approach in a model-driven process", ERTS
10. Wiels V, Delmas R, Doose D, Garoche P-L, Cazin J, Durrieu G (2012) Formal verification of critical aerospace software. AerospaceLab Journal, Issue 4
11. Varet A, Larrieu N (2011) New methodology to develop certified safe and secure aeronautical software – an embedded router case study. 30th digital avionics systems conference (DASC), Seattle, Washington, USA
12. Varet A, Larrieu N (2012) Design and Development of an embedded aeronautical router with security capabilities. Integrated Communication, Navigation and Surveillance Conference (ICNS), Washington DC, Colombia, USA
13. Izerrouken N, Thirioux X, Pantel M, Strecker M. Design and development of an embedded aeronautical, certifying an automated code generator using formal tools: preliminary experiments in the GeneAuto Project. European Congress on Embedded Real-Time Software (ERTS), 2010, Toulouse
14. Klochkov Y, Gazizulina A, Golovin N(2016) Assessment of organization development speed based on the analysis of standards efficiency. Second international symposium on stochastic models in reliability engineering, Life Science and Operations Management (SMRLO), pp. 530–53

Discrete-Time Framework for Determining Optimal Software Release and Patching Time

Anshul Tickoo, P. K. Kapur, A. K. Shrivastava, and Sunil K. Khatri

1 Introduction

As software are everywhere in our daily life, reliability becomes one of its most important characteristics. Software reliability is very important for measurement of the performance of a software system. However, software is usually developed by human work; introduction of faults into the software during its development process is inevitable, despite great advancements in programming technology [17]. Therefore, it is essential to apply necessary tools and technologies for developing highly reliable software products.

While software testing, the faults that lead to failures are detected and removed; hence, the reliability of software improves. Various SRGMs have been developed to characterize the reliability growth phenomenon in the software testing process. These models help to determine the reliability growth by testing with time [5, 19, 22, 25]. Most of the developed SRGMs are nonhomogeneous Poisson process (NHPP) based [16, 22, 25].These software reliability models are of two types: continuous-time and discrete-time models. Continuous-time models use CPU time or calendar time for measurement. While the discrete models make use of the total test cases executed as a measurement unit. A test case is represented by a test run which may either be executed in seconds or minutes or hours or days or weeks or months.

A. Tickoo (✉)
Amity School of Engineering, Amity University, Noida, UP, India
e-mail: anshultickoo@hotmail.com

P.K. Kapur • A.K. Shrivastava
Amity Center for Interdisciplinary Research, Amity University, Noida, UP, India
e-mail: pkkapur1@gmail.com; kavinash1987@gmail.com

S.K. Khatri
Amity Institute of Information Technology, Amity University, Noida, UP, India
e-mail: sunilkkhatri@gmail.com

© Springer Nature Singapore Pte Ltd. 2018

P.K. Kapur et al. (eds.), *Quality, IT and Business Operations*, Springer Proceedings in Business and Economics, DOI 10.1007/978-981-10-5577-5_11

The failure data sets which are observed in real life are most of the time discrete in nature; therefore, discrete models help to get a better fit than continuous-time models. Due to limited real-life application of continuous-time model, in our paper a discrete model is used to obtain the software release and patch time. Available literature in the above field focuses on cost minimization, and maximizing reliability [15, 18, 28], and determination of the optimum release time [26]. In our current work we have formulated a cost model to obtain the optimum release and patch time of the software.

From the developer's perspective, determining testing stop time and release time of the software is crucial. The software company along with building a good-quality product has to make sure that it is delivered in time. In such a scenario of early release, a major threat that hovers over firms is releasing unreliable software with large number of undetected bugs lying dormant in the software. In such an environment, newer ways of testing are quite soothing, and alongside greater consideration is given toward post-release debugging. It has now become a common practice for software firms to release the software early but continue removing the bugs to improve its reliability and hence overall quality.

In the literature, the optimal testing stop time coincides the optimal time of software release [10, 11, 17, 27]. This approach can be improved upon by continuing the fault removal process even after the release for some specified time period .In order to make sure that less number of errors is faced by the user during operational phase, software firms continue debugging even after release and release patches to fix the bugs [1, 7, 9]. After each patch release, reliability of the software increases. This is to say, software updating can also be regarded as an important approach to improving software reliability. After the release of the software, the companies keep providing patches to users. If detected faults are not removed in the operational phase, then the failure occurrence rate remains fixed, while if the detected faults during the testing phase got rid of successfully, then the reliability increases. The latter is more in practice [20]. This can be seen in Figs. 1 and 2 below.

Usually updates are released by the developer against one or several reported failures; therefore, the update release is strongly related to the discovery of faults

Fig. 1 Software reliability growth without patching

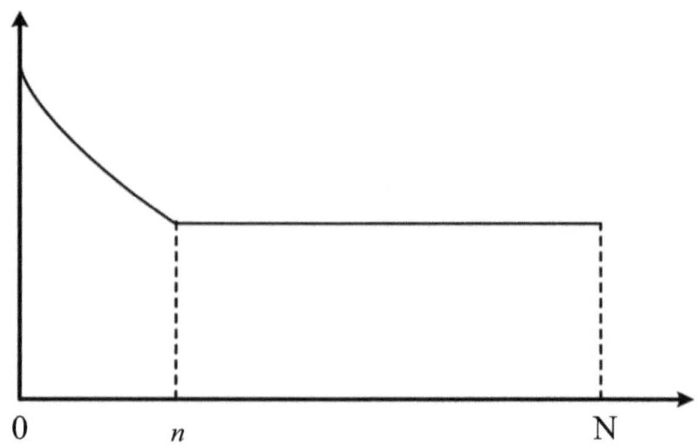

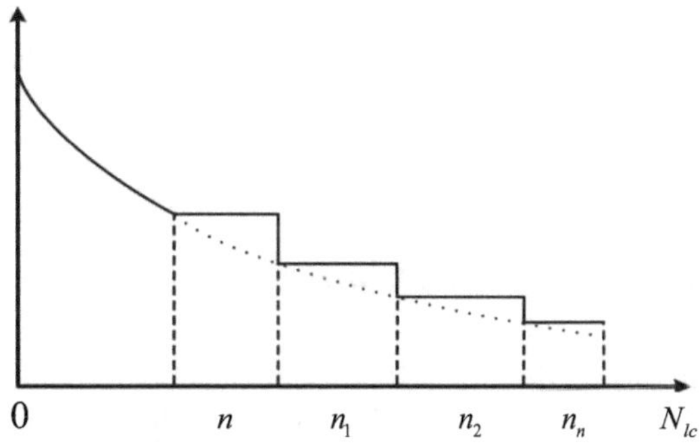

Fig. 2 Software reliability growth with patching

[5, 6]. Recently Zachariah [32] proposed a cost model determining optimal testing time of software based on failure size. One particular topic related to software update management is security patch management, which has received some attention. The patch release problems from perspectives of both vendor and firm were studied by Cavusoglu et al. [4].This work was extended by considering a nonhomogeneous bug detection process in their work [21, 23, 24]. Arora et al. [2, 3] have shown that it is beneficial to release software early and patch it later.

Quite recently, a study suggested by Jiang et al. [13, 14] in the similar direction examined a comprehensive scenario of software release policy with post-release testing. Jiang et al. [14] justify the post-release testing practice for large-scale enterprise-level information system. However, their modeling framework is based on a very strong assumption that the bug detection/correction process follows exponential distribution and does not model for any other distribution, whereas our proposed framework presents a generalized formulation of the detection/correction process for post-release phenomenon. We also provide a framework based on which further research can be developed by relaxing the assumption of perfect debugging in case of bug detection process. Furthermore, from the work proposed by Kapur et al. [17] and Pham [27] extended the prior literature by proposing models to simultaneously evaluate optimal release time and warranty period of the software but did not consider post-release testing phenomenon to minimize cost.

Today software firms release early and update the software by releasing patches. In the literature related to cost modeling for single and multiple releases of software by Pham [27] and Kapur et al. [17, 19, 29] it was assumed that testing and operational phases are influenced by one distribution function, which is quite contrary to today's testing practices followed by practitioners. Huang et al. [10, 11] and Zhao et al. [33] proposed that there is a change in fault detection rate from testing to operational phase. Accordingly, we consider a more realistic scenario that debugging process, pre- and post-release of software, takes place differently, and, therefore, we propose a generalized cost framework with different distribution functions for various phases of a software cycle. In the paper, the fault detection process is strongly assumed to follow discrete logistic distribution function.

In the rest of the paper, Sect. 2 describes the model assumptions, notations, and model formulation. In Sect. 3 modeling of the cost function is done. In Sect. 4, numerical illustration is provided to verify the proposed work. Finally conclusion is in Sect. 5.

2 Model Formulation

In this section we first describe the notations and assumptions of the applied SRGM. Then, we discuss the discrete-time SRGM formation.

2.1 *Notations (Table 1)*

Table 1

$m(n)$	Expected number of failures that occur by the nth test run
a	Initial fault content in the software when the testing starts
$m(N_{\mathrm{lc}})$	Number of faults removed in software life cycle
b	Failure detection/correction rate
$F(n)$	Distribution functions for fault correction
n	Optimum number of test runs required for release of the software (Here we have assumed that one test run is executed in 1 week)
n_i	Test runs required for release of ith patch of the software
c_1	Cost of testing per test run
c_2	Fault detection and removal cost before the release of the software
c_3	Cost of debugging a fault reported by the user in pre-patching phase
c_4	Cost of debugging a fault by developer reported by user in post-patching period

2.2 *Assumptions*

The following are the basic assumptions for the proposed model:

1. The detection process during the entire life cycle of the software product follows NHPP.
2. The detection of each bug is not dependent on the detection of other faults.
3. Faults remaining in the software cause failures in the software during execution.
4. An instantaneous debugging effort is undertaken on failure observation to discover the source of failure in order to remove it.
5. The faults remaining in the software affect the failure rate equally.
6. The fault removal is perfect.
7. Numbers of dormant faults in the software are finite.
8. Software life cycle is finite.

2.3 Discrete-Time SRGM Formation with Logistic Learning Function

As discussed in Sect. 1, NHPP-based SRGMs are categorized in two groups: continuous-time models and discrete-time models [8, 17].In the SRGMs discussed in literature, a constant fault removal rate is assumed. However, practically with the progress of the testing process, learning of the testing team increases, and hence here we assume that the fault removal rate follows a logistic learning function. Here we have formulated a discrete model to obtain the software release and patch time using a discrete logistic function.

The model discussed here [17] has incorporated the testing team learning process into the SRGM. The difference equation for this model is as below:

$$\frac{m\,(N+1)-m(N)}{\delta}=\frac{b.\,(a-m(N))}{1+\beta(1-b\delta)^{N+1}},\tag{1}$$

here δ refers to a constant time interval.

The mean value function obtained from the Eq. (1) with the initial condition m $(n=0)=0$ and assuming $\delta=1$ is:

$$m(n)=\frac{a\,(1-(1-b)^{n})}{1+\beta(1-b)^{n}}=a.F(n)\tag{2}$$

Software reliability for this discrete model is represented as below:

$$R\,(h|N)=e^{(-\{m(N+h)-m(N)\})}\tag{3}$$

Here h represents the additional test runs that are required for discovery of faults during software testing.

3 Cost Model Development

In this case the debugging is continued after the software release as well in order to improve the reliability of the software and hence minimize the threat of software failure in the field. Here we consider that the developer continues to invest their resources in software testing post-release by debugging the faults detected by the user.

Here, the software life cycle is classified in three phases, viz., prerelease testing phase, post-release pre-patching phase, and post-patching phase, denoted by $[0,n]$, $[n,n_1]$, and $[n_1,N_{\mathrm{lc}}]$, respectively, as shown in Fig. 3.

Here $n_1 = n + n_1'$ and n_1' are the test runs between software release and first patch.

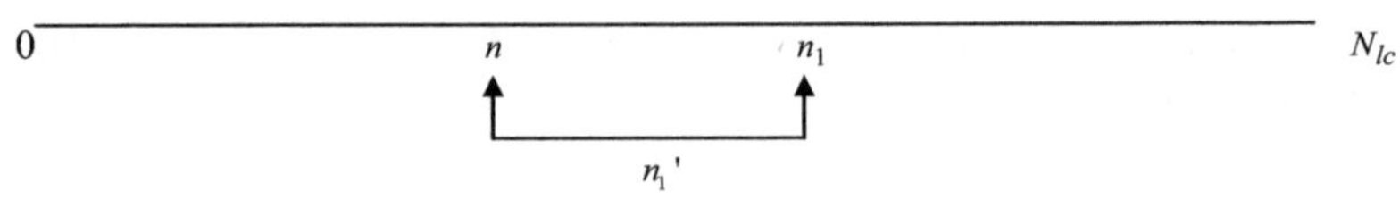

Fig. 3 Single patch release

In order to determine the optimal test runs required to release the software, we make use of those costs that impact the concerned parameters. The total cost is the sum of cost of testing and the cost sustained by removal of faults in each phase which are described below.

(a) Cost of testing

The per unit testing cost refers to the cost of testing, test case generation, planning, test case execution, and testing cost analysis. It also includes CPU hours consumed in the process of testing by the developer.

$$\text{Per unit testing cost is given by } c_1 N. \tag{4}$$

(b) Phase 1: $[0, n]$ (prerelease testing phase)

In phase 1 testing team works toward detection and correction of faults, and the total faults discovered and removed in this phase are represented by

$$m(n) = a.F_1(n), \tag{5}$$

where $F_1(n)$ is the rate of fault removal from the software in the interval $[0, n]$.

$$\text{Cost sustained by the company in this phase is given by } c_2 \cdot m(n) \tag{6}$$

(c) Phase 2: $[n, n_1]$ (post-release before patching phase)

This phase represents the period in which the software has already been released in the market. Now only users detect the faults and developers remove the faults detected by the users. The faults discovered and removed in this period in this phase are represented by

$$m(n_1 - n) = (a - m(n)) \cdot F_2(n_1 - n) = a(1 - F_1(n)).F_2(n_1 - n). \tag{7}$$

Here $a(1 - F_1(n))$ are the faults that remain undetected in prerelease testing phase, and $F_2(n_1 - n)$ is the rate of fault removal from the software in the post-release phase $[n, n_1]$.

$$\text{Cost sustained by company in this phase is given by } c_3.m(n_1 - n) \tag{8}$$

(d) Phase 3: $[n_1, N_{\mathrm{lc}}]$ (post-patching phase)

In this interval $[n_1, N_{\mathrm{lc}}]$, only users detect the faults. These detected faults are removed by the developers till the completion of software life cycle. In this interval, users face the failures due to faults which are not detected till the release of the patch.

The faults discovered and removed in this period are represented by:

$$\begin{aligned} m\left((N_{\mathrm{lc}} - n_1\right) &= \left(a - m\left(n_1\right)\right)\right) \cdot F_3\left(N_{\mathrm{lc}} - n_1\right) \\ &= a\left(1 - F_1(n)\right) . \left(1 - F_2\left(n_1 - n\right)\right) . F_3\left(N_{\mathrm{lc}} - n_1\right) \end{aligned} \tag{9}$$

where n_1 is the optimum patch release time and N_{lc} is the software life cycle. Here $a . (1 - F_1(n)) . (1 - F_2(n_1 - n))$ denotes the faults which were not detected in the first and second phase, and $F_3(N_{\mathrm{lc}} - n_1)$ is the rate at which faults are removed from the software in this post-patching phase, i.e., interval $[n_1, N_{\mathrm{lc}}]$.

Cost sustained by the company in this phase is given by

$$c_4 . m\Big(\left(N_{\mathrm{lc}} - n_1\right) . \tag{10}$$

It is important to note here that, since at some point management has to bring a new version of the software in the market, hence we do not release any patch in this last interval [n1, Nlc].

After combining the cost sustained in each, the total cost sustained by the company is given by

$$C\left(n, N_{\mathrm{lc}}\right) = c_1 . n + c_2 . m(n) + c_3 . m\left(n_1 - n\right) + c_4 m\left(N_{\mathrm{lc}} - n_1\right) . \tag{11}$$

Also, based on the argument in Sect. 1, the rate of detection of testers is b, and r is the ratio of fault detection rate under customer's usage with respect to tester's testing in the previous phase; then, by the above argument, rate of detection of user is $r1 . b$ in $[n, n_1]$ and $r2 . b$ in $[n_1, N_{\mathrm{lc}}]$.

Hence, F_1, F_2, and F_3 in (5), (7), (9), and (11) are given by $F_1(n) = \frac{(1-(1-b)^n)}{1+\beta(1-b)^n}$,

$$F_2\left(n_1 - n\right) = \left(1 - \left(\frac{(1+\beta) . (1-b)^{n_1 - n}\Big)}{1 + \beta(1-b)^{n_1 - n}}\right)^{r1}\right),$$

$$\text{and } F_3\left(N_{\mathrm{lc}} - n_1\right) = \left(1 - \left(\frac{(1+\beta) . (1-b)^{N_{\mathrm{lc}} - n_1}\Big)}{1 + \beta(1-b)^{N_{\mathrm{lc}} - n_1}}\right)^{r2}\right).$$

3.1 Objective

We have selected budget and reliability as two constraints. In this problem the objective is to minimize the cost $C(n, N_{\text{lc}})$ and maximize software reliability to get the optimum software release and patch time.

Minimize: $C(n, N_{\text{lc}})$
Subject to: $\text{R} >= \text{R}_0$
$C(n, N_{\text{lc}}) <= \text{C}_\text{B}$

The reliability can be represented by $R(h|\,N) = e^{(-\{m(N+h)-m(N)\})}$, and it should be maximized. The management decides a reliability level R_0 which has to be achieved. Here, C_B is the total budget of the firm.

4 Numerical Illustration

Determination of optimum test runs required for releasing the software is an important implementation of software reliability models. Various release time problems in literature have been studied for different SRGMs [12, 19, 30, 31]. In this paper we determine the optimum number of test runs required when software is ready to be released. If software testing is done for long duration, large number of software faults are removed which increases reliability. However, it may cause financial loss and cause delay in the software release. In this section we obtain the optimum release and patch time of the software with the constraint to keep the total software cost minimum and fulfill the reliability requirement.

We assume logistic distribution in each phase, given by $F(n) = \frac{(1-(1-b)^n)}{1+\beta(1-b)^n}$. For the estimation of parameters, the failure data set of Wood [29] has been used. SPSS software has been applied for parameter estimation. Estimated values of parameters are given in Table 2.

With the help of testing team experience, we assume the following cost parameters as:

Per unit testing cost, $C_1 = 40$; fault detection and removal cost before the release of the software, $C_2 = 30$; debugging cost in the pre-patching phase, $C_3 = 70$; debugging cost in post-patching period, $C_4 = 70$; and software life cycle, $N_{\text{lc}} = 100$. Management decision of the reliability requirement is $R_0 = 0.94$, and the budget allotted is $C_B = 5000$.

Also let $r1 = 0.5$ and $r2 = 0.5$. Note that $a = m(N_{\text{lc}})$.

Table 2 Estimation of model parameters

Model parameters	
a	110.83
b	0.17
β	1.2

Now the faults detected with the corresponding cost for each phase is described below.

Phase 1: $[0, n]$ (prerelease testing phase)

In phase 1 testing team works toward detection and correction of faults, and the faults discovered and removed in this period are represented by

$$m(n) = a.F_1(n)$$

where $F_1(n)$ is the rate of fault removal in the interval $[0, n]$.

Cost sustained by the company in this phase is given by

$$c_2 \cdot m(n) = a.F_1(n) = c_2 \cdot \frac{(1-(1-b)^n)}{1+\beta(1-b)^n}. \tag{12}$$

Phase 2: $[n, n_1]$ (post-release pre-patching phase)

This phase represents the period in which the software has already been released in the market. Now only users detect the faults and developers remove the faults detected by the users. The faults discovered and removed in this period are represented by

$$m(n_1-n) = (a-m(n)) \cdot F_2(n_1-n) = a(1-F_1(n)).F_2(n_1-n)$$
$$= a.\left(1-\frac{(1-(1-b)^n)}{1+\beta(1-b)^n}\right) \cdot \left(1-\left(\frac{(1+\beta).(1-b)^{n_1-n}}{1+\beta(1-b)^{n_1-n}}\right)^{r1}\right). \tag{13}$$

Cost sustained by the company in this phase is given by

$$c_3.a.\left(1-\frac{(1-(1-b)^n)}{1+\beta(1-b)^n}\right) \cdot \left(1-\left(\frac{(1+\beta).(1-b)^{n_1-n}}{1+\beta(1-b)^{n_1-n}}\right)^{r1}\right). \tag{14}$$

Phase 3: $[n_1, N_{\mathrm{lc}}]$ (post-patching phase)

In this interval $[n_1, N_{lc}]$, only users detect the faults. These detected faults are removed by the developers till the completion of software life cycle. In this interval, users face the failures due to faults which are not detected till the release of the patch.

The faults discovered and removed in this period are represented by

$$m\left(N_{\mathrm{lc}}-n_1\right)=\left(a-m\left(n_1\right)\right)\Big)\cdot F_3\left(N_{\mathrm{lc}}-n_1\right)$$
$$=a\left(1-F_1(n)\right).\left(1-F_2\left(n_1-n\right)\right).F_3\left(N_{\mathrm{lc}}-n_1\right)$$
$$=a.\left(1-\frac{\left(1-(1-b)^n\right)}{1+\beta(1-b)^n}\right).\left(1-\left(1-\left(\frac{(1+\beta).(1-b)^{n_1-n}}{1+\beta(1-b)^{n_1-n}}\right)^{r1}\right)\right).$$
$$\left(1-\left(\frac{(1+\beta).(1-b)^{N_{\mathrm{lc}}-n_1}}{1+\beta(1-b)^{N_{\mathrm{lc}}-n_1}}\right)^{r2}\right).$$

Cost sustained by the company in this phase is given by

$$c_4.a.\left(1-\frac{\left(1-(1-b)^n\right)}{1+\beta(1-b)^n}\right).\left(1-\left(1-\left(\frac{(1+\beta).(1-b)^{n_1-n}}{1+\beta(1-b)^{n_1-n}}\right)^{r1}\right)\right).\left(1-\left(\frac{(1+\beta).(1-b)^{N_{\mathrm{lc}}-n_1}}{1+\beta(1-b)^{N_{\mathrm{lc}}-n_1}}\right)^{r2}\right). \quad (15)$$

Total cost sustained in this case is given by

$$C\left(n,N_{\mathrm{lc}}\right)=c_1.n+c_2.m(n)+c_3.m\left(n_1-n\right)+c_4 m\left(N_{\mathrm{lc}}-n_1\right). \quad (16)$$

Using the estimated parameter values in Eq. (15) and minimizing total cost in Eq. (16) after incorporating the budget and reliability constraint, by using Maple software, we get optimal test runs required for releasing the software as 19, i.e., $n=19$ weeks, optimal test runs for first patch release as 1, i.e., 1 week, and optimal cost as $c(n,N_{\mathrm{lc}})=4357$. Figure 4 shows the graph for cost for the proposed model.

5 Conclusions

Software firms are bringing the up-gradations/add-ons in the software early, due to increased demand for new features. This implies early software release, and hence more bugs are still lying dormant in the system. In order to handle this situation, the concept of patch release became quite useful in which most of the software developers post-release of the software would continue testing the software and release small programs to fix those remaining bugs in the system. Here, we consider a realistic scenario that debugging process, followed by the testers and developers, pre- and post-release of software, takes place differently, and, therefore, we propose

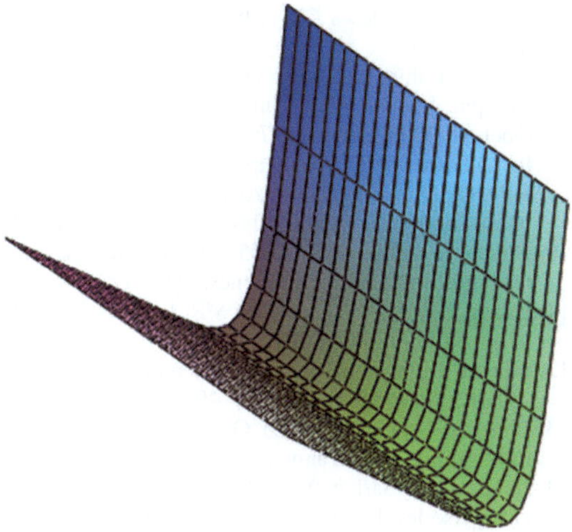

Fig. 4 Cost-time graph

a generalized cost framework with distinct distribution functions for various phases of a software life cycle in our paper.

This paper models a framework where development and testing team develops, tests, releases the software, and, during the operational phase, releases updates to upgrade the software. In this paper a discrete-time model has been formulated to determine the optimal test runs required for software release and patch release with the cost and reliability constraints. We have verified the model using a numerical example using real-life data set using logistic distribution function. The result is encouraging and fairly accurate. Future research may consider more advanced SRGMs which can characterize imperfect debugging and change-point phenomena, etc.

Acknowledgment Authors express their deep sense of gratitude to the Founder President of Amity University, Dr. Ashok K. Chauhan, for his keen interest in promoting research in the Amity University and always been an inspiration for achieving greater heights.

References

1. Apple (2015) https://www.apple.com/softwareupdate. Accessed 24 Sept 2015
2. Arora A, Jonathan P (2006) Caulkins and Rahul Telang "Research note: sell first, fix later: impact of patching on software quality". Manag Sci 52(3):465–471
3. Arora A, Telang R, Xu H (2008) Optimal policy for software vulnerability disclosure. Manag Sci 54:642–656
4. Cavusoglu H, Zhang J (2008) Security patch management: share the burden or share the damage? Manag Sci 54:657–670
5. Chatterjee S, Singh JB (2014) A NHPP based software reliability model and optimal release policy with logistic–exponential test coverage under imperfect debugging. Int J Syst Assur Eng Manag Volume 5, Issue 3, pp 399–406

6. Dalal SR, Mallows CL (1988) When should one stop testing software? J Am Stat Assoc 83:872–879
7. Dey D, Lahiri A, Zhang G Optimal policies for security patch management INFORMS J Comput, Vol. 27, No. 3, Summer 2015, pp 462–477.
8. Goel AL, Okumoto K (1979) Time-dependent error-detection rate model for software reliability and other performance measures. IEEE Trans Reliab R-28:206–211
9. HP (2015) http://www.zdnet.com/hp-to-begin-charging-for-firmware-updates-and-service-packs-for-servers-7000026110. Accessed 24 July 2015
10. Huang C-Y, Lyu MR (2005) Optimal release time for software systems considering cost, testing-effort, and test efficiency. IEEE Trans Reliab 54:583–591
11. Huang CY, Kuo SY Yen, Lyu MR, Lo Jung Hua. Quantitative software modeling from testing to operation, ISSRE 2000
12. Jain M, Priya K (2005) Software reliability issues under operational and testing constraints. Asia-Pac J Oper Res 22(1):33–49
13. Jiang S, Sarkar S (2003) Optimal Software Release Time with Patching considered. In: Proceedings of 13th annual workshop information technologies and systems. Seattle, pp 61–66
14. Jiang, Z., S. Sarkar, V. S. Jacob (2012) Post-release testing and software release policy for enterprise-level systems. Inform Syst Res 23(3), Part 1 of 2: 635–657
15. Kapur PK, Shrivastava AK (2015) When to release and stop testing of a software: a new insight, international conference on reliability, infocom technology and optimization (trends and future directions), held during 2–4 Oct 2015 at Amity University, Noida, Uttar Pradesh, pp 1–6
16. Kapur PK, Garg RB, Kumar S (1999) Contributions to hardware and software reliability. World Scientific Publishing Co. Ltd, Singapore
17. Kapur PK, Pham H, Gupta A, Jha PC (2011) Software reliability assessment with OR applications. Springer, London
18. Kapur PK, Singh VB, Singh O, Singh JNP (2013) Software release time based on different multi-attribute utility functions. Int J Reliab, Qual Saf Eng 20(4):1350012. 15 pages
19. Kapur PK, Khatri SK, Tickoo A, Shatnawi O (2014) Release time determination depending on number of test runs using multi attribute utility theory. Int J Syst Assur Eng Manag (IJSAEM) 5(2):186–194. ISSN: 0975-6809
20. Li X, Li YF, Xie M, Ng SH (2011) Reliability analysis and optimal version-updating for open source software. Inf Softw Technol 53:929–936
21. Luo C, Okamura H, Dohi T (2015) Optimal planning for open source software updates. Proc IMechE Part O. doi:10.1177/1748006x15586507
22. Musa JD, Iannino A, Okumoto K (1987) Software reliability: Measurement, prediction, applications. Mc Graw Hill, New York
23. Okamura H, Dohi T, Osaki S (2001) A reliability assessment method for software products in operational phase – proposal of an accelerated life testing model. Electron Commun Jpn 84:25–33
24. Okamura H, Tokuzane M, Dohi T (2009) Optimal security patch release timing under non-homogeneous vulnerability-discovery processes. In: Proceedings of the 20th international symposium on software reliability Engineering (ISSRE '09), Mysuru, India, 2009, pp 120–128
25. Okumoto K, Goel AL (1979) Time dependent error detection rate model for software reliability and other performance measures. IEEE Trans Reliab R-28(3):206–211
26. Ompal Singh, Kapur PK, Anand A (2012) A multi attribute approach for release time and reliability trend analysis of a software. Int J Syst Assur Eng Manag (IJSAEM) 3(3):246–254. ISSN: 0975-6809
27. Pham H (2006) System software reliability. Springer-Verlag, London
28. Singh O, Kapur PK, Shrivastava AK, Kumar V (2015) Release Time Problem with Multiple Constraints. Int J Syst Assur Eng Manag 6(1):83–91
29. Wood A (1996) Predicting software reliability. IEEE Comput 29:69–77
30. Yang B, Xie M (2000) A study of operational and testing reliability in software reliability analysis. Reliab Eng Syst Safe 70:323–329

31. Yang B, Hu H, Jia L (2008) A study of uncertainty in software cost and its impact on optimal software release time. IEEE Trans Softw Eng 34:813–825
32. Zachariah B (2015) Optimal stopping time in software testing based on failure size approach. Ann Oper Res. doi:10.1007/s10479-015-1959-5
33. Zhao J, Liu H, Cui G, Yang XZ (2005) Software Reliability Growth Model from Testing to Operation. ICSM

EFA-FTOPSIS-Based Assessment of Service Quality: Case of Shopping Websites

Vivek Agrawal, Akash Agrawal, and Anand Mohan Agrawal

1 Introduction

Technology is narrowing the gap between physical and online shopping environment. The customers are becoming very selective in choosing the products. This is easily possible in online shopping. Consumers are keen to search online shopping, online booking, online financial transactions, etc., and their offerings.

It raises the Internet users and growth in electronic commerce (e-commerce). Enterprises are attempting to gain a competitive advantage by using e-commerce for interacting with the customers [33]. These types of businesses are commencing to realize that success of any business is not only the low price of the product, but service quality of their website is equally important. But service quality is an intangible and theoretical construct that is not easily elucidated and evaluated.

In India, the growth has been forecasted in the online retail market from 2012 to 2018 (in billion US dollars) as shown in Fig. 1.

This shows that there is a need of research which can explore the factors affecting the service quality of online shopping sites so that sites can be compared and ranked.

V. Agrawal (✉)
IBM, GLA University, Mathura, Uttar Pradesh, India
e-mail: vivek.agrawal@gla.ac.in

A. Agrawal
Quality Council of India, Delhi, India
e-mail: akashranu@gmail.com

A.M. Agrawal
GLA University, Mathura, Uttar Pradesh, India
e-mail: provc@gla.ac.in

© Springer Nature Singapore Pte Ltd. 2018
P.K. Kapur et al. (eds.), *Quality, IT and Business Operations*, Springer Proceedings in Business and Economics, DOI 10.1007/978-981-10-5577-5_12

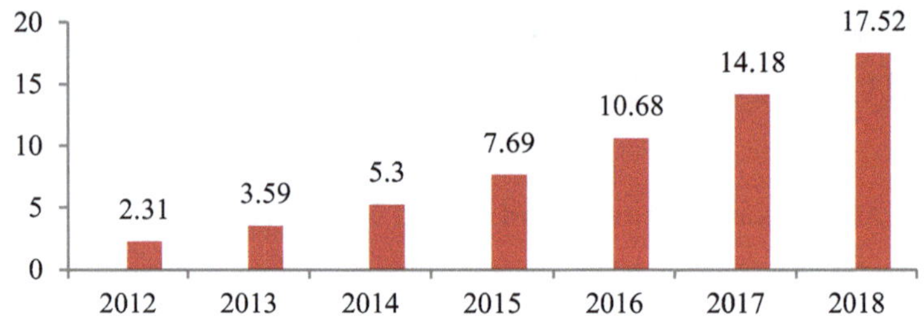

Fig. 1 Growth in sales in online retail from 2012 to 2018 (Source: Ref. [40])

2 Literature Review

Literature in domain of service quality and e-service quality has been reviewed for the present study and has been categorized as follows:

(a) Service quality
(b) Service quality measurement
(c) E-service quality
(d) E-Service quality measurement

2.1 *Service Quality*

Service quality has drawn a major attention of researchers; as a result, numerous studies are available in the literature. Gummesson [15] was first to suggest the concept of service quality and its strong association to the trust and perception. Oliver [25] conceptualized the service quality based on his disconfirmation model. Thus, service quality is generally considered to be a measure of how soundly the level of service delivery matched the expectation of customer.

The previous studies have also revealed that there is less managerial control over the service quality due to higher involvement of customers in the process [26]. Many researchers considered service quality as economies of services as it is playing a significant role in economic environment. Thus, Table 1 presents the select compilation of definitions on service quality.

2.2 *Service Quality Measurement*

There are various measures available to measure the service quality; SERVQUAL is the most popular scale for measuring service quality used by various researchers among all other measures [5, 9, 32].

Table 1 Definitions of service quality

S. no.	Authors	Definition of service quality
1	[13]	The outcome of an evaluation process, where the consumer compares his expectations with the service he perceives he has received
2	[26]	The comparison between customer expectations and perceptions of service
3	[10]	Based on the customers' perceptions of how well the service matches their needs and expectations
4	[30]	It is a function of the difference in scores or gaps between expectations and perception
5	[33]	The attitude or belief about the excellence of degree of service offered in service provider location

This area of research has been very rich in terms of basic concepts and models [13], applications [1], linkages with customer satisfaction, customer loyalty, and profitability.

The previous studies have [17, 19, 31, 34, 35, 39] well documented the critique and application in different contexts and also highlight that subject of service quality is not generic and its measurement varies with respect to specific service as shown in Table 2.

2.3 E-Service Quality

The studies in the field of service quality have been admired for more than three decades, but recently it has been applied to the e-commerce environment [19]. The beginning of "e-service" emerged upon the expansion of Internet applications [23]. Information technology (IT) is used by e-commerce organizations to gain a competitive advantage around the world. IT is used to elaborate the interaction with customers more friendly. It is an efficient means at minimal cost to expand a vast market share. With the increase of e-service acceptance in business environment, the significance of measuring e-service quality in the virtual world has been acknowledged.

With more use of e-services in business, many studies have been conducted to better understand its dynamics [6]. The previous studies have focused on various conceptual definitions of e-service quality [36] which have been presented in Table 3.

2.4 E-Service Quality Measurement

SERVQUAL instrument is used to measure the traditional service quality of the company according to the five dimensions: tangibles, reliability, responsiveness,

Table 2 Compilation of measures of service quality

[13]	[14]	[26]	[21]	[28]	[8]
				Reliability	Reliability
Technical quality	Recovery	Credibility	Physical quality (physical product + physical environment)	Responsiveness	Comfort
Functional quality	Attitude and behavior	Access		Tangibles	Features
		Reliability			
Corporate image	Accessibility and flexibility	Communication	Interactive quality (interaction with persons and equipments)	Access	Personal attention
	Reputation and credibility	Understanding the customer			
	Professionalism and skills reliability and	Courtesy		Knowing the customers	
				Assurance	
		Competence	Corporate quality		
	Trustworthiness	Responsiveness	Process quality		
		Tangibles	Output quality		
		Security			

Source: Modified from [32]

Table 3 Definitions of e-service quality

S No.	Authors	Definitions
1	[27]	Effectiveness and efficiency of online browse, online purchase, and delivery of goods and service
2	[11]	The degree to which an electronic service is able to effectively and efficiently fulfil relevant customer needs
3	[2]	The entire stages of a customer's interactions with the Internet, website

Table 4 E-service quality instrument

SN	Author	Instrument	Dimensions
1	[22]	WebQual	Visual appeal, integrated communication, business processes, informational fit to task, interaction, trust, response time, design, intuitiveness, and substitutability
2	[38]	SITE-QUAL	Processing speed, ease of use, aesthetic design, and security
3	[4]	WebQual	Usability, design, information, trust, and empathy
4	[37]	eTailQ	Website design, reliability/fulfilment, privacy/security, and customer service
5	[27]	E-S-QUAL	Efficiency, system availability, fulfilment, and privacy
		E-RecS-QUAL	Responsiveness, compensation, and contact
6	[20]	Revised SERVQUAL	Website design, reliability, responsiveness, trust, and personalization

empathy, and assurance. The measurement of e-service quality emerged on the basis of SERVQUAL. For measuring the e-service quality, Gefen [12] combined the SERVQUAL five dimensions into three. But various researchers also realized that there is a need of another instrument for measuring e-service quality and SERVQUAL cannot be considered for measuring the e-service quality [27]. Since then, many researches are addressed in account of e-service quality and explored different measurements (different dimensions) for measuring e-service quality. Some of them are presented in Table 4.

After an extensive literature review, an instrument called E-S-QUAL and E-RecS-Qual was developed to measure the service quality of online shopping websites by Parasuraman [27]. According to his study, service quality has provided online services an effective and efficient way for online browse, online purchase, and delivery of goods and service. In the present research, context E-S-QUAL has been adopted to measure the service quality of online shopping sites.

Taking insights and gap from the literature, the following objectives are framed:

- To explore the factors of service quality affecting the service quality of online shopping websites
- To propose the methodology for comparing the performance of online shopping sites

3 Research Methodology

In the present study, E-S-QUAL scale was adopted for measuring the service quality of online shoppers. The responses were collected from the respondents on 5-point Likert scale ranging from 1 (strongly disagree) to 5 (strongly agree). A total of 320 questionnaires were circulated in North India region, and 232 were received in the decided time period. Out of this, 157 questionnaires were used, and the rest were eliminated because they did not have any online shopping experience. The number of responses was considered suitable for exploratory factor analysis as suggested ratio (1:5) according to Hair [16]. The reliability of the instrument was found to be 0.698, which is to be considered acceptable according to Nunnally [24]. The value of KMO was 0.732 which shows the sample is adequate [3]. Five factors were identified on the basis of their eigenvalue and factor loading, since it is more than 1 and 0.5, respectively. These five factors were named as efficiency (F1), system availability (F2), fulfilment (F3), contact (F4), and privacy (F5). On the basis of this, expert opinions were taken to develop a methodology for making the comparison of online shopping sites by using fuzzy technique for order performance by similarity to ideal solution (FTOPSIS).

4 Fuzzy TOPSIS

TOPSIS is a multi-criteria decision-making technique which is used for ranking and comparing the alternatives among various alternatives by the numerical evaluations and calculations with respect to certain attributes/criterion. In this technique, weights will be specified for each criterion for measuring the relative importance which is felt by the decision maker. The basic principle of the fuzzy TOPSIS is that the chosen alternative should have the shortest distance from the positive ideal solution and the farthest distance from the negative-ideal solution in a geometrical (i.e., Euclidean) sense [18].

This study has considered three shopping sites as SS-I, SS-II, and SS-III. For the rating purpose of these three shopping sites, as suggested by Saaty and Vargas [29], five customers were selected as experts.

This group of five customers was asked to assign the weights to the three shopping sites. The steps for the methodology using FTOPSIS are as follows.

Step 1: Defining Fuzzy Decision Matrix
Five customers as experts are selected assessing the selection of shopping sites, since Saaty and Vargas [29] suggest that three to seven experts are suitable.

In the linguistic language, the item weightings are assessed from Table 5 by the team of decision makers and converted into TFN (shown in Table 6).

The linguistic ratings and values (Table 7) are expressed in exact numerical values [7]. This rating of alternatives is formulated based on decision makers' judgments as shown in Table 7 after conversion of linguistic terms into TFN.

Table 5 Linguistic variable set

Linguistic terms		
Weightings	Ratings	Triangular fuzzy numbers (TFN)
Very low (VL)	Very poor (VP)	(0,0,1)
Low (L)	Poor (P)	(0,1,3)
Medium low (ML)	Medium poor (MP)	(1,3,5)
Medium (M)	Fair (F)	(3,5,7)
Medium high (MH)	Medium good (MG)	(5,7,9)
High (H)	Good (G)	(7,9,10)
Very high (VH)	Very Good (VG)	(9,10,10)

Table 6 Decision makers' judgment on item weightings

	F1	F2	F3	F4	F5
DM_1	7,9,10	3,5,7	3,5,7	7,9,10	7,9,10
DM_2	5,7,9	3,5,7	3,5,7	5,7,9	3,5,7
DM_3	5,7,9	7,9,10	5,7,9	7,9,10	5,7,9
DM_4	3,5,7	7,9,10	7,9,10	7,9,10	7,9,10
DM_5	7,9,10	5,7,9	3,5,7	5,7,9	7,9,10

Table 7 Decision makers' judgments on rating for alternative (shopping sites)

		F1	F2	F3	F4	F5
DM_1	**SS-I**	1,3,5	3,5,7	5,7,9	5,7,9	3,5,7
	SS-II	5,7,9	5,7,9	7,9,10	7,9,10	3,5,7
	SS-III	5,7,9	7,9,10	5,7,9	5,7,9	5,7,9
DM_2		**F1**	**F2**	**F3**	**F4**	**F5**
	SS-I	3,5,7	3,5,7	7,9,10	1,3,5	7,9,10
	SS-II	5,7,9	3,5,7	5,7,9	7,9,10	1,3,5
	SS-III	7,9,10	5,7,9	5,7,9	7,9,10	5,7,9
DM_3		**F1**	**F2**	**F3**	**F4**	**F5**
	SS-I	3,5,7	1,3,5	3,5,7	3,5,7	1,3,5
	SS-II	1,3,5	5,7,9	3,5,7	1,3,5	1,3,5
	SS-III	1,3,5	5,7,9	7,9,10	7,9,10	5,7,9
DM_4		**F1**	**F2**	**F3**	**F4**	**F5**
	SS-I	5,7,9	7,9,10	7,9,10	3,5,7	7,9,10
	SS-II	1,3,5	1,3,5	1,3,5	7,9,10	5,7,9
	SS-III	1,3,5	3,5,7	1,3,5	5,7,9	1,3,5
DM_5		**F1**	**F2**	**F3**	**F4**	**F5**
	SS-I	1,3,5	1,3,5	7,9,10	1,3,5	5,7,9
	SS-II	5,7,9	7,9,10	1,3,5	7,9,10	5,7,9
	SS-III	1,3,5	1,3,5	7,9,10	5,7,9	3,5,7

Table 8 Averaged frequency weightings and ratings of three shopping sites

	F1	F2	F3	F4	F5
Weights	5.4, 7.4, 9	5, 7, 8.6	4.2, 6.2, 8	6.2, 8.2, 9.6	5.8, 7.8, 9.2
SS-I	2.6, 4.6, 6.6	3, 5, 6.8	5.8, 7.8, 9.2	2.6, 4.6, 6.6	4.6, 6.6, 8.2
SS-II	3.4, 5.4, 7.4	4.2, 6.2, 8	3.4, 5.4, 7.2	5.8, 7.8, 9	3, 5, 7
SS-III	3, 5, 6.8	4.2, 6.2, 8	5, 7, 8.6	5.8, 7.8, 9.4	3.8, 5.8, 7.8

Table 9 Normalized fuzzy decision matrix

	F1	F2	F3	F4	F5
SS-I	0.2708, 0.4792, 0.6875	0.3125, 0.5208, 0.7083	0.6042, 0.8125, 0.9583	0.2708, 0.4792, 0.6875	0.4792, 0.6875, 0.8542
SS-II	0.3778, 0.6, 0.8222	0.4667, 0.6889, 0.8889	0.3778, 0.6, 0.8	0.6444, 0.8667, 1	0.3333, 0.5556, 0.7778
SS-III	0.3125, 0.5208, 0.7083	0.4375, 0.6458, 0.8333	0.5208, 0.7292, 0.8958	0.6042, 0.8125, 0.9792	0.3958, 0.6042, 0.8125

Step 2: Formulating the Complex Fuzzy Decision Matrix
The fuzzy item weightings and fuzzy decision matrix are formulated by converting the linguistic terms into TFN [Tables 6 and 7]. After TFN, convert this into complex decision matrix (Table 8) by using following formulas:

$$\tilde{a}_{\mathrm{ij}} = 1/\mathrm{t}\left[\tilde{a}_{ij}^{1} + \tilde{a}_{ij}^{2} + \ldots\ldots.. + \tilde{a}_{ij}^{t}\right], i = 1, 2, \ldots\ldots.., s; j = 1, 2, \ldots$$

$$\tilde{w} = 1/\mathrm{t}\left[\tilde{w}_{i}^{1} + \tilde{w}_{i}^{2} + \ldots\ldots\cdots + \tilde{w}_{i}^{t}\right], i = 1, 2, \ldots.., s$$

Step 3: Normalizing the Complex Fuzzy Decision Matrix
The fuzzy decision matrix now is normalized (Table 9) by using the following formulas:

$$\tilde{\mathrm{r}}_{ij} = \left(\frac{a_{lij}}{a_{uij}^{*}}, \frac{a_{mij}}{a_{uij}^{*}}, \frac{a_{uij}}{a_{uij}^{*}}\right) \; i \in \mathrm{B}$$

$$\tilde{\mathrm{r}}_{ij} = \left(\frac{a_{\bar{li}}}{a_{uij}}, \frac{a_{\bar{li}}}{a_{mij}}, \frac{a_{\bar{li}}}{a_{lij}}\right) \; i \in \mathrm{C}$$

Step 4: Construction of Weighted Normalized Fuzzy Decision Matrix
With the normalized fuzzy numbers, now construct the weighted normalized fuzzy decision matrix (Table 10) by using the formula:

$$\tilde{\mathrm{v}}_{ij} = \tilde{\mathrm{w}}_{\mathrm{I}} * \mathrm{r}_{ij}, i = 1, 2, 3 ----- s, j = 1, 2, 3, \ldots\ldots\ldots\ldots\ldots, n$$

Table 10 Weighted normalized fuzzy decision matrix

	F1	F2	F3	F4	F5
SS-I	1.4625, 3.5458, 6.1875	1.5625, 3.6458, 6.0917	2.5375, 5.0375, 7.6667	1.6792, 3.9292, 6.6	2.7792, 5.3625, 7.8583
SS-II	2.04, 4.44, 7.4	2.3333, 4.8222, 7.6444	1.5867, 3.72, 6.4	3.9956, 7.1067, 9.6	1.9333, 4.3333, 7.1556
SS-III	1.6875, 3.8542, 6.375	2.1875, 4.5208, 7.1667	2.1875, 4.5208, 7.1667	3.7458, 6.6625, 9.4	2.2958, 4.7125, 7.475

Table 11 Distance measurement of alternatives

Alternative	d*	d^-
SS-I	29.62822	24.03577
SS-II	27.36385	26.97287
SS-III	27.37666	26.64665

Table 12 Closeness coefficient (CC) of shopping sites

Alternative	CC
SS-I	0.44789
SS-II	0.4964
SS-III	0.49324

Step 5: Calculate the FPIS and FNIS

$$A^* = [(1,1,1),(1,1,1),(1,1,1),(1,1,1),(1,1,1),(1,1,1),(1,1,1)]$$

$$A^- = [(0,0,0),(0,0,0),(0,0,0),(0,0,0),(0,0,0),(0,0,0),(0,0,0)]$$

Step 6: Calculate the Distance (Tables 11 and 12)
The distance for each alternative form FNIS and FPIS can be calculated according to the following formulas:

$$d_j^* = \sum_{i=1}^{s} d\left(\tilde{v}_{ij}\tilde{v}_i^*\right) j = 1, 2, \ldots\ldots s$$

$$d_j^- = \sum_{i=1}^{s} d\left(\tilde{v}_{ij}\tilde{v}_i^-\right) j = 1, 2, \ldots\ldots s$$

Step 7: Calculate the Closeness Coefficient of Each Alternative Using the Following Formula

$$\mathrm{CC}_i = \frac{d_j^-}{d_j^* + d_j^-} = j = 1, 2 \ldots .n$$

According to the closeness coefficient, three online shopping sites can be ranked as SS-II > SS-III > SS-I, from highest to the lowest. The results indicate that SS-II is outstanding one. SS-I performs worse than SS-II and SS-III. So SS-I has some more gaps to improve their service quality to their customers while purchasing online through his sites.

5 Discussion and Conclusion

In today's scenario, the uses of online shopping sites are increasing day by day; thus, it requires more improvement in their services in an effective and efficient way. Improved service quality can reduce so many problems. In this paper, an integrated approach of EFA- FTOPSIS and E-S-QUAL scale is presented to evaluate the service quality, by the online shopping sites, which they are providing to their online customers. The present methodology can be applied for ranking more online shopping websites. Online companies can also use this methodology for comparing their performance with their competitors and hence make enhancement in providing the services to their online customers.

6 Limitations and Future Research

The practical difficulties have confined the study only for measuring the service quality of online shopping sites; future research may focus on the measuring quality of service in other service sectors. The dimensions and items of service quality have been only taken from E-S-QUAL. These items and factors/criterions can be varying. Another limitation for this research is number of respondents. Future research can be conducted by more number of respondents to get more generalized result.

As this case has been solved by using Fuzzy TOPSIS approach, so it is recommended that future research can be conducted in the same area by considering more items and factors with same methodology or by some other approach like analytical hierarchy process (AHP), analytical network process (ANP), etc.

References

1. Aktas A, Akin AA, Cizel B (2003) Tourist profile research: Antalya region example 2001. Tour Rev 58(1):34–40
2. Al-Nasser M, Yusoff RZ, Islam R, Alnasser A (2013) E-service quality and its effect on consumers' perceptions trust. Am J Econ Bus Adm 5(2):47–55
3. Bajpai N (2011) Business research methods. Pearson Publication, Delhi
4. Barnes S, Vidgen RT (2002) An integrative approach to the assessment of e-commerce. J Electron Commerce Res 3(3):114–126

5. Bebko CP, Garg RK (1995) Perceptions of responsiveness in service delivery. J Hosp Mark 9(2):35–45
6. Burgess L (2014), A conceptual model of B2B online service quality. In: Hasan H (ed) Being practical with theory: a window into business research, pp 87–90
7. Chen SJ, Hwang CL (1992) Fuzzy multiple attribute decision making: methods and applications. Springer, Berlin
8. Dabholkar PA, Shepherd CD, Thorpe DI (2000) A comprehensive framework for service quality: an investigation of critical conceptual and measurement issues through a longitudinal study. J Retail 76(2):131–139
9. Einasto O (2014) Investigating e-service quality criteria for university library: a focus group study. New Libr World 115(1/2):4–14
10. Ennew C, Waite N (2007) Financial services marketing: an International guide to principles and practice. Elsevier, Oxford
11. Fassnacht M, Koese I (2006) Quality of electronic services: conceptualizing and testing a hierarchical model. J Serv Res 9(1):19–37
12. Gefen D (2002) Customer loyalty in e-commerce. J Assoc Inf Syst 3(1):27–51
13. Grönroos C (1984) A service quality model and its marketing implications. Eur J Mark 18(4):36–44
14. Grönroos C (1988) Service quality: the six criteria of good service quality. Rev Bus 9(3):10–13
15. Gummesson E (1979) The marketing of professional services—an organizational dilemma. Eur J Mark 13(5):308–318
16. Hair JF, Black WC, Babin BJ, Anderson RE (2015) Multivariate data analysis, 7th edn. Pearson, Edinburg
17. Haywood-Farmer J (1988) A conceptual model of service quality. Int J Oper Prod Manag 8(6):19–29
18. Hwang CL, Yoon KP (1981) Multiple attribute decision making: methods and applications. Springer-Verlag, New York
19. Ladhari R (2008) Alternative measures of service quality: a review. Manag Serv Qual 18(1): 65–86
20. Lee G, Lin H (2005) Customer perceptions of e-service quality in online shopping. Int J Retail Distrib Manag 33(2):161–176
21. Lehtinen U, Lehtinen JR (1991) Two approaches to service quality dimensions. Serv Ind J 11(3):287–305
22. Loiacono ET, Watson RT, Goodhue DL (2000) WebQual: a web site quality instrument. Working Paper No. 2000-126-0, University of Georgia, Athens
23. Mary L, O'Loughlin D (2008) An observation analysis of e-service quality in online banking. J Financ Serv Mark 13(2):164–178
24. Nunnally J (1978) Psychometric theory. McGraw-Hill, New York, NY
25. Oliver RL (1980) A cognitive model of the antecedents and consequences of satisfaction decisions. J Mark Res 17(4):460–469
26. Parasuraman A, Zeithaml VA, Berry LL (1985) A conceptual model of service quality and its implications for the future research. J Mark 49(1):41–50
27. Parasuraman A, Zeithaml VA, Malhotra A (2005) E-S-Qual: a multiple-item scale for assessing electronic service quality. J Serv Res 7(3):213–233
28. Rosen LD, Karwan KR (1994) Prioritizing the dimensions of service quality. Int J Serv Ind Manag 5(4):39–52
29. Saaty TL, Vargas LG (1994) Decision making in economic, political, social and technological environment with the analytic hierarchy process. RWS Publications, Pittsburgh
30. Santhiyavalli G, Sandhya B (2011) Service quality evaluation in select commercial banks. IUP J Oper Manag 10(1):43–62
31. Seth N, Deshmukh SG, Vrat P (2005) Service quality models: a review. Int J Qual Reliab Manag 22(9):913–949
32. Seth N, Deshmukh SG, Vrat P (2006) A framework for measurement of quality of service in supply chains. Supply Chain Manag 11(1):82–94

33. Shabbir MF, Aslam DH, Capusneanu S, Barbu CM, Tanveer MA (2012) Perceived service quality of Islamic and non Islamic banks operating in Pakistan. Am J Sci Res 51:27–36
34. Soteriou AC, Stavrinides Y (2000) An internal customer service quality data envelope analysis model for bank branches. Int J Bank Market 18(5):246–252
35. Sweeney JC, Soutar GN, Johnson LW (1997) Retail service quality and perceived value. J Consum Serv 4(1):39–48
36. Sylvie R, Ina F (2010) A new measure of e-service quality in France. Int J Retail Distrib Manag 38(7):497–517
37. Wolfinbarger M, Gilly MC (2003) eTailQ: dimensionalizing, measuring, and predicting etail quality. J Retail 79(3):183–198
38. Yoo B, Donthu N (2001) Developing a scale to measure the perceived quality of an internet shopping site (Sitequal). Q J Electron Commerce 2(1):31–46
39. Zeithaml VA (1988) Consumer perceptions of price, quality, and value: a means-end model and synthesis of evidence. J Mark 52(3):2–22
40. http://www.statista.com/statistics/289770/india-retail-e-commerce-sales/

Finding Efficiency in Data Envelopment Analysis Using Variable Reduction Technique

Seema Gupta, K. N. Rajeshwari, and P. C. Jha

1 Introduction

Charnes, Cooper, and Rhodes (1978) which originated the DEA as a technique described it as a *mathematical programming model applied to observational data that provides a new way of obtaining empirical estimates of relations – such as the production functions and/or efficient production possibility surfaces – that are cornerstones of modern economies* [4]. Data envelopment analysis (DEA) is a relatively new data-oriented approach for evaluating the performance of a set of peer entities called decision-making units (DMUs) which convert multiple inputs into multiple outputs. The definition of a DMU is generic and flexible [16]. Over the years, DEA has been used in a great variety of application for evaluating the performances in different contexts and of different kinds of entities engaged in many different activities. The domain of application is as smaller as different departments of a company or university or bank branches to as large as different states and countries on global scale. Different sets of DMUs include but not limited to hospitals, academic institutions, airlines, bank branches, cities, business firms, countries, regions, etc. DEA requires very few assumptions and hence opened up the possibilities for use in different cases which cannot be evaluated by other techniques because of complexities and assumptions involved [7].

As the number of inputs and outputs increase in numbers, the dimension of the linear programming solution space also increases. Thus, the discriminating power of DEA decreases. As given in [13], total number of input and output variables should

S. Gupta (✉) • K.N. Rajeshwari
School of Mathematics, Devi Ahilya University Indore, Indore, India
e-mail: seemadavv@gmail.com

P.C. Jha
Department of Operational Research, University of Delhi, New Delhi, India

© Springer Nature Singapore Pte Ltd. 2018

P.K. Kapur et al. (eds.), *Quality, IT and Business Operations*, Springer Proceedings in Business and Economics, DOI 10.1007/978-981-10-5577-5_13

be less than one third of the number of DMUs in the analysis. The initial approaches toward variable reduction were applied when the variables in the study are found to be highly correlated, and in the set of highly correlated variables, few can be highly correlated. [12] has used canonical correlation to compute weights for input set and output set so that correlation of weighted input vector to weighted output vector is maximum for the given set. Then a comparison is made between efficiency scores by DMUs and correlation weights. Reference [20] compared the DEA with principal component analysis (PCA), and two approaches are used in aggregating multiple inputs and multiple outputs in the evaluation of decision-making units (DMUs). PCA, a multivariate statistical method, combines new multiple measures defined by the inputs/outputs. Both methods are applied to three real-world data sets to characterize the economic performance of Chinese cities, and he got consistent and mutually complementary results. He further used nonparametric statistical tests to validate the consistency between the rankings obtained from DEA and PCA.

Reference [8] described a multivariate statistical approach to find out variable reduction among highly correlated variables with least loss of information. The method also described which variable should be retained. They have chosen the set of most informative variables by calculating variance-covariance matrix and trace matrix. In their approach, they have normalized the variables to have a mean value of zero and a variance value of one. By doing so they affirmed that all the input and output variables are considered at equal preference level. Reference [9] applied an approach toward variable reduction as decomposing the original problem into smaller subproblems. Once these smaller problems with fewer inputs and outputs are solved, the solution is used as initial approximation to the original problem. Lexicographic parametric programming along with dimensional decomposition procedure is used to increase the computational efficiency. The production possibility set is specified based on the information about the existing input and output data values which consists of all the input and output combinations.

Reference [14] provides a methodology based upon Shannon's entropy formula, for combining the efficiency results of different DEA models. An application of the same is illustrated by considering numerical example of educational departments of a university in Iran. They talked about different models as CCR, BCC, FDH (FDH-CRS, FDH-VRS, FDH-NDRS, and FDH-NIRS), SBM, RAM, cost models, FG, ST, etc., and different approaches as input-oriented and output-oriented views or optimistic and pessimistic views. Reference [10] discussed the four approaches, namely, efficiency contribution measure (ECM), principal component analysis (PCA), a regression-based test, and bootstrapping method for variable specification and reduction by comparing them. They analyzed three-input, one-output production process for at least 300 observations under these methods, and method for variable selection is being guided based on the results.

There are many real applications found for variable reduction technique in DEA [2, 5, 11, 15, 17, 19]. For example, [18] have used the model for hotel chain and 23 public libraries in Tokyo. Reference [3] have used the model for real data set of 30 provinces in China. Reference [6] have used the model for 24 of Taiwan's commercial banks. Reference [1] have used the variable reduction model for National Gas Company (NGC) of Iran with 14 large branches in 13 provinces. The proposed model in this paper based on variable reduction technique in DEA can also be used in these and different real-life applications to calculate the efficiency scores and discriminating among similar entities.

Rest of the paper is arranged in the following manner. Section 2 talks about the existing CCR model and then details the proposed methodology. Section 3 describes the data used in the study. Next section details the result followed by conclusion section. The article completed with acknowledgement and references.

2 Methodology

The basic model for DEA is constant return to scale model being given in Banker, Charnes, and Cooper in 1978, known as CCR model. There are models for formulating self-efficiency in case of multiple inputs and multiple outputs.

2.1 CCR Model

Suppose there are n numbers of DMUs, each having m number of inputs and s number of outputs. Let index for DMU is j, for input is i, and for output is r.

Thus, we have $j = 1, 2,3, \ldots n$ DMUs each using $i = 1,2, \ldots m$ inputs x_{ij} and delivering $r = 1,2, \ldots s$ outputs y_{rj}.

The efficiency score of a DMU, say k, under consideration is calculated by taking ratio of weighted output to weighted input of DMU k, subject to the constraint that values of optimal weights for inputs and outputs for DMU k should be such that it keeps efficiency score of all of the DMUs within definition of efficiency, i.e., between 0 (minimum) and 1 (maximum).

Mathematically the formulation is

$$\max \quad \frac{\sum_{r=1}^{s} \mu_{rk} y_{rk}}{\sum_{i=1}^{m} \vartheta_{ik} x_{ik}}$$

$$\text{s.t.} \quad \frac{\sum_{r=1}^{s} \mu_{rk} y_{rj}}{\sum_{i=1}^{m} \vartheta_{ik} x_{ij}} \leq 1 \qquad \text{for } j = 1, 2, \ldots, n$$

$$\mu_{rk} \geq 0 \qquad \text{for } r = 1, 2, \ldots, s$$

$$\vartheta_{ik} \geq 0 \qquad \text{for } i = 1, 2, \ldots, m$$

The above formulation is fractional programming problem which is converted to equivalent linear programming problem using Charnes-Cooper method as

$$\text{put } u_{rk} = t\, \mu_{rk}$$

$$\text{put } v_{rk} = t\,\vartheta_{rk}$$

$$\text{where } t = \left[\sum_{i=1}^{m} \vartheta_{ik} x_{ik}\right]^{-1}$$

Thus, the fractional programming problem can be written in linear form as

$$\begin{aligned}
\max \quad & \sum_{r=1}^{s} u_{rk} y_{rk} \\
\text{s.t.} \quad & \sum_{i=1}^{m} v_{ik} x_{ik} = 1 \\
& \sum_{r=1}^{s} u_{rk} y_{rj} - \sum_{i=1}^{m} v_{ik} x_{ij} \le 1 \qquad \text{for } j = 1, 2, \ldots, n \\
& u_{rk} \ge 0 \qquad \text{for } r = 1, 2, \ldots, s \\
& v_{ik} \ge 0 \qquad \text{for } i = 1, 2, \ldots, m
\end{aligned}$$

Traditional DEA model differentiate between efficient and inefficient DMUs based on their calculated efficiency value. A DMU is efficient if its efficiency value is one. However, there are cases where this differentiation becomes difficult with large number of inputs and outputs, in comparison with number of DMUs. In such scenario, most of DMUs becomes efficient since calculated efficiency value comes out to be 1. As the number of inputs and outputs increases, the dimensionality of observation space is increased, and thus the number of orthogonal directions increases. This gives increase in Euclidean distance between the observation, and thus more and more DMUs become efficient. Thus, discrimination power of DEA is reduced. Hence, variable reduction technique is used in DEA model to aggregate some of the inputs and outputs so that rule of thumb is satisfied.

The proposed approach to find the efficiency scores in DEA problem comprised of following steps.

2.2 Algorithm

1. Assign countin = number of inputs, countout = number of outputs, countdmu = number of DMUs.
2. Check the rule of thumb with countin, countout, and countdmu as countdmu $\ge$ max {countin $\times$ countout, 3(countin + countout)}.

3. If the thumb rule is satisfied, exit and apply standard DEA model, otherwise, go to step 4.
4. Form the input and output matrix as

$$\underset{\text{Input Matrix}}{\begin{bmatrix} x_{11} & x_{12} & \cdots & x_{1n} \\ x_{21} & x_{22} & \cdots & x_{2n} \\ \vdots & \vdots & \cdots & \vdots \\ x_{m1} & x_{m2} & \cdots & x_{mn} \end{bmatrix}_{m\times n}} \underset{\text{Output Matrix}}{\begin{bmatrix} y_{11} & y_{12} & \cdots & y_{1n} \\ y_{21} & y_{22} & \cdots & y_{2n} \\ \vdots & \vdots & \cdots & \vdots \\ y_{s1} & y_{s2} & \cdots & y_{sn} \end{bmatrix}_{s\times n}}$$

5. Normalize the input and output matrix.
6. For the normalized input matrix, calculate the correlation coefficient for each pair of inputs.
7. Put the correlated value into another matrix named as input correlated matrix.

$$\begin{bmatrix} c_{11} & c_{12} & \cdots & c_{1m} \\ & c_{22} & \cdots & c_{2m} \\ \vdots & \vdots & \cdots & \vdots \\ & & \cdots & c_{mm} \end{bmatrix}$$

8. Repeat steps 6 and 7 with output matrix to get output correlation matrix.

$$\begin{bmatrix} d_{11} & d_{12} & \cdots & d_{1s} \\ & d_{22} & \cdots & d_{2s} \\ \vdots & \vdots & \cdots & \vdots \\ & & \cdots & d_{ss} \end{bmatrix}$$

9. Choose the highest correlated value among the entries of input (output) correlation matrix, and then replace these two correlated input values with a new input value as average of these two input (output) values (i.e., we are replacing with weighted sum of equal weights to them).
10. Check rule of thumb equation as given in step 2.
11. Repeat steps from 5 to 10 till rule of thumb is satisfied.
12. Find efficiency of each using CCR-DEA model.

3 Data

We have considered a case study of 13 renewable energy technologies included in the renewable energy plan approved by the Spanish government on 2005 as given in [11] (Table 1).

Table 1 Input and output matrix

DMU	i1	i2	i3	o1	o2	o3	o4
D1	0.937	1	1.47	0.5	2.35	20	1.93
D2	0.937	1	1.47	1	2.35	20	3.22
D3	1.5	1	1.51	2.5	2.35	20	9.65
D4	0.7	1.5	1.45	0.5	3.1	25	0.47
D5	0.601	2	0.7	2	2	25	0.26
D6	5	2.5	0.6	3.5	2	25	0.26
D7	1.803	2	4.2	5	2.59	25	0.48
D8	1.803	1	7.11	0.5	7.5	15	2.52
D9	1.803	1	5.42	0.5	7.5	15	2.52
D10	1.803	1	5.42	0.5	7.5	15	2.52
D11	1.803	1	2.81	0.5	7.5	15	2.52
D12	0.856	1	4.56	5.6	7.5	20	4.84
D13	1.503	1.5	2.51	0.2	7	20	5.91

The inputs are:

Input 1: Investment ratio (euro × 10^3/Kw)
Input 2: Implement period (years)
Input 3: Operating and maintenance costs (euro × 10^3/Kwh)

The outputs are:

Output 1: Power (MW × 10^3)
Output 2: Operating hours (hours × 10^3/year)
Output 3: Useful life (years)
Output 4: Tons of CO_2 avoided ($tCO_2 \times 10^6$/year)

The first table shows the values of inputs and outputs for each of the DMUs. Table 2 shows the efficiency scores of DMU with the basic CCR model. Since all but one DMUs have efficiency score one, so the discrimination among them is difficult. Also we see that the number of DMU is only 13 which are far less than maximum of the product of number of outputs and number of inputs and trice the sum of number of inputs and number of outputs. Thus, the thumb rule is not satisfied in this case. Hence, we proceed for calculating the efficiency scores using the proposed model. Table 3 shows the normalized input and output matrix which is required for further calculations.

4 Result

The input correlation matrix is formed by inserting the correlation coefficient values at the respective places in the input correlation matrix. Thus, we get first-level correlation coefficient matrix for inputs which is given in Table 4. Similarly output correlation matrix is obtained and results are put in Table 5.

Table 2 Efficiency scores of DMU with basic CCR model

DMU	Efficiency calculated with basic CCR model without variable reduction
D1	1
D2	1
D3	1
D4	1
D5	1
D6	1
D7	0.78
D8	1
D9	1
D10	1
D11	1
D12	1
D13	1

Table 3 Normalized input and output matrix

DMU	i1	i2	i3	o1	o2	o3	o4
D1	0.044515	0.057143	0.037471	0.02193	0.038374	0.076923	0.052022
D2	0.044515	0.057143	0.037471	0.04386	0.038374	0.076923	0.086792
D3	0.071262	0.057143	0.038491	0.109649	0.038374	0.076923	0.260108
D4	0.033256	0.085714	0.036962	0.02193	0.050621	0.096154	0.012668
D5	0.028552	0.114286	0.017843	0.087719	0.032658	0.096154	0.007008
D6	0.237541	0.142857	0.015294	0.153509	0.032658	0.096154	0.007008
D7	0.085657	0.114286	0.107061	0.219298	0.042293	0.096154	0.012938
D8	0.085657	0.057143	0.181239	0.02193	0.122469	0.057692	0.067925
D9	0.085657	0.057143	0.13816	0.02193	0.122469	0.057692	0.067925
D10	0.085657	0.057143	0.13816	0.02193	0.122469	0.057692	0.067925
D11	0.085657	0.057143	0.071629	0.02193	0.122469	0.057692	0.067925
D12	0.040667	0.057143	0.116238	0.245614	0.122469	0.076923	0.130458
D13	0.071405	0.085714	0.063982	0.008772	0.114304	0.076923	0.159299

Table 4 Correlation coefficient matrix for inputs at first level

Inputs	Input1	Input2	Input3
Input1	1	0.5371	−0.04904
Input2		1	−0.44543
Input3			1

From Table 4, it is clear that highest correlation exists among input1 and input2. Thus, we combine the input1 and input2 to get a new input12 which is the equal weighted sum of these two inputs. From Table 5, we observe that a highest correlation coefficient value comes for output2 and output3. Thus, we combine output2 and output3 to get a new output23.

Table 5 Correlation coefficient matrix for outputs at first level

Outputs	Output1	Output2	Output3	Output4
Output1	1	−0.24111	0.494462	0.010632
Output2		1	−0.78986	0.202265
Output3			1	−0.32887
Output4				1

Table 6 Correlation coefficient matrix for outputs at second level

Outputs	Output1	Output23	Output4
O1	1	0.538271	0.010632
O23		1	−0.31529
O4			1

Table 7 Comparative efficiency scores from basic CCR model and proposed model

DMU	Original efficiency scores	Modified efficiency scores
DMU1	1	0.886575
DMU2	1	0.954665
DMU3	1	1
DMU4	1	0.979019
DMU5	1	1
DMU6	1	1
DMU7	0.78	0.644814
DMU8	1	0.402079
DMU9	1	0.447981
DMU10	1	0.447981
DMU11	1	0.578715
DMU12	1	1
DMU13	1	0.711143

Now the number of outputs is three and number of inputs is two. We again check the thumb rule as

$$\text{countdmu} \geq \max\{\text{countin} \times \text{countout}, 3\,(\text{countin} + \text{countout})\}$$

in which countdmu is 13, countin is 2, and countout is 3, since 13 is less than trice the sum of countin and countout which is 15. Thus again apply correlation formula for inputs and outputs at second level. Since the inputs are already reduced to two, hence in this example, we form only output correlation matrix at second level as given as in Table 6. Since this table is obtained at second level, it is thus named so. From Table 6, we observed that output1 and output23 are highly correlated; thus, we calculate a new output by combining these two as output123.

Again we check the thumb rule which is now satisfied as $13 \geq \{4, 12\}$. Thus, we proceed to the next step which is calculating the efficiency scores of DMUs using these new input and output matrices. We use the CCR model for this. Thus, for each of 13 DMUs, 13 input-oriented DEA models are formulated and then solved for weights and efficiency scores. The results are given in the following Table 7.

Table 8 Comparative efficiency scores from basic CCR model and proposed model

DMU	Modified efficiency scores	Efficiency scores from models in prior art	
		Model 1	Model 2
DMU1	0.886575	0.88	1
DMU2	0.954665	0.91	1
DMU3	1	1	1
DMU4	0.979019	0.79	0.88
DMU5	1	0.68	0.67
DMU6	1	0.6	0.54
DMU7	0.644814	0.62	0.36
DMU8	0.402079	0.58	0.53
DMU9	0.447981	0.7	0.65
DMU10	0.447981	0.7	0.65
DMU11	0.578715	0.82	1
DMU12	1	1	0.8
DMU13	0.711143	0.72	0.86

We can observe the significant difference in the efficiency scores calculated after the variable reduction technique is applied. Before applying the technique, all but one was efficient, but now only four DMUs are efficient and rests are inefficient with different efficiency scores. Thus, we also observe that discrimination power of DMU is significantly enhanced.

Table 8 gives the comparative results obtained from the proposed model and those found in prior art. We can observe that our modified efficiency scores are very much comparable with the given models. Model 1 which is an improvement over model 2 is very much similar to our model results; thus, the validity of the model is also established.

5 Conclusion

From the case study, it is clear that initially, without variable reduction, except one DMU, the rest all have efficiency score 1. After applying the variable reduction, we can see the difference in efficiency scores as now only four DMUs are efficient and nine are inefficient. The difference among inefficient DMUs is also evident from the results. By using the weights obtained, we can find the scope for improvement of inefficient units. The results obtained here are comparable with the deviation models given in prior art, thus establishes the validity of the results. We can further find cross-efficiency to rank the DMUs; if required we can also work toward extending the model in cases involving uncertainty.

References

1. Amirteimoori A, Despotis DK, Kordrostami S (2014) Variables reduction in data envelopment analysis. Optimization 63(5):735–745
2. Bhattacharjee S (2012) Efficiency dynamics and sustainability of the Indian IT-ITeS industry: an empirical investigation using DEA. IIMB Manage Rev 24(4):203–214
3. Bian Y, Yang F (2010) Resource and environment efficiency analysis of provinces in China: a DEA approach based on Shannon's entropy. Energ Policy 38(4):1909–1917
4. Charnes A, Cooper WW, Rhodes E (1978) Measuring the efficiency of decision making units. Eur J Oper Res 2(6):429–444
5. Dyson RG, Allen R, Camanho AS, Podinovski VV, Sarrico CS, Shale EA (2011) Pitfalls and protocols in DEA. Eur J Oper Res 132(2):245–259
6. Hsiao B, Chern CC, Chiu CR (2011) Performance evaluation with the entropy-based weighted Russell measure in data envelopment analysis. Expert Syst Appl 38(8):9965–9972
7. Jain RK, Natarajan R (2015) A DEA study of airlines in India. Asia Pac Manage Rev
8. Jenkins L, Anderson M (2003) A multivariate statistical approach to reducing the number of variables in data envelopment analysis. Eur J Oper Res 147(1):51–61
9. Korhonen PJ, Siitari PA (2009) A dimensional decomposition approach to identifying efficient units in large-scale DEA models. Comput Oper Res 36(1):234–244
10. Nataraja NR, Johnson AL (2011) Guidelines for using variable selection techniques in data envelopment analysis. Eur J Oper Res 215(3):662–669
11. San Cristóbal JR (2011) A multi criteria data envelopment analysis model to evaluate the efficiency of the renewable energy technologies. Renew Energ 36(10):2742–2746
12. Sengupta JK (1990) Tests of efficiency in data envelopment analysis. Comput Oper Res 17(2):123–132
13. Sinuany-Stern Z, Friedman L (1998) DEA and the discriminant analysis of ratios for ranking units. Eur J Oper Res 111(3):470–478
14. Soleimani-Damaneh M, Zarepisheh M (2009) Shannon's entropy for combining the efficiency results of different DEA models: method and application. Expert Syst Appl 36(3):5146–5150
15. Sueyoshi T, Yuan Y (2015) China's regional sustainability and diversified resource allocation: DEA environmental assessment on economic development and air pollution. Energ Econ 49:239–256
16. Thianjaruwatthana MW (2009) Technical efficiency and its determinants of regional hospitals in Thailand. Doctoral dissertation, Chulalongkorn University
17. Toloo M, Babaee S (2015) On variable reductions in data envelopment analysis with an illustrative application to a gas company. Appl Math Comput 270:527–533
18. Wagner JM, Shimshak DG (2007) Stepwise selection of variables in data envelopment analysis: procedures and managerial perspectives. Eur J Oper Res 180(1):57–67
19. Wang D, Li S, Sueyoshi T (2014) DEA environmental assessment on US industrial sectors: investment for improvement in operational and environmental performance to attain corporate sustainability. Energ Econ 45:254–267
20. Zhu J (1998) Data envelopment analysis vs. principal component analysis: an illustrative study of economic performance of Chinese cities. Eur J Oper Res 111(1):50–61

Fire Safety Experimental Investigations of Time to Flashover as a Function of Humidity in Wood

Ajit Srividya, Torgrim Log, and Arjen Kraaijeveld

1 Introduction

Fire safety is of great concern in Norway since it is rich in wooden constructions increasing the chances of fire hazard due to proximity of the buildings, the dry climate in winters, and lighting of candles and firewood in winter time. Torgrim Log [1] has presented the investigations on the fire at Laerdalsoyri in January 2014 and shown that if values of fuel moisture content (FMC) in wood fall below 5%, wood products would ignite faster than wood with more normal FMC values. The equilibrium moisture content (EMC) in wood is a function of air temperature and relative humidity in air. The FMC will, due to diffusion process, approach the EMC value. Log [1] showed theoretically that this was indeed the case in the previously mentioned Laerdalsoyri fire. When flashover occurs quickly and everything inside the house gets involved, then it would be uncontrollable even by the fire brigade. In dry and windy weather, the risk of uncontrollable fire spread to adjacent and distant structures, as well as vegetation and garden trees, hedges, etc., is very high. Jing Xin and Chong Fu Huang [2] presented the analyses of fire situation in China from 1991 to 2010 and found that the number of fires was more frequent during winter months and fires were more frequent during the weekend. Also electrical failures and improper use of fire in daily life were major causes of fire incidents. Annemarie Poulsen et al. [3] have presented two full-scale room fire tests for the study of onset of flashover by varying the fire loads representing seven different commercial applications and two noncombustible linings. They have found that for fire loads composed of pure wood/celluloses, the onset of flashover occurred about the same time as fire spread irrespective of linings and at significantly higher room temperature (725 °C).

A. Srividya (✉) • T. Log • A. Kraaijeveld
Stord/Haugesund University College, Haugesund, Norway
e-mail: asvidya88@gmail.com; Torgrim.Log@hsh.no; Arjen.Kraaijeveld@hsh.no

© Springer Nature Singapore Pte Ltd. 2018

P.K. Kapur et al. (eds.), *Quality, IT and Business Operations*, Springer Proceedings in Business and Economics, DOI 10.1007/978-981-10-5577-5_14

Aiping Chen et al. [4] established a model with theories of fire dynamics and relevant parameters of combustible lining materials for predicting the hot gas layer temperature during pre-flashover and theoretically analyzed the effects of lining materials on the likelihood of flashover. Zhaozhi Wang et al. [5] proposed a method known as TTFO model to predict time to flashover in ISO 9705 room corner fire tests (full scale) based on material density, time to ignition, and heat release collected from cone calorimeter tests and given different equations to different group of materials. J. Francis and A. P. Chen [6] grouped flashover definitions into three categories, one mathematical, one visual, and the last depending on diction (the ISO definition), and suggested that the most convenient definition of flashover for experimentalists is a visual one that is based on observation of flames projecting from the vent opening. Aiping Chen et al. [7] studied the effects of four factors (thermal inertia of lining materials, ventilation factor of door openings, heat release rate of fuel, and internal dimensions of enclosure) on likelihood of flashover and developed a method by correlating a vast amount of data gained in small-scale and large-scale experiments to predict whether the flashover occurs or not in enclosure fires. Chi-Ming Lai et al. [8] presented a review of the characteristics of prediction models of flashover. Based on their experimental investigations in full-scale room fires, they presented a method to predict flashover on the energy filling concept by comparing the room volume and heat release rate (HRR) at flashover with previous data on similar building materials and fire loads. Based on their study, Vytenis Babrauskas et al. [9] concluded that the predicted values of HRR at flashover using existing correlations provide lower-bound estimates and actual HRR at flashover may be significantly higher and highlighted the importance of understanding the causal factors leading to flashover. Richard D. Peacock et al. [10] presented a review of the techniques to predict onset of flashover, and the various predictions give estimates that are commensurate with the precision of available experimental data.

Various studies on understanding the onset of flashover and methods to predict flashover have been presented earlier, and the importance of the FMC on the onset of flashover has been analyzed theoretically by Log [1]. In the current study based on experimental investigations, the focus is on understanding the effect of low humidity levels in wood on the onset of flashover and to propose simple models to predict time to flashover for known humidity levels in wood constructions.

2 Experimental Setup and Data Collection

The experiments were conducted on approximate equivalent of ¼ ISO rooms constructed of wood with a door opening. Nine wooden rooms were constructed and stored at different locations to achieve different fuel moisture contents (FMCs). Referring to the door opening side of box as the front side, a hole is drilled in a corner of the noncombustible (light weight concrete) floor at the back corner for placement of the initial burner which was filled with methanol. A stack of thermocouples were placed in the smoke layer height to record the temperatures

Thermocouple Layout		
Back Plane		
Top Layer	13 12 11	Cassette 1
Middle Layer	16 15 14	Cassette 1
Bottom Layer	09 03 12	Cassette 1
Front Plane		
Top Layer	3 2 1	Cassette 1
Middle Layer	7 5 4	Cassette 1
Bottom Layer	10 9 8	Cassette 2
Front of door		
Top	7	Cassette 2
Middle	11	Cassette 2
Bottom	8	Cassette 2

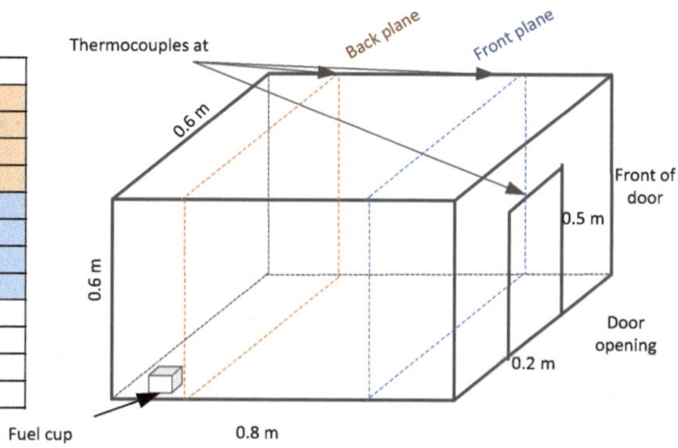

Fig. 1 Approx. equiv. of 1/4 ISO wooden room with thermocouple layout

inside the room at the back end and the front end of the room. Three thermocouples were placed in front of the door opening in the lower side, middle, and at the top end. Mass reduction of methanol, mass reduction of wooden one-fourth-scale ISO room, and temperatures within the room and outside the opening were measured. Spread of flame on the walls and through the door opening was recorded by a video camera. Before igniting the methanol, relative air humidity and air temperature in the laboratory are noted. The room used for experiment and layout of thermocouples is shown in Fig. 1.

3 Prediction of Flashover

The ISO 13943 definition states, "Flashover is a stage of fire transition to a state of total surface involvement in a fire of combustible materials within an enclosure." It is not easy to quantify flashover process in terms of measurable physical parameters, and hence based on the various researches, a working definition has been formulated and presented in detail by Peacock [10]. In the present experimental study, the visual definition of flashover has been adopted, i.e., "The time when the first flames were seen to emerge out of the opening is recorded as time to flashover."

4 Analysis of Experimental Flashover Data

Onset of flashover for the nine small-scale rooms have been recorded when the first flames were seen coming out of the opening and presented in Table 1. The temperatures at the door opening at onset of flashover are also presented in Table 1. Heat released (Q_C) based on the mass loss of wood and heat radiated (Q_E) based on the average temperature at the front side of box are evaluated. The values of heat

Table 1 Data recorded for humidity and temperatures and heat values evaluated based on data

S. No /fuel gm	Humidity in air H_a %	Temp. of air T_a°C	Humidity of ISO room, FMC/H_b %	Flashover F_o sec	Temp. at opening T_o°C	Q_C – HRR KW	Q_E KW/m^2
1/120	42.5	14.5	7.52[a]	411	433.45	182.636	216.44
2/120	40	17.7	8.05[b]	355	410.12	174.94	135.738
3/120	42.5	14.5	9.03[a]	409	451.65	194.976	260.363
4/120	43.9	18	9.07[b]	454	380.64	160.007	127.067
5/120	43.9	18	9.16[c]	709	405.91	182.816	178.455
6/120	43.9	18	9.36[b]	408	427.232	190.873	182.987
7/150	42.5	14.5	9.52[b]	562	443.63	217.85	265.456
8/120	43.9	18	10.59[b]	635	444.815	236.479	242.629
9/120	30.8	11.2	11[a]	782	451.804	228.959	242.611

[a]First time, [b]Second time, [c]Third time

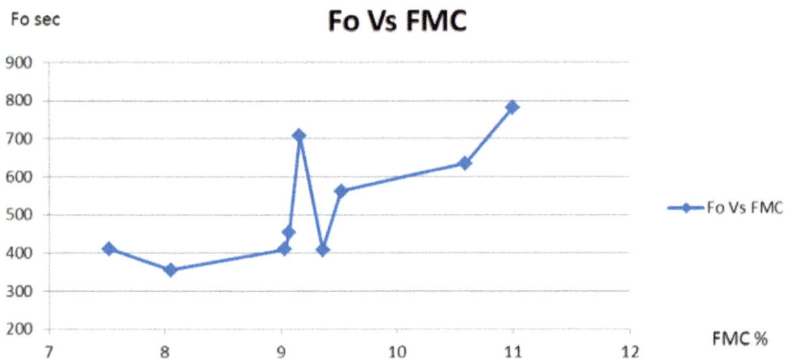

Fig. 2 Plot of time to flashover vs. FMC of the test rooms

released and heat radiated at the time of onset of flashover are presented in Table 1. The plot of time to onset of flashover as a function of fuel moisture content (FMC) is shown in Fig. 2.

The data obtained on the 5th room seems to deviate considerably and hence is eliminated from the set for further analysis. This may be because this box was burnt twice earlier with too smaller quantity of fuel. In that previous test, flashover was not observed. After increasing fuel, quantity onset of flashover was observed during third experimentation. This may have depleted some volatiles from the wooden structure. Further, onset of flashover was observed during second experimentation for the rooms 2,4,6,7, and 8. These boxes were not burned much during first experimentation, and the rooms were remade by reversing the sides, and hence the observations on onset to flashover have been assumed as similar to first experimentation like the observations made for rooms 1, 3, and 9. For further analysis and discussion on onset of flashover, the observation for the fifth box is

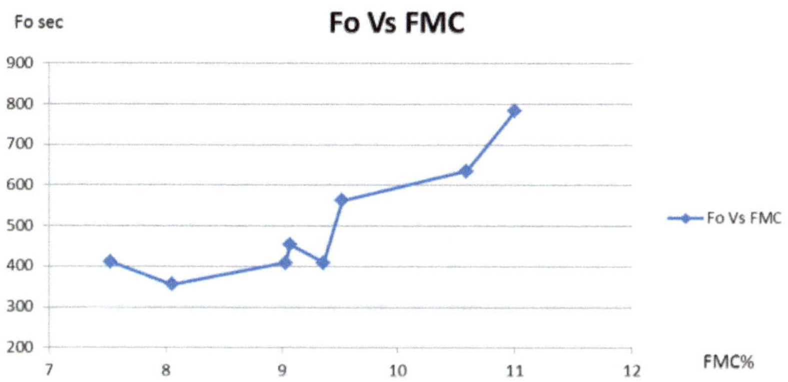

Fig. 3 Plot of time to flashover vs. FMC after eliminating the outlier

eliminated. The plot of flashover as a function of humidity of the rooms (after eliminating the outlier, room 5) is shown in Fig. 3.

5 Observations

It is observed that the onset of flashover occurs very quickly when humidity levels in the wooden rooms are low. The trend observed in Fig. 3 gives a clear indication of increase in onset of flashover as humidity levels increase in wood. This clearly implies that the lower humidity levels in wood would lead to onset of early flashover, and this results in less time for taking preventive actions for spread of fire. Further, at onset of flashover, the temperatures at the door opening recorded are ranging from 380 °C to 450 °C as presented in Table 1. These are consistent with reported literature [10]. Heat released and heat radiated are evaluated using the following expressions:

$$\text{Heat released, } Q_C = (dm/dt)^* \chi^* \, \Delta H_C KW$$

where dm/dt is rate of loss of mass of wood and, for fuel methanol, $\chi = 0.983$ and $\Delta H_C = 19.83$ KJ/g [11].

$$\text{Heat radiated, } Q_E = \sigma^* \left[(T_{av} + 273)\,4 - (T_a + 273)\,4 \right] / 1000 \text{ KW}$$

Here, T_{av} is the average temperatures noted on the front stack of thermocouples, T_a is the temperature of air, and "σ" is the Stefan-Boltzmann constant and is 5.67*10–8 $W/m^2.K^4$ [1]. At flashover, the variation of the heat released (Q_C) and heat radiated (Q_E) as a function of humidity in wooden room is presented in Table 1.

A comparison of various parameters (flashover, heat radiated, heat released, and temperature at flashover at door opening) with respect to humidity of wooden room

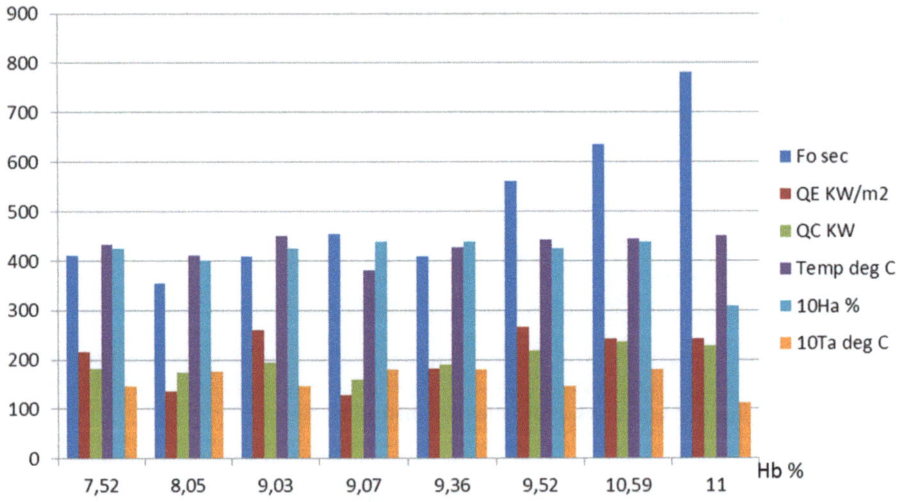

Fig. 4 Comparison of parameters with respect to humidity of wooden room

is shown in Fig. 4. The temperature in the room and humidity in air are also shown. The temperature and humidity in air are almost at the same levels except in the last data set. Hence, it has been assumed that the data on flashover for different levels of humidity in wooden boxes is only a function of the humidity level in wooden boxes.

The data for the lowest humidity level is point 1, i.e., $H_b = 7.52$ and $F_o = 411$ s. $H_a = 42.5$ and $T_a = 14.5$ °C. Plots of the temperatures w.r.t time are shown in Fig. 5. It can be seen in Fig. 5 that temperatures rise very rapidly once the flashover (at 411 s) is reached. Also rise in heat released and heat radiated is rapid once flashover is reached.

Based on the experiments conducted, it has been observed that the onset of flashover occurs earlier when the humidity levels in the wooden rooms are lower. An increasing trend of flashover is observed with increasing values of humidity, and mathematical models are developed based on the data collected to predict flashover when the humidity levels in the wood are known.

6 Mathematical Model to Predict Flashover

Onset of flashover (F_o) as a linear function of FMC in wooden room (FMC) is carried out, and the mathematical linear relation with correlation of 72.2% is presented below and shown in Fig. 6:

$$F_o = -514.5 + 109.7\ \text{FMC}$$

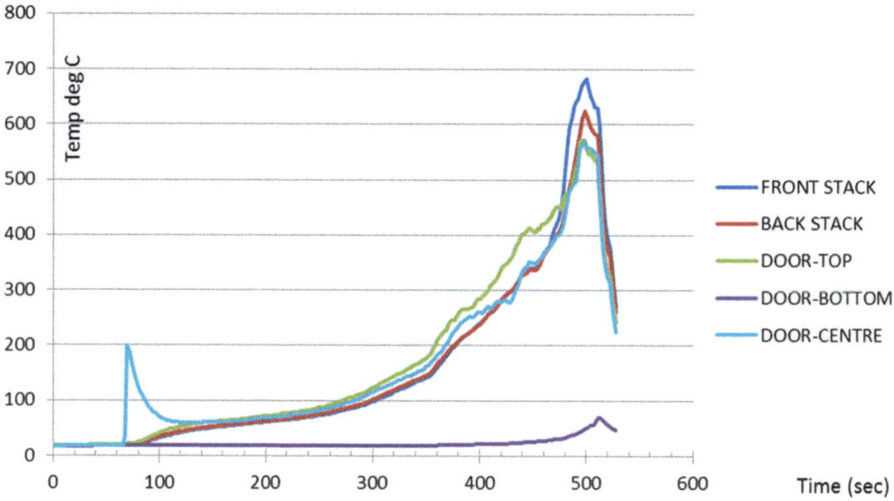

Fig. 5 Temperature at various locations of wooden room with time

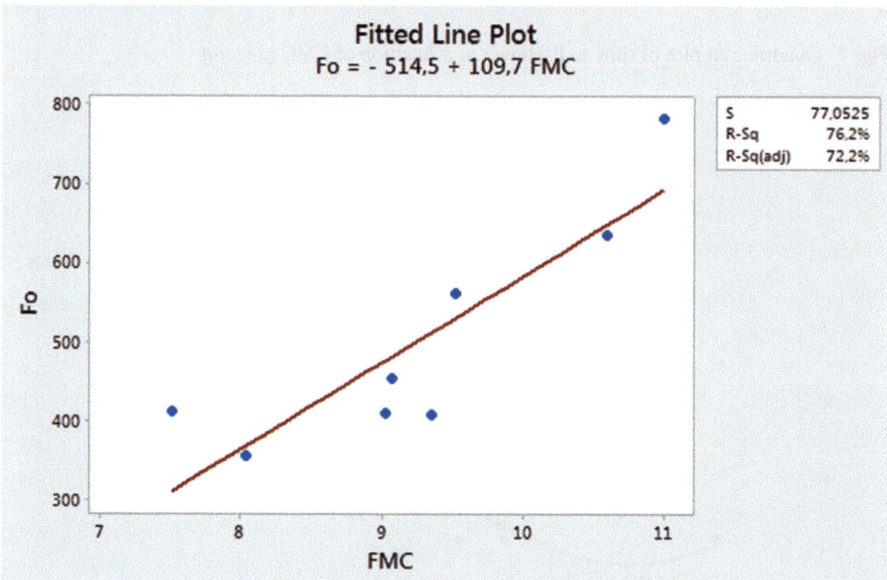

Fig. 6 Linear fit plot of time to flashover as a function of FMC in wood

Polynomial fits (quadratic and cubic) have been tried, and it is found that the correlation for quadratic fit is 88% and cubic fit is 91.7% and shown in Figs. 7 and 8 with the following expressions.

The regression equation for quadratic fit is

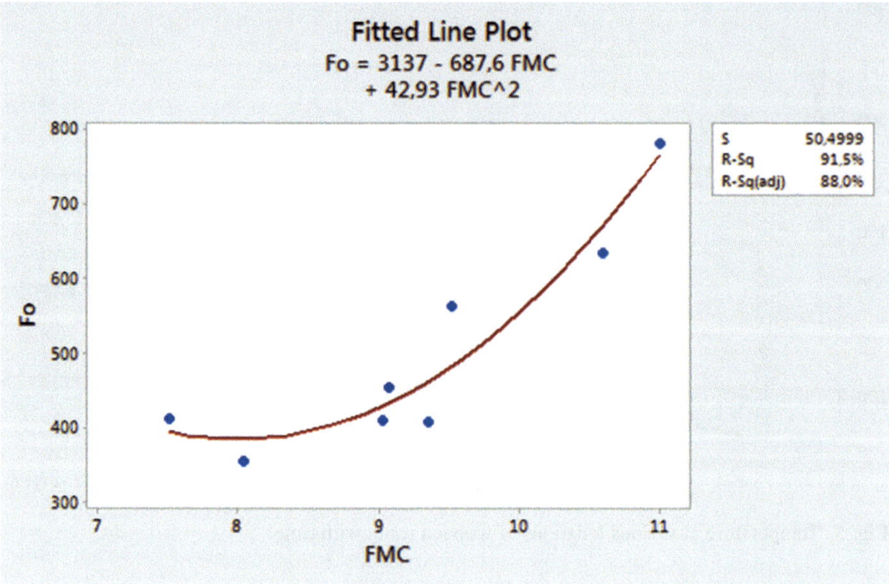

Fig. 7 Quadratic fit plot of time to flashover as a function of FMC in wood

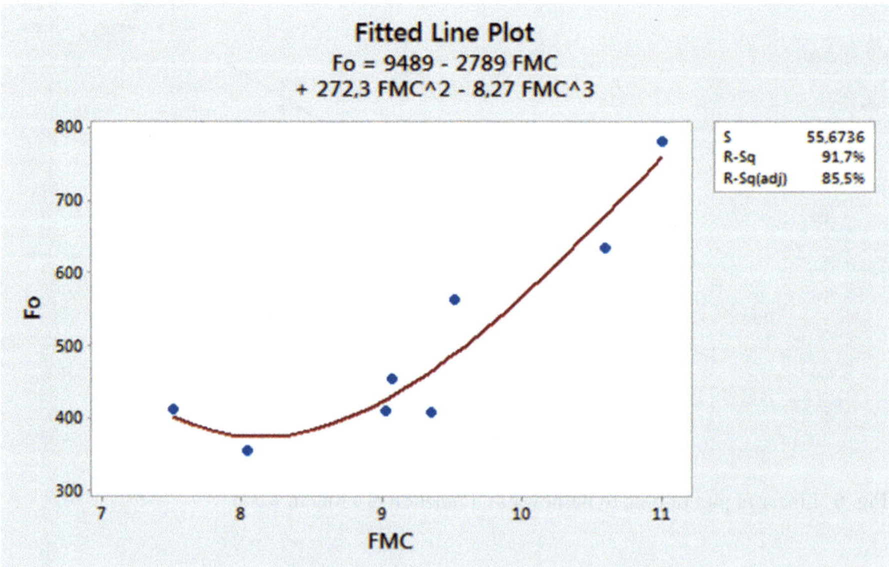

Fig. 8 Cubic fit plot of time to flashover as a function of FMC in wood

$$F_o = 3137–687.6\text{FMC} + 42.93\text{FMC}^2$$

The regression equation for cubic fit is

$$F_o = 9489 - 2789\ \text{FMC} + 272.3\ \text{FMC}^2–8.27\ \text{FMC}^3$$

Both quadratic and cubic fits have a higher correlation of about 86% and hence would give a better prediction of the onset of flashover values for the given FMC level in wood.

Due to limited range of FMC levels studied, the scatter in the time to flashover versus FMC may result in erroneous trends when using higher-order regression equations. It is quite unlikely that even lower FMC than studied in the present work will give an increase in time to flashover, as indicated in Figs. 7 and 8 for FMC < .5%. For now, the linear regression is probably the most reliable if slight extrapolation to lower FMC is needed to conclude regarding time to flashover for a larger range of FMC values.

7 Conclusion

Experiments have been carried out on approximate equivalent of one-fourth ISO rooms constructed of wood with a door opening, and time to flashover is recorded when the first flames escape through the vent (door) opening. The wooden rooms were maintained at different conditions to achieve desired fuel moisture content (FMC) levels. When temperature and humidity in the air is maintained almost at the same level, it is demonstrated that onset of flashover occurs very fast when the FMC level in the wood is low. Further, the temperatures rise very rapidly with onset of flashover with higher heat release and increased radiated heat. The lower humidity levels that tend to be the fact in winter time can lead to quick development of fire, and this has to be borne in mind, and necessary precautions are to taken. Based on the experimental data gathered, the mathematical models developed can be used to get an idea of the onset of flashover for the humidity levels in wood which can help in being alert for control of fire development and spread of fire; thus, risk of fire accidents can be reduced. The limitation of the model is that it is based on a small-scale rooms and also some of the rooms were reused for conducting the experiment. More experiments, especially with rooms of a larger FMC value range, need to be conducted to establish models for better prediction of onset of flashover as a function of FMC in wood.

Acknowledgments I would like to thank Dr. Maria-Monika Metallinou Log for her kind support in arranging funds for the experimental work. Thanks are also due to Håkon Gjøvåg Olaisen and Anders Hole for their support in conducting the experiments and Adithya Thaduri for his timely support.

References

1. Log T 2015 Cold climate fire risk; a case study of Laerdalsoyri fire, January 2014. Fire Technol 1–19
2. Xin J, Huang CF (2014) Fire Risk Assessment of Residential Buildings Based on Fir Statistics from China. Fire Technol 50:1147–1161
3. Poulsen A, Jomaas G, Bwalya A (2013) Evaluation of the onset of flashover in room fire experiments. Fire Technol 49:891–905
4. Chen A, Liang Z, Lu S (2013) Study on flashover in a single chamber lined with combustible materials. Fire Mater 37:358–373
5. Wang Z, Hu X, Jia F, Galea ER (2013) A two-step method for predicting time to flashover in room corner test fires using cone calorimeter data. Fire Mater 37:457–473
6. Francis J, Chen AP (2012) Observable characteristics of flashover. Fire Saf J 51:42–52
7. Chen A, Yang S, Dong X (2011) Studies of the combined effects of some important factors on the likelihood of flashover. Fire Mater 35:105–114
8. Lai C-M, Ho M-C, Lin T-H (2010) Experimental Investigations of fire spread and flashover time in Office Fires. J Fire Sci 28:278–302
9. Babrauskas V, Peacock RD, Reneke PA (2003) Defining flashover for fire hazard calculations: part II. Fire Saf J 38:613–622
10. Peacock RD, Reneke PA, Bukowski RW, Babrauskas V (1999) Defining flashover for fire hazard calculations. Fire Saf J 32:331–345
11. Drysdale D (2011) An introduction to fire dynamics. Wiley

Fixing of Faults and Vulnerabilities via Single Patch

Yogita Kansal, Uday Kumar, Deepak Kumar, and P. K. Kapur

1 Introduction

In today's competitive market, technology developers are struggling more with the software security problems rather than with its development problems. The consequences of security problems range from inconvenience to life threatening. Most of the security issue arises due to the defects residing in the software specifications. When a demand of new software comes to the technical organization, the developer is bounded to fulfill the requirements of customer with respect to its organization's policy. A customer demands for high-quality software in no time, whereas a developing organization desires for appropriate profit margins and goodwill in the market. In software engineering, software experiences several verification phases in its whole life cycle. Prior to release, testing techniques are applied so as to discover and debug the defects immediately.

It may be noted that at the beginning of a new project, the management team provides some resource (time, budget, manpower, etc.) constraint to the developing team. The developers are then restricted to achieve the task within the provided resource limits. A Gantt or PERT chart is maintained for resource and time management among all phases of software development life cycle (SDLC). The testing phase consumes a large amount of resources among all phases. If the testing

Y. Kansal (✉) • D. Kumar
Amity Institute of Information Technology, Amity University, Noida, India
e-mail: ykansal35@gmail.com; deepakgupta_du@rediffmail.com

U. Kumar
Lulea University of Technology, Luleå, Sweden
e-mail: Uday.Kumar@ltu.se

P.K. Kapur
Amity Centre for Interdisciplinary Research, Amity University, Noida, UP, India
e-mail: pkkapur1@gmail.com

© Springer Nature Singapore Pte Ltd. 2018
P.K. Kapur et al. (eds.), *Quality, IT and Business Operations*, Springer Proceedings in Business and Economics, DOI 10.1007/978-981-10-5577-5_15

goes prolonged, then the organization pays a big penalty to the customers, while hasty testing may leave some defects in the software [23, 24]. Therefore, when a developing team should stop, testing is first decided by the management.

The job of a developing team doesn't end up with software release rather the problem becomes more complex after its release. The severe defects are seen to be observed more at customer's site. Under scrutiny, developing team classifies the defects into faults (simple defects) and vulnerability (security defects). These defects are interdependent, i.e., faults are the cause of vulnerability and vulnerability is the effect of multiple faults. As stated by Schultz et al. [1], vulnerability is security defect that enables an attacker to bypass the software system. In software engineering, Kapur et al. [2] defined software fault or bug as an accidental condition that causes a functional unit to fail to perform its required function due to some error made by the software designers. Additionally, software faults are the errors which arise due to misinterpretation of specifications, design errors, etc., while software vulnerability is due to a gap left behind while developing the software. After software delivery, these defects can cause calamitous situation for users and developers too. Thus, the eventual solution for the defects is to provide patch timely. However the patch is developed by the developing team and the distribution is done by the vendor [3]. Generally, the patch for faults and vulnerabilities is developed and delivered separately [4] which increases the development cost, time, and manpower.

Some of the evidence for the statement is provided via prior art and is elaborated below.

The literature survey states that Arora, Telang, and Xu [5] provided a framework to assess the vulnerability disclosure time and discussed the stress handled by vendors to provide fixes just after the disclosure. Choi et al. [6] modeled the management's decisions that focused on quality of the software such that the vulnerabilities could be reduced. Whether the vendor should disclose vulnerabilities or apply patches is also focused here. August and Tunca [7] examined the efforts of external users in providing patches for vulnerabilities. Telang and Wattal [8] illustrated a case study that describes how the vulnerability disclosure is affecting the market share of the vendors especially due to unavailability of patches at the time of disclosure. Okamura et al. [9] assessed the optimal patch release time on the basis of stochastic modeling without considering developer's and vendor's perspective. Recently, Kansal et al. [21] developed a model to assess the number of patches released successfully.

Various researchers have also put their skills in fault patching. Cavusoglu et al. [10] investigated the security patch release and management problems from both vendor and firm perspectives. They developed a game-theoretic model to study the strategic interaction between the vendor and firm in balancing the costs and benefits of patch management. Subsequently, Okamura et al. and Luo et al. [11] have considered a nonhomogeneous fault identification process with patch/update management. Dey et al. [12] developed a cost framework for solving the optimal release time problems for patch management. Jiang et al. [13] proposed a software scheduling method that highlights the pros and cons of early release and continuing testing after release of software. Kapur et al. [2] provide solution for software release time decision problems only when faults are considered. Kapur and Shrivastava [14] also discuss about the software release and testing stop time without bothering about its patch. Recently Kansal et al. [22] have proposed a cost framework to solve the optimal release time problem with the consideration of warranty period.

The above mentioned research work discusses more about the vulnerability patch and fault patch separately. Although the resources required for fault removal are less as compared to the vulnerability removal, but if we combine the time and manpower for both the processes, it will cost much higher to the developers.

In an organization teams are divided into an efficient manner so as to manage the whole loop easily. The appointed teams are developers for assisting new projects, debuggers (including or excluding testers) for running projects, and technical supporters for maintaining customer relationships with the organization. When any customer observes instability in the system, the problem was first reported to the vendor, and then the problem is disclosed to the developing team. The developing team analyzes the problem and provides a solution for the same in form of patch. If a fault and vulnerability are reported simultaneously, then naturally priority should be given to vulnerability patch as it can be more exploitable. Further faults are the cause of vulnerability; thus, neglecting faults is also not a feasible solution. Moreover, it doesn't seem economically feasible to spend so much effort in building different patches for quality improvement. The study also proves that half of the time, manpower and cost are required by an organization after the release of software.

Although the literature suggests multiple optimization solution for patch development of faults and vulnerabilities independently, none of the researchers focuses on reducing the developer's effort by introducing vulnerability and fault patch as a single unit. In the present work, it is assumed that a single patch is provided for fixing the faults and vulnerabilities jointly.

After considering a single patch for both faults and vulnerabilities, when to release that patch is further decided by the management. Early patching may leave some faults or vulnerabilities behind, while later patching may harm the organization with respect to its market opportunity. As it is the most critical decision, management needs to have an optimal model for evaluating the release.

According to the developer's perspective, a trade-off needs to be maintained between time and money. The developing and debugging processes both consume high resources. The amount of time spends in any phase correspondingly affect the allocated cost and resources. A developer faces a gradual loss when any failure is observed or when any variation is seen in the allocation process.

Thus in the current work, a cost model is proposed for evaluating software and patch (for both faults and vulnerabilities) release time so that the total developer's cost could be minimized. To find the solution of release time problem, it is assumed that users are not familiar with the software; thus, the defect detection rate of the user is obviously less than the testers. In operational phase customer reports defects to the developing team. The developers then figure it out whether it is a fault or vulnerability. Generally, a significant amount of vulnerabilities are found after the software release through external users [4]. In proposed model, some constants are taken so as to differentiate discovered faults with discovered vulnerabilities.

After finalizing the type of defect, a repair process will start. For determining the cost incurred by developers for debugging the fault and vulnerability, it may also be noted that the fault detection rate and vulnerability detection rate of the user are different. As discovering vulnerability is the most complex task in itself, the vulnerability detection rate of the user is very small as compared to the fault detection rate [15]. The change in detection rate also leads to different distribution of defects.

The cost of discovering any type of defect by customers after testing may lead developers with burden of debt. The developer faces a large amount of loss in terms of capital and reputation. An organization needs to hire more employees on a monthly basis, and as time goes on, they discover fewer bugs so the cost per defect discovered climbs steadily. Moreover, at some threshold, perhaps a defect removal starts to cost more than the damage that defect could do.

In our paper we model two cases. In the first case, we investigate how a single patch for both faults and vulnerabilities can become more effective from the developer's perspective, while in the second case we release two patches, one specifically for vulnerability and the other for both. The cases will also help in validating the model. Different distributions have been used for faults and vulnerabilities of each interval.

Rest of the paper is organized as follows. Section 2 discusses the model assumptions and formulation of the proposed cost model. In this study we are presenting two cases. In the first case, we are assessing the optimal release time for single patch. In the second case, we are evaluating the optimal release time for two patches, i.e., first patch for removing faults and vulnerabilities at same time and second patch for removing vulnerabilities specifically. Section 3 elaborates the cost model with the detailed description of the two cases. In Sect. 4, numerical example is provided to test the proposed model with the help of real-time data set. Finally conclusion has been drawn in Sect. 5.

2 Model Formulation

In this section, software reliability growth model is formulated. Our first step is to provide some assumptions for model clarity. The next step is to narrate the software testing and updating policies in order to minimize the total cost.

2.1 Assumptions

The basic assumptions used throughout the paper are as follows:

- Defect discovery and debugging process follow the nonhomogeneous Poisson process (NHPP) in the whole software life cycle.

- The detection of each defect in a software system is independent of the detection of others.
- During testing there is no distinction between faults and vulnerabilities.
- Software is prone to failures due to the faults and vulnerabilities residing in the software.
- All faults and vulnerabilities are debugged perfectly.
- Total numbers of faults and vulnerabilities lying dormant in the software are finite.
- The fault detection rate and vulnerability detection rate of the user are always different.
- In operational phase (after delivering last patch), no patch will be released for remaining and newly discovered defects; rather, the reported defects are removed by the developing team at the company end, and later a new upgraded version of the software will be released.

2.2 *Devising the Model*

Several software reliability growth models and vulnerability discovery models have been developed to detect and debug the faults and vulnerabilities, respectively. Among them NHPP-based models are proven to be quite successful tool in the decision-making process. The hazard rate approach is used for deriving the mean value function of a cumulative number of faults and vulnerabilities removed and represented as

$$\frac{dm(t)}{dt} = \left(\frac{f(t)}{1 - F(t)}\right) \cdot (a \cdot p_i - m(t)) \qquad \text{for } i = 1, 2$$

where p_1 is the proportion of discovered faults, while p_2 is the proportion of discovered vulnerabilities. Also, $\frac{f(t)}{1-F(t)}$ is the defect detection rate.

A mean value function (the expected number of faults and vulnerabilities detected) is expressed as follows:

$$m(t) = a \cdot p_i \cdot F(t) \qquad \text{for } i = 1, 2 \tag{1}$$

The proportion of faults and vulnerabilities must be $p_1 + p_2 = 1$. Different distribution functions will help in gaining different MVF.

3 Cost Model

In this section the formulated cost model is represented with two cases. The first case discusses about the model with a single patch release for fixing both faults and vulnerabilities together. The second case argues about the vulnerability prioritiza-

Fig. 1 Software life cycle

tion by providing a model that releases two patches (first patch is specifically for vulnerability and other one is for both faults and vulnerabilities together).

3.1 *Case 1: Model with Single Patch Release (M1)*

In this case, the life cycle of the software, i.e., $[0, Tlc]$, is divided in three phases, i.e., before release $[0, \tau]$, after release $[\tau, \tau 1]$, and after patching $[\tau 1, Tlc]$ as shown in Fig. 1 where τ represents the software release time, $\tau 1$ represents the patch (for fixing faults and vulnerabilities together) release time, and *Tlc* is the software support cycle.

Per unit testing: During this model, per unit testing cost is given by

$$c_1 \cdot \tau,$$

where c_1 is the cost of testing per unit time.

First Interval $[0, \tau]$ In this phase, testers are working independently from the developing team. While testing, they detect the defects and report to the debuggers for immediate removal process. The defect distinction process is not done in this interval as detection and correction process is executed by the organization itself (no external entities, i.e., customers or users are involved). Every fault and vulnerability is considered to be severe and equal by the debuggers. Thus, the total number of defects residing in the system before software release is given by

$$m(\tau) = a \cdot F_1(\tau) \tag{2}$$

where a is the actual number of defects and $F_1(\tau)$ is the distribution rate or removal rate in $[0, \tau]$. The total cost incurred in this phase is only due to the tester's report, and at the developer's end, thus it is given by

$$C_{0,\tau} = c_2 \cdot m(\tau) \tag{3}$$

where c_2 represents the cost of removing the defects in the testing phase.

Second Interval $[\tau, \tau 1]$ After handing over the software, a job of intensive observation of faults and vulnerabilities is left with customers or users [2, 15]. It may also be noted that users are not that much capable of distinguishing between a fault and a vulnerability. But as discussed earlier the cost of discovering vulnerability is much higher than a fault which is ultimately incurred by the developers. Thus for developer's convenience, from this interval the defect distinction process starts for evaluating the debugging cost for faults and vulnerabilities separately. The total number of faults discovered and removed in this interval is represented as:

$$m_f\left(\tau 1-\tau\right)=a\cdot p1\cdot\left(1-F_1\left(\tau\right)\right)\cdot F_2\left(\tau 1-\tau\right) \quad (4)$$

The total number of vulnerabilities discovered and removed in this interval is represented as:

$$\Omega\left(\tau 1-\tau\right)=a\cdot p2\cdot\left(1-F_1\left(\tau\right)\right)\cdot F_3\left(\tau 1-\tau\right) \quad (5)$$

where $F_2(\tau 1-\tau)$ is the fault removal rate, whereas $F_3(\tau 1-\tau)$ is the vulnerability removal rate. The number of residual faults of preceding interval $[0,\tau]$ which are reported in current interval is represented as $a\cdot p1\cdot(1-F_1(\tau))$, while the number of residual vulnerabilities is represented as $a\cdot p2\cdot(1-F_1(\tau))$. The cost incurred after software release for debugging a fault is lesser than the cost of vulnerability removal. The total cost incurred in this phase is represented as

$$C_{\tau,\tau 1}=c_3\cdot m_f\left(\tau 1-\tau\right)+c_4\cdot\Omega\left(\tau 1-\tau\right) \quad (6)$$

where c_3 is the cost of debugging a fault and $c_4$is the cost of debugging a vulnerability.

Third Interval $[\tau 1, Tlc]$ After delivering the maximum number of patches, the software saturation period starts. In this last interval, no more patch is introduced rather the defects are removed for new version. The total number of faults discovered and removed in this interval is represented as:

$$m_f\left(Tlc-\tau 1\right)=a\cdot p1\cdot\left(1-F_1\left(\tau\right)\right)\cdot\left(1-F_2\left(\tau 1-\tau\right)\right)\cdot F_4\left(Tlc-\tau 1\right) \quad (7)$$

The total number of vulnerabilities discovered and removed in this interval is represented as:

$$\Omega\left(Tlc-\tau 1\right)=a\cdot p2\cdot\left(1-F_1\left(\tau\right)\right)\cdot\left(1-F_3\left(\tau 1-\tau\right)\right)\cdot F_5\left(Tlc-\tau 1\right) \quad (8)$$

where $F_4(Tlc-\tau 1)$ is the fault removal rate and $F_5(Tlc-\tau 1)$ is the vulnerability removal rate in $[\tau 1, Tlc]$. The number of residual faults and vulnerabilities from the testing and patching period of software is represented as $a\cdot p1\cdot(1-F_1(\tau))\cdot(1-F_2(\tau 1-\tau))$ and $a\cdot p2\cdot(1-F_1(\tau))\cdot(1-F_3(\tau 1-\tau))$, respectively. Hence, the total cost incurred in this phase is represented as

$$C_{\tau 1,Tlc}=c_5\cdot m_f\left(Tlc-\tau 1\right)+c_6\cdot\Omega\left(Tlc-\tau 1\right) \quad (9)$$

where c_5 is the cost of debugging fault and c_6 is the cost of debugging vulnerability.

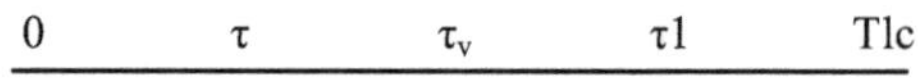

Fig. 2 Software life cycle

By Eqs. 3, 6, and 9, the total amount of money spent in debugging the defects is represented as

$$Total\_Cost_1 = C_{0,\tau} + C_{\tau,\tau 1} + C_{\tau 1,Tlc} \tag{10}$$

3.2 *Model for Multiple Patch Release (M2)*

When vulnerability prioritization is considered, patch specifically for vulnerability is released first and thereafter patch for faults are taken care of. The idea to release vulnerability patch is to ensure that the risks generated through vulnerability exploitation are mitigated on time.

If the developing team finds any vulnerability in the list of reported defects, then they must develop a patch for vulnerabilities first (as vulnerability is more harmful than the faults) and afterward faults are judged. Hence a combine patch for fixing both vulnerability and faults together will be released after delivering the vulnerability patch.

In this case, life cycle of the software (let's suppose $[0, Tlc]$) is fragmented into four phases, i.e., before software release $[0, \tau]$, before first patch release $[\tau, \tau_v]$, before second patch release $[\tau_v, \tau 1]$, and after patching $[\tau 1, Tlc]$ as shown in Fig. 2 where τ_v represents the patch release time for fixing vulnerabilities.

First Interval $[\mathbf{0}, \boldsymbol{\tau}]$ In this interval the total number of defects discovered is given by

$$m(\tau) = a \cdot F_1(\tau) \tag{11}$$

where a is the initial number of defects and $F_1(\tau)$is the defect distribution or removal rate.

Cost incurred in this phase by the developing team is given by

$$C_{0,\tau} = c_2 \cdot m(\tau) \tag{12}$$

where c_2 represents the cost of debugging in the testing phase.

Second Interval $[\boldsymbol{\tau}, \boldsymbol{\tau_v}]$ Since priority has been given to vulnerabilities, no faults are considered simultaneously within this interval. The reported faults are taken care of after mitigating the risks associated with the reported vulnerabilities. The total number of vulnerabilities discovered is

$$\Omega(\tau_v - \tau) = a \cdot p2 \cdot (1 - F_1(\tau)) \cdot F_2(\tau_v - \tau) \tag{13}$$

where $F_2(\tau_v - \tau)$ is the vulnerability removal rate. The number of residual vulnerabilities arriving from testing phase is represented as $a \cdot p2 \cdot (1 - F_1(\tau))$. The total cost incurred in this phase for debugging vulnerabilities is given by

$$C_{\tau,\tau_v} = c_3 \cdot \Omega\,(\tau_v - \tau) \tag{14}$$

where c_3 is the cost of debugging per vulnerability.

Third Interval $[\tau_v, \tau 1]$ Though the users are reporting the faults, we haven't released any patch for faults in preceding interval $[\tau, \tau_v]$. Hence, the faults of preceding and current interval are jointly cured here with the reported vulnerabilities. In this phase, patch for fixing all the faults of $[\tau, \tau_v]$, $[\tau_v, \tau 1]$, and vulnerabilities of $[\tau_v, \tau 1]$ is released at $\tau 1$. The total number of vulnerabilities discovered is represented as

$$\Omega\,(\tau 1 - \tau_v) = a \cdot p2 \cdot (1 - F_1\,(\tau)) \cdot (1 - F_2\,(\tau_v - \tau)) \cdot F_3\,(\tau 1 - \tau_v) \tag{15}$$

where $F_3(\tau 1 - \tau_v)$ is the vulnerability removal rate and $a \cdot p2 \cdot (1 - F_1(\tau)) \cdot (1 - F_2(\tau_v - \tau))$ is the number of residual vulnerabilities. The total cost incurred in this phase for debugging vulnerabilities is given by

$$C_{\tau_v,\tau 1} = c_4 \cdot \Omega\,(\tau 1 - \tau_v) \tag{16}$$

where c_4 is the cost of debugging a vulnerability. Before going further, we have to understand the difference between $a \cdot p1 \cdot (1 - F_1(\tau)) \cdot (1 - G_1(\tau_v - \tau)) \cdot F_4(\tau 1 - \tau_v)$ and $a \cdot p1 \cdot (1 - F_1(\tau)) \cdot F_4(\tau 1 - \tau)$. The first term where $a \cdot p1 \cdot (1 - G_1(\tau_v - \tau))$ are the remaining numbers of faults from $[\tau, \tau_v]$ denotes the total number of bugs detected during $[\tau_v, \tau 1]$ when faults are considered with vulnerabilities in interval $[\tau, \tau_v]$, but though we are considering faults in this interval, it becomes invalid in our proposed model. Thus, the second term is used here to calculate the total number of removed faults and is represented as

$$m_f\,(\tau 1 - \tau) = a \cdot p1 \cdot (1 - F_1\,(\tau)) \cdot F_4\,(\tau 1 - \tau) \tag{17}$$

where $F_4(\tau 1 - \tau)$ is the fault removal rate and $a \cdot p1 \cdot (1 - F_1(\tau))$ is the number of residual faults. The total cost incurred in this phase for debugging faults is given by

$$C_{\tau,\tau 1} = c_5 \cdot m_f\,(\tau 1 - \tau) \tag{18}$$

where c_5 is the cost of debugging per fault.

Fourth Interval $[\tau 1, Tlc]$ The total number of faults discovered and removed in this interval is represented as

$$m_f\,(Tlc - \tau 1) = a \cdot p1 \cdot (1 - F_1\,(\tau)) \cdot (1 - F_4\,(\tau 1 - \tau)) \cdot F_5\,(Tlc - \tau 1) \tag{19}$$

The total number of vulnerabilities discovered and removed in this interval is represented as

$$\Omega\,(Tlc-\tau 1)=a\cdot p2\cdot(1-F_1\,(\tau))\cdot(1-F_2\,(\tau_v-\tau))\cdot\ (1-F_3\,(\tau 1-\tau_v))\\ \cdot F_6\,(Tlc-\tau 1) \tag{20}$$

where $F_5(Tlc-\tau 1)$ is the fault removal rate and $F_6(Tlc-\tau 1)$ is the vulnerability removal rate. The number of residual faults is represented as $a\cdot p1\cdot(1-F_1(\tau))\cdot(1-F_4(\tau 1-\tau))$, while the residual vulnerabilities is represented as $a\cdot p2\cdot(1-F_1(\tau))\cdot(1-F_2(\tau_v-\tau))\cdot(1-F_3(\tau 1-\tau_v))$. The total cost incurred in this phase is represented as

$$C_{\tau 1,Tlc}=c_6\cdot m_f\,(Tlc-\tau 1)+c_7\cdot\Omega\,(Tlc-\tau 1) \tag{21}$$

where c_6 is the cost of debugging faults, whereas c_7 is the cost of debugging vulnerability. In the last phase when software is at obsolete position, the cost of debugging per vulnerability decreases for the purpose of user's attention on new software [16].

By Eqs. 12, 14, 16, 18, and 21, the total amount of money spent in debugging the defect is represented as

$$Total\_Cost_2=C_{0,\tau}+C_{\tau,\tau_v}+C_{\tau_v,\tau 1}+C_{\tau,\tau 1}+C_{\tau 1,Tlc} \tag{22}$$

4 Case Study

In this section, a numerical is illustrated for model validation purpose. We have chosen past defect data set described by Yoshira Tohma [17] which consists of 535 defects which were detected during the testing of 109 days. We have considered that the data set contains both faults and vulnerabilities.

On the basis of assumptions that all the defects are identically and independently distributed, we have used exponential model that is preferable for faults (SRGM, Goel-Okumuto model [18]) and vulnerabilities (VDM, Rescorla exponential model [19]) detection/removal process. Since we had not contrasted between faults and vulnerabilities before software release, thus we have used same distribution throughout the paper.

The defect detection rate of users is always less than that of testers. Let the defect detection rate of tester be b; thus, user fault and vulnerability detection rates are $f\cdot b$ and $v\cdot b$, respectively, where f is the proportion of fault detection rate with respect to testers and v is the proportion of vulnerability detection rate with respect to testers.

Thus, the mean value function for exponential model is $m(t)=a\cdot p1\cdot(1-\exp(-b\cdot t))$ for faults and $m(t)=a\cdot p2\cdot(1-\exp(-b\cdot t))$ for vulnerabilities. After estimating the data set through Statistical Package for the Social Sciences (SPSS), we get initial number of defects $a=713.889$ and defect detection rate of tester $b=0.015$.

4.1 Model for Single Patch Release

From Eq. (2), the total number of defects removed in prerelease phase, i.e., in $[0, \tau]$, is given by

$$m(\tau) = a \cdot (1 - \exp(-b \cdot \tau)) \tag{23}$$

therefore, by Eqs. (3) and (23), we can calculate the $Cost_{0,\tau}$.

From Eqs. (4) and (5), the total number of faults and vulnerabilities removed, respectively, in $[\tau, \tau 1]$ is given by

$$\begin{aligned} m_f(\tau 1 - \tau) = a \cdot p1 \cdot (\exp(-b \cdot \tau)) \cdot \\ (1 - \exp(-b \cdot f \cdot (\tau 1 - \tau))) \end{aligned} \tag{24}$$

$$\begin{aligned} \Omega(\tau 1 - \tau) = a \cdot p2 \cdot (\exp(-b \cdot \tau)) \cdot \\ (1 - \exp(-b \cdot v \cdot (\tau 1 - \tau))) \end{aligned} \tag{25}$$

therefore, by Eqs. (6), (24), and (25), we can calculate the $Cost_{\tau,\tau 1}$.

From Eqs. (7) and (8), the total number of faults and vulnerabilities removed, respectively, in $[\tau 1, Tlc]$ is given by

$$\begin{aligned} m_f(Tlc - \tau 1) = a \cdot p1 \cdot (\exp(-b \cdot \tau)) \cdot (\exp(-b \cdot f \cdot (\tau 1 - \tau))) \\ \cdot (1 - \exp(-b \cdot f \cdot (Tlc - \tau 1))) \end{aligned} \tag{26}$$

$$\begin{aligned} \Omega(Tlc - \tau 1) = a \cdot p2 \cdot (\exp(-b \cdot \tau)) \cdot (\exp(-b \cdot v \cdot (\tau 1 - \tau))) \\ \cdot (1 - \exp(-b \cdot v \cdot (Tlc - \tau 1))) \end{aligned} \tag{27}$$

therefore, by Eqs. (9), (26), and (27), we can calculate the $Cost_{\tau 1, Tlc}$. The total cost for the single-patch model can be calculated by putting all the values in Eq. (10), i.e.,

$$\begin{aligned} \boldsymbol{Total\_Cost}_1 = {} & c_1 \cdot \tau + c_2 \cdot a \cdot \left(1 - e^{-b \cdot \tau}\right) + c_3 \cdot a \cdot p1 \cdot \left(e^{-b \cdot \tau}\right) \cdot \left(1 - e^{-b \cdot f \cdot (\tau 1 - \tau)}\right) \\ & + c_4 \cdot a \cdot p2 \cdot \left(e^{-b \cdot \tau}\right) \cdot \left(1 - e^{-b \cdot v \cdot (\tau 1 - \tau)}\right) \\ & + c_5 \cdot a \cdot p1 \cdot \left(e^{-b \cdot \tau}\right) \cdot \left(e^{-b \cdot f \cdot (\tau 1 - \tau)}\right) \cdot \left(1 - e^{-b \cdot f \cdot (Tlc - \tau 1)}\right) \\ & + c_6 \cdot a \cdot p2 \cdot \left(e^{-b \cdot \tau}\right) \cdot \left(e^{-b \cdot v \cdot (\tau 1 - \tau)}\right) \cdot \left(1 - e^{-b \cdot v \cdot (Tlc - \tau 1)}\right) \end{aligned} \tag{28}$$

4.2 Model for Multiple Patch Release (M2)

From Eq. (11), the total number of defects removed in prerelease phase, i.e., in $[0, \tau]$, is given by

$$m(\tau) = a \cdot (1 - \exp(-b \cdot \tau)) \tag{29}$$

therefore, by Eqs. (12) and (29), we can calculate the $Cost_{0,\tau}$.

From Eq. (13), the total number of vulnerabilities removed in $[\tau, \tau_v]$ is given by

$$\Omega(\tau_v - \tau) = a \cdot p2 \cdot (\exp(-b \cdot \tau)) \cdot (1 - \exp(-b \cdot v \cdot (\tau_v - \tau))) \tag{30}$$

therefore, by Eqs. (14) and (30), we can calculate the $Cost_{\tau,\tau_v}$.

From Eq. (15), the total number of vulnerabilities removed in $[\tau_v, \tau 1]$ is given by

$$\begin{aligned}\Omega(\tau 1 - \tau_v) = a \cdot p2 \cdot (\exp(-b \cdot \tau)) \cdot (\exp(-b \cdot v \cdot (\tau_v - \tau))) \\ \cdot (1 - \exp(-b \cdot v \cdot (\tau 1 - \tau_v)))\end{aligned} \tag{31}$$

therefore, by Eqs. (16) and (31), we can calculate the $Cost_{\tau_v,\tau 1}$.

From Eq. (17), the total number of faults removed in $[\tau, \tau 1]$ is given by

$$\begin{aligned}m_f(\tau 1 - \tau) = a \cdot p1 \cdot (\exp(-b \cdot \tau)) \\ \cdot (1 - \exp(-b \cdot f \cdot (\tau 1 - \tau)))\end{aligned} \tag{32}$$

therefore, by Eqs. (18) and (32) we can calculate the $Cost_{\tau,\tau 1}$.

From Eqs. (19) and (20), the total number of faults and vulnerabilities removed, respectively, in $[\tau 1, Tlc]$ is given by

$$\begin{aligned}m_f(Tlc - \tau 1) = a \cdot p1 \cdot (\exp(-b \cdot \tau)) \cdot (\exp(-b \cdot f \cdot (\tau 1 - \tau))) \\ \cdot (1 - \exp(-b \cdot f \cdot (Tlc - \tau 1)))\end{aligned} \tag{33}$$

$$\begin{aligned}\Omega(Tlc - \tau 1) = a \cdot p2 \cdot (\exp(-b \cdot \tau)) \cdot (\exp(-b \cdot v \cdot (\tau_v - \tau))) \\ \cdot (\exp(-b \cdot v \cdot (\tau 1 - \tau_v))) \cdot (1 - \exp(-b \cdot v \cdot (Tlc - \tau 1)))\end{aligned} \tag{34}$$

therefore, by Eqs. (21), (33), and (34), we can calculate the $\mathrm{Cost}_{\tau 1, Tlc}$. The total cost for the two-patch model can be calculated by putting all the values in Eq. (22), i.e.,

$$
\begin{aligned}
Total\_Cost_2 = {} & c_1 \cdot \tau + c_2 \cdot a \cdot \left(1 - e^{(-b\cdot\tau)}\right) + c_3 \cdot a \cdot p2 \cdot e^{(-b\cdot\tau)} \cdot \left(1 - e^{(-b\cdot v\cdot(\tau_v - \tau))}\right) \\
& + c_4 \cdot a \cdot p2 \cdot \left(e^{(-b\cdot\tau)}\right) \cdot \left(e^{(-b\cdot v\cdot(\tau_v - \tau))}\right) \cdot \left(1 - e^{(-b\cdot v\cdot(\tau 1 - \tau))}\right) \\
& + c_5 \cdot a \cdot p1 \cdot \left(e^{(-b\cdot\tau)}\right) \cdot \left(1 - e^{(-b\cdot f\cdot(\tau 1 - \tau))}\right) \\
& + c_6 \cdot a \cdot p1 \cdot \left(e^{(-b\cdot\tau)}\right) \cdot \left(e^{(-b\cdot f\cdot(\tau 1 - \tau))}\right) \cdot \left(1 - e^{(-b\cdot f\cdot(Tlc - \tau 1))}\right) \\
& + c_7 \cdot a \cdot p2 \cdot \left(e^{(-b\cdot\tau)}\right) \cdot \left(e^{(-b\cdot v\cdot(\tau_v - \tau))}\right) \cdot \left(e^{(-b\cdot v\cdot(\tau 1 - \tau))}\right) \\
& \cdot \left(1 - e^{-b\cdot v\cdot(Tlc - \tau 1)}\right)
\end{aligned}
\tag{35}
$$

Now to find the optimal software and patch release time, both the cases of the single-patch model and two-patch model are considered. Suppose $f = 0.5$, $v = 0.3$, $p1 = 0.9$, $p2 = 1 - 0.9 = 0.1$, and $Tlc = 1000$.

On optimizing Eqs. (28) and (35) by Maple software for the cost function of single-patch model (M1) with cost values $c_1 = 350, c_2 = 260, c_3 = 300, c_4 = 2400$, $c_5 = 310$, and $c_6 = 2300$ and two-patch model (M2) with cost values $c_1 = 350$, $c_2 = 260$, $c_3 = 2400$, $c_4 = 2400$, $c_5 = 300$, and $c_6 = 310$ $c_7 = 2300$, we get the optimal values of release time, patch release time, and developer's cost as shown in Table 1.

As proved from the above results, optimal release time in the case when the model for single patch release is considered is the same as when the model for multiple patch release is considered. Consequently, we propose that both the models are efficient in terms of cost, software release time, and patch release time. The first patch release time is scattered in the two-patch model for the sake of vulnerability prioritization.

The evident table also proves that when patch optimization is done separately for faults and vulnerabilities then the total developer's cost achieved is 7% larger than the proposed models. Also, when no vulnerabilities are considered, the first patch release time is very large, whereas when no faults are considered, the first patch release time is very small, which is not economical and ethical.

The most significant alarming situation arises in the presence of any vulnerability in the software and delaying its fix due to any circumstances is not beneficial for developers [4, 20].

In M1, the developer has to wait for approximately 135 days (4.5 months) to deliver a patch which may lead into economical and physical loss due to

Table 1 Obtained results

Model	Optimal release time	Optimal first patch release time		Optimal cost
M1	134.11	135.15		259972.9
M2	134.11	86.75	48.39	259972.9

Table 2 Phase-wise description for model for single patch release

Phase	Mean value function	No. of defects removed
$[0, \tau]$	$m(\tau)$	618.39
$[\tau, \tau 1]$	$m_f(\tau 1 - \tau)$	54.75
	$\Omega(\tau 1 - \tau)$	4.35
$[\tau 1, Tlc]$	$m_f(Tlc - \tau 1)$	31
	$\Omega(Tlc - \tau 1)$	5

Table 3 Phase-wise description for model for multiple patch release

Phase	Mean value function	No. of defects removed
$[0, \tau]$	$m(\tau)$	618.39
$[\tau, \tau_v]$	$m_v(\tau_v - \tau)$	3
$[\tau_v, \tau 1]$	$\Omega(\tau 1 - \tau_v)$	1.2
$[\tau, \tau 1]$	$m_f(\tau 1 - \tau)$	54.72
$[\tau 1, Tlc]$	$m_f(Tlc - \tau 1)$	31
	$\Omega(Tlc - \tau 1)$	5

vulnerability exploitation. The problem is overcome in model M2, by delivering one patch specifically for vulnerability. In M2, the developer has to wait only for approximately 87 days (2.9 months) for first patch release and thereafter within 48 days patch for fixing faults and vulnerabilities together is delivered.

The validation of model M1 becomes simpler after looking into Table 2 which states the number of faults and vulnerabilities detected and removed in each interval.

From the above table we conclude that there are four vulnerabilities which have been reported in $[\tau, \tau 1]$ interval which signifies if these vulnerabilities are remotely exploitable then prolonged waiting may cause something disastrous.

Let's look at Table 3 description of model M2.

From the above table we conclude that three highly exploitable vulnerabilities were removed within $[\tau, \tau_v]$ interval, i.e., these vulnerabilities are removed before any black hat user tries to breach the security of the software. The number of faults found within $[\tau, \tau 1]$ of model M1 and $[\tau, \tau 1]$interval of model M2 is equal which denotes if the time interval is fragmented for the sake of vulnerability prioritization, then it will not affect faults, but the severity of vulnerability can definitely be reduced.

In the presented demonstration, though the time gap between the last patch and support cycle is very large, i.e., $1000 - 134 - 135 = 731$ days when M1 is considered and $1000 - 134 - 87 - 48 = 731$ days when M2 is considered, the number of vulnerabilities and faults still exists in the software. Hence, more patches are delivered later to make it a bug-free secure software. We can introduce multiple patches until the time frame between last patch and software support cycle goes down to minimum.

5 Conclusions

In this paper we have proposed a mathematical model to determine the optimal software release time and the software patching time so as to minimize the developer's effort. Historically, it has been observed that 40% of the developer's resources are consumed after the release of software. The main reason of resource consumption is the failures which were detected post release. The failure repair process requires patches which take large amount of money, time, and manpower which interrupt developers in focusing on a new project. The complications arise more when fault patch and vulnerability patch are considered separately.

Thus, for improving the optimization, we have proposed a novel method to fix both faults and vulnerabilities together via a single patch. The paper consists of two models M1 and M2; despite of releasing same patch, the models are still said to be different. For vulnerability prioritization, we have distributed the patching time with one more patch in model M2. This is done to meet the challenges associated with vulnerabilities risk exposure to a large extent.

As per the demonstrated results, we can conclude that it is economically beneficial to fix both faults and vulnerabilities together through a single patch rather than fixing them separately. We had distributed the time in model M2, but it has also proven that it will not affect the cost, time, and manpower when comparing with M1 though the possibility of exploitation is definitely be reduced.

If vulnerabilities are reported earlier than faults then model M2 is considered otherwise M1.

This phenomenon can help software firms in acquiring good position in the market and gives opportunity of getting higher sales due to optimal software release. The technical organizations can aid these models to forecast the software release time and budget.

References

1. Schultz Jr EE, Brown DS, Longstaff TA Responding to Computer Security Incidents. Lawrence Livermore National Laboratory. ftp://ftp.cert.dfn.de/pub/docs/csir/ihg.ps.gz, July 23 1990
2. Kapur PK, Pham H, Gupta A, Jha PC (2011) Software reliability assessment with OR applications. Springer, London
3. Ven K, Mannaert H (2008) Challenges and strategies in the use of open source software by independent software vendors. Inf Softw Technol 50(9):991–1002
4. Alhazmi OH, Malaiya YK (2005, November) Modeling the vulnerability discovery process. In 16th IEEE International Symposium on Software Reliability Engineering (ISSRE'05). IEEE, pp 10–pp
5. Arora A, Telang R, Xu H (2004) Optimal time disclosure of software vulnerabilities. In: conference on information systems and technology, Denver CO, October (pp 23-2)
6. Choi JP, Fershtman C (2005) Internet security, vulnerability disclosure and software provision
7. August T, Tunca TI (2006) Network software security and user incentives. Manag Sci 52(11):1703–1720

8. Telang R, Wattal S (2005) Impact of vulnerability disclosure on market value of software vendors: an empirical analysis. In: Proceedings of the fourth annual Workshop on Economics and Information Security (WEIS'06)
9. Okamura H, Tokuzane M, Dohi T (2009). Optimal security patch release timing under non-homogeneous vulnerability-discovery processes. In: 2009 20th International Symposium on Software Reliability Engineering. IEEE, pp 120–128
10. Cavusoglu H, Cavusoglu H, Zhang J (2008) Security patch management: share the burden or share the damage? Manag Sci 54(4):657–670
11. Luo C, Okamura H, Dohi T (2016) Optimal planning for open source software updates. Proc Inst Mech Eng Part O J Risk Reliab 230(1):44–53
12. Dey D, Lahiri A, Zhang G (2015) Optimal policies for security patch management. INFORMS J Comput 27(3):462–477
13. Jiang Z, Sarkar S, Jacob VS (2012) Postrelease testing and software release policy for enterprise-level systems. Inf Syst Res 23(3-part-1):635–657
14. Kapur PK, Shrivastava AK (2015, September). Release and testing stop time of a software: a new insight. In: 2015 4th International Conference on Reliability, Infocom Technologies and Optimization (ICRITO) (Trends and Future Directions). IEEE, pp 1–7
15. Alhazmi OH, Malaiya YK, Ray I (2007) Measuring, analyzing and predicting security vulnerabilities in software systems. Comput Secur 26(3):219–228
16. Foundations, C (1997) Protecting America's infrastructures. The Report of the
17. Tohma Y et al 1990 Parameter estimation of the hyper-geometric distribution model for real test/debug data. Tokyo Institute of Technology
18. Lyu MR (1996) Handbook of software reliability engineering, 222. IEEE Computer Society Press, Los Alamitos
19. Rescorla E (2005) Is finding security holes a good idea? IEEE Secur Priv 3(1):14–19
20. Nizovtsev D, Thursby M (2005). Economic analysis of incentives to disclose software vulnerabilities. In WEIS
21. Kansal Y, Kumar D, Kapur PK, Vulnerability Patch Modeling, Communicated
22. Kansal Y, Singh G, Kumar U, Kapur PK Optimal release and patching time of software with warranty. Int J Syst Assur Eng Manag, 1–7
23. Kapur PK, Garg RB, Kumar S (1999) Contributions to hardware and software reliability. World Scientific
24. Kapur PK, Pham H, Gupta A, Jha PC (2011) Software reliability assessment with OR applications. Springer, London

LC-ELM-Based Gray Scale Image Watermarking in Wavelet Domain

Rajesh Mehta and Virendra P. Vishwakarma

1 Introduction

Digital watermarking offers an effective way out to multimedia security and digital right management in the era of emerging internet and multimedia technologies. Digital watermarking [1–3] is the procedure of inserting the bits of watermark in the cover signal in an imperceptible mode. The main issues in any of the watermarking schemes are imperceptibility, robustness, security, and payload [2–4]. Digital image watermarking is divided into robust and fragile watermarking. For copyright protection and copy protection, applications where high degree of robustness is prerequisite come under the class of robust watermarking, whereas fragile watermarking is generally used for tamper detection (integrity proof) applications. The key focus of all the researchers is to design a robust watermarking scheme which shows resistance toward common signal processing operations and geometric alterations [2–4]. In the existing research methods of digital image watermarking, it is observed that imperceptibility and resistance toward attacks called robustness are more on frequency domain methods [9–12, 14]. In addition to the use of transform domain methods, the applications of artificial neural methods and their variants called machine learning approaches are successfully working onto watermarking applications [5, 7, 9–12] by many researchers to minimize the trade-off between the issues of watermark approaches. In this paper, a newly designed

R. Mehta (✉)
Department of Computer Science, Amity School of Engineering and Technology, New Delhi, India
e-mail: rajesh2010usit@gmail.com

V.P. Vishwakarma
University School of Information and Communication Technology, USICT, Guru Gobind Singh Indraprastha University, New Delhi, India
e-mail: virendravishwa@rediffmail.com

© Springer Nature Singapore Pte Ltd. 2018
P.K. Kapur et al. (eds.), *Quality, IT and Business Operations*, Springer Proceedings in Business and Economics, DOI 10.1007/978-981-10-5577-5_16

machine learning algorithm called LC-ELM by Y. Qu [6] is employed onto image watermarking application. The generality performance of LC-ELM on regression and classification using synthetic datasets which are obtained from UCI repository and against noisy datasets is already examined [6]. With the help of the work presented in this paper, the learning ability and high generality characteristics of LC-ELM onto image watermarking for improving the robustness are examined.

In this paper, a new gray scale watermarking method using LC-ELM, fuzzy entropy, and wavelet decomposition transformed domain is presented. Firstly, the cover (original) image is decomposed using discrete wavelet transform to get approximate and detail subbands. Secondly, select the nonoverlapping blocks of approximate subband based on the randomness (fuzzy entropy) to embed the bits of the watermark. Thirdly, the approximate wavelet coefficients (approximate subband) of the desired blocks are extracted to generate the image dataset for the purpose of training of LC-ELM. Then, a scrambled image of binary watermark bits is inserted into the predicted value (output) attained through the well-trained LC-ELM. The performance in terms of learning capability of the characteristics of image and robustness against attacks of the LC-ELM are examined by the experimental results. The security of the watermark is achieved by Arnold transformation along with imperceptibility and robustness.

The rest of the paper is organized as follows. The related work such as DWT and LC-ELM are discussed in preliminaries in Sect. 2. Proposed gray scale watermarking scheme is described in Sect. 3. The experimental work, their results along with the discussion, is enlightened in the Sect. 4. In Sect. 5, finally the conclusion is described.

2 Preliminaries

In this section, the preliminaries related to the method proposed in this work such as discrete wavelet transform, Arnold transformation, and local coupled extreme learning machine are described.

2.1 Discrete Wavelet Transform (DWT)

The characteristics of discrete wavelet transform such as spatial frequency localization, multiresolution property, etc. make it a very efficient tool in pattern recognition and image processing applications [16]. Approximate and detailed subbands represented by low–low and low–high, high–low, high–high frequency are obtained after one-level decomposition of the original image. The coefficients obtained of approximate subband are low frequency or contain the maximum energy of the signal (coarse level coefficients), and the coefficients of detailed subband are called finest scale wavelet coefficients since the coefficients containing the energy compactness property are robust when attacks are performed as compared to detailed subband coefficients. Therefore, approximate subband coefficients are selected to embed the watermark. It is found in the literature that detailed coefficients which are less robust are inserting the watermark into it which degrade the quality of signed image.

2.2 Arnold Transformation

The periodicity feature of Arnold transformation [15] has made this transformation as widely used in many applications to scramble the squared image. The 2D Arnold transformation of a square image is given as

$$\begin{bmatrix} x_t \\ y_t \end{bmatrix} = \begin{bmatrix} 1 & p \\ p & pq+1 \end{bmatrix} \begin{bmatrix} x_{t-1} \\ y_{t-1} \end{bmatrix} \bmod M \tag{1}$$

where x_t and y_t are the coordinates of scrambled image corresponding to x_{t-1} and y_{t-1} after tth iteration,M is the height of image being processed, and p and q are positive integers ($p = 1, q = 1$).The original position of (x, y) coordinates gets back after T_M iterations due to its periodicity property [15].

2.3 Local Coupled Extreme Learning Machine (LC-ELM)

An artificial neural network (ANN) approach of problem solving is derived from the human brain processing, in which feed forward neural network with single hidden layer (SLFN) architecture which is most commonly applied for solving any nonlinear problem [13]. The most popular algorithm under supervised learning category is back propagation (BP) which is iterative and gradient descent-based learning, which leads to slow convergence, local minima, and over-fitting problems. Extreme learning machine (ELM) [13] provides the non-iterative solution to overcome the slow convergence problem. ELM is simple, fast, and efficient approach having wide applications in regression and classification problems. For further improvement in the performance of ELM, which incorporates ELM training and structural strategies, local coupled ELM (LC-ELM) was proposed by Y. Qu. [6]. The performance of LC-ELM in case of classification and regression problems was investigated on synthetic datasets obtained from UCI repository, which reveals the efficacy of LC-ELM over traditional ELM. In the present investigation, LC-ELM is applied onto image watermarking application to solve the nonlinear regression problem. The mathematical formulation of LC-ELM is presented as

A set of training examples $\{p_q, t_q, q = 1,2,..., Q\}$ is used to train SLFN having the number of hidden layer neuron as N. Here, p_q is the qth sample for training and

t_q is the respective target vector. Consider F: $\Re \rightarrow \Re$ to be a real valued function as activation function of hidden layer, so that $g(\boldsymbol{w}_i{}^T . \boldsymbol{p}_q + b_i)$ is the output of the ith neuron in hidden layer with weight given as

$$w_i = [w_{i1}, w_{i2}, \ldots, w_{is}, \ldots, w_{im}]^T \tag{2}$$

where w_{is} be the weight assigned to sth input and ith neuron of the hidden layer. The bias b_i is element of $\Re$, and input vector $\boldsymbol{p}_q \in \Re^{LM}$ represents the qth feature vector of training set. The inner product of $\boldsymbol{w}_i$ and $\boldsymbol{p}_q$ is represented by $\boldsymbol{w}_i{}^T . \boldsymbol{p}_q$. The SLFN output is given by

$$f(p_q) = \sum_{i=1}^{N} \beta_i F\left(w_i^T . p_q + b_i\right) = o_j, \ \ \forall j = 1, 2, \ldots, L; \tag{3}$$

$$\text{where} \ \ \beta_i = [\beta_{i1}, \beta_{i2}, \ldots, \beta_{ik}]^T \in R^k \tag{4}$$

is the weight vector connecting the ith neuron in hidden layer with the kth neuron of the output layer, O is the predicted output vector of SLFN, and jth output layer neuron is represented by O_j. The linear activation function is selected for output layer neurons. The SLFN can approximate these Q input samples with zero error, if there exist $\boldsymbol{\beta}_i$, $\boldsymbol{w}_i$, and b_i such that

$$\sum_{i=1}^{N} \beta_i g\left(w_i^T . p_q + b_i\right) = t_j, \forall j = 1, 2, \ldots, L \tag{5}$$

The above set of L equations can be rewritten compactly as

$$H\beta = T \tag{6}$$

$$\text{where } H = \begin{bmatrix} F\left(w_1^t . p_1 + b_1\right) \ldots F\left(w_L^t . p_1 + b_L\right) \\ \cdots\cdots \\ \cdots\cdots \\ F\left(w_1^t . p_q + b_1\right) \ldots F\left(w_L^t . p_q + b_L\right) \end{bmatrix}, \tag{7}$$

$$\beta = \begin{bmatrix} \beta_1^T \\ . \\ . \\ \beta_n^T \end{bmatrix} \ \text{and} \ T = \begin{bmatrix} t_1^t \\ . \\ . \\ t_L^t \end{bmatrix} \tag{8}$$

If infinitely differentiable activation function is selected in the hidden layer of SLFN, then the least squares solution $\boldsymbol{\beta}$ ^ of the linear system given by (6) with minimum norm of output weights $\boldsymbol{\beta}$ is obtained [13]. By this process the training error is least and may be reached by the solution $\boldsymbol{\beta}$ ^ given as

$$\beta^{\wedge} = H^{!}T \tag{9}$$

where $H^{!}$ denotes the Moore-Penrose inverse [13] of H. The least squares solution $\hat{\beta}$ of the linear system $H\beta{=}T$ is obtained by the training procedure of SLFN. This is generated due to fixed input weights and biases, and because of the linear activation function at the output layer, the one and only one smallest norm least squares solution of $H\beta{=}T$ is $\hat{\beta} = H^{!}T$.

Local Coupled ELM (LC-ELM) ELM enjoys the full connectivity from input region to feature region, that is, a fully coupled structure is utilized. This generates the computation cost in proportion to the input space and network size. In *LC-ELM*, the computational cost is reduced with the help of local coupling proportion to fuzzy membership proportional to similarity between the associated address (hidden neuron) and given input. In *LC-ELM*, the similarity relation (*SR*) between input and search space is defined with the help of kernel function [6]. Based on this relation, a fuzzy membership (*FM*) function is derived. The output of hidden layer neurons which is given by (5) for ELM is now modified with the help of *FM* as

$$\begin{aligned} &\textstyle\sum_{i=1}^{N} \beta_i g\left(w_i^T . p_q + b_i\right) \times \\ &FM\left(SR\left(p_q, a_i\right)\right) = t_j, \forall j = 1, 2, \ldots, L \end{aligned} \tag{10}$$

where SR(p_q, a_i) is the distance between input and the *i*th hidden neuron (with the address a_i). The various forms of *FM* function [6] are utilized such as Gaussian function, sigmoid and reverse sigmoid, etc.

Algorithm LC-ELM
For a given training set and *N* number of hidden layer neurons of *LC-ELM*,

1. Input weights $\boldsymbol{w}_i$ and bias b_i are randomly allocated, for $i = 1,2,\ldots, N$.
2. The hidden layer output is computed using (10), which is the modified matrix *H*.
3. Evaluate the output weight matrix $\boldsymbol{\beta}\,\hat{}$ using (9) after computing $H^{!}$.

3 The Proposed Watermarking Scheme

Let Im g = {Im g(x, y):$1 \leq x \leq M$, $1 \leq y \leq N$} be the gray scale image of size 512×512 with bit depth *8*. The watermark of size $N1 \times N2$ is a binary image which is used in our method. The proposed watermark embedding and extracting scheme is given as the following:

$LF_{r-2,s-2}$				$LF_{r-2,s+2}$
	$LF_{r-1,s-1}$		$LF_{r-1,s+1}$	
		$LF_{r,s}$		
	$LF_{r+1,s-1}$		$LF_{r+1,s+1}$	
$LF_{r+2,s-2}$				$LF_{r+2,s+2}$

Fig. 1 Selected wavelet coefficients of each block to form dataset

3.1 Embedding Watermark

1. Approximate and detail subband decomposition of cover image is obtained using one-level DWT with *Haar* as wavelet.
2. Divide the approximate subband into blocks of size 5×5 as shown in Fig. 1; the fuzzy entropy [8] of every selected block is calculated. The blocks are grouped into nonincreasing order. The blocks obtained are nonoverlapping.
3. Select first m no. of blocks, and image dataset of significant wavelet coefficients as shown in Fig. 1 of each selected block is formed. The main property of selected wavelet coefficients is that they are perceptually significant because of their energy compactness and robustness to image operations as explained in experimental results and discussion section.
4. The LC-ELM is trained with the help of image dataset DS formed using the selection of prominent features from the designated blocks

$$\mathrm{DS} = \left\{ \begin{array}{l} (x_i, y_i) \in R^8 \times R : i = 1, 2, \ldots, m \\ = \left\{ \begin{array}{l} \left(\begin{array}{l} LF_{r-2,s-2}, LF_{r-2,s+2}, LF_{r-1,s-1}, \\ LF_{r-1,s-1}, LF_{r+1,s-1}, LF_{r+1,s+1}, \\ LF_{r+2,s-2}, LF_{r+2,s+2} \end{array} \right), \\ LF_{r,s} \end{array} \right\} \end{array} \right\} \quad (11)$$

where $LF_{r,s}$ as depicted in Fig. 1 is the target and its diagonal elements (neighbors) are supplied as input vector to *LC-ELM*. Thus a dataset $DS = \{(x_i, y_i : i = 1,2,\ldots,m)\}$ of size $m \times l$ is generated. Here, the total number of blocks is m, and the dimension of feature vector is $l (l = 8)$ for *LC-ELM*. From correlation property, the central coefficient $LF_{r,s}$ can be approximated by the selected elements (diagonal neighbor elements). Using the dataset DS, selected (odd number) samples are used to train the *LC-ELM* with (2) (to compute weights and biases input) and (9) (to calculate weights output), that is, $DS = \{(x_i, y_i : i = 1,3,5,\ldots,m)\}$ acts as input to train the *LC-ELM* corresponding to target vector $y_i = \{LF_{r,s} : i = 1,3,5,\ldots,m\}$.

5. The predicted value corresponding to even number of samples is obtained through the well-trained *LC-ELM* using (9) analogous to target vector $y_i = \{LF_{r,s} : i = 2,4...,m\}$. The scrambled bits of watermark computed using Arnold transformation [15] are embedded into the predicted value of LC-ELM, according to the following rule:

$$\begin{aligned} &\text{if } wm\_bit = 1 \\ &LF\_Wm_{r,s} = \max\left(LF_{r,s}, LF'_{r,s} + \delta\right) \\ &\text{else} \\ &LF\_Wm_{r,s} = \min\left(LF_{r,s}, LF'_{r,s} - \delta\right) \end{aligned} \tag{12}$$

where $LF'_{r,s}$ is the predicted value of trained *LC-ELM*, the modified central value is $LF\_Wm_{r,s}$, the scrambled watermark bit is *wm_bit*, and δ is the strength of the watermark which is obtained after performing a number of experiment ($\delta = 15$).

6. The selected blocks which are used for embedding the watermark are with the original block of approximate subband. Then watermarked (signed) image is reconstructed using inverse DWT. Peak signal-to-noise ratio (PSNR) demarcated by (14) is used to find the perceptual quality of the signed image. High PSNR value justifies that that there is no loss in the perceptual quality of the signed (watermarked) image.

3.2 Watermark Extraction

The watermark is extracted from the signed image as follows:

1. Decompose the watermarked image into approximate and detail subband using one-level DWT.
2. The *m* blocks used in embedding the watermark are chosen using the block index, and step 4 of watermark embedding is applied to generate the watermarked dataset.
3. From this generated dataset, only even number of samples is picked which is supplied to trained *LC-ELM* to obtain the predicted value (output) denoted by $LF'_{r,s}$ of each selected block corresponding to desired output $y_i = \{LF_{r,s} : i = 2,4...,m\}$ of the signed (watermarked) image and recover the binary bits as extracted watermark:

$$SW'_m = \begin{Bmatrix} 1 & \text{if } LF_{r,s} > LF'_{r,s} \\ 0 & \text{otherwise} \end{Bmatrix} \tag{13}$$

4. Extracting the scrambled watermark sequence in a vector form in step 3 is reshaped into the scrambled watermark, and then anti-Arnold transformation is applied to this scrambled watermark to get the original watermark. Then bit

error rate (BER) using (16) is computed between the original and recovered watermark.

4 Results and Discussions

The validity of the watermarked embedding and extracting procedure described in Sect. 3 is tested on gray scale images *Lena*, *Pepper*, and *Cameraman* of size 512×512 with bit depth 8 shown in Fig. 2 through experimental results. The binary watermark of size 32×32 bits shown in Fig. 3 is used in our experiments. The hidden node parameter of *LC-ELM* is randomly selected which lies in the range of [−1, 1], and address of these nodes is given a value which lies in between [0, 1]. The radial basis function is used as the hidden layer activation function of *LC-ELM*. The Gaussian function $F(x) = \exp\left(-\frac{x^2}{r}\right)$ with $r = 0.5$ is used as fuzzy membership function in *LC-ELM*. The resemblance between input and hidden nodes of *LC-ELM* is evaluated using the Gaussian kernel given by $G(x, y) = \exp\left(-\frac{\|x-y\|^2}{\theta}\right)$ with $\theta = 0.5$. The optimal value of r and θ is found by the repeating a no. of experiments on different textured images. The nodes in the hidden layer for performing all the experiments are set to 8. The learning capability of *LC-ELM* and its generalization performance corresponding to the abovementioned parameters are demonstrated by the quality of watermarked image shown in Fig. 4, and the resistance against different kinds of image operations is listed in Table 1 on all the test images which verifies the robustness of the proposed approach.

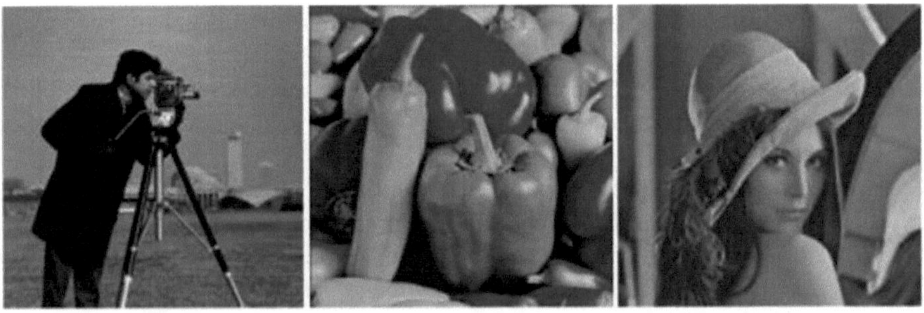

Fig. 2 The original (**a**) *Cameraman*, (**b**) *Pepper*, and (**c**) *Lena* images

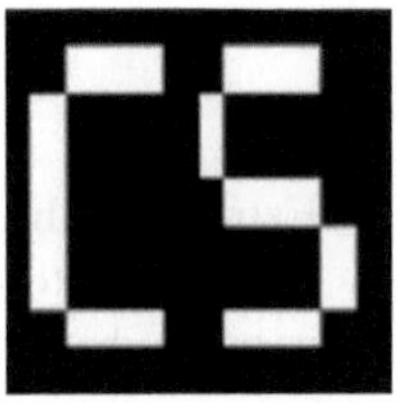

Fig. 3 Original watermark

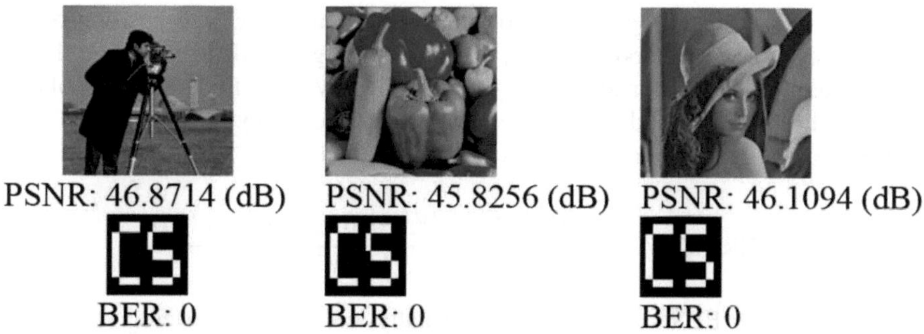

Fig. 4 Watermarked version images corresponding to Fig. 2 and extracted watermark against without attack case

Table 1 BER value of extracted watermark against attacks

Attacks/images	Cameraman	Pepper	Lena
Attack-free	0	0	0
Median filtering	0.1602	0.1357	0.1734
Average filtering	0.1084	0.1309	0.1377
Scaling	0.0371	0.0184	0.0605
JPEG (QF = 80)	0.002	0.0039	0
Cropping from top left	0.0244	0.0137	0.0195
Blurring	0.0049	0.0059	0.0049
Contrast enhancement	0.0098	0.0068	0.0088
Addition of Gaussian noise	0.0195	0.0195	0.041
Addition of salt-and-pepper noise	0.0508	0.0508	0.0996

4.1 Results for Imperceptibility and Robustness

The watermark imperceptibility and resemblance between the original and signed (watermarked) image are measured by PSNR given as

$$\mathrm{PSNR} = 10\log_{10}\frac{255^2}{\mathrm{MSE}} \quad (dB) \tag{14}$$

where MSE is the mean square error between the original and the distorted image and is defined as

$$\mathrm{MSE} = \frac{\sum_{x=1}^{M}\sum_{y=1}^{N}\left(\mathrm{Im}\,g\,(x,y) - \mathrm{Im}\,g'\,(x,y)\right)^2}{M \times N} \tag{15}$$

where Im $g(x, y)$ and $\mathrm{Img}'(x,y)$ denote the pixel value at (x, y)th position of the host and signed (watermarked) image, respectively. High PSNR value implies better visual quality of the signed image. BER is employed to find the resemblance between the original and extracted watermark denoted by W_org and W_ext, respectively against a series of attacks which is given by

$$\mathrm{BER}\left(W, W^{*}\right) = \frac{\sum_{i=1}^{N1} \sum_{j=1}^{N2} W\_\mathrm{org}\,(i,j) \otimes W\_ext\,(i,j)}{N1 \times N2} \tag{16}$$

The lower the BER value, the better visually will be the recovered watermark against image operations. The images after embedding watermark along with recovered watermark against without attack equivalent to all test images are revealed in Fig. 4. From Fig. 4, it is perceived that high PSNR value of all watermarked version images shows the good learning ability of image characteristics of *LC-ELM*. The zero BER value corresponding to without attack of all test images proves the accurate extraction of watermark.

4.2 Results for Robustness

Different kinds of image operations such as median filtering, average filtering, JPEG compression, scaling, cropping, Gaussian blurring, addition of Gaussian noise and addition of salt-and-pepper noise, etc. are performed to examine the robustness by experiments. The results of robustness on the images are reported in Table 1. The visual quality of distorted watermarked image Pepper along with extracted watermark against attacks is shown in Table 2. From Table 1, it is concluded that the lower BER value against all the attacks verifies that the extracted watermark is recognizable against the operations. The experimental results depicted in Table 2 reveal the generalization performance of the LC-ELM onto image watermarking applications.

The main points of attraction of the proposed scheme include (1) the utilization of LC-ELM as regression improves the robustness against attacks as verified by the experimental results, which proves its good generalization performance with low computational cost; (2) selection of approximate subband (low-frequency subband) which provides the imperceptibility of watermark and more robustness to the proposed scheme; and (3) fuzzy entropy [8] which is sensitive to image variations; it is used for selecting smooth nonoverlapping blocks and discards blocks with redundant data.

Table 2 Visual quality of watermarked images after attacks and corresponding extracted watermark of Pepper image

Attacks	Images	Extracted watermark
Median filtering		
Average filtering		
Scaling (512-256-512)		
JPEG (QF = 80)		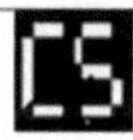
Cropping from top left		
Gaussian blurring		
Contrast enhancement		
Addition of Gaussian noise		
Addition of salt-and-pepper noise		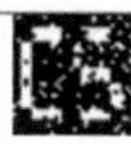

5 Conclusion

The applicability of LC-ELM onto image watermarking for copyright protection applications is discussed in this work. In the proposed scheme, LC-ELM is used as regression mode onto image watermarking to generate the nonlinear relationship among the inputs and targets. The well learning ability of the characteristics of image and the generality performance against attacks of LC-ELM are examined by the experimental results. Low-frequency wavelet coefficients of the blocks selected using fuzzy entropy are carried out to train the LC-ELM, and then with the help of trained LC-ELM, watermark is embedded. Fuzzy entropy [11] is sensitive to image variations; it is used for selecting smooth nonoverlapping blocks and discards blocks with redundant data. Experimental results imply that the method discussed in the work is highly robust against image processing operations and is applicable for copyright protection applications.

References

1. Cox IJ, Kilian J, Leighton FT, Shamoon T (1997) Secure spread spectrum watermarking for multimedia. IEEE Trans Image Process 6(12):1673–1687
2. Kutter M, Peticolas FAP (1999) A fair benchmark image watermarking system. In: Proceedings of Electronic Imaging 99, Security and Watermarking of multimedia contents, vol. 3657, SanJose, pp. 226–239
3. Moulin P, Mincak M (2002) A framework for evaluating the data-hiding capacity of image sources. IEEE Trans Image Process 11(9):1029–1042
4. Chang CC, Hwang KF, Hwang MS (2000) A digital watermarking scheme using human visual effects. Informatica 24:505–511
5. Mehta R, Rajpal N (2013) General regression neural network based image watermarking using fractional DCT-II transform. In Proceedings of IEEE international Conference on Image Information Processing (ICIIP), pp. 340–345
6. Qu Y (2014) Local coupled extreme learning machine. Neural Comput Appl. doi:10.1007/s00521-013-1542-4
7. Shen RM, Fu YG (2005) A novel image watermarking scheme based on support vector regression. J Syst Softw 78:1–8
8. Kumar RR, Das RR, Mishra VN, Dwivedi R (2011) Fuzzy entropy based neuro-wavelet identifier-cum- quantifier for discrimination of gases/odors. IEEE Sensors J 11(7):1548–1555
9. Wen XB, Zhang H (2009) A new watermarking approach based on probabilistic neural network in wavelet domain. Soft Computing 13:355–360
10. Mishra A, Goel N (2014) Novel gray scale image watermarking using extreme learning machine. Int J Sci Eng Res 6(5):737–744
11. Mehta R, Rajpal N, Vishwakarma VP (2015) Robust image watermarking scheme in lifting wavelet domain using GA-LSVR hybridization. Int J Mach Learn Cybern. doi:10.1007/s13042-015-0329-6
12. Mehta R, Rajpal N, Vishwakarma VP (2015) A robust and efficient image watermarking scheme based on Lagrangian SVR and lifting wavelet transform. Int J Mach Learn Cybern. doi:10.1007/s13042-015-0329-z
13. Huang GB, Zhu QY, Siew CK (2004) Extreme learning machine: a new learning scheme of Feedforward Neural Networks. In Proceedings of IEEE International joint conference on Neural Networks, pp. 985–990
14. Gonalez RC, Woods RE, Eddins SL (2005) Digital image processing using MATLAB. Pearson Education
15. Wu L, Deng W, Zhang J, He D (2009) Arnold transformation algorithm and antiArnold transformation algorithm. In: Proceedings of 1st International Conference on Information Science and Engineering (ICISE). pp. 1164–1167
16. Mallat S (1989) The theory for multiresolution signal decomposition: the wavelet representation. IEEE Trans Pattern Anal Mach Intell 11(7):654–693

Implementing and Evaluating R-Tree Techniques on Concurrency Control and Recovery with Modifications on Nonspatial Domains

Rucche Sharrma and Amit Gupta

1 Introduction

A spatial database is a collection of objects that span the same underlying space, where each object may have an arbitrary extent. Usually, spatial databases are very large in data volume. Consequently, a spatial database is organized using spatial access methods that provide efficient access and flexible manipulation of the data. There are many ways to represent and organize a set of objects inside a data structure [1]. These representations have been reviewed earlier [2].

One way is to decompose a spatial object into disjoint cells. Thus more than one entity inside the data structure is used to depict the spatial object. Some example representations include a partition of the spatial object into a collection of convex blocks (the cell tree [3]), a collection of square blocks at predetermined positions (the quadtree [4]), or a collection of rectangles (the R+ − tree [5]). An alternative way is to represent a spatial object by only one entity inside the data structure, e.g. by a point in higher dimensions just like depicting an n-dimensional polygon with k boundary points by a point in nk dimensions, and then store it in a point data structure (e.g. the grid file [6]) or by some conservative approximation of the object as in the case of representing the same polygon by its minimum enclosing rectangle (e.g. the R-tree [7]), with R-tree being the most popular approach.

R. Sharrma (✉)
SOCIS, IGNOU, Delhi, India
e-mail: rucchesharrma@gmail.com

A. Gupta
Maharaja Agarsen Institute of Technology (MAIT), G.G.S. Indraprastha University, Delhi, India
e-mail: amitgupta21@gmail.com

© Springer Nature Singapore Pte Ltd. 2018
P.K. Kapur et al. (eds.), *Quality, IT and Business Operations*, Springer Proceedings in Business and Economics, DOI 10.1007/978-981-10-5577-5_17

2 R-Tree

Spatial database systems utilize a multidimensional index structure, called R-tree [7]. The concept of R-trees states that a minimum bounding box is used to index a spatial object and contains the object. A minimum bounding box is used to represent the data in R-trees and the data can be of any dimension. All the children are bounded by the parent node. One node can contain multiple objects inside it. Actual objects are represented in the leaves but are stored separately on the disk. It is always height balanced with height as log n.

The R-tree [7] is a height-balanced tree that stores the spatial data by using the concept of a minimum bounding (or enclosing) rectangle. Objects are grouped into hierarchies of enclosing rectangles. The R-tree is similar to a B-tree with key values in the B-tree nodes replaced by bounding rectangles. However, unlike a B-tree, the bounding boxes of different nodes of the R-tree may overlap. Unlike other spatial access methods [6, 9–11], R-trees are not limited to storing multidimensional points, but can also store multidimensional spatial objects, which are denoted by their minimal bounding box.

Even now, R-tree structure modifications such as page split or deletion pose a challenge to concurrency and recovery. However, there are more issues with respect to R-tree, mainly:

- Lack of ordering on R-tree keys
- Minimum bounding box resizing
- To implement the above, two questions must be addressed to improve efficiency:
    - How to identify that the node/child has split
    - How to cap the extent to move right

In case of R-trees, when multiple subtrees are traversed down, it can be possible that the same node is visited twice if traversal is done too far to the right. Hence, the second question is very important.

Over the years many improving variations of R-trees have come up which are offering suggestions on cost models. Our proposed approach tries to improve upon this.

3 R-Link Tree

The various refinements and concurrency optimizations that were earlier intended for B-trees, have not provided much improvisation on R-trees. One such variation is the B-link tree [12], where the nodes (siblings) at each level are interconnected using a right link and are used to manage the splits through traversal across the links. In recent times, this process has demonstrated the highest degree of concurrency in comparison with other locking protocols of B-trees [13] and also averts holding locks at the time of input-output operations. This concept in B-link trees works on the principle that key spaces are maintained in a linear order, which makes it irrelevant for R-trees.

Inspired from Yao and Lehman's research, another deviation from R-trees has been developed called R-link trees [14]. Just like B-link trees [15], it proposes a high degree of concurrency and similar locking behavior. Each page is allocated a sequence number which orders the keys linearly and assists in traversal of sibling links in deciding when and how to do the same. A new characteristic of link-style trees is that without a separate reorganization phase, the deletion algorithm removes an empty node.

R-link trees handle unexpected splits almost like B-link trees and traverse across to child nodes at the same level using the link pointers and hold only a few locks at one time for every operation. However, the conceptual design difference between the two is that there is no linear order among spatial keys. Unexpected splits are handled differently in both, where the actual keys of the search are used to settle them in B-link trees, but in the case of R-link, the sequence number allocated to the node is used for the same. Theoretically, the degree of concurrency achieved using R-link trees is equivalent to the B-link tree, the best B-tree algorithm. During traversal to the leaf node, in R-link tree, at one time only, a single node is locked, hence no lock coupling. Invalid pointers created due to node deletion are maintained using a generation number. This is done by allocating and tracking a generation number of when the node deletion takes place. This method ensures that during tree traversal, at the time of online node deletion, there is no need of using lock coupling. Concurrently, many updates of the index are possible as during any delete or insert operation, maximum of only two nodes need to be locked. Also, to ensure that this happens, logical undo needs to be implemented in the recovery mechanism. R-link trees are competent of supporting high throughput and low response times whenever the load increases. This is comparable with R-tree's lock-coupling concurrency mechanism.

4 Proposed Approach: NS Link Tree

This paper proposes a new variant of R-tree, NS link tree. Instead of R-tree which traditionally uses minimum bounding rectangle and R-link trees which use a system of sequence numbers assigned to each node with linked tree components at every level, the research introduced NS link tree (nonspatial) with minimum bounding circle and adding relevant nonspatial data at each point to reduce the number of query results to the database.

In [8], the minimum bounding box, the convex hull, the minimum bounding 4- and 5-corner, the rotated boxes, the ellipses, and the circles were selected and investigated. The result is very interesting. The convex hull has the best approximation quality (124%), followed by the 5-corner (133%), the 4-corner (144%), the rotated box (162%), and the ellipse (168%). Compared to the minimum bounding box (193%), clear gains of approximation quality were obtained in all. In this research minimum bounding circle (MBC) is selected.

4.1 Concurrency Control

Concurrency control can be designed in different manners. It depends on the volume and type of data and the application it is used for. Here the concurrency control is implemented in the following manner:

- The Operations are prioritized as first-come, first-served.
- Simultaneous operations can occur without any problem, if their respective MBCs (minimum bounding circles) don't clash.
- If MBCs are clashing, then the operation which is received first is executed first.
- The remaining operations on the same MBC will have to wait till the previous one is complete.
- Hence, no locking of the entire branch takes place like R-tree and only the active node is locked.

4.2 Recovery Mechanism

Just like concurrency, recovery mechanism can be designed in different manners depending on the type of application where it will be used and the type of data. In this implementation, a separate log file is created. In the log file, record of every transaction (insert and delete) is maintained. Every entry made is log file constitutes of the address of the node where insertion and deletion are taking place along with the data to be inserted/deleted. In this implementation the log file stores 100 transactions. However, the same can be altered depending on the application where it is used. If the tree has crashed, this can be used to reconstruct the tree or undo the change.

4.3 Algorithm

The following algorithms are used in the implementation:

Add_datapoint()

1. Chooseleaf() will give the leaf node in which the data point is to be inserted.
2. Insert() will insert the data point in the leaf node selected and split it if needed.

Chooseleaf()

1. Choose the mbc in which there is minimum increase in area to insert the data point. In case of conflict, choose the one having smaller area.
2. Change the attributes of the selected mbc.

3. Recursively run chooseleaf() on child node of the selected mbc till a leaf node is found.
4. Return the leaf node.

Insert() If the node is not full:

1. Insert data point/mbc.

   Else: //in case of split

1. If the node is root node, create a new node and add the mbc of the node which is to be split.
2. Find the points/mbc to be grouped together and obtain two sets of the same (using clustering or sorting based).
3. Update the previous node and its mbc with one set of points.
4. Create another node with the second set of mbc/points and create its mbc.
5. Recursively run this function again to add the mbc of newly created node to the parent node.

Sorting-Based Algorithm for Grouping

1. Sort the M + 1 data points or the centroids of the mbc on the basis of their x and y coordinate separately.
2. Find the sum of area of the mbc formed by the smallest two points and the largest two points on x axis.
3. Do step 2 for the y axis also.
4. Compare the sum of area obtained in steps 2 and 3, and the one which is smaller will give the two sets required.
5. Add the remaining points one by one to either the set containing the smallest two points or the largest two points based upon in which set the increase in area of the mbc is smaller. In case the increase is same, choose the one with smaller area of mbc.
6. Return the two sets.

Clustering-Based Algorithm

1. Select k points(=2) as initial centroids.
2. Do until centroids do not change:

   (i) Form k clusters by assigning each point to its closest centroid using Euclidean distance formula.
   (ii) Recompute the centroid of each cluster.

   Distance formula: sqrt[(x2−x1)^2 + (y2−y1)^2 + (z2−z1)^2 ....].

Search()

- Do until leaf node is not reached:

  1. For each mbc of the node, check whether the search query overlaps the area of the mbc.

2. For all overlapping mbcs, invoke search() on the subtree of that mbc.

- If leaf node is reached, for each data point in the leaf node, use the nonspatial attribute to check if the data point is qualified for the query or not.

4.4 Implementation

For implementation, Python language is used. The environment and computing resources are not standardized and may vary as per the scalability of the database.

Publically available data set of airports all across the world is used in this implementation. It contains over 6600 records. The structure of the data is as follows:

Each entry in the database comprises of the following information:

- *Airport ID*: Unique open flight identifier for an airport.
- *Name*: Name of airport, which may or may not contain the *City* name.
- *City*: Main city served by airport that can be spelled differently from *Name*.
- *Country*: Country or territory where airport is located.
- *IATA/FAA*: 3-letter FAA code, for airports located in *Country* "United States of America." 3-letter IATA code, for all other airports. Blank if not assigned.
- *ICAO*: 4-letter ICAO code. Will show blank if it is not assigned.
- *Latitude*: Decimal degrees, usually to six significant digits. Negative is south and positive is north.
- *Longitude*: Decimal degrees, usually to six significant digits. Negative is west and positive is east.
- *Altitude*: In feet.
- *Time zone*: Hours offset from UTC. Fractional hours are expressed as decimals, e.g. India is 5.5.
- *DST*: Daylight savings time. One of E (Europe), A (US/Canada), S (South America), O (Australia), Z (New Zealand), N (none), or U (unknown).

The relevant nonspatial element used in the implementation for this data is the field, DST. This provides categorization of all records in the database (Table 1).

4.5 Results

The performance of the NS link tree over an R-tree is measured by the number of links generated to search into the database and illustrated through these equations:

Final links to database for search:

$$\text{In case of R} - \text{tree} : N^{*}M \tag{1}$$

$$\text{In case of NS link tree (nonspatial)} : S^{*}N^{*}M \tag{2}$$

Table 1 Shows the results of the tests for many search queries

Query	Links generated by R-tree	Links generated by NS link tree
Search all "E" airports between the coordinates (0,0) to (20,40)	124	5
Search all "U" airports between the coordinates (0,0) to (20,40)	124	97
Search all "A" airports between the coordinates (−100,−100) to (10,0)	657	5
Search all "Z" airports between the coordinates (30,50) to (30,50)	193	0
Search all "N" airports between the coordinates (60,23) to (75,60)	78	43
Search all "U" airports between the coordinates (−10,−10) to (−10,−30)	16	4
Search all "A" airports between the coordinates (−23,−20) to (−45,−50)	42	0

Where:

N: No. of leaf nodes which are overlapping with the search query area.
M: No. of data points a leaf node can store.
S: Percentage of records in the search area fulfilling the nonspatial conditions. This is out of N*M. S < =1 always.

Values returned by Eq. 2 will always be less than or equal to Eq. 1. To simplify:

- If the number of nonspatial categories is Z, then considering the homogeneous distribution of data according to nonspatial categories in the search area, on an average the resultant number of links to the database would be 1/Zth of the number of links obtained by R-trees.
- In case of nonhomogeneous distribution, the resultant may further reduce or may not be 1/Zth but will always be less than R-tree.
- The only scenario when the two results will be equal is when there is only one category of data in the search area which makes it the worst case scenario. However, that may only be possible if the search area is very narrow or the data is inadequate.

And so the list continues.

The table clearly shows that the number of links generated through NS link is much less than the ones generated through a regular R-tree (Fig. 1).

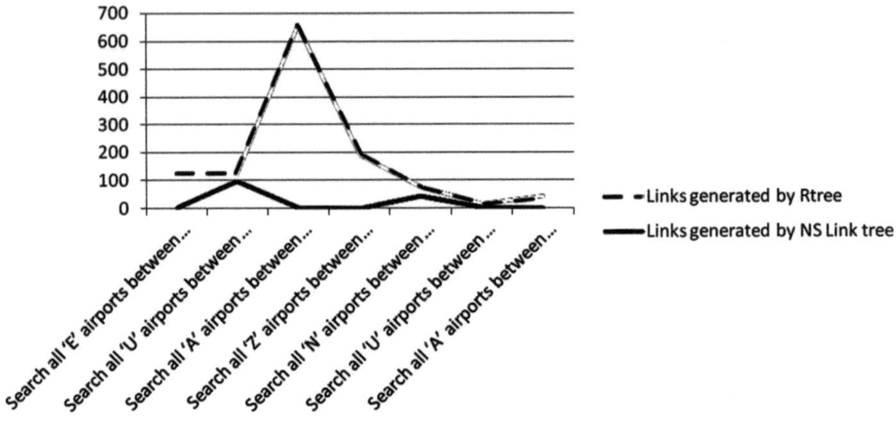

Fig. 1 Shows the pictorial representation of the difference between the number of links generated through R-tree and NS link tree

5 Summary

NS link trees are an extension on R-trees and support high concurrency. They are better than R-trees. Incorporating the related nonspatial data in the NS link tree with the minimum bounding circle significantly improves the number of links to the database for searching and thus speeds up the performance.

Some performance study may analyze that the resultant overhead in keeping a nonspatial element for every node in memory will increase the overhead of the tree. However, the same can be considered as a trade-off in terms of lesser records to be searched in the database, and if the database is very large, then this actually works in favor.

Acknowledgment I would like to thank Dr. Amit Gupta for inspiring me and having faith in me to complete this. I would also like to acknowledge the contribution of one of my former students, Parteek Singhal, Software Engineering, Delhi Technological University, whose questions on the concepts made me brainstorm further into the topic. He was also the person who worked on the implementation and improvised the entire code.

References

1. Kriegel HP, Heep P, Heep S, Schiwiets M, Schneider R (1992) An access method based query processor for spatial database systems. In: Gambosi G, Scholl M, Six H-W (eds) Geographic Databose management systems. Workshop Proceedings, Copri, Italy, Mug 1991, pages 19-t-211, Springer-Verlag, Berlin
2. Walid G, Aref, HS (1994) A cost model for query optimization using R-Trees. CIKM'94 Workshop 12/94, Gaithersburg, MD, USA @ 1994 ACM

3. Gunther O (1989). The design of the cell tree: an object oriented index structure for geometric databases. In Proceedings of the Fifth IEEE' International Conference on Dots Engineering, pp 593-005, Los Angeles
4. Klinger A (1971) Patterns and search statistics. In: Rustagi JS (ed) Optimising methods in statistics. Academic Press, New York, pp 303–337
5. Faloutsos C, Sellis T, Roussopoulos N (1997) Analysis of object oriented spatial access methods. In Proceedings of the 1987′ ACM SIGHCD International Conference on Management of Dots, pages 420–439, San Francisco
6. Nievergelt J, Hinterberger H, Sevcik KC (1984) The grid file: an adaptable, symmetric multikey file structure. ACM Trans Dotobose Syst 9(1):38–T1
7. Guttman A (June 1984) R-trees: a dynamic index structure for spatial searching. In Proceedings of the 1984 ACM SIGMOD International Conference on Management of Doro, pp 47–57, Boston
8. Brinkhoff T, Kriegel H-P, Schneider R Comparison of approximations of complex objects used for approximation-based query processing in spatial database systems. Institute for Computer Science, University of Munich, Germany
9. Bentley JL (1975) Multidimensional binary search trees used for associative searching. CACM 18(9):509–517
10. Robinson JT (1981) The K-D-B-Tree: a search structure for large multidimensional dynamic indexes. Proc. 1981 ACM SIGMOD Conf., pp 10–18
11. Lomet D, Salzberg B (1990) The hB-tree: a Multiattribute indexing method with good guaranteed performance. ACM TODS 15(4):625–685
12. Lehman P, Yao S (1981) Efficient locking for concurrent operations on B-Trees. ACM TODS 6(4)
13. Johnson T, Shasha D (1993) The performance of current B-tree algorithms. ACM TODS 18(1):51–101
14. Banks D, Kornacker M, Stonebraker M (June 1994) High-concurrency locking in R-Trees. Sequoia 2000 Technical report 94/56
15. Kornacker M Douglas Banks: high-concurrency locking in R-Trees. Proceedings of the 21st VLDB Conference. Proceedings of the 21st VLDB Conference

Inventory Decisions for Imperfect Quality Deteriorating Items with Exponential Declining Demand Under Trade Credit and Partially Backlogged Shortages

Chandra K. Jaggi, Prerna Gautam, and Aditi Khanna

1 Introduction

Recent advancements and developments in the field of inventory management have driven plenty of researchers toward expanding their horizon in exploring various realistic inventory situations, with the motto to make a better and versatile model that deals with real-time system. Thus, it is mandatory to build inventory models that exhibit the real inventory situation. Out of many aspects, deterioration is one such factor which should not be snubbed for making a realistic inventory model, because there is loss of inventory due to deterioration and the value of the inventory depends upon worth of the item at the time of evaluation. Commonly, deterioration process can be explained in terms of product damage, decay, spoilage, evaporation, pilferage, obsolescence, etc. which ultimately declines the utility of the original product. At first, [1] investigated an inventory model assuming exponential decay with deterioration. Reference [2] further explored the model for items that deteriorate and the rate of deterioration to follow Weibull distribution. Afterward, a number of research articles appeared in different journals like [3–5], etc. that were summarized by [6]. References [7, 8] also contributed in the field of deterioration.

Recently, [9] provided a review of inventory literature that considered deterioration. It is believed that the assumption of taking demand rate to be constant does not hold in day-to-day practice. In fact, inventory products (viz., electronic products, fashionable goods, etc.) face fluctuations in their demand patterns oftentimes. A number of products experience expanding demand through their growth stage. However, the demand of few items tends to decrease with the advent of attractive and better products which influences the preference of customers. Also, the physical loss

C.K. Jaggi • P. Gautam • A. Khanna (✉)
Department of Operational Research, Faculty of Mathematical Sciences, University of Delhi, Delhi, 110007, India
e-mail: ckjaggi@yahoo.com; prerna3080@gmail.com; dr.aditikhanna.or@gmail.com

© Springer Nature Singapore Pte Ltd. 2018

P.K. Kapur et al. (eds.), *Quality, IT and Business Operations*, Springer Proceedings in Business and Economics, DOI 10.1007/978-981-10-5577-5_18

of materials and decreasing customer assurance in the product quality negatively affect the demand. This scenario provoked researchers to propose inventory models that incorporated deterioration and varying demand patterns.

The two popular types of varying demands which were considered by research experts were continuous and discrete time. Under the category of inventory models with continuous time, mostly the demand patterns are assumed to vary either linearly or exponentially. Reference [3] developed an instantaneous replenishment inventory model with deterioration and demand to be proportional to time. The first one to propose an inventory model incorporating demand rate to be decreasing exponentially was done by [4].Reference [8] presented a generalized version of [4] by implementing both exponentially inclining and declining demand. References [10, 11] presented a deterministic model with deterioration where demand decreases exponentially.

Later, [12] exhibited optimal method recommended by [10] to be independent of the rate of demand. Reference [13] proved "Newton's method" incorporated in [11] to be inadequate and applied it again with suitable changes. Reference [14] developed a scenario considering deterioration and exponentially decreasing demand.

Further, it remains worthless to consider the quality of items to be perfect as it is difficult to procure/produce perfect items. This way it becomes necessary to incorporate a screening process so as to distinguish the good-quality and bad-quality items. Motivated with this, plenty of research in the field of imperfect quality was carried out. Reference [15] incorporated the impact of defectives in the economic order quantity model. Reference [16] considered the time lag "between the in-control and out-of- control state is exponential," and the nonconforming items can be reworked at some cost.

Reference [17] extended the work done for imperfect quality and contradicted with the findings of [16] that the lot size decreases with increase in the number of defectives. Reference [18] investigated the subject further by considering the defective proportion to be random. Reference [19–21] developed some interesting work in this area.

Furthermore, trade credit is necessary for both, the supplier and retailer in terms of financial benefits. It is a well-known strategy to elevate the sales and acquire profits. Reference [22–26] presented some of the pioneer work in this category, and since then plenty of research was done in this field, which was later summarized by [27]. Reference [28] presented an optimal replenishment model considering items of imperfect quality with trade credit.

The abovementioned models were developed under the assumption that shortages are not permissible. However, in common practice, stockout cannot be avoided due to various supply and demand suspicions. Therefore, it becomes reasonable to incorporate the shortages in inventory model. References [29, 30] developed models for deteriorating items with time-varying and linear trend in demand. Reference [31] developed optimal inventory model with shortages and imperfect quality items. Reference [32] incorporated imperfect quality items with backlogging and obtained closed-form expressions. Further, [33, 34] discussed economic ordering policies for defective items with shortages and trade credit. Recently, [35] developed an inventory model that explores the defective and deteriorating items under trade credit environments with allowable shortages and demand to be dependent on price.

As mentioned earlier, nearly all researchers in the past have considered that shortages are fully backlogged. But in a realistic environment, only some customers wait for the item during the shortage period. Therefore, it becomes pertinent to include the opportunity cost owing to lost sales in the modeling framework. Also, several researchers have used fixed backlogging rates while developing the inventory models. In some cases, for instance, in the inventory system for clothing industry, the duration of waiting time for the subsequent replenishment acts as a prime factor to decide whether the demand will be backlogged or not. It has been observed that the more the waiting period, the smaller is the pace of backlogging and also holds conversely. Thus, the rate of backlogging is not constant but is a function of waiting time for the subsequent replenishment. In the literature [36–38] have developed the optimal replenishment model with shortages. Throughout the interval of shortages, the backlogging rate is dynamic and is a function of waiting time interval.

In the present study, the inventory model is formulated to cater the problem of the retailer that deals with imperfect-quality deteriorating products whose demand decreases exponentially in trade credit environment. The shortages have been taken into consideration with partially backlogging. Also, the rate of screening is considered to be more as compared to demand rate. This is because it helps in meeting the demand and back orders with perfect-quality products in parallel to a screening process. The model mutually optimizes the shortage point and cycle length in order to maximize the retailer's overall profit function. The model is appropriate for items such as fashionable and seasonal products, electronic equipments, and mobile phones. These items face variations in their demand rate due to the introduction of new technology that eventually alters customer's preferences.

2 Assumptions and Notations

1. The demand rate, $R(t)$, is known and decreases exponentially. $R(t) = \begin{cases} Ae^{-\gamma t}; I(t) > 0 \\ D; I(t) \leq 0 \end{cases}$
   where $A(> 0)$ is initial demand, and $\gamma(0 < \gamma < \theta)$ is a constant representing the declining demand rate.
2. The defectives, α, are considered, with p.d.f. $f(\alpha)$
3. Fraction θ of inventory gets deteriorated per unit time.
4. A fixed delay period M is offered to the retailer by the supplier.
5. Process of screening and demand continues at the same time, but $\lambda > D$
6. Defectives are free from deterioration.

7. During stock-out time, the rate of backlogging varies and depends upon the length of the waiting time for subsequent replenishment. The fraction of customers who are willing to accept backlogging at time t will decrease with the waiting time $(T - t)$ waiting for the succeeding replenishment. Hence, backlogging rate is given as $\frac{1}{1+\delta(T-t)}$ when inventory is negative. The backlogging parameter δ is a positive constant, $t_1 \le t \le T$.

K	Ordering cost
c	Retailer's purchasing price
h	Holding cost
θ	Inventory loss due to deterioration
α	Average proportion of defectives
λ	Rate of screening
β	Screening cost
t_s	Time of screening
Q	Quantity ordered
T	Length of inventory cycle
t_1	Point when shortages start to build
D	Rate of demand
I_e	Interest made
I_p	Interest payable
B	Maximum permissible demand to be backlogged
p	selling cost
$f(\alpha)$	Probability density function of α
$E(\alpha)$	$= \int_a^b \alpha f(\alpha)\, d\alpha, 0 < a < b < 1$
M	Credit period
C_2	Shortage cost
C_s	Salvage cost
C_L	Cost of lost sales per unit
$I_1(t)$	Level of inventory for $(0,t_s)$
$I_2(t)$	Level of inventory for (t_s,t_1)

3 Mathematical Model

An inventory scenario is considered with size of lot Q which is delivered to the retailer, purchase cost is c per unit, and cost of ordering is K. The received lot is assumed to contain defectives, α with p.d.f., $f(\alpha)$ After receiving the consignment, full inspection of the lot is carried out at λ rate. The defectives separated from the good-quality lot are retained in stock and are sold out as solo batch at a reduced price C_s, where $C_s < c$. The rate of demand, $R(t)$, is deterministic and decreases exponentially, $R(t) = Ae^{-\gamma t}$, where A is the opening demand and $\gamma(0 < \gamma < \theta)$ is a constant which represents the decreasing demand rate, θ proportion deteriorated, t_s time of screening, and T the cycle length. During time interval (0, t_s) the inventory level $I_1(t)$ depletes because of the combined effect of deterioration and demand. At time t_s, screening is performed, and the level of inventory is reduced by the number of defectives secluded, αQ and the back orders B, the backlogged demand which gets satisfied from the inventory. For the time interval (t_s, t_1), the level of inventory $I_2(t)$ decreases owing to demand and deterioration and reaches zero at time t_1 Also, for the interval (t_1, T), partial backlogging is made.

Let $I_1(t)$ be the inventory level for period $0 \le t \le t_s$.

Differential equation for interval (0, t_s) is

$$\frac{dI_1(t)}{dt} + \theta I_1(t) = -Ae^{-\gamma t}, 0 \le t \le t_s \tag{1}$$

The above differential equation is solved with the boundary condition, t=0, $I_1(0) = Q$

$$I_1(t) = \frac{A}{\theta - \gamma}\left(e^{-\theta t} - e^{-\gamma t}\right) + Qe^{-\theta t} \tag{2}$$

With the completion of screening, the number of defectives at time t_s is αQ and backlogged demand is B.

Further, the inventory in effect at $t = t_s$, after secluding defectives and backorders, is

$$I_{\text{eff.}}(t_s) = \frac{A}{\theta - \gamma}\left(e^{-\theta t_s} - e^{-\gamma t_s}\right) + Qe^{-\theta t_s} - \alpha Q - B \tag{3}$$

Now, let $I_2(t)$ be level of inventory for ($t_s \le t \le t_1$).

Differential equation for period (t_s, t_1) is

$$\frac{dI_2(t)}{dt} + \theta I_2(t) = -Ae^{-\gamma t}, t_s \le t \le t_1 \tag{4}$$

The above differential equation is solved with the boundary condition $t = t_s$, $I_2(t_s) = I_{eff.}(t_s)$, given by

$$I_2(t) = \frac{A}{\theta - \gamma}\left(e^{-\theta t} - e^{-\gamma t}\right) + Qe^{-\theta t} - \alpha Qe^{\theta(t_s - t)} - Be^{\theta(t_s - t)} \tag{5}$$

The equation is solved for Q with condition $t = t_1$, $I_2(t_1) = 0$ (Fig. 1)

$$Q = \frac{\frac{A}{\theta-\gamma}\left(e^{-\theta t_1} - e^{-\gamma t_1}\right) - Be^{\theta(t_s - t_1)}}{\alpha e^{\theta(t_s - t_1)} - e^{-\theta t_1}} \tag{6}$$

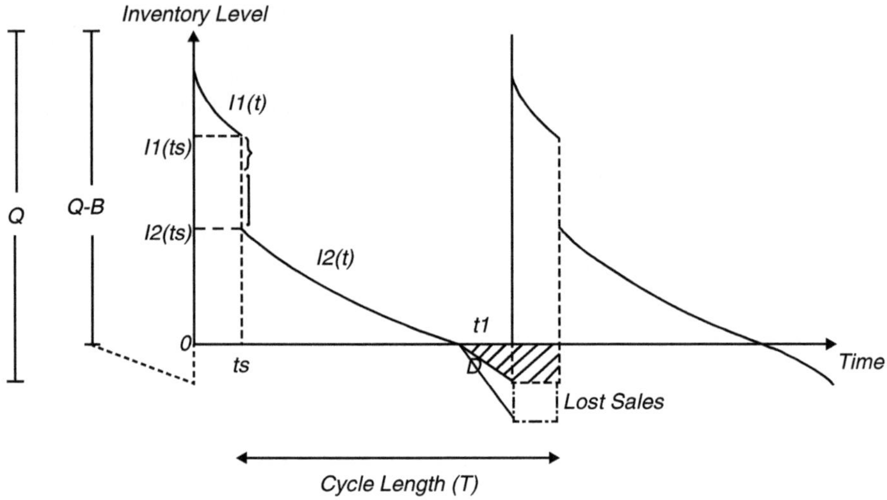

Fig. 1 Level of inventory over time T

$$t_s = \frac{\frac{A}{\theta-\gamma}\left(e^{-\theta t_1} - e^{-\gamma t_1}\right) - Be^{-\theta t_1}}{\lambda\alpha e^{-\theta t_1} - \lambda e^{-\theta t_1} + B\theta e^{-\theta t_1}} \tag{7}$$

During the shortage period $[t_1, T]$, demand at time t is partly backlogged as $\frac{1}{1+\delta(T-t)}$. The differential equation is given as

$$\frac{dB(t)}{dt} = -\frac{D}{1+\delta(T-t)}, t_1 \leq t \leq T \tag{8}$$

$$B(t) = \frac{D}{\delta}\{\ln[1+\delta(T-t)] - \ln[1+\delta(T-t_1)]\} \tag{9}$$

Let $t = T$ in (8); the maximum demand which is backlogged can be obtained as

$$B(t) = \frac{D}{\delta}\ln[1+\delta(T-t_1)] \tag{10}$$

The model incorporated trade credit; hence, pertaining to delay period, three different possible cases are discussed for the profit expression of the retailer π_j, $j = 1, 2, 3$, *viz.*

Case 1: $0 \leq M \leq t_s$
Case 2: $t_s \leq M \leq t_1$
Case 3: $t_1 \leq M \leq T$

Following are the components of the retailer's total profit expression (Fig. 2):

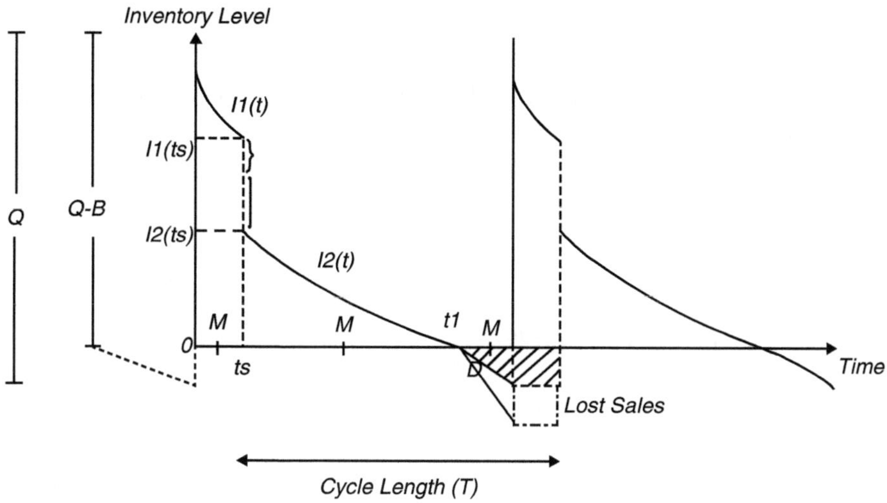

Fig. 2 Level of inventory over time T

$$\begin{aligned}\pi_j(t_1, T) = \; & \text{Sales Revenue} - \text{Ordering cost} - \text{Purchase cost} \\ & -\text{HoldingCost} - \text{Screening Cost} - \text{Shortage cost} - \text{Cost of lost sales} \\ & +\text{Interest Earned} - \text{Interest Paid}\end{aligned} \tag{11}$$

All the components of retailer's total profit are discussed below:

1. The total revenue from sales is obtained by selling good-quality and imperfect-quality items:

$$= p(1-\alpha)Q + C_s\alpha Q \tag{12}$$

2.

$$\text{Ordering cost} = K \tag{13}$$

3.

$$\text{Purchase cost} = cQ \tag{14}$$

4.

$$\text{Screening cost} = \beta Q \tag{15}$$

5.

$$\text{Shortage cost} = C_2 D\left\{\frac{T-t_1}{\delta} - \frac{1}{\delta^2}\ln\left[1+\delta(T-t_1)\right]\right\} \tag{16}$$

6.

$$\text{Cost of lost sales} = C_L D\left\{(T - t_1) - \frac{1}{\delta}\ln\left[1 + \delta\,(T - t_1)\right]\right\} \tag{17}$$

7. Cost of holding for the time interval 0 to t_s and t_s to t_1

$$=h\left[\int_0^{t_s} I_1(t)dt + \int_{t_s}^{t_1} I_2(t)dt\right]$$

$$=h\left[\begin{array}{l}\left\{\frac{A}{\theta-\gamma}\left[\frac{1}{\gamma}\left(e^{-\gamma t_s}-1\right)-\frac{1}{\theta}\left(e^{-\theta t_s}-1\right)\right]-\frac{Q}{\theta}\left(e^{-\theta t_s}-1\right)\right\}\\ +\left\{\begin{array}{l}\frac{A}{\theta-\gamma}\left[\frac{1}{\gamma}\left(e^{-\gamma t_1}-e^{-\gamma t_s}\right)-\frac{1}{\theta}\left(e^{-\theta t_1}-e^{-\theta t_s}\right)\right]\\ -\frac{Q}{\theta}\left(e^{-\theta t_1}-e^{-\theta t_s}\right)+\frac{\alpha Q e^{\theta t_s}}{\theta}\left(e^{-\theta t_1}-e^{-\theta t_s}\right)+\frac{Be^{\theta t_s}}{\theta}\left(e^{-\theta t_1}-e^{-\theta t_s}\right)\end{array}\right\}\end{array}\right] \tag{18}$$

To determine expressions for interest to be paid and earned, three possible cases are derived:

Case 1: $t_s \le M \le t_1$

For this case, interest can be earned from the revenue generated by selling the items up to the delay period, M, though the accounts require settlement at M, which further requires the arrangement of money at some identified interest rate so as to get the remaining stocks financed for the time period M to T.

8.

$$\text{Interest earned} = pI_e\int_0^M R(t)tdt = pI_e\left[\left(e^{-\gamma M} - 1\right)\left(\frac{-M}{\gamma} - \frac{1}{\gamma^2}\right)\right] \tag{19}$$

9.

$$\text{Interest payable} = cI_p\left[\int_M^{t_s} I_1(t)dt + \int_{t_s}^{t_1} I_2(t)dt\right]$$

$$= cI_p\left[\begin{array}{l}\left\{\frac{A}{\theta-\gamma}\left[\frac{1}{\gamma}\left(e^{-\gamma t_s}-e^{-\gamma M}\right)-\frac{1}{\theta}\left(e^{-\theta t_s}-e^{-\theta M}\right)\right]-\frac{Q}{\theta}\left(e^{-\theta t_s}-e^{-\theta M}\right)\right\}\\ +\left\{\begin{array}{l}\frac{A}{\theta-\gamma}\left[\frac{1}{\gamma}\left(e^{-\gamma t_1}-e^{-\gamma t_s}\right)-\frac{1}{\theta}\left(e^{-\theta t_1}-e^{-\theta t_s}\right)\right]-\frac{Q}{\theta}\left(e^{-\theta t_1}-e^{-\theta t_s}\right)\\ +\frac{\alpha Q e^{\theta t_s}}{\theta}\left(e^{-\theta t_1}-e^{-\theta t_s}\right)+\frac{Be^{\theta t_s}}{\theta}\left(e^{-\theta t_1}-e^{-\theta t_s}\right)\end{array}\right\}\end{array}\right] \tag{20}$$

Substitute the values from (12) to (20) in (11), the total profit expression per unit time per cycle for Case 1, $\pi_1(t_1, T)$.

$$
\begin{aligned}
\pi_1(t_1, T) = \tfrac{1}{T} [&p(1-\alpha)Q + C_s\alpha Q - K - cQ - \beta Q \\
&-C_2 D\left\{\tfrac{T-t_1}{\delta} - \tfrac{1}{\delta^2}\ln[1+\delta(T-t_1)]\right\} \\
&-C_L D\left\{(T-t_1) - \tfrac{1}{\delta}\ln[1+\delta(T-t_1)]\right\} \\
&-h\left[\left\{\tfrac{A}{\theta-\gamma}\left[\tfrac{1}{\gamma}(e^{-\gamma t_s}-1) - \tfrac{1}{\theta}\left(e^{-\theta t_s}-1\right)\right] - \tfrac{Q}{\theta}\left(e^{-\theta t_s}-1\right)\right\}\right. \\
&+\left\{\tfrac{A}{\theta-\gamma}\left[\tfrac{1}{\gamma}(e^{-\gamma t_1}-e^{-\gamma t_s}) - \tfrac{1}{\theta}\left(e^{-\theta t_1}-e^{-\theta t_s}\right)\right] - \tfrac{Q}{\theta}\left(e^{-\theta t_1}-e^{-\theta t_s}\right)\right. \\
&+\tfrac{\alpha Q e^{\theta t_s}}{\theta}\left(e^{-\theta t_1}-e^{-\theta t_s}\right) + \tfrac{B e^{\theta t_s}}{\theta}\left(e^{-\theta t_1}-e^{-\theta t_s}\right)\}] \\
&+pI_e\left[\left(e^{-\gamma M}-1\right)\left(\tfrac{-M}{\gamma} - \tfrac{1}{\gamma^2}\right)\right] - cI_p[\{\tfrac{A}{\theta-\gamma}\left[\tfrac{1}{\gamma}\left(e^{-\gamma t_s}-e^{-\gamma M}\right)\right. \\
&\left.\left.-\tfrac{1}{\theta}\left(e^{-\theta t_s}-e^{-\theta M}\right)\right] - \tfrac{Q}{\theta}\left(e^{-\theta t_s}-e^{-\theta M}\right)\right\} \\
&+\left\{\tfrac{A}{\theta-\gamma}\left[\tfrac{1}{\gamma}(e^{-\gamma t_1}-e^{-\gamma t_s}) - \tfrac{1}{\theta}\left(e^{-\theta t_1}-e^{-\theta t_s}\right)\right.\right. \\
&\left.\left.-\tfrac{Q}{\theta}\left(e^{-\theta t_1}-e^{-\theta t_s}\right)\right] + \tfrac{\alpha Q e^{\theta t_s}}{\theta}\left(e^{-\theta t_1}-e^{-\theta t_s}\right) + \tfrac{B e^{\theta t_s}}{\theta}\left(e^{-\theta t_1}-e^{-\theta t_s}\right)\right\}]]
\end{aligned} \tag{21}
$$

Case 2: $\pi_1(t_1, T)$

For the Case 2, one can make interest over the revenue made from sales till M; also, one will earn interest for the shortages which are met during $(M - t_s)$ and due to the sale of defective items during $(M - t_s)$.

10.

$$
\begin{aligned}
&\text{Interest earned} = pI_e\left[\textstyle\int_0^M R(t)tdt + B[(M-t_s)]\right] + C_s\alpha Q I_e(M-t_s) \\
&= pI_e\left[\left(e^{-\gamma M}-1\right)\left(\tfrac{-M}{\gamma} - \tfrac{1}{\gamma^2}\right)\right] + pI_e B[(M-t_s)] + C_s\alpha Q I_e(M-t_s)
\end{aligned} \tag{22}
$$

11.

$$
\text{Interest payable} = cI_p\left[\textstyle\int_M^{t_s} I_2(t)dt\right]
$$

$$
= cI_p\left[\begin{array}{l}\tfrac{A}{\theta-\gamma}\left[\tfrac{1}{\gamma}\left(e^{-\gamma t_1}-e^{-\gamma M}\right) - \tfrac{1}{\theta}\left(e^{-\theta t_1}-e^{-\theta M}\right)\right] - \tfrac{Q}{\theta}\left(e^{-\theta t_1}-e^{-\theta M}\right) \\ + \tfrac{\alpha Q e^{\theta t_s}}{\theta}\left(e^{-\theta t_1}-e^{-\theta M}\right) + \tfrac{B e^{\theta t_s}}{\theta}\left(e^{-\theta t_1}-e^{-\theta M}\right)\end{array}\right] \tag{23}
$$

Substitute the values from (12) to (18), (22), and (23) in (11), the total profit expression per unit time per cycle for Case 2, $\pi_2(t_1, T)$.

$$\begin{aligned}\pi_2(t_1,T) = \tfrac{1}{T}\Big[& p(1-\alpha)Q + C_s\alpha Q - K - cQ - \beta Q - C_2 D \\ & \left\{\tfrac{T-t_1}{\delta} - \tfrac{1}{\delta^2}\ln\left[1+\delta(T-t_1)\right]\right\} \\ & - C_L D\left\{(T-t_1) - \tfrac{1}{\delta}\ln\left[1+\delta(T-t_1)\right]\right\} \\ & - h\Big[\left\{\tfrac{A}{\theta-\gamma}\left[\tfrac{1}{\gamma}\left(e^{-\gamma t_s}-1\right) - \tfrac{1}{\theta}\left(e^{-\theta t_s}-1\right)\right] - \tfrac{Q}{\theta}\left(e^{-\theta t_s}-1\right)\right\} \\ & + \left\{\tfrac{A}{\theta-\gamma}\left[\tfrac{1}{\gamma}\left(e^{-\gamma t_1}-e^{-\gamma t_s}\right) - \tfrac{1}{\theta}\left(e^{-\theta t_1}-e^{-\theta t_s}\right)\right]\right. \\ & - \tfrac{Q}{\theta}\left(e^{-\theta t_1}-e^{-\theta t_s}\right) + \tfrac{\alpha Q e^{\theta t_s}}{\theta}\left(e^{-\theta t_1}-e^{-\theta t_s}\right) \\ & + \tfrac{Be^{\theta t_s}}{\theta}\left(e^{-\theta t_1}-e^{-\theta t_s}\right)\}] + pI_e\left[\left(e^{-\gamma M}-1\right)\left(\tfrac{-M}{\gamma}-\tfrac{1}{\gamma^2}\right)\right] \\ & + pI_e B\left[(M-t_s)\right] + C_s\alpha Q I_e(M-t_s) \\ & - cI_p\Big[\left\{\tfrac{A}{\theta-\gamma}\left[\tfrac{1}{\gamma}\left(e^{-\gamma t_1}-e^{-\gamma M}\right) - \tfrac{1}{\theta}\left(e^{-\theta t_1}-e^{-\theta M}\right)\right]\right. \\ & \left. - \tfrac{Q}{\theta}\left(e^{-\theta t_1}-e^{-\theta M}\right)\right\} + \tfrac{\alpha Q e^{\theta t_s}}{\theta}\left(e^{-\theta t_1}-e^{-\theta M}\right) \\ & + \tfrac{Be^{\theta t_s}}{\theta}\left(e^{-\theta t_1}-e^{-\theta M}\right)\}]\Big]\end{aligned} \tag{24}$$

Case 3: $t_1 \le M \le T$

The Case 3 depicts the state in which interest payable is zero, and besides earning interest in previous case, one is earning interest for the demand met for the time period $(M - t_1)$.

12.

$$\begin{aligned}\text{Interest earned} &= pI_e\int_0^{t_1} R(t)t\,dt + pI_e(M-t_1)\int_0^{t_1} R(t)dt \\ &\quad + pI_e B(M-t_s) + C_s\alpha Q I_e(M-t_s) \\ &= pI_e\left[\left(e^{-\gamma M}-1\right)\left(\tfrac{-t_1}{\gamma}-\tfrac{1}{\gamma^2}\right)\right] - \tfrac{pI_eA}{\gamma}\left(e^{-\gamma t_1}-1\right)(M-t_1) \\ &\quad + pI_e B(M-t_s) + C_s\alpha Q I_e(M-t_s)\end{aligned} \tag{25}$$

13.

$$\text{Interest payable} = 0 \tag{26}$$

Substitute the values from (12) to (18), (25), and (26) in (11), the total profit expression per unit time per cycle for Case 3. $\pi_3(t_1, T)$ becomes

$$
\begin{aligned}
\pi_2(t_1,T) = \tfrac{1}{T}\Big[& p(1-\alpha)Q + C_s\alpha Q - K - cQ - \beta Q - C_2 D \\
& \left\{\tfrac{T-t_1}{\delta} - \tfrac{1}{\delta^2}\ln\left[1+\delta(T-t_1)\right]\right\} \\
& - C_L D\left\{(T-t_1) - \tfrac{1}{\delta}\ln\left[1+\delta(T-t_1)\right]\right\} \\
& - h\Big[\left\{\tfrac{A}{\theta-\gamma}\left[\tfrac{1}{\gamma}\left(e^{-\gamma t_s}-1\right) - \tfrac{1}{\theta}\left(e^{-\theta t_s}-1\right)\right] - \tfrac{Q}{\theta}\left(e^{-\theta t_s}-1\right)\right\} \\
& + \Big\{\tfrac{A}{\theta-\gamma}\left[\tfrac{1}{\gamma}\left(e^{-\gamma t_1}-e^{-\gamma t_s}\right) - \tfrac{1}{\theta}\left(e^{-\theta t_1}-e^{-\theta t_s}\right)\right] \\
& - \tfrac{Q}{\theta}\left(e^{-\theta t_1}-e^{-\theta t_s}\right) + \tfrac{\alpha Q e^{\theta t_s}}{\theta}\left(e^{-\theta t_1}-e^{-\theta t_s}\right) \\
& + \tfrac{Be^{\theta t_s}}{\theta}\left(e^{-\theta t_1}-e^{-\theta t_s}\right)\Big\}\Big] + pI_e\left[\left(e^{-\gamma t_1}-1\right)\left(\tfrac{-t_1}{\gamma}-\tfrac{1}{\gamma^2}\right)\right] \\
& - \tfrac{pI_eA}{\gamma}\left(e^{-\gamma t_1}-1\right)(M-t_1) \\
& + pI_eB(M-t_s) + C_s\alpha QI_e(M-t_s)\Big]
\end{aligned}
\tag{27}
$$

Case 4: $T \leq M$

The expressions for the interest made and payable overlap with that of previous case.

Hence, effectively there are three different cases for the retailer's total profit expression per cycle, $\pi(t_1, T)$,

$$
\pi(t_1,T) = \begin{cases} \pi_1(t_1,T), & 0 \leq M \leq t_s, \ \text{case 1} \\ \pi_2(t_1,T), & t_s \leq M \leq t_1, \ \text{case 2} \\ \pi_3(t_1,T), & t_1 \leq M \leq T, \ \text{case 3} \end{cases} \tag{28}
$$

4 Solution Procedure

The aim is to determine the optimal values of t_1 and T that maximize the total profit expression, $\pi_i(t_1, T)$; consequently, the conditions necessary for optimal $\pi_i(t_1, T)$ are

$$
\frac{\partial\left[\pi_i(t_1,T)\right]}{\partial t_1} = 0, \ \text{and} \tag{29}
$$

$$
\frac{\partial\left[\pi_i(t_1,T)\right]}{\partial T} = 0 \ \text{where} \ i = 1,2,3 \tag{30}
$$

Further, in order to establish the concavity of expected total profit expression, below are the sufficient conditions which should be satisfied:

$$
\left(\frac{\partial^2\left[\pi_i(t_1,T)\right]}{\partial t_1 \partial T}\right)^2 - \left(\frac{\partial^2\left[\pi_i(t_1,T)\right]}{\partial t_1{}^2}\right)\left(\frac{\partial^2\left[\pi_i(t_1,T)\right]}{\partial T^2}\right) \leq 0, \ \text{and} \tag{31}
$$

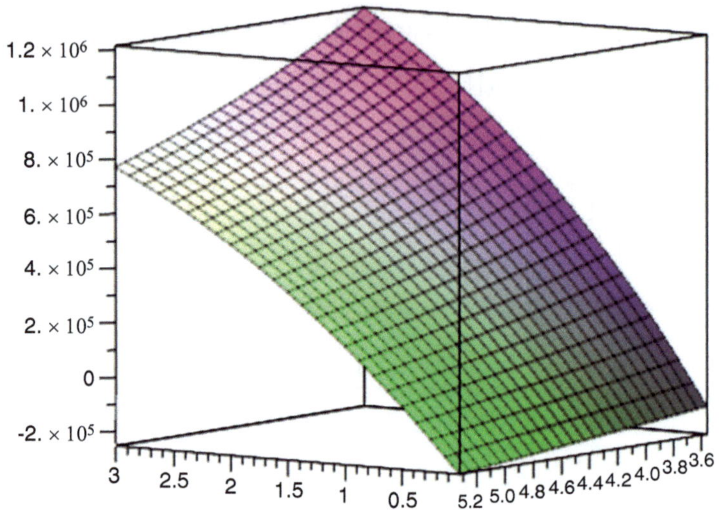

Fig. 3 Concavity of expected total profit function (Case 1)

$$\left(\frac{\partial^2\left[\pi_i\left(t_1,T\right)\right]}{\partial t_1{}^2}\right)\le 0,\quad \left(\frac{\partial^2\left[\pi_i\left(t_1,T\right)\right]}{\partial T^2}\right)\le 0. \tag{32}$$

Due to the complexity of these derivatives, it becomes challenging to show the concavity mathematically; therefore the optimality of all the profit functions is proven with the help of graphs; different scenarios are depicted in Figs. 3, 4, and 5 below.

Algorithm

1. Determine $t_1^* = t_{11}$ (say) and $T = T_1$ (say) from (29) to (30). By making use of the values of t_1 and T, obtain value of Q and B from (6) and (7). If $0 \le M \le t_1$ profit expression can be calculated from (28), otherwise set $\pi_1(t_1, T) = 0$.
2. Determine $t_1^* = t_{12}$ (say) and $T = T_2$ (say) from (29) to (30). By making use of the values t_1 and T, calculate the value of Q and B from (6) and (7). If $t_s \le M \le t_1$ profit expression can be obtained from (28), else set $\pi_2(t_1, T) = 0$.
3. Determine $t_1^* = t_{13}$ (say) and $T = T_3$ (say) from (29) to (30). Now using the value of t_1 and T, calculate the value of Q and B from (6) to (7). If $t_1 \le M \le T$, the expected total profit can be obtained from (28), else set $\pi_3(t_1, T) = 0$.
4. Comparing the obtained values of expected profit for all the three cases, choose the optimal values of t_1 and T that are connected to the optimal expected total profit.

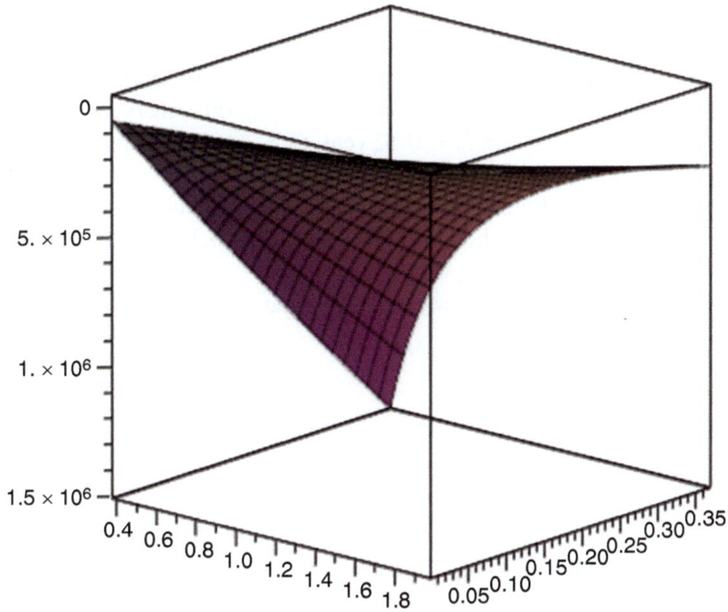

Fig. 4 Concavity of expected total profit function (Case 2)

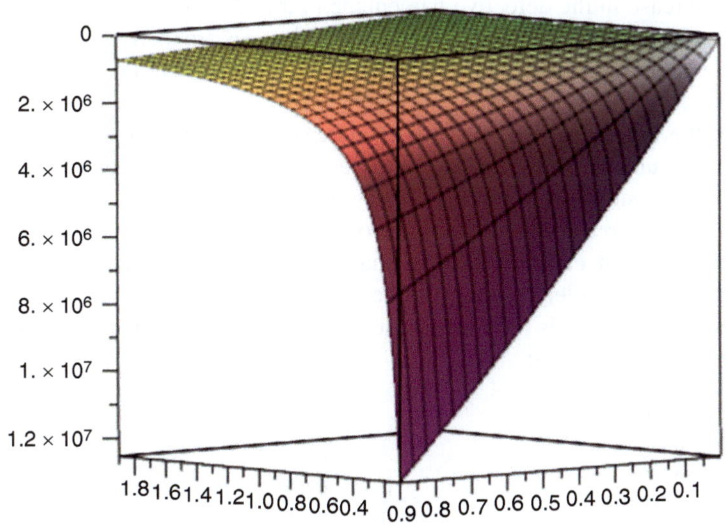

Fig. 5 Concavity of expected total profit function (Case 3)

Numerical Example

Example: An example is devised to validate the model by using the data below:

D = 800 units/year, K = \$1000 per cycle, c = \$250/unit, p = \$400/unit, h = \$100/unit/year, $\theta = 0.1$, $\alpha = 0.08$, λ = 48,000 units/year, β = \$15/unit, M = 0.75 year, I_e = 0.10 year, I_p = 0.12 year, C_s = \$100/unit, and C_2 = \$150/year, A = 1200, C_L = \$200/year, $\gamma = 0.07$, $\delta = 0.02$.

By applying the developed algorithm, the following results are obtained: optimal order level $t_1^* = 0.369$ units, the back order level $T^* = 0.380$ units, and the expected total profit $\pi(t_1, T)$ \$122976.5.

5 Sensitivity Analysis

Managerial Insights

- Table 1 indicates the decreasing value of parameter of backlogging δ, that is, an increasing rate of backlogging tends to increase the size of order that ultimately contributes to more profits. As an increase in the rate of backlogging indicates backlogged demand to be more, hence forward from the ordered quantity, a big share gets used up in order to satisfy demand which is backlogged that lessens the primary inventory level for the cycle and hence the carrying costs.
- With increase in the defective percentage (Table 2), the profit decreases and the order level increases considerably. As the defectives are increasing in the ordered lot, there are more items that are salvaged at a discounted price, resulting in lower profits.
- When the inventory carrying cost is increasing (Table 3), the back order level increases and the order level decreases and results into lower profits. It is advisable to stock less in the inventory in this case and back order more of the demand so as to avoid large holding costs.
- From Table 4 it can be observed that with increase in the credit period, the total profit of retailer increases as he is able to earn more interest on the sales revenue that increases total profit. Hence trade credit has the positive impact on the inventory modeling.

Table 1 Optimal values obtained by varying backlogging parameter δ

δ	t_1	T	Q	B	Profit
0.07	0.649	0.926	1095	219.47	113012.2
0.06	0.648	0.930	1098	223.33	113130.8
0.05	0.647	0.933	1101	227.35	113253.1
0.04	0.646	0.937	1104	231.51	113379.3
0.03	0.645	0.941	1107	235.84	113509.6
0.02	0.644	0.945	1111	240.34	113644.2

Table 2 Optimal values obtained by varying defective percentage α

α	t_1	T	Q	B	Profit
0.01	0.648	0.950	1037	240.24	115306.1
0.02	0.648	0.949	1047	240.26	115086.0
0.04	0.647	0.948	1068	240.30	114629.4
0.06	0.645	0.947	1089	240.32	114149.4
0.08	0.644	0.945	1111	240.34	113644.2

Table 3 Optimal values obtained by varying the inventory carrying cost h

h	t_1	T	Q	B	Profit
25	1.124	1.194	1553	55.41	144323.3
50	0.892	1.063	1329	136.61	131016.7
100	0.644	0.945	1111	240.34	113644.2
125	0.568	0.913	1048	275.55	107575.4
150	0.508	0.889	999	303.90	102585.4
200	0.42	0.855	930	346.52	94817.1

Table 4 Optimal values obtained by varying credit period M

M	t_1	T	Q	B	Profit
0.07	0.970	1.723	1938	597.19	107515.4
0.08	0.974	1.729	1944	599.53	108241.6
0.09	0.977	1.735	1951	601.87	108966.5

Table 5 Optimal values obtained by varying initial demand A

A	t_1	T	Q	B	Profit
1000	0.577	0.968	973	311.34	102720.3
1050	0.597	0.967	1010	294.52	105295.8
1100	0.615	0.963	1045	277.06	107977.6
1150	0.631	0.955	1079	259.00	110761.1
1200	0.644	0.945	1111	240.34	113644.2
1250	0.656	0.933	1142	221.09	116626.6

- Table 5 indicates that the rise in the initial demand increases the order level, and back orders decrease as a huge amount of inventory is used initially to satisfy the inceptive demand resulting in lesser holding cost that ultimately contributes to increase the total profit of the firm.

6 Conclusion

The primary aim of this paper is to formulate an optimal inventory strategy keeping in mind the retailer who has to cope up with imperfect quality and product that deteriorates with time. The demand of the item is taken to be exponentially decreasing. This is due to the fact that more and more new products are being introduced in the market. Furthermore, the paper also investigates the impact of trade credit in the inventory system. Trade credit is a well-known promotional activity that suppliers usually implement in today's competitive business environment. Moreover, the shortages are considered to be partially backlogged, and the backlogging rate has an inverse relation to the waiting time for the next replenishment. To jointly optimize the inventory cycle length and the shortage point, an algorithm has been used in order to maximize the overall profit function. The validity of the model is illustrated using the numerical example, and further the sensitivity analysis has been performed to comprehend the significance of different model parameters that holds vital managerial implications.

The developed model has future application and can be explored further by incorporating some other practical scenarios, viz., investment in preservation technology, varying holding costs, etc.

Acknowledgment The first and last author would like to acknowledge the support of the Research Grant No.: RC/2015/9677, provided by the University of Delhi, Delhi, India, for conducting this research.

References

1. Ghare PM, Schrader GF (1963) A model for exponentially decaying inventory. J Ind Eng 14(5):238–243
2. Covert RP, Philip GC (1973) An EOQ model for items with Weibull distribution deterioration. AIIE Trans 5(4):323–326
3. Dave U, Patel LK (1981) (T, S i) policy inventory model for deteriorating items with time proportional demand. J Oper Res Soc 32(2):137–142
4. Hollter RH, Mak KL (1983) Inventory replenishment policies for deteriorating items in a declining market. Int J Prod Res 21(6):813–836
5. Sachan RS (1984) On (T, S i) policy inventory model for deteriorating items with time proportional demand. J Oper Res Soc 35(11):1013–1019
6. Raafat FF, Wolfe PM, Eldin HK (1991) An inventory model for deteriorating items. Comput Ind Eng 20(1):89–94
7. CHUNG KJ, TING PS (1994) On replenishment schedule for deteriorating items with time-proportional demand. Prod Plan Control 5(4):392–396
8. Hariga MA, Benkherouf L (1994) Optimal and heuristic inventory replenishment models for deteriorating items with exponential time-varying demand. Eur J Oper Res 79(1):123–137
9. Goyal SK, Giri BC (2001) Recent trends in modeling of deteriorating inventory. Eur J Oper Res 134(1):1–16
10. Wee HM (1995) A deterministic lot-size inventory model for deteriorating items with shortages and a declining market. Comput Oper Res 22(3):345–356
11. Wee HM (1995) Joint pricing and replenishment policy for deteriorating inventory with declining market. Int J Prod Econ 40(2):163–171
12. Benkherouf L (1998) Note on a deterministic lot-size inventory model for deteriorating items with shortages and a declining market. Comput Oper Res 25(1):63–65
13. Chung KJ, Tsai SF (1999) A solution procedure to determine inventory replenishment policies for deteriorating items in a declining market. J Inf Optim Sci 20(1):1–15
14. Su CT, Lin CW, Tsai CH (1999) A deterministic production inventory model for deteriorating items with an exponential declining demand. Opsearch-New Delhi 36:95–106

15. Porteus EL (1986) Optimal lot sizing, process quality improvement and setup cost reduction. Oper Res 34(1):137–144
16. Rosenblatt MJ, Lee HL (1986) Economic production cycles with imperfect production processes. IIE Trans 18(1):48–55
17. Salameh MK, Jaber MY (2000) Economic production quantity model for items with imperfect quality. Int J Prod Econ 64(1):59–64
18. Papachristos S, Konstantaras I (2006) Economic ordering quantity models for items with imperfect quality. Int J Prod Econ 100(1):148–154
19. Maddah B, Jaber MY (2008) Economic order quantity for items with imperfect quality: revisited. Int J Prod Econ 112(2):808–815
20. Maddah B, Moussawi L, Jaber MY (2010) Lot sizing with a Markov production process and imperfect items scrapped. Int J Prod Econ 124(2):340–347
21. Jaggi CK, Mittal M, Khanna A (2013) Effects of inspection on retailer's ordering policy for deteriorating items with time-dependent demand under inflationary conditions. Int J Syst Sci 44(9):1774–1782
22. Kingsman BG (1983) The effect of payment rules on ordering and stockholding in purchasing. J Oper Res Soc 34(11):1085–1098
23. Goyal SK (1985) Economic order quantity under conditions of permissible delay in payments. J Oper Res Soc 36(4):335–338
24. Davis RA, Gaither N (1985) Optimal ordering policies under conditions of extended payment privileges. Manag Sci 31(4):499–509
25. Aggarwal SP, Jaggi CK (1995) Ordering policies of deteriorating items under permissible delay in payments. J Oper Res Soc 46(5):658–662
26. Chu P, Chung KJ, Lan SP (1998) Economic order quantity of deteriorating items under permissible delay in payments. Comput Oper Res 25(10):817–824
27. Soni H, Shah NH, Jaggi CK (2010) Inventory models and trade credit: a review. Control Cybern 39(3):867–882
28. Chung KJ, Huang YF (2006) Retailer's optimal cycle times in the EOQ model with imperfect quality and a permissible credit period. Qual Quant 40(1):59–77
29. Hariga M (1995) An EOQ model for deteriorating items with shortages and time-varying demand. J Oper Res Soc 46(3):398–404
30. Chakrabarti T, Chaudhuri KS (1997) An EOQ model for deteriorating items with a linear trend in demand and shortages in all cycles. Int J Prod Econ 49(3):205–213
31. Wee HM, Yu J, Chen MC (2007) Optimal inventory model for items with imperfect quality and shortage backordering. Omega 35(1):7–11
32. Chang HC, Ho CH (2010) Exact closed-form solutions for "optimal inventory model for items with imperfect quality and shortage backordering". Omega 38(3):233–237
33. Jaggi CK, Goel SK, Mittal M (2013) Credit financing in economic ordering policies for defective items with allowable shortages. Appl Math Comput 219(10):5268–5282
34. Khanna A, Mittal M, Gautam P, Jaggi CK (2016) Credit financing for deteriorating imperfect quality items with allowable shortages. Decis SciLett 5(1):45–60
35. Khanna A, Gautam P, Jaggi CK (2017) Inventory modeling for deteriorating imperfect quality items with selling price dependent demand and shortage backordering under credit financing. Int JMath Eng Manag Sci 2(2):110–124
36. Chang HJ, Dye CY (1999) An EOQ model for deteriorating items with time varying demand and partial backlogging. J Oper Res Soc 50(11):1176–1182
37. Ouyang W, Cheng X (2005) An inventory model for deteriorating items with exponential declining demand and partial backlogging. Yugoslav J Oper Res 15(2):277–288
38. Dash BP, Singh T, Pattnayak H (2014) An inventory model for deteriorating items with exponential declining demand and time-varying holding cost. A J Oper Res 4(01):1

Maintenance in the Era of Industry 4.0: Issues and Challenges

Uday Kumar and Diego Galar

1 Introduction to Industry 4.0

Industry 4.0 based on its characteristics of smart systems and Internet-based solutions is already declared as the fourth generation of industrial activity (Lasi et al. [17]). In the eighteenth to nineteenth century, we observed the first revolution with production getting mechanized. The most distinct characteristic of this revolution was marked with production moving away from local workshops to huge factories, and birth of the working class took place. With second revolution during the last century, production became electrified with processes getting standardized with exemplary model marked by the Ford's assembly line. In late 1960s came the digitization of production marking the third revolution with introduction to programmable logic controllers (PLC).

With unprecedented growth observed in ICT in the last 2–3 decades, the fourth industrial revolution is observed based on ICT and data-driven decision-making processes. Industry 4.0 is characterized by two important features: intelligent factories based on IoT (Internet of Things) (Ashton [1]) and cyber-physical systems. Lee [19] suggested that cyber-physical systems are embedded systems of integrated digital components, primarily designed to monitor and control physical devices. The embedded systems are then networked based on Internet technology and create what is nowadays called as "Internet of Things". With ever-increasing competition around, there are distinctive signs of organizations, systems, human beings, etc. that are all increasingly getting interconnected, instrumented and intelligent.

This has led to improved quality of services, new savings, enhanced resource utilization and efficiency. This has also facilitated the development of the new trends in manufacturing technology popularly known as Industry 4.0 based on

U. Kumar (✉) • D. Galar
Luleå University of Technology, Luleå, Sweden
e-mail: Uday.Kumar@ltu.se; diego.galar@ltu.se

© Springer Nature Singapore Pte Ltd. 2018

P.K. Kapur et al. (eds.), *Quality, IT and Business Operations*, Springer Proceedings in Business and Economics, DOI 10.1007/978-981-10-5577-5_19

the capability of industrial Internet. Industry 4.0 provides foundation for the next generation of manufacturing technologies based on the use of advanced information logistic analytics to transform the current state-of-the-art manufacturing platforms into a network of collaborative machine communities' seamlessly manufacturing and delivering goods and services in a planned way. Currently, Germany is leading transformation towards fourth industrial revolution (Industry 4.0) grounded on the capability industrial Internet. Industry 4.0 symbolizes the current trend of automation and data exchange in manufacturing sector striving to adopt and adapt the new and emerging technologies to achieve new level of effectiveness and efficiency. Industry 4.0 concept essentially includes cyber-physical systems, IoT and cloud computing. It creates what has been called a "smart factory".

1.1 Maintenance 4.0 Within Industry 4.0

In maintenance, Industry 4.0 find its application in designing of self-learning and smart system that helps predict failures, diagnose and trigger maintenance schedules. In order to extract specific and relevant information, these smart systems are highly demanded for data access, for quality and also for the use of multiple sources of data (Lee et al. [20]). Development of intelligent maintenance systems based on cyber-physical approach, for failure detection, providing diagnostics and prognostics, has been the core focus on several research projects (Syed et al. [27], Sankavaram et al. [26] and Kroll et al. [16]). Till now, such systems are applied in manufacturing and process industries; however, future of Maintenance 4.0 seems highly relevant in a lot of other areas.

Based on the use of advanced technology, Maintenance 4.0 does predictive analytics and suggests feasible solution, with major application in Industry 4.0 and especially on those maintenance aspects that deals with collection of data, its analysis and visualization and asset decision-making. It also handles another common vulnerable aspect in asset management: prognosis, a better asset status forecasting. Probability of mission accomplishment by the asset is checked by remaining useful life estimation for any operation/maintenance service [7].

To create new services based on exploited ontologies, [2] suggested the importance of "big data" and its application to diverse sources of information and therefore discovery of knowledge accomplished. The "big data" adoption paves the ground for efficient operation and maintenance policies for the future and would surely bridge the gap between them. In this paper the application of Maintenance 4.0, from a system's perspective, is described to suggest its positive effects on technology, organization and operations.

In summary, the foundation of Industry 4.0 is built around the concepts of interconnectivity, instrumentation and intelligence for the assets by the means of successfully proven technologies such as industrial Internet, cloud computing or industrial Internet of Things (IoT).

2 Instrumentation and Interconnection

2.1 Internet of Things

The Internet of Things (IoT) is a collective term for the developments that means that machinery, vehicles, goods, appliances, clothes and other things and creatures (including humans) are equipped with tiny sensors and computers. IoT could be defined as a dynamic network infrastructure with self-configuring capabilities based on standard interoperable communication protocols where physical and virtual things have identities, physical attributes, and virtual personalities and use intelligent interfaces and are seamlessly integrated into the information network.

In short the Internet of Things (IoT) is the network of everything or physical objects—devices, vehicles, buildings and other items embedded with electronics, software, sensors and network connectivity—that enables these objects to collect and exchange data.

2.2 Industrial Internet

Industrial Internet can be defined as the new and emerging technologies for managing interconnected machines and systems between its physical assets and computational capabilities [29] .

The Industrial Internet of Things (IIoT) is the use of Internet of Things (IoT) technologies in manufacturing, incorporating machine to machine communication, big data analytics, and harnessing of the sensor data and robotics and automation technologies that have existed in industrial settings for years.

Industrial Internet comprises three elements (Fig. 1):

- Intelligent machines: New ways of connecting the word's myriad of machines, facilities, fleets and networks with advanced sensors, controls and software applications.
- Advanced Analytics: Harnessing the power of physics-based analytics, predictive algorithms, automation and deep domain expertise in material science, electrical engineering and other key disciplines required to understand how machines and larger systems operate.
- People: connecting people, whether they be at work in industrial facilities, offices, hospitals or on the move, at any time to support more intelligent design, operations, maintenance as well as higher quality service and safety. Connecting and combining these elements offers new opportunities across firms and economies.

Fig. 1 Elements of industrial Internet

To collect and analyse huge amount of data to improve machine performance and invariably increase the efficiency of the systems and networks that link them, industrial Internet begins with embedding sensors and other complex instrumentation in a network of machines in the order of complexity. Eventually, data itself becomes "intelligent" enough to know which users it finally needs to reach.

The three main components of this concept are intelligent devices, intelligent systems and digital instrumentation wherein the first step in this revolution of industrial Internet is industrial machines.

To ensure that myriad of machines is more intelligent, several forces like costs of deployment (IoT, instrumentation), computing power and advanced analytics (big data analytics) are put in place.

2.3 Cyber-Physical Architecture

Baheti and Gill [3] defined cyber-physical systems (CPS) as transformative technologies that can be used to manage networked systems between their physical assets and computational capabilities. Lee et al. [21] proposed a five-level CPS structure, called 5C architecture (Fig. 2), demonstrating a step-by-step procedure to develop and deploy a CPS at a manufacturing setup.

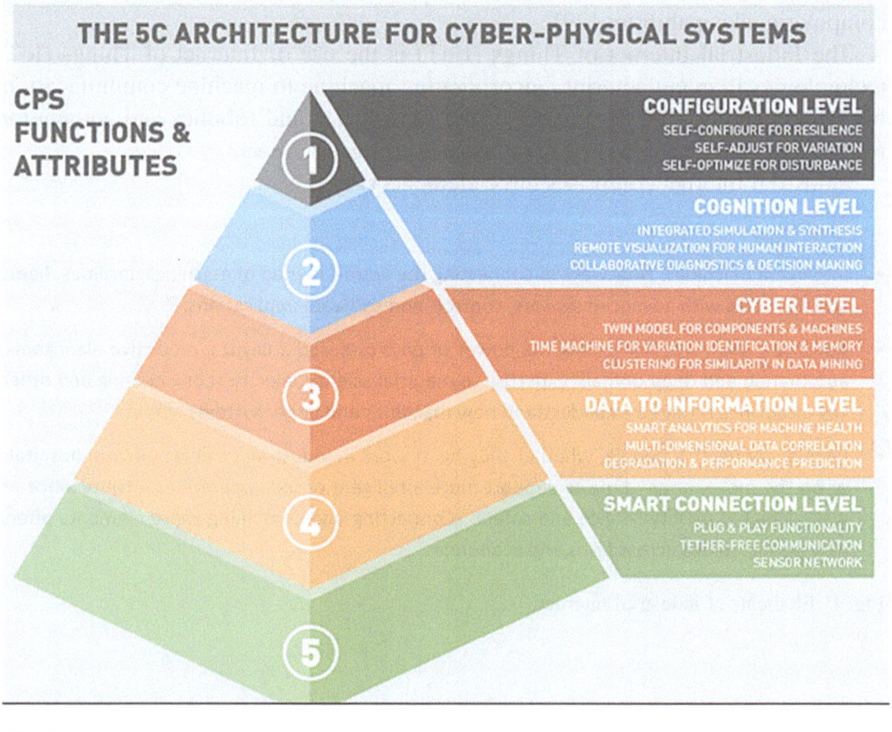

Fig. 2 Architecture of CPS [21]

- Smart connection: While developing a CPS application, the first step is to acquire accurate and reliable data from machines. This data may be directly measured through sensors or is accessed through an ERP, MES, CMM or SCM.
- Data→information conversion: The real task in any CPS application is to obtain meaningful information from the data. In recent years, extensive focus has been applied to develop these algorithms specifically for prognostics and health management applications. By calculating health value, estimated remaining useful life, etc., the second level of CPS architecture brings context awareness to machines.
- Cyber: The cyber level acts as central information hub in this architecture. Information is being pushed to it from every connected machine to form the machine network.
- Cognition: Implementing CPS upon this level generates a thorough knowledge of the monitored system. Presentation of the acquired knowledge to expert users supports the correct decision.
- Configuration: The configuration level is the feedback from cyberspace to physical space and acts as supervisory control to make machines self-configure and self-adaptive. This stage acts as resilience control system (RCS) to apply the corrective and preventive decisions, which has been made in cognition level, to the monitored system.

2.4 Cloud Computing

According to National Institute of Standards and Technology definition of cloud computing V15, dated October 7, 2009 [25], cloud computing is a model for enabling convenient, on-demand network access to a shared pool of configurable computing resources (e.g. networks, servers, storage, applications and services) that can be rapidly provisioned and released with minimal management effort or service provider interaction. Talking in terms of a cloud for industrial services, cloud computing can be seen as a broad array of Web-based services aimed at allowing users to obtain a wide range of functional capabilities on a "pay-as-you-go" basis that previously required tremendous hardware and software investments and professional skills to acquire. Cloud computing is the realization of the earlier ideals of utility computing without the technical complexities or complicated deployment worries.

Therefore, cloud is considered as a combination of hardware, networks, space, services and interfaces that offers computing as a service to its users. Cloud comes as a solution for maintenance as a medium to connect highly dispersed data across various repositories. In cloud-based solutions, the user need not to know anything about the underlying technology as the collection and distribution of data done through various applications are dispersed throughout the network at several locations.

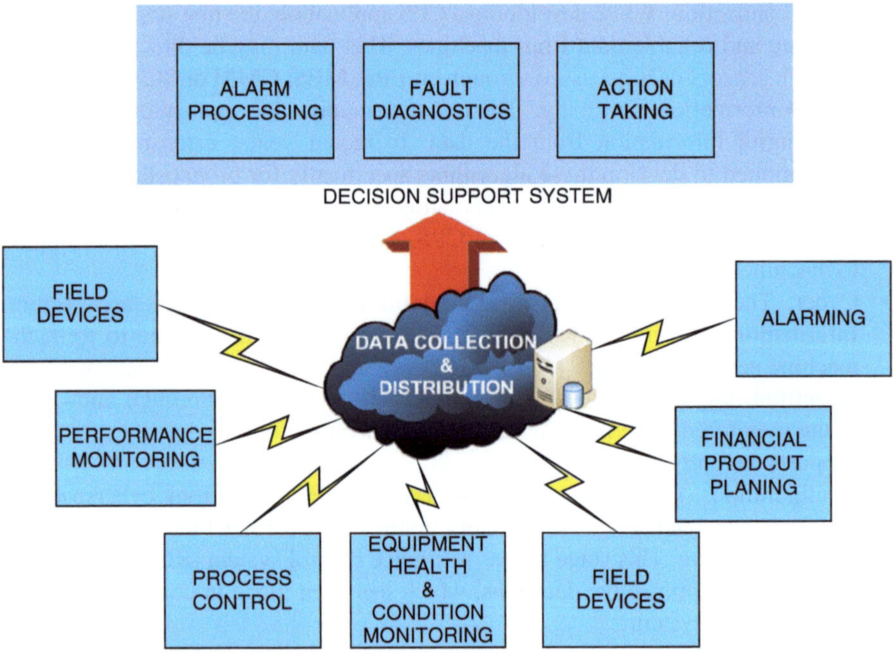

Fig. 3 Services provided by the asset cloud

Figure 3 showcases the block diagram of how data flow and communication are associated with or used by the asset cloud, highlighting the data collection and distribution system, wherein data is received from numerous data sources.

Here the cloud can help maintain and store these large data repositories in a central data store and also provide user with an interface to carry out powerful and robust analysis tool simply because of its ability to integrate various maintenance inputs, layout, conditions, weather, inventory, etc.

3 Maintenance Data and Knowledge Discovery

Latest technologies have emerged in the last 10 years to capture, process and visualize huge volumes of data. Nowadays, big data analytics use is fairly common in business sector with most of their information structured and came from the same source, like banking processes. In banking, there is substantial number of IT services (tools) available to meet the customer needs. Chen et al. [5] summarized some of the successful implementation of big data applications. In 2009, Google was able to obtain timely information about the flu pandemic using big data analytics to surpass the value generated by any of the disease prevention centres. Working on the same lines, models are built to predict spread of influenza and its origin destination. In 2008, Microsoft bought a company who used to predict trends of airline ticket prices. Connecting this system to the big data search engine resulted in saving almost US $50 per ticket per passenger, with prediction accuracy of 75%. But, such cases are only exception, as most businesses, vulnerable towards adopting big data concept, are still to embrace it yet. This has been either due to the lack of specific tools required for big data implementation or the huge cost involved in bringing key stakeholders together.

Chen et al. [5] defined that the traditional IT and software/hardware tools are incapable of perceiving, acquiring, managing and processing big data as a data set within a tolerable time limit. Earlier, Zikopoulos and Eaton [30] proposed big data model on 3Vs: volume, variety and velocity. Later on, Lomotey and Deters [23] extended the 3Vs model to 5Vs and added value and veracity. Analytics-as-a-service tool was proposed by Lomotey and Deters [23] for knowledge discovery in big data. A key process in knowledge discovery, data mining, is designed for schema- and structure-oriented data storages. Manyika et al. [24] report that US medical industry may surpass US $300 billion, improvement of over 60% of retailer's profit; EU can save over 100 billion EUR by improving the efficiency of government operations, all this with the effective and creative usage of big data.

3.1 Big Data in Maintenance

In asset management, data can be described using 5Vs (Lomotey and Deters [23]; Zikopoulos and Eaton [30]). With a velocity of several thousands of samples per second for every measuring point, data from accelerometers of acoustic sensors can be assimilated. With myriad of such points, we obtain big data. Maintenance data, some structured and others not, such as comments from texts based on performed maintenance or failure reports, adding to that variety of data formats depicting the variety of data in asset management, clearly leads with a dire need to assess and manage this uncertainty of data called the veracity. It is of utmost importance to derive this true potential from such data in asset management. Finally, to understand the true value of the data, it is imperative to know how data enables managers with effective decision-making capabilities in maintenance and also how to choose the most cost-effective means of processing data.

In asset management, data mining with big data can help discover knowledge in terms of identifying newer patterns and relations not visible upfront. The big data approach enables combining of contextual information in decision support systems of maintenance (Galar et al. [8]). Discovery of the root causes of failure is one such example of useful knowledge, which can provide input for design improvement, along with enhanced maintenance planning.

To support an effective maintenance decision-making process needs a trusted DSS based on knowledge discovery. The process of knowledge discovery will essentially consist of data acquisition, to obtain relevant data and manage its content; data transition, to communicate the collected data; data fusion [11], to compile

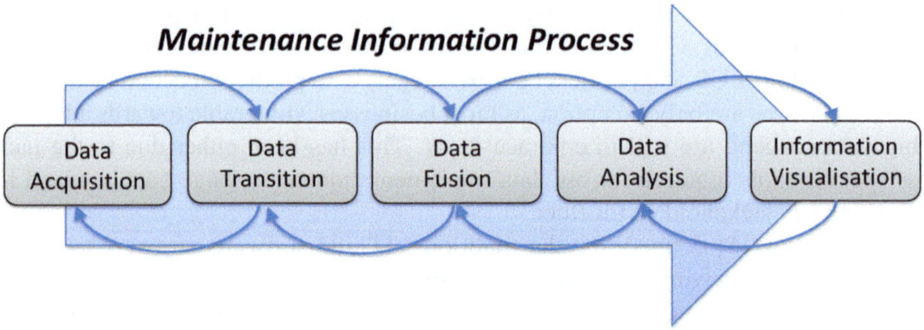

Fig. 4 A generic maintenance information process

data and information from different sources; data mining, to analyse data to extract information and knowledge; and information extraction and visualization, to support maintenance decision, as shown in Fig. 4. Figure 4 illustrates a maintenance decision based on real-time data using data fusion and big data analytics and context sensing to get real-time decisions and solutions for maintenance problems.

The integration of data, recorded from a multiple-sensor system, together with information from other sources to achieve inferences, is known as data fusion [10].

Data fusion is a prerequisite when handling data from heterogeneous sources or from multiple sensors. Knowledge discovery when applied for maintenance decision support uses eMaintenance concept [18] for integrating the data mining and knowledge discovery. To get the right decision for the context sensing is a must as illustrated in Fig. 4. However, several challenges are encountered while developing an eMaintenance solution for industrial application [15], like challenges related to (1) architecture, (2) organization, (3) content and contextual, (4) infrastructural and (5) integration.

Using the advanced technologies of predictive analytics, Maintenance 4.0 provides effective feasible decisions for Industry 4.0 with major thrust on collection, analysis and visualization of data.

3.2 *eMaintenance Solutions for Industry 4.0*

The ongoing industrial digitalization provides enormous capabilities for industry to collect vast amount of data and information (i.e. industrial big data), from various processes and data sources such as operation, maintenance and business processes. However, having accurate data and information available is one of the prerequisites in maintenance knowledge discovery (Fig. 5). Besides collecting data and information, another challenge is to understand the patterns and relationships of these data useful and relevant for maintenance decisions.

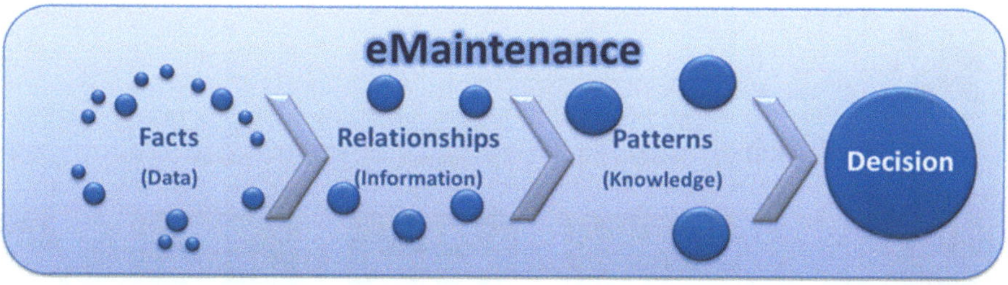

Fig. 5 eMaintenance solution

To deal with the challenges arising out of high volume of data generated by machines in Industry 4.0 scenario, big data and advanced tools are developed and implemented so that data can be systematically processed into information and facilitate decision-making with more information in real time. The concept is captured within the framework of eMaintenance solutions. Since there is no standard definition, we define eMaintenance:

eMaintenance is a concept that connects all the stakeholders, integrates their requirements and facilitates optimal decision-making on demand or in real time to deliver the planned and expected function and services from the assets and minimizes the total business risks.

3.3 *eMaintenance Challenges*

Maintenance process [14] is facilitated by eMaintenance which represents the services that are aimed at managing information related to maintenance. These services can be used during all lifecycle phases of a system under different purposes of preparation, execution, assessment and knowledge management related to maintenance (Fig. 6). Hence, it's imperative to design appropriate eMaintenance solution considering a holistic perspective and based on appropriate methodologies, strategies and technologies. However, this process of designing is a very challenging task and few of these challenges are:

Organizational Challenges
These challenges mainly focus on enterprise resource management related aspects like (1) organizations restructuring for those involved in maintenance, (2) resource planning (e.g. spare part, material, etc.), (3) information management, (4) management of heterogeneous organizations and (5) knowledge management.

Architectural Challenges
Challenges dealing with the issues of the architecture of eMaintenance solutions, like (1) developing framework for eMaintenance development, (2) developing models for distributed processing and analysis of data, (3) service model development

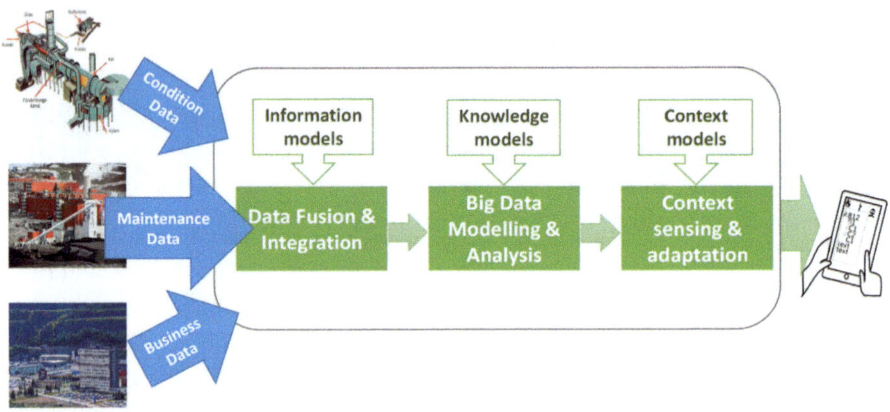

Fig. 6 Maintenance decision process incorporating knowledge discovery for decision-making in maintenance

for distributed data analysis, (4) developing prognostic tool-based models, (5) model development for visualization of relevant data that supports interaction between human and machine, and (6) developing model for dispersed data storage capability.

Infrastructural Challenges
When services, according to SOA, are developed and implemented, infrastructural challenges arise to address to the issues pertaining to providing necessary tools and technologies required to meet the service needs and requirements. Some of these challenges include (1) wired/wireless network infrastructure, (2) service and user authentication, (3) mechanism for safety and security, (4) maintainability of eMaintenance services, (5) availability performance management and tracing and tracking mechanism and (6) establishment of documentation and archiving mechanism.

Content and Contextual
There are those challenges that are connected with the data sourced from eMaintenance services, like (1) establishing appropriate ontology through which data from data sources (e.g. process, product, condition monitoring and business data) are integrated smoothly, 2) providing quality assurance mechanism so as to increase decision-making quality, (3) providing mechanism to establish user's current situation so as to adapt information to user's context, (4) mechanism to manage uncertainty in data sets, (5) mechanism for describing various context and (6) for pattern recognition.

Integration Challenges
Coordination, integration and orchestration of services and data managed by eMaintenance solution raises integration challenges like (1) service management, interaction and interactivity, (2) management of configurations and (3) enablement of integration capability across a multiplatform and technologies.

3.4 Data Mining and Knowledge Discovery

Within Maintenance 4.0, the process of automating the maintenance decision support and through knowledge discovery forms important parts of the decision support system looking for failure-free operations. The process of knowledge discovery will essentially consist of data acquisition from intelligent devices, to obtain relevant data and manage its content; data transition, to communicate the collected data; data fusion, to compile data and information from different sources; data mining, to analyse data to extract information and knowledge; and information extraction and visualization, to support maintenance decision in real time.

In asset management, big data with the help of data mining can help discover newer patterns and relations which are otherwise not available of the surface. Combining the contextual information in DSS of maintenance is possible with this big data approach (Galar et al. [8]). Conceptually, KDD refers to a multiple-step process that can be highly interactive and iterative in the following [6, 28].

Artificial intelligence techniques possess advance knowledge management which includes knowledge acquisition, repositories, discovery and distribution. With knowledge acquisition, tacit and explicit knowledge from domain experts is captured, while repositories formalize knowledge acquisition outcomes, and integrate it in distributed corporate environment. Knowledge discovery and mining approaches explore relationships and trends in the knowledge repositories to create new knowledge [22].

In data integration, multiple sources or multiple data types can be integrated. By integrating multiple sources of the same data types, it is possible to compare parameters and by that retrieve an indication on the quality of the data; integrating multiple data types enables a more thorough analysis utilizing the relation between the data types and their behaviour. The last step, data visualization, is the interface between the user and the cloud by means of Web service (Fig. 7).

This includes any interaction from the user. As previously mentioned, this turns into quite a complex task due to the varying character of the customers and their users. The information to visualize and the relevancy in the information will differ a lot depending on customer characteristics.

4 Maintenance Analytics

The ongoing industrial digitalization provides enormous capabilities for industry to collect vast amount of data and information (i.e. industrial big data), from various processes and data sources such as operation, maintenance and business processes. However, having accurate data and information available is one the prerequisites in maintenance knowledge discovery. Beside the collecting data and information, another puzzle is to understand the patterns and relationships of these data useful and relevant for maintenance decisions.

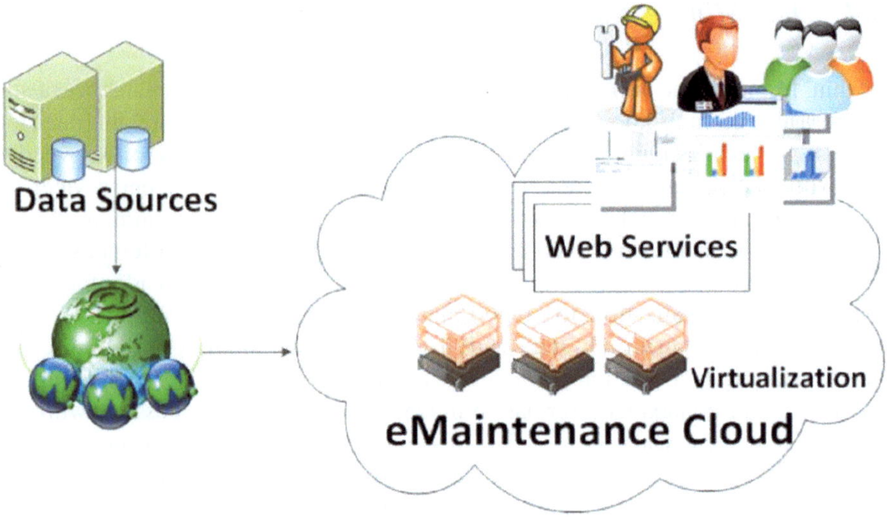

Fig. 7 eMaintenance cloud as a service provider by means of Web services

Hence, the major objective of this paper is to suggest a knowledge delivery concept, *Maintenance Analytics* (MA), in maintenance with prime focus on Industry 4.0. The concept for Maintenance Analytics (MA) is based on three interconnected time-lined phases, which aim to facilitate maintenance actions through enhanced understanding of data and information. Major focus of MA is towards knowledge discovery in maintenance. Based on four time-related perspectives, MA tries to address the discovery, understanding and communication process of maintenance data. These time-related perspectives match with the determination of the past, present and future state of an asset summarized by Gartner [9] in four questions, as it can be seen in Fig. 8. What happened, why it happened, what will happen and how can we make it happen are the issues involving the determination of the state of an asset. The questions are ordered by the value of the information given by each of them, in such a way that the former has the less value and the latter the higher value. Nevertheless, obtaining this valuable information requires more and more resources as the difficulty to achieve the goals proposed by the questions is higher. The last question will be in the spotlight in the coming future in order to decide how to take advantage of a future opportunity or mitigate a future risk, getting information about the implications of each decision option. The selection of the best option, based on some given parameters, will provide a meaningful tool for improving maintenance planning and production scheduling.

- Maintenance descriptive analytics (monitoring) focuses to discover and describe what happened in the past and why something happened. In this phase, access to data related to system operation, system condition and expected condition is highly important. Another important aspect in order to understand the relation-

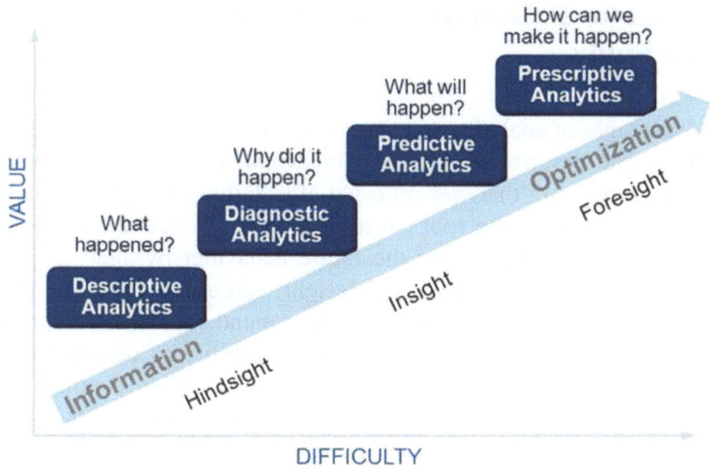

Fig. 8 The way to prescriptive analysis (Gartner [9])

ship of events and states during the descriptive analytics is time and time frame associated with each specific log.

- Maintenance diagnostic analytics: It explains the possible reasons for faults or failures, i.e. the why and the where question, since diagnosis is defined by EN13306 [4] as the fault detection, identification and localization.
- Maintenance predictive analytics focuses to estimate what will happen in the future. The maintenance predictive analytics phase of MA aims to answer "what will happen in the future?" but also "why will it happen?" In this phase the outcome from "maintenance descriptive analytics" is used. Additionally, in this phase, availability of reliability data and maintainability data is necessary beside the data used in descriptive phase. In addition, in order to predict upcoming failure and fault, there is a need to provide business data such as planned operation and planned maintenance to this phase.
- Maintenance prescriptive analytics which addresses what need to be done next. The maintenance prescriptive analytics phase of MA aims to answer "What needs to be done?" When dealing with maintenance analytics (MA), provision of appropriate information logistics is essential. The main aim of information logistics is to provide just-in-time information to targeted users and optimization of the information supply process, i.e. making the right information available at the right time and at the right point of location [12, 13]. Solutions for information logistics need to deal with (I) time management, which addresses "when to deliver"; (II) content management, which refers to "what to deliver"; III) communication management, which refers to "how to deliver"; and (IV) context management, which addresses "where and why to deliver" [12, 13].

5 The Need for Prescriptive Analytics in Maintenance: A Case Study

There are four stages of analytics that vary depending on difficulty, value and trends of technology. The descriptive analytics (hindsight) provides what has happened based on measuring asset that was reflected after failure. Diagnostic analytics can provide the reason behind the root cause of failure. Predictive analytics (insight) in the present provides can predict the future behaviour by analysing remaining useful life. The prescriptive analytics (foresight) can assess the recommendations provided by the predictive analytics and recommend measures for corrective or preventive maintenance actions. This has the capability to design the operation and maintenance according to our requirements that adapts continually without excessive user intervention thus acts as a back end for Industry 4.0 systems. Next figures show the natural deterioration and restoring process of an asset with the corresponding thresholds. Looking upon all mentioned before, next figure (Fig. 9) summarizes the three potential scenarios:

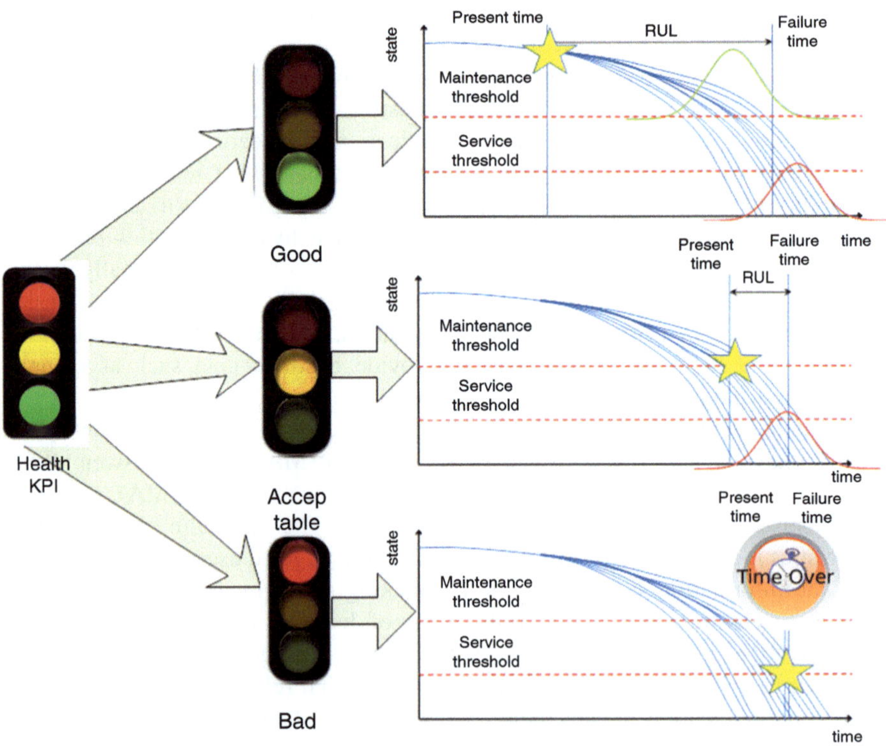

Fig. 9 Maintenance scenarios

On the one hand, there is the health of the component, which will be represented by a traffic light. Each of these states is related to a different scenario in terms of maintenance. It must be remarked that the diagnosis phase, i.e. the answer to "what is happening", may be able to distinguish between three different health states, limited by two thresholds: maintenance threshold and service threshold (depicted Fig. 9). The first to be understood as a warning limit when maintenance personnel must start considering to deploy a maintenance action. The last is formally equivalent to RUL, when the component and consequently the machine get a failure and the service is interrupted.

In the first scenario, when neither the maintenance threshold nor the service threshold has been crossed, two figures may be relevant for operators and maintainers:

- Remaining time to get the maintenance threshold
- Remaining time to get the service threshold

In this first phase of the life span, the system does not suggest to perform any maintenance action, since the component is considered in an early safety stage, and the RUL estimation dictates that the risk of failure is still far (Fig. 10).

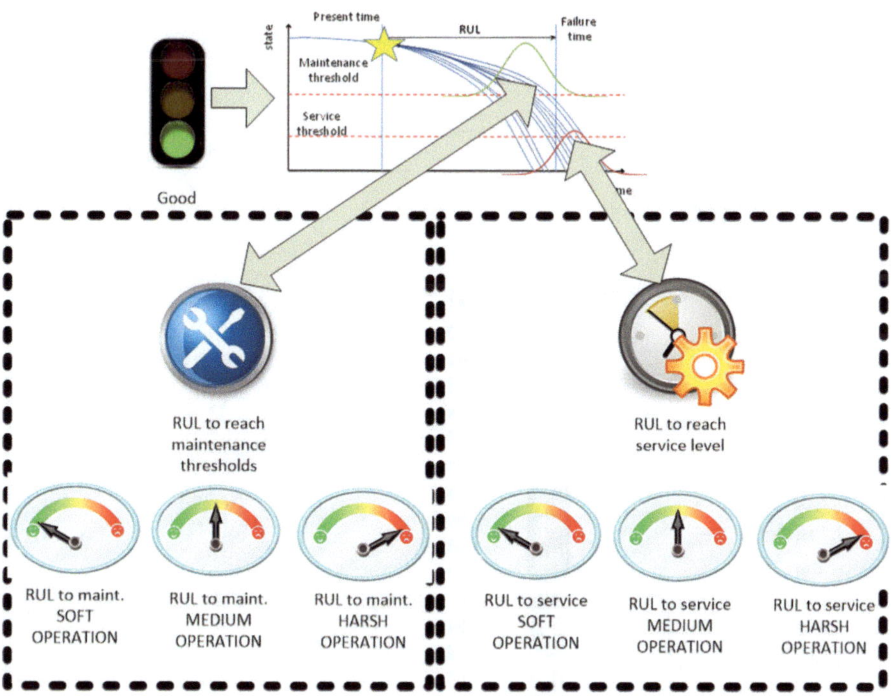

Fig. 10 Good health scenario

Once the maintenance threshold has been crossed, the component gets into a risky stage where a maintenance action should be performed in a near future, in order to avoid a failure.

In this case, the RUL estimation to get the service threshold should be presented to the end user as a result; it means that this would be the time to failure if the maintenance personnel do not perform any maintenance action ("do nothing" option), but the end user may also be interested in the consequences of taking one or another maintenance action (preventive or corrective). It is in this point where the RUL restoration parameter must be considered to decide which health state would have the component after a maintenance action.

The challenge is to know the real condition of the asset, i.e. the RUL consumed in each threshold and therefore to know if the RUL restoration of each maintenance action is a fact and the maintenance threshold has been crossed back restoring the asset to the healthy condition. If so, the component will get a good health condition; if not, it may remain in the risky situation. This mentioned situation may happen when performing a preventive maintenance action; while applying corrective maintenance in such situation, it is considered that a recovery up to the good health condition is feasible. The reason is that after a corrective maintenance action, the improvement is high enough to cross back the maintenance threshold but not the PM action (Fig. 11).

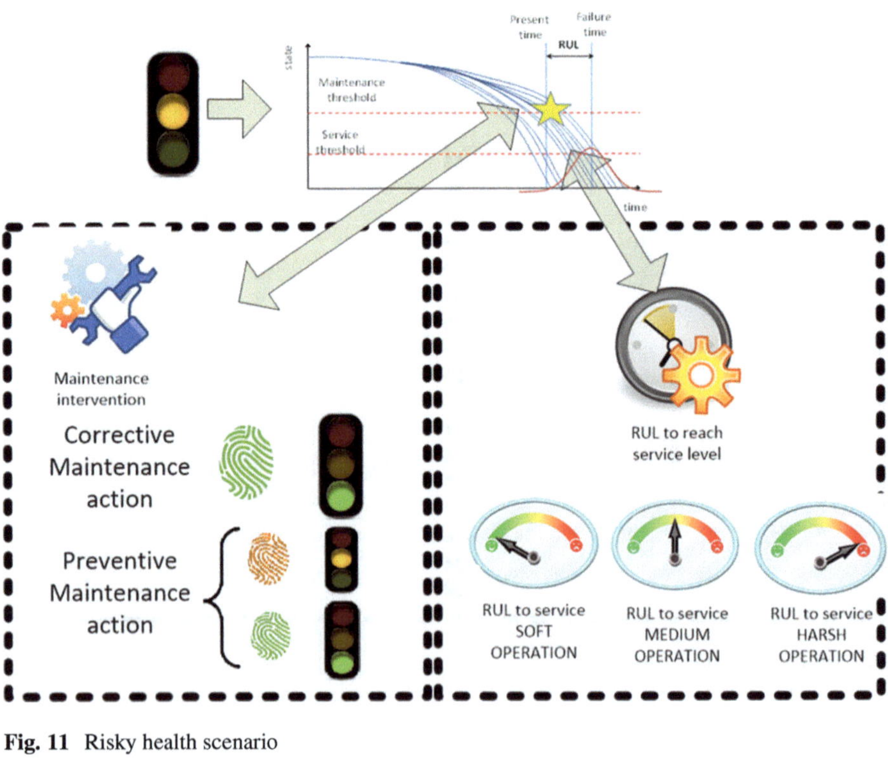

Fig. 11 Risky health scenario

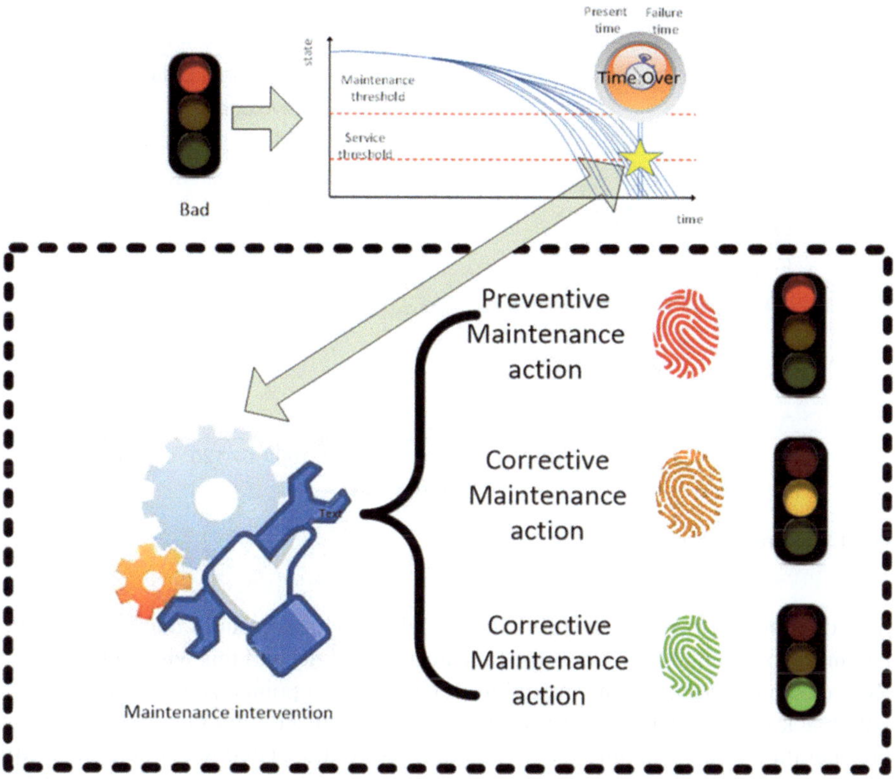

Fig. 12 Faulty health scenario

Finally, the last scenario to be covered is the one where the component reaches the service threshold. After this point, the only solution to recover the component's health is to perform a corrective action (many times reactive). As it happened in the previous scenario, the recovery situation would get the good or risky condition depending on the threshold values (Fig. 12).

However prediction of RUL is no longer the information requested by the user in a predictive analytics approach since this information even valuable may be considered incomplete for the decision-makers. In the RUL visualization of Fig. 13, two different aspects of the prognosis techniques are shown. Since prognostics deals with predicting the future behaviour of engineering systems, there are several sources of uncertainty which influence such future prediction, and therefore, it is rarely feasible to obtain an estimate of the RUL with complete precision.

In fact, it is not even meaningful to make such predictions without computing the uncertainty associated with RUL. In the case of prescriptive analytics, the uncertainty can be meaningful for the user since one of the uncertainty sources comes from the lack of knowledge in the operation of the machine. Indeed, the most intuitive way to show a component's degradation is drawing the evolution over

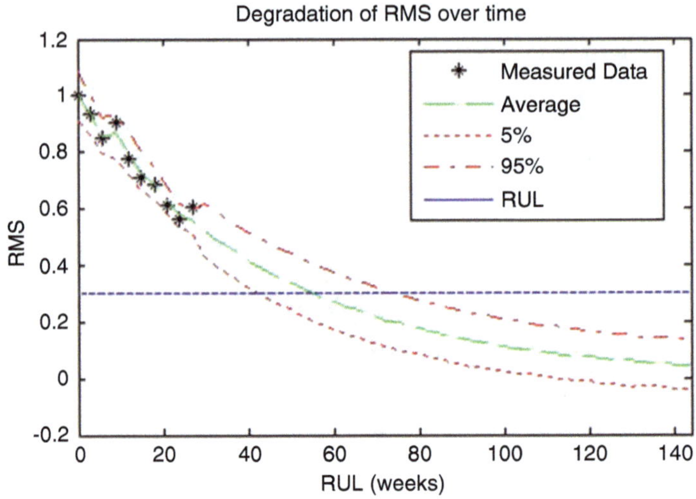

Fig. 13 Degradation over time

time of some performance/health index. In Fig. 13, the threshold that indicates the beginning of the faulty region is the blue horizontal line, and this will determine the RUL of the component. The evolution of the real data points over time is depicted with black stars. It can be observed that red lines as uncertainty measures regarding the operation of the machine provide different RUL associated with operations hard and soft against the medium or normal operation mode considered in the RUL estimation.

This estimation provides to operators and maintainers the information to schedule maintenance actions but not only that. Since the different operational profiles provide different RUL estimates, then operators and maintainers have the opportunity to decide different operational sequences in order to open up maintenance windows whenever convenient for the business. These alternatives far away from static RUL predictions are the prescriptive analytics requested by O&M departments and asset managers in Industry 4.0.

6 Concluding Remarks

The foundation of Maintenance solution for manufacturing sector adopting Industry 4.0 vision is founded on the capability of big data analytics (descriptive, predictive and prescriptive analytics). Traditional view of viewing data as record keeping has been transformed to an asset capable of driving business improvement. However "big data" like any other data needs to be cleaned, consistent and secured. The biggest challenge in the big data analytics is dealing the missing or corrupted multidimensional data. There is a need for high-performance computing power and algorithms to process the massive quantities of real-time data.

Key trigger for the big data approach in asset management has been IoT. It is a paradigm where everyday objects are connected to the Internet. In this approach physical asset equipped with multiple sensors can produce large volume of data. Those data could be used through the knowledge discovery to build, for example, asset degradation prediction model that can be employed for optimization in the form of a maintenance decision support system. A data-driven approach is not as accurate as physics-based approaches; however it does not require excessive knowledge about underlying physical processes.

The proposed framework described in this article implies a progress beyond the state of the art in the application of SOA ICT technologies within Industry 4.0 and specifically Maintenance 4.0. eMaintenance solutions grounded on maintenance analytics and built on intelligent, instrumented and well-managed IT infrastructure will enable innovation and transformation in pursuit of implementation of Industry 4.0. In this regard, the directions identified along the paper for further research can be summarized as follows:

- Real-time knowledge discovery algorithms from plethora of heterogeneous asset data sources may be derived. The idea is to ensure privacy-preserved processing, appropriate and effective selection of features and instances, discretization, compression of data, etc.
- With the help of a virtualization layer in between, the data acquisition process and data analytics, derived from vast cross-domain acquisition of data sources, may be considered. This shall enable newer solutions to be associated with database capabilities and to integrate heterogeneous data sources
- Using open-interface gateways with monitoring systems to enabling big data communications in order to provide timestamp and position synchronization, priority protocols for real-time transmission of information, and heterogeneous communication support, including mobility and aggregation.

References

1. Ashton K (2009) That 'internet of things' thing. RFiD J 22(7):97–114
2. Baglee D, Marttonen S (2015) The need for Big Data collection and analyses to support the development of an advanced maintenance strategy. In: Proceedings of the international conference on data mining (DMIN). The Steering Committee of The World Congress in Computer Science, Computer Engineering and Applied Computing (WorldComp)
3. Baheti R, Gill H (2011) Cyber-physical systems. Impact Control Technol 12:161–166
4. CEN EN (2001) EN 13306: maintenance terminology
5. Chen M, Mao S, Zhang Y, Leung VC (2014) Big data: related technologies, challenges and future prospects. Springer, Heidelberg
6. Fayyad UM, Uthumsamy R (1995) Preface. In: Proceedings of the first international conference on knowledge discovery and data mining. AAAI Press, New York

7. Galar D, Kumar U, Lee J, Zhao W (2012) Remaining useful life estimation using time trajectory tracking and support vector machines. J Phys Conf Ser 364(1):012063. IOP Publishing
8. Galar D, Thaduri A, Catelani M, Ciani L (2015) Context awareness for maintenance decision making: a diagnosis and prognosis approach. Measurement 67:137–150
9. Gartner Report (2012) Forecast: enterprise software markets. Worldwide, 2011–2016, 4Q12 Update. Accessed on 18 Mar 2013 through http://www.gartner.com/resId=2054422
10. Hall D, Llinas J (2001) Handbook of multisensor data fusion. CRC Press LLC, Boca Raton, pp 17–26
11. Han J, Cai Y, Cercone N, Huang Y (1992) DBLEARN: a knowledge discovery system for large databases. In: International conference on information and knowledge management, Baltimore, MD
12. Haseloff S (2005) Context awareness in information logistics
13. Heuwinkel K, et al. (2003) Information logistics and wearable computing. Distributed computing systems workshops, 2003. In: Proceedings of the 23rd international conference on IEEE, 2003
14. IEV (2016) International electrotechnical vocabulary (IEV). Available at: http://www.electropedia.org/. Accessed 18 Mar 2016
15. Karim GD, Westerberg J, Kumar U (2016) Maintenance analytics- the new know in maintenance, Accepted in IFAC AMEST 2016, October 19–21, 2016 Biarritz, France
16. Kroll B, Schaffranek D, Schriegel S, Niggemann O (2014) System modeling based on machine learning for anomaly detection and predictive maintenance in industrial plants. In: 2014 IEEE Emerging Technology and Factory Automation (ETFA), pp 1–7
17. Lasi H et al (2014) Industry 4.0. Bus Inf Syst Eng 6(4):239
18. Lee J (2003) E-manufacturing–fundamental, tools, and transformation. Robot Comput Integr Manuf 19(6):501–507
19. Lee EA (2008) Cyber physical systems: design challenges. 2008 11th IEEE international symposium on object and component-oriented real-time distributed computing (ISORC). IEEE, 2008
20. Lee J, Kao H-A, Yang S (2014) Service innovation and smart analytics for industry 4.0 and big data environment. Proc CIRP 16:3–8
21. Lee J, Bagheri B, Kao HA (2015) A cyber-physical systems architecture for industry 4.0-based manufacturing systems. Manuf Lett 3:18–23
22. Liu D, Ke C (2007) Knowledge support for problem-solving in a production process: a hybrid of knowledge discovery and case based reasoning. Expert Syst Appl 33:147–161
23. Lomotey RK, Deters R (2014) Towards knowledge discovery in big data. Service Oriented System Engineering (SOSE), IEEE 8th International Symposium on, 7–11 April 2014, pp 181–191
24. Manyika J, et al. (2011) Big data: the next frontier for innovation, competition, and productivity
25. NIST (2009) Definition of cloud computing v15. NIST, Editor
26. Sankavaram C, Kodali A, Pattipati K (2013) An integrated health management process for automotive cyber-physical systems. In: International conference on computing, networking and communications, Workshops Cyber Physical System, pp 82–86
27. Syed B, Pal A, Srinivasarengan K, Balamuralidhar P (2012) A smart transport application of cyber-physical systems: road surface monitoring with mobile devices. In: Sixth international conference on sensing technology, pp 8–12
28. Wang H (1997) Intelligent agent-assisted decision support systems: integration of knowledge discovery, knowledge analysis, and group decision support. Expert Syst Appl 12(3):323–335
29. Wang L, Wang G (2016) Big data in cyber-physical systems, digital manufacturing and industry 4.0
30. Zikopoulos P, Eaton C (2011) Understanding big data: analytics for enterprise class hadoop and streaming data. McGraw-Hill Osborne Media, New York

Modeling Fault Detection Phenomenon in Multiple Sprints for Agile Software Environment

Prabhanjan Mishra, A. K. Shrivastava, P. K. Kapur, and Sunil K. Khatri

1 Introduction

During the past two decades, we have seen projects getting either over budget or over scheduled too often and do not provide the value to the clients and their end users. We also understand that IT projects are often too complex that require thorough appreciative of customer's goals, challenges, customer needs, and expectations. As we note that the customer's needs evolve with change in their business strategies very often, thus the conventional approach (waterfall) to software development is too rigid to cater to these needs as it mandates that the requirements get frozen before the design and development starts. The time to market is the "key" as the marketers can rapidly gauge the interest and find out if there's a real demand for specific feature or function in the software. In order to meet these challenges, more and more software projects are adapting "Agile" [1].

Agile is a way of developing software (web applications, web sites, mobile apps) aiming to deliver high-quality working software frequently and consistently and, at the same time, minimizing the project overhead, thus increasing business value. Marketers, stakeholders, and (most importantly) customers are consulted at each step with further changes being made in response to testing, feedback, and business prioritization, before beginning a new sprint. This ensures that teams can evaluate and respond to any change in customer requirements.

In February 2001, "the Agile Manifesto" [2, 3, 4] was created to find alternative approaches to software development. The team studied the existing software

P. Mishra (✉) • S.K. Khatri
Amity Institute of Information Technology, Amity University, Noida, UP, India
e-mail: prabm@yahoo.com; sunilkkhatri@gmail.com

A.K. Shrivastava • P.K. Kapur
Amity Center for Interdisciplinary Research, Amity University, Noida, UP, India
e-mail: kavinash1987@gmail.com; pkkapur1@gmail.com

© Springer Nature Singapore Pte Ltd. 2018
P.K. Kapur et al. (eds.), *Quality, IT and Business Operations*, Springer Proceedings in Business and Economics, DOI 10.1007/978-981-10-5577-5_20

development processes, which were considered to be inflexible, quite intense, and heavily focused on documentation. The outcome of the study is summarized in the manifesto, which values:

- *Individuals and interactions over processes and tools* – Projects get build around inspired individuals. They should be provided the required support and environment and be trusted to get the work done.
- *Working software over comprehensive documentation* – The focus should be on delivering working software frequently and in preferably shorter timescale.
- *Customer collaboration over contract negotiation* – Key representative from the business and software developers needs to work together throughout the project.
- *Responding to change over following a plan* – The change in the requirement is welcome even at later stages in the development cycle.

There are a number of agile development methodologies [5] that are extensively used in software projects; the most common ones are Extreme Programming (XP), Dynamic Systems Development Method (DSDM), Feature-Driven Development (FDD), and Scrum. All methodologies engage in iterative model which emphasizes on incremental delivery of working software in short time intervals, which are called time-boxed iterations.

Extreme Programming (XP) – This is also referred as pair programming, where the emphasis is more on the peer reviews where one member reviews the code of another developer and vice versa, thus ensuring the highest quality of the deliverables.

Feature-Driven Development (FDD) – The FDD is one of the agile methodologies developed by Jeff De Luca and supported by Jon Kern and Stephen Palmer. FDD is based on a model having short iterations [6]. The process is initiated by forming an overall model followed by a sequence of 2-week "design by feature" and "build by feature" iterations. FDD recommends the practices such as "build regularly" and "ownership of components."

Dynamic Systems Development Method (DSDM) – DSDM focuses on "fit for business purpose" as the key criteria for delivery and acceptance. Requirements are planned and deliveries are done in short, fixed-length time boxes, referred as iterations.

Scrum – The term is borrowed from the "Rugby formation" where "the players huddle closely together, attempting to advance down the playing field after an accidental infraction" [7]. Scrum is the most widespread method for implementing agile in software projects due to its nature of simplicity and flexibility. Scrum emphasizes empirical feedback and advocates team self-management. Scrum has three roles:

(i) Product owner – The person has to have the vision for the product and authority to take decisions and should be available throughout the development period. The product owner has the responsibility to communicate the vision continuously and set the priorities to the team. He also manages the product backlog.

(ii) Scrum master – The scrum master's role is to act as a facilitator for the product owner and the project team. The scrum master is not supposed to manage the team; his primary job is to remove any obstructions that are hampering the team from achieving its sprint goals. This leaves the team remains focused on the sprint tasks.
(iii) Team – Ideally, a team comprises of 4–9 resources that include the developers, UI designers, architects, and testers. The team is self-disciplined and has the autonomy and responsibility to carry out the sprints tasks, thus meeting the sprints goals.

The deliverables which are "shippable incremental product" in Scrum are delivered in cycles called **sprints**. Typically, one sprint comprises of 2–4 weeks. Each sprint starts with sprint planning meeting, and in the meeting, the product owner and the development team agree upon the scope of work that will get accomplished during the sprint [8, 9]. The development team themselves determines how much work can be realistically accomplished during the sprint, whereas the product owner defines the acceptance criteria for the sprint. Scrum master help the team and the product owner to determine the duration of the sprint. Once the duration is finalized, all future sprints have to follow the same duration. Once a sprint is started, the product owner should keep himself aside and let the team do their work. During the sprint, the team gets together for a daily stand-up meeting, which is facilitated by the scrum master to discuss progress and any issues that are blocking the progress. The presence of the project owner is optional and he may attend the meeting as an observer and should not participate unless he has to provide some clarifications. The product owner is not supposed to request any changes during sprint. At the end of the sprint, the team demonstrates the completed work to the product owner and the product owner applies the defined criteria to ascertain the sprint deliverables, and he can either accept or reject the work based on the established criteria [1] (Fig. 1).

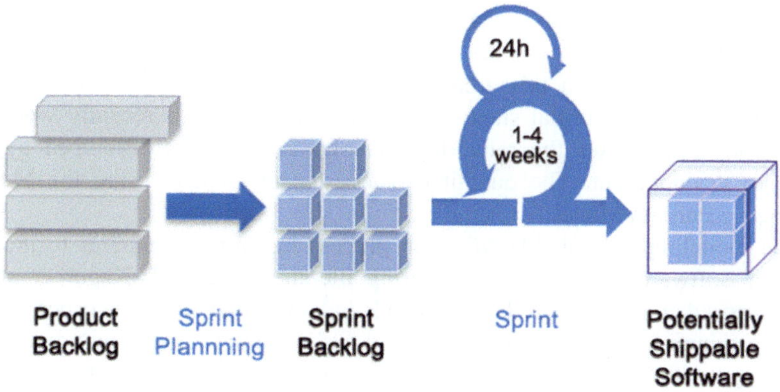

Fig. 1 Scrum process

2 Proposed Framework

In our proposed work, we intend to develop a framework to establish a mathematical model which depicts the trends of fault detection in agile software development. We have chosen Scrum methodology for the study and have gathered the data for multiple sprints.

The modeling of fault detection process in sprint is similar to the modeling done in case of multi-upgradation of software. The basic difference between the two can be understood as following:

In case of software upgradations, companies come up with multiple versions of the released software by adding new functionalities to the previous releases, whereas in case of sprint, it has a predefined period during which specific work is completed and the same is available for review. Once the product is in production, companies upgrade software by eliminating the dependencies of earlier versions and introduce new features for better user experience. There is always a risk attached with every upgrade as this could lead to new defects which may cause software failures. Hence there is a need to consider the defects that were detected from the previous releases along with the testing of the new code. The very phenomenon may lead to an increase in the failure rate during the upgrade. Failure rate tends to decrease as the testing team detects and the developers rectify the detected defects before the release. There is an increase in the failure rate due to the addition of new software features, which in turn increases the software complexity [10]. The objective of the software companies is to provide an upgrade along with enhancing the software reliability, which can be achieved by either redesigning or reusing of some modules by applying better engineering practices to reduce software failure rates.

As we see more and more projects are now being delivered in agile environment, but as the delivery methodology is relatively new as compared to conventional methods, the aim of our study is to assess the quality of various sprints of the entire life cycle of the software development. We would like to apply the similar approach (as described above in case of multiple releases) in case of sprints, where the outstanding faults from earlier sprints are removed together with the faults of current sprint.

The research done in this area (modeling of fault detection phenomenon in sprints) is still in its initial stages. Yamada et al. [11] have recently worked on prediction methods for the reliability of software products developed by agile software development. The quantitative assessment is done using the Moranda Geometric Poisson Model.

The modeling of fault identification process in sprints is similar to the work done in Software Reliability Engineering literature [12, 13]. Kapur et al. [10, 14] proposed a multi-upgradation model and extended it by incorporating the impact of imperfect debugging. Kapur et al. [15] also presented a two-dimensional multi-upgradation modeling. In the current paper, we have presented a generalized framework to model fault detection phenomenon for multiple sprints in agile software development environment. The paper is organized as follow: Section III describes the modeling framework of the proposed model. Section IV discusses the modeling of multiple sprints. In section V the numerical example is provided. Finally conclusion is drawn in section VI.

3 Derivation of SRGM with Generalized Modified Weibull Distribution

3.1 Basic Notations

$F_i(t)$	is the fault distribution function for ith sprint.
$m_i(t)$	is the expected number of faults removed in time $(0, t]$ for ith sprint, $i = 1, 2, 3, 4$.
a_i	is the expected initial fault content lying dormant in ith sprint, $i = 1, 2, 3, 4$, $a_i > 0$.
t_i	is the testing time of ith sprint.
$\alpha_i, \gamma_i, \lambda_i, \beta_i$	are the four parameters of the flexible GWM distribution prior to and after the change point, $i = 1, 2, 3, 4$.

3.2 Basic Assumptions

The following are the assumptions made for SRGM where the number of discoverable bugs is finite:

(A1) As soon as a software failure occurs, the fault causing this failure is instantly removed without introducing any new fault.
(A2) Fault detection phenomenon follows nonhomogeneous Poisson process (NHPP).
(A3) All the faults are mutually independent.
(A4) The expected number of faults identified in the time interval $(t, t + \Delta t)$ is directly proportional to the average number of outstanding faults.
(A5) Each fault is equally likely to cause software failure.
(A6) Likelihood of the existence of the remaining faults of ith sprints to $(i + 2)$th sprint is negligible.

By applying the hazard rate approach [16] in deriving the mean value function of cumulative number of faults removed, we have:

$$\frac{dm(t)}{dt} = \left(\frac{f(t)}{1 - F(t)}\right) \cdot (a - m(t)) \tag{1}$$

where $\frac{f(t)}{1-F(t)}$ is the fault detection rate. In solving Eq. (1) under the initial condition of $m(0) = 0$, we get

$$m(t) = a \cdot F(t) \quad (2)$$

Here, $F(t)$ is the cumulative distribution function.

The "bathtub"-shaped failure curves are very advantageous in carrying out the survival analysis of software. The four parameters of flexible GWM are used to model bathtub-shaped distributions. Also it is capable to model monotone and non-monotone failure curves, which are very common in reliability problems. This is the reason behind selecting this distribution for modeling fault detection phenomenon in our paper. The density function of flexible $GMW(\alpha, \gamma, \lambda, \beta)$ distribution is given by

$$f(t) = \frac{\alpha\beta t^{\gamma-1}(\gamma + \lambda t)\exp\{\lambda t - \alpha t^{\gamma}\exp(\lambda t)\}}{[1 - \exp\{-\alpha t^{\gamma}\exp(\lambda t)\}]^{1-\beta}}, t > 0 \quad (3)$$

and cumulative distribution function (CDF) is given by

$$F(t) = [1 - \exp\{-\alpha t^{\gamma}\exp(\lambda t)\}]^{\beta} \quad (4)$$

where $\alpha > 0, \beta > 0, \gamma \geq 0$ and $\lambda \geq 0$.

Here αcontrols the scale of distributions and the shape is controlled by γ, β. The parameter λ is an accelerating factor in the imperfection time, and it works as a factor of fragility in the survival of the individual when the time increases. The GMW distribution provides great flexibility with various forms of well-known distributions, i.e., exponential, Rayleigh, Weibull, modified Weibull [17], etc. We have used this CDF for modeling fault detection phenomenon in multiple sprints for agile software development process. Now on using Eq. (2), we get the mean value function

$$m(t) = aF(t) = a[1 - \exp\{-\alpha t^{\gamma}\exp(\lambda t)\}]^{\beta} \quad (5)$$

4 Modeling Fault Detection Phenomenon in Successive Sprints

Companies today face challenges to keep up to the ever increasing demands of its customers, and in order to out beat the competitors, companies are constantly applying innovation to remain relevant in the business. Thus, time to market has become an important factor; therefore, firms are adopting iterative development using multiple sprints.

The objective of the company is to ensure that the defects get detected at an early stage; therefore, the testing team gets engaged right from the first sprint. During the first sprint, the testing team prepares the test cases; therefore, actual defects start getting reported from subsequent sprints. In the early sprint, the defects encountered are primarily related to design phase as the actual coding starts from onward sprints.

4.1 Sprint 1

The outcome of each sprint is a potentially shippable product which in itself a complete feature that could be tested. Thus, during the sprints the testing team tests the deliverables of the previous sprint and the objective is to remove as many defects as possible, but we cannot be sure if all the defects would be detected; the testing team may detect only a finite number, which is less than the total number of existing defects in a particular coded functionality.

The number of defects getting detected during testing gets removed perfectly, which is represented by the mathematical equation as follows:

$$m_1(t) = a_1 F_1(t); 0 < t < t_1 \tag{6}$$

where F_1 (t) is the cumulative distribution function with respect to first release.

4.2 Sprint 2

New features in terms of user stories are added to the existing software which may result in introducing some new defects; thus, it becomes more imperative to get the code thoroughly tested so that the overall software quality could be enhanced. In sprint 2, testing starts right at the beginning of t_1. In this duration, there are defects from sprint 2 as well as $a_1. (1-F_1 (t_1))$; the remaining fault content of the first sprint interacts with new detection/correction rate. As a result of this, a portion of faults which were not detected during the testing of the first sprint now get acknowledged during this sprint. Along with previous sprint's defects, faults that are identified due to the enhancement of the features are also removed during the testing with new detection proportion, i.e., $F_2(t-t_1)$. The finite faults removed during (t_1, t_2) can be written mathematically as

$$m_2(t) = a_2 F_2 (t - t_1) + a_1 (1 - F_1 (t_1)) F_2 (t - t_1); t_1 < t < t_2 \tag{7}$$

4.3 ith Sprint

During subsequent sprints, new features and functionality get added to the code base of the previous sprint, and the testing is carried out for ith sprint. We consider faults generated in *i*th sprint and remaining number of faults from the just previous $(i-1)$th sprint. The new functionality keeps getting added in every sprint till the desired product features for a release are attained. The mathematical expression for this fault detection process can be given as

$$m_i(t) = a_i F_i\,(t - t_{i-1}) + a_{i-1}\,(1 - F_{i-1}\,(t_{i-1}))\,F_i\,(t - t_{i-1}) \tag{8}$$

$$t_{i-1} < t < t_i$$

5 Numerical Illustration

For numerical illustration purpose, we have collected sprint data of 7 months during which 287 faults were identified. Each sprint is of 4 weeks (1 month). The actual numbers of faults in the eight sprints are 10, 23, 19, 67, 66, 70, and 32, respectively. For estimating the parameters of the proposed model, we have used statistical package for social sciences (SPSS). The parameter estimation values with the coefficient of determination values (R^2) are given in Table 1. Figures 2, 3, 4, 5, 6, and 7 represent the goodness of fit curves for each sprint.

From Fig. 2, we can see that our proposed model does not present the same trend as we observe for subsequent sprints, the reason being the data for the sprint 1 comprises of the faults related to the design phase and not the faults related to coding phase.

Table 1 Parameter estimation

	Parameters					
Sprints	a_i	α_i	γ_i	λ_i	β_i	R^2
1	14.809	0.221	0.694	0.006	5.613	0.966
2	40.641	0.173	0.596	0.009	3.590	0.979
3	20.818	0.003	0.000	0.229	0.469	0.978
4	109.537	2.615	0.045	0.012	37.697	0.985
5	76.505	0.409	0.271	0.022	2.833	0.983
6	68.985	0.632	0.167	0.041	2.924	0.974
7	49.039	0.060	0.588	0.030	0.973	0.971

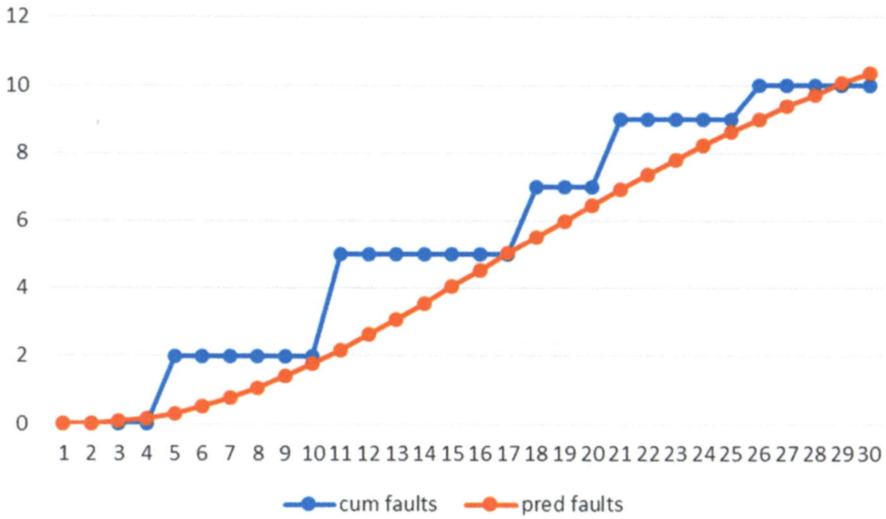

Fig. 2 Goodness of fit curve for sprint 1

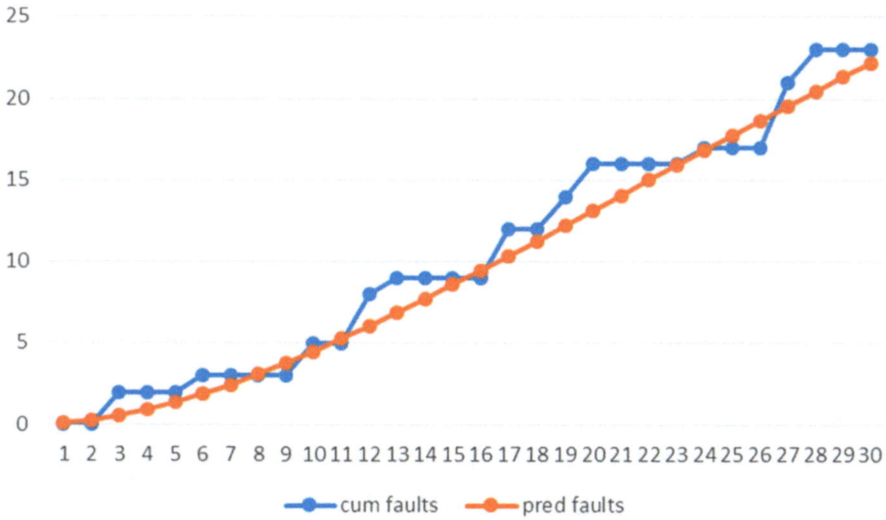

Fig. 3 Goodness of fit curve for sprint 2

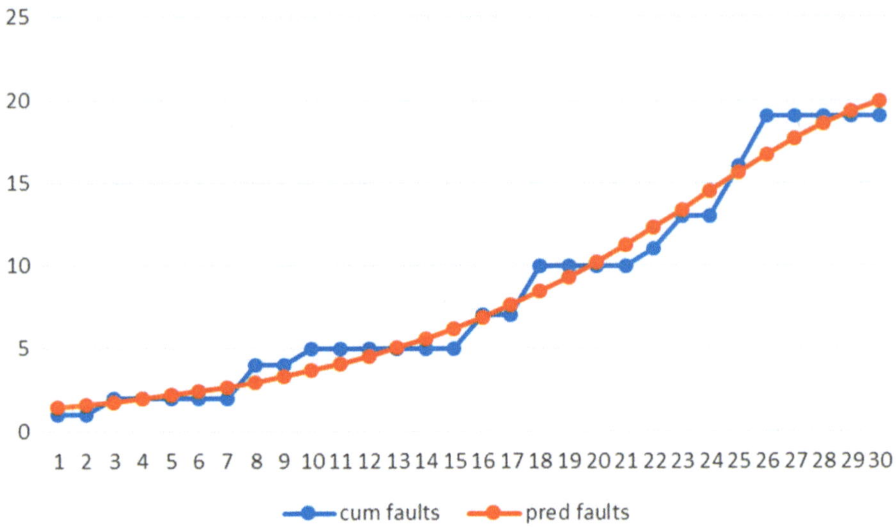

Fig. 4 Goodness of fit curve for sprint 3

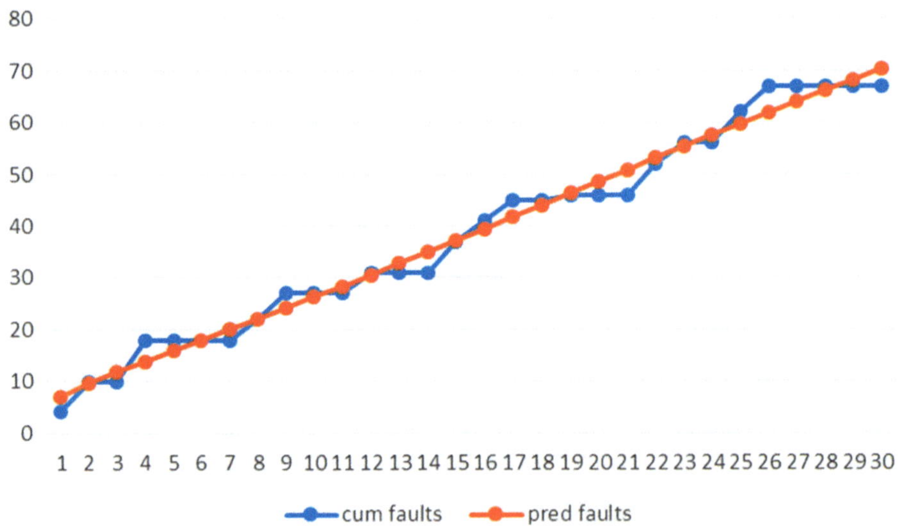

Fig. 5 Goodness of fit curve for sprint 4

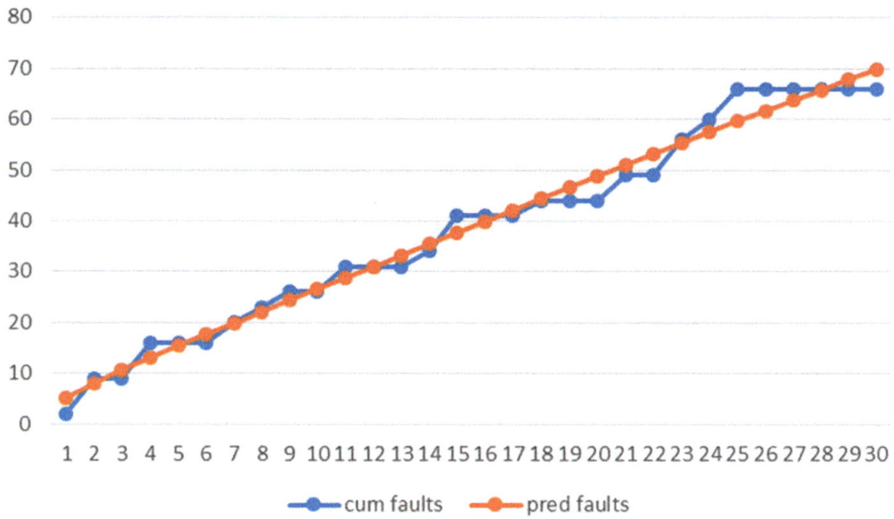

Fig. 6 Goodness of fit curve for sprint 5

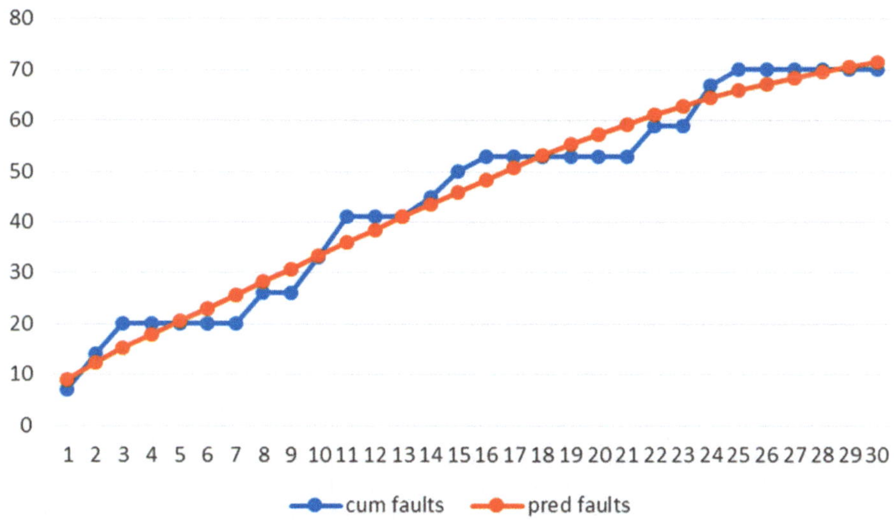

Fig. 7 Goodness of fit curve for sprint 6

6 Conclusion

Nowadays, the practice of delivering software using Scrum software methodology is becoming very prevalent; thus, there is a need to establish the trend of faults found during multiple sprints. We have proposed a generalized framework to find out the fault identification process during various sprints. From numerical illustration, we see that the data fits well on all the sprints except for first sprint.

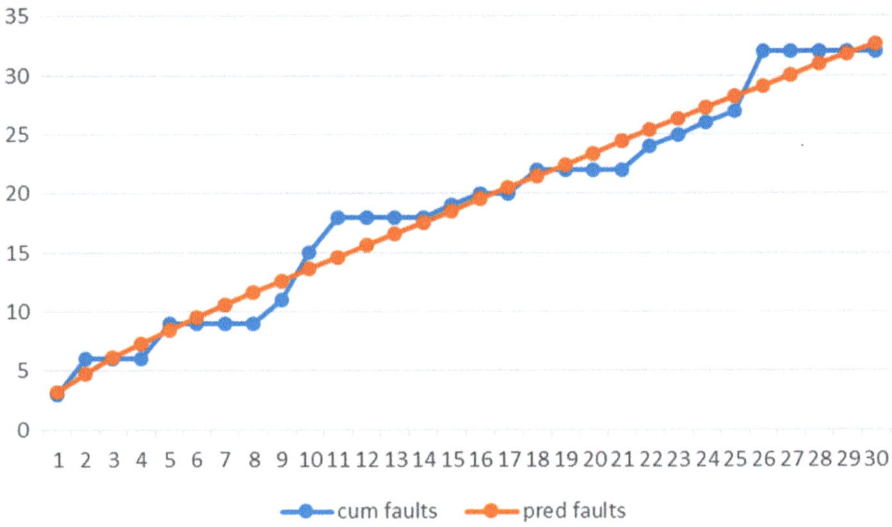

Fig. 8 Goodness of fit curve for sprint 7

For the future study, we can extend this model to include the phenomenon of fault identification process by taking into consideration testing resource consumption to derive the testing efficiency using our proposed model.

References

1. Sandra M, David F (2000) Journey towards agility: the agile wheel explored. TQM Mag 12(2):137–143
2. https://www.agilealliance.org/agile101/the-agile-manifesto/
3. Agile Maturity Model (AMM): a software process improvement framework for agile software development practices. Int J Software Eng 01/2009;2(1)
4. Dybå T, Dingsøyr T (2008) Empirical studies of agile software development: a systematic review. Inform Softw Technol 50(9–10):833–859
5. Miguel Morales R, Oktaba H, Pino FJ, Orozco MJ Applying agile and lean practices in a software development project into a CMMI organization. In: Product-focused software process improvement, Volume 6759 of the series Lecture notes in computer science. Springer, Berlin, pp 17–29
6. Jawadekar WS (2004) Software engineering. McGraw-Hill Education (India) Pvt Limited
7. Hossain E (2009) Using scrum in global software development: a systematic literature review. CSE, University of New South Wales, Sydney, NSW, Australia. Hye-young Paik, Babar. Global Software Engineering (2009) ICGSE 2009. Fourth IEEE international conference on, pp 175–184, 16 July 2009
8. Dingsøyr T, Dybå T, Moe NB (2007) Agile software development: current research and future directions. Springer, Berlin
9. Sharp H, Robinson H (2010) Three 'c's of agile practice: collaboration, coordination and communication. In: Agile software development: current research and future directions. Springer, Berlin, pp 61–85

10. Kapur PK, Tandon A, Kaur G (2010) Multi up-gradation software reliability model. 2nd international conference on Reliability, Safety and Hazard, (ICRESH-2010), 468–474
11. Yamada S (2015) Software quality analysis for agile development. 4th international conference on reliability, Infocom Technologies and Optimization (ICRITO) (Trends and Future Directions)
12. Pham H (2006) System software reliability. Springer, London
13. Kapur PK, Pham H, Gupta A, Jha PC (2011) Software reliability assessment with OR applications. Springer, London
14. Kapur PK, Garmabaki AHS, Singh J (2010) Multi up-gradation software reliability model with imperfect debugging. Int J Syst Assur Eng Manag 1(4):299–306
15. Kapur PK, Pham H, Aggarwal AG, Kaur G (2012) Two dimensional multi-release software reliability modeling and optimal release planning. IEEE Trans Reliab 61(3):758–768
16. Kapur PK, Pham H, Anand S, Yadav K (2011) A unified approach for developing reliability growth models in the presence of imperfect debugging and error generation. IEEE Trans Reliab 60(1)
17. Lai CD, Xie M, Murthy DNP (2003) A modified Weibull distribution. IEEE Trans Reliab 52:33–37

Optimal Price and Warranty Length for Profit Determination: An Evaluation Based on Preventive Maintenance

Adarsh Anand, Shakshi Singhal, Saurabh Panwar, and Ompal Singh

1 Introduction

Due to rapidly evolving technologies, the global market is being flooded with various new products. For companies to sustain in today's constantly changing business environment, they need to provide better services to their customers. Consumerism as a factor in the marketplace is an advent which will require increasing attention and adjustment on the part of producers wishing to maintain or improve their competitive situations in the face of expanding consumer awareness [1]. The intent of manufacturing organization is to strengthen their position in the market and to earn a maximum profit by increasing the sale of its product which can be achieved only if a customer is satisfied with the quality of their product.

To increase the sales of a product and to distinguish it with a competitor's product, warranty and maintenance serve as important marketing attributes for any organization. Murthy and Blischke [2] espoused two important roles of warranty. Firstly, warranty guarantees a protection against defective products for consumers and against excessive claims by customers for manufacturers and secondly warranty acts as a promotional tool for product differentiation by manufacturers. Within the marketing coverage, maintenance is commonly carried out for warranty service. Maintenance actions are classified into two strategies, which are *corrective maintenance* (CM) and *preventive maintenance* (PM). Preventive maintenance is used to control the deterioration process leading to failure of a system, and corrective maintenance restores the system to its operational state through corrective actions

A. Anand (✉) • S. Singhal • S. Panwar • O. Singh
Department of Operational Research, University of Delhi, Delhi, 110007, India
e-mail: adarsh.anand86@gmail.com; sakshi.singhal5@gmail.com; saurabhpanwar89@yahoo.com; drompalsingh@live.com

© Springer Nature Singapore Pte Ltd. 2018

P.K. Kapur et al. (eds.), *Quality, IT and Business Operations*, Springer Proceedings in Business and Economics, DOI 10.1007/978-981-10-5577-5_21

after a failure [3]. PM actions are scheduled actions and performed before the product fails. PM can be perfect which restores the product to "as good as new" state or can be imperfect, which restores the product to a state which lies between "as good as new" and "as bad as old."

Offering warranty also leads to the loss in revenue earned due to the cost of repairing the failed items over the warranty period. This loss can be minimized by providing preventive maintenance during the warranty period. Though by providing preventive maintenance, the manufacturer also incurs some maintenance cost. So, it is efficacious for the manufacturer to perform preventive maintenance only if the reduction in the cost of servicing the warranty is higher than the additional cost incurred with preventive maintenance. Preventive maintenance actions are executed either to lower the probability of a failure or to extend the useful life of the product.

Consumer buying behavior analysis is a first step in recognizing the customer's attitudes, preferences, intentions, and decisions regarding the purchase of the product. Understanding the rate of adoption of a product in the market is another important step which enables the firm to enhance their profit in the long run. Diffusion theory helps in explaining how a product is adopted and spread in the marketplace.

Diffusion research is the branch of marketing that aims to answer the important questions of modeling the sales cycle of new products [4]. Diffusion research outlines the conditions which increase or decrease the likelihood of adoption of a new product in the social system. Although the study of diffusion was originated in sociology and anthropology, the marketing and consumer behavior theorists have embraced the general paradigm of diffusion in their fields to expound new product acceptance and diffusion over time. The diffusion modeling was first introduced in early 1960, and since then, it has become the prominent topic of academic and practical research. The most widely accepted diffusion model was given by Bass [5] who proposed that the innovation diffusion can be brought to two information channels: innovation or external influence and imitation or internal influence. Despite its good fit to historical data, it lacks in explaining the adopter's decision-making process and does not completely account for market diversity. Therefore, in our study, we consider a two-dimensional technology diffusion innovation model proposed by Kapur et al. [6] which combines the adoption time of technological diffusion and price of the technology product.

The purpose of the current study is to understand the significance of warranty in sales of the product and in the overall profit maximization of the firm. The impact of providing preventive maintenance on the warranty within a warranty period is also investigated. In the proposed model, the warranty price and warranty length of the product act as a decision variable to maximize the firm's profit and for determining the optimum maintenance level.

2 Literature Review

2.1 Prior Literature on Innovation Diffusion Modeling

The introduction of new product in the market is the key dimension of a company's business strategy. Thus, the innovation diffusion modeling has been a very beneficial process. Several models on diffusion and adoption of innovation were developed in the literature since the 1960s. Some of the earliest models were given by [5, 7], and [8]. However, among these researches, Bass model [5] is the most well known and the simplest diffusion model [6]. Even though many studies have been carried out in later years on the usefulness of the Bass model, it has limited applicability. The model emphasized only on one dimension of diffusion: either to examine the individual's adoption of an innovation or to explain the time path of adoption of technologies that typically follows an S-shaped curve. The diffusion model must incorporate marketing variables such as price, and also the marketing decisions must be made jointly so that the ultimate goal of the firm, i.e., maximum net profit, can be achieved. Kapur et al. [15] considered price as one of the dimensions of the diffusion of an innovation other than the time. In their study, they have combined the adoption time of technological diffusion and price of the product to depict the sales pattern.

2.2 Previous Study of Sales Model Incorporating Product Warranty

Apart from continuation time and price, warranty also acts as a significant marketing variable that uplifts the sales of the product and eventually the profit levels. Berke and Zaino [9] defined warranty as a contractual obligation offered by the manufacturer in correspondence with the sale of the product. Chun and Tang [10] dealt with warranty model for the free-replacement warranty policy that determines the optimal warranty price for a given warranty period. Wu et al. [11] developed a sales rate function with a positive discount rate and a separable sales rate function with a zero discount rate to determine the optimal price and warranty length. Ladany and Shore [12] in their study assumed that a longer warranty period increases the selling price of the product. In their study, the manufacturer's lower specification limit (LSL) coincides with the offered warranty period, and the optimal lower specification limit was determined so as to maximize the expected revenue per item. The model given by [4] treated optimal warranty length and price of goods as decision variables and finds overall maximized profit for producers.

2.3 Literature on Connecting Preventive Maintenance and Warranty

The concept of preventive maintenance has been examined rigorously in the marketing literature. Article by [13] provides an impressive review in this area. Chun and Lee [18] proposed a model of a system with an increasing failure rate and subjected to periodic preventive maintenance actions during and after the warranty period. Later, in Djamaludin et al. [3], the warranty cost model provides a linkage between the warranty and preventive maintenance. The literature review on maintenance models given by [14] overviews numerous research models that deliver strong interlink between warranty and preventive maintenance. An effective preventive maintenance strategy has a significant impact on the warranty servicing cost to the manufacturer as it reduces the product failure rate and increases the useful life of the product. Therefore, in our proposed framework, we have integrated the concept of preventive maintenance and product's warranty to obtain the optimal warranty length and price of the goods. Our aim is to maximize the firm's overall profit using optimal maintenance level.

3 Mathematical Modeling

3.1 Notations

The following notations are used for modeling purpose:

x:	The value of the product
t:	The continuation time of the product in the market
p:	The innovation rate
q:	The imitation rate
r_s:	The revenue per unit
η:	The output elasticity
$\overline{N}$:	Initial potential adopters or market size
LSL:	Lower specification limit or warranty length
USL:	Upper specification limit
m:	Preventive maintenance level
$C(t, N(t, r_s))$:	Production cost function
$C_m(t)$:	Preventive maintenance cost per unit time
λ:	Inverse scale parameter for Weibull distribution
β:	Shape parameter for Weibull distribution
$f(t; m)$:	Failure rate function of the product with maintenance level m

In the proposed profit maximization problem, we consider a two-dimensional diffusion model to evaluate the sales pattern of the innovation given by [15]. The model combines the continuation time of the product in the market and the price of the technology product, i.e., sales rate is dependent not only on time, but price is another important aspect that governs the pace of adoption.

The mathematical form of the effort function can be expressed using Cobb-Douglas production function [16] as:

$$x\,(t, r_s) = t^{\eta} {r_s}^{(1-\eta)} \tag{1}$$

The differential equation of rate of change of sales due to the value of the product is given as:

$$n(x) = \frac{dN(x)}{dx} = p\left(\overline{N} - N(x)\right) + \frac{qN(x)}{\overline{N}}\left(\overline{N} - N(x)\right) = \left(p + \frac{qN(x)}{\overline{N}}\right)\left(\overline{N} - N(x)\right) \tag{2}$$

With the initial condition at $x(0,0)=0, N(0,0)=0$, Eq. (2) can be solved to get cumulative sales till time t as:

$$N\,(t, rs) = \overline{N}\left(\frac{1 - e^{-(p+q)t^{\eta}.rs^{1-\eta}}}{1 + \frac{q}{p}e^{-(p+q)t^{\eta}.rs^{1-\eta}}}\right) \tag{3}$$

After the formulation of two-dimensional modeling, we will now emphasize on the details of proposed model development.

3.2 Product Preventive Maintenance Policy

One of the most important objectives of preventive maintenance is to improve the functional reliability of the product as it reduces the number of breakdowns of the product.

We assume that preventive maintenance services are carried out very frequently so that it can be considered to be continuous over time. The maintenance level m is bounded by $0 \le m \le M$ where M denotes the upper limit. Higher value of m represents greater maintenance effort, i.e., more often inspection or product evaluation. We also assume the maintenance level to be constant throughout the product life cycle.

As $f(t; m)$ denotes the failure rate function of the product with maintenance level m, therefore, for $m = 0$, we have:

$$f\,(t; 0) = f_0(t) \tag{4}$$

If $m > 0$, then the failure rate is given by:

$$f\,(t; m) = f_m(t) \tag{5}$$

Also, $[f_0(t) - f_m(t)]$ is an increasing function of m that signifies that as the value of m increases the value of failure rate reduces.

3.3 Product Failure

We consider failure distribution, $F(t)$ to follow the Weibull distribution with shape parameter β. Let λ_m denotes the inverse scale parameter with maintenance level m and λ_0 be the inverse scale parameter with no maintenance level. We use the concept of imperfect maintenance, i.e., after each PM action, the state of the product lies between "as good as new" and "as bad as old" depending on the level of PM activities. The preventive maintenance effect is formulated by the change in the scale parameter. The functional relationship between λ_m and λ_0 is of the form:

$$\lambda_m = \lambda_0 \left(\frac{M - m}{M} \right)^{\gamma} \tag{6}$$

where $0 \leq m \leq M$ and $\gamma > 0$. This implies that failure rate increases in m.

The failure distribution function is given by:

$$F(t; m) = 1 - e^{-(\lambda_m t)^{\beta}} \tag{7}$$

3.4 Profit Model

It is assumed that the expected profit depends on the following factors: (1) sale volume, (2) unit producing cost, (3) sale price, (4) expected warranty cost, and (5) maintenance cost. We consider an optimization problem where the objective is to maximize the firm's profit and making price and warranty length as the decision variables. The proposed profit function is given as:

Profit = Sales Revenue – Expected Warranty Cost – Preventive Maintenance Cost – Production Cost

Revenue Function Suppose that the selling price of the product is r_s, then the expected revenue earned per product is given as:

$$J = r_s N(t, r_s) \tag{8}$$

Warranty Cost Warranties define the responsibility of the manufacturer in the occurrence of early product failure. The warranty period is the time duration in which if a sold-out product does not give satisfactory performance to the customer as offered by the retailer, then the manufacturer will provide replacement or repair of one or some parts of the product. The manufacturer provides a warranty in the beginning period, and it is assumed that warranty period coincides with the lower specification limit (LSL) of the product. Lower specification limit is a time duration during which a product is supposed to perform to the best of its capabilities. Also, manufacturer expects the product to last before its upper specification limit (USL) which constitutes the product useful life.

By providing warranty on products, there is a loss of revenue in each period of the product life cycle, i.e.:

- Loss in revenue due to the replacement of the product within the warranty period
- Post upper specification limit period, revenue loss due to non-failure of the product, and the producer losing in terms of repeat sales

The effect of the warranty cost may be expressed in terms of loss in revenue per unit produced. We assume:

- The company incurs costs of repairing the failed products in the form of warranty cost during the warranty period where $(r_L)r_S$, $(r_L < 1)$ represents the lower revenue per unit earned during the warranty period, i.e., $(1 - r_L)$ is the relative loss per unit for items with lifetime below the LSL due to servicing the warranty.
- The company also faces loss in revenue if the lifetime of the product exceeds USL. In this case, $(r_U)r_S$, $(r_U < 1)$ represents the lower revenue per unit earned by the manufacturer, i.e., $(1 - r_U)$ is the relative loss per unit of these items due to non-failure of the product and the producer lose due to no repeat purchase.

Therefore, the expected loss of revenue due to warranty is given as:

$$Z = r_S.(1 - r_L)\int_0^{\text{LSL}} f(x)dx + r_S.(1 - r_U)\int_{\text{USL}}^{\infty} f(x)dx \tag{9}$$

Loss in revenue in terms of the cumulative distribution function F is given as:

$$Z = r_S\left[(1 - r_L)F(\text{LSL}) + (1 - r_U)(1 - F(\text{USL}))\right] \tag{10}$$

Preventive Maintenance Cost Let $C_m(t)$ be the preventive maintenance cost per unit time with maintenance level m. It is assumed that maintenance cost increases linearly with maintenance level, i.e.:

$$C_m(t) = C_{m_0}{}^*m \tag{11}$$

where C_{m_0} is a constant.

Manufacturer incurs the maintenance cost during the warranty period, i.e., till LSL. So, the total preventive maintenance cost for the manufacturer of a product is:

$$P_m(t) = C_m(t)^*\text{LSL} \tag{12}$$

Production Cost Consider Q denotes the quantity produced by the manufacturer which depends upon the demand of the product, expressed in terms of number of units sold, $N(t, r_s)$, and on the product's sale price r_s. The relationship between these factors is expressed by the Cobb-Douglas type production function as given in [4]:

$$Q(N, r_s) = \delta N^{\alpha} {r_s}^{1-\alpha} \tag{13}$$

where δ and α are positive parameters.

Additionally, we assume C_0,C_1 and C_2 to be the setup cost, per unit production cost, and per unit inventory carrying cost, respectively. The total production cost comprising of fixed and the variable component at any time t and sales price r_S is given as:

$$C(t, N(t, r_s)) = [C_0 + C_1 \{Q(N, r_S)\} + C_2 \{Q(N, r_S) - N(t, r_S)\}] \tag{14}$$

4 Optimization Problem

The net profit Π earned by the manufacturer from selling of the product under warranty and preventive maintenance at any time t and at per unit price r_S is given as:

$$\begin{aligned} \Pi(t, r_S) = {} & r_s N(t, r_s) - r_S . [(1 - r_L) F(L) \\ & + (1 - r_U) . (1 - F(U))] . N(t, r_S) - m C_{m_0} L . N(t, r_s) \\ & - \left[C_0 + C_1 \{Q(N, r_S)\} + C_2 \left\{Q(N, r_S) - N\left(t, r_S\right)\right\}\right] \end{aligned} \tag{15}$$

For ease of use, (LSL, USL) is replaced by (L, U). Given the parameters in the above profit equation and also the distribution of the lifetime of the product, Π becomes a function of t and r_S. It is also assumed that the sales price of the product increases linearly with the warranty period, i.e.,

$$r_s = a + b.L \tag{16}$$

The effect of maintenance often leads to the useful life of the product being extended. Thus, it is considered that USL increases linearly in m.

$$U = U_0 + b_1 m \tag{17}$$

where U_0 is the constant useful life of the product in the absence of preventive maintenance.

The aim of the problem is to find the optimal value of L that can be extracted by maximizing Π and by determining optimal r_S, which when substituted in (16) provides optimal L. Thus, our optimization problem becomes:

$$\begin{aligned} \text{Max}\,\Pi\,(t, r_S) = {} & r_s N\,(t, r_s) - r_S.\left[(1 - r_L)\,F(L) + (1 - r_U).(1 - F(U))\right]. \\ & N\,(t, r_S) - mC_{m_0}\left(\tfrac{r_s - a}{b}\right).N\,(t, r_s) \\ & - \left[C_0 + C_1\,\{Q\,(N, r_S)\} + C_2\left\{Q\,(N, r_S) - N\big(t, r_S\big)\right\}\right] \end{aligned} \tag{18}$$

Subject to:

$$N\,(t, r_s) \geq N_0 \;\; \text{(Demand Constraint)} \tag{19}$$

$$Q\,(N, r_s) \leq Q_0 \;\; \text{(Production Constraint)} \tag{20}$$

$$Q\,(N, r_s) \geq N\,(t, r_s) \;\; \text{(No Shortages)} \tag{21}$$

where N_0 and Q_0 are nonnegative constants.

5 Numerical Illustration

The parameter estimates of the two-dimensional diffusion model which is validated on DRAM sales data sets are given in Table 1. The lifetime distribution function is assumed to follow a Weibull distribution. Then, the cumulative Weibull distribution function with variable L is given as:

$$F(L) = 1 - e^{-(\lambda_m L)^\beta} = 1 - e^{-\left(\lambda_m \frac{r_S - a}{b}\right)^\beta} \tag{22}$$

and the distribution function with variable U is:

$$F(U) = 1 - e^{-(\lambda_m U)^\beta} = 1 - e^{-(\lambda_m\{U_0 + b_1 m\})^\beta} \tag{23}$$

Table 1 Parameter estimates of two-dimensional sales model

Parameters	Estimates
$\overline{N}$	315.679
p	0.003
q	0.994
η	0.67

Table 2 Profit model parameters

Parameters	Values
λ_0	0.04
p	0.003
q	0.994
β	2
C_0	4
C_1	2
C_2	0.5
C_{m_0}	0.02
a	3
b	0.4
b_1	2
U_0	24
r_L	0.2
r_U	0.9
δ	4
α	0.603
N_0	250
Q_0	300

Table 3 Profit function results

m	λ_m	U	r_s	L	Profit
0	0.04	24	9.13	15.33	1339
1	0.036	26	9.13	15.33	1354
2	**0.032**	**28**	**9.13**	**15.33**	**1360**
3	0.028	30	9.13	15.33	1355
4	0.024	32	9.13	15.33	1338
5	0.02	34	9.13	15.33	1306

The formulated optimization problem for profit maximization model is solved using the LINGO software [17], and numerical data are given in Table 2. Results obtained using these values of parameters in maximization problem are given in Table 3.

6 Findings

From Table 3, it can be seen that the optimal maintenance level is obtained at $m = 2$ where the profit function is maximum. The firm should quote sales price of the product as Rs 9.13 and should provide warranty length of duration 15.33 months. Also, the firm should provide preventive maintenance in order to reduce failure of the product during the warranty period and thereby reducing loss in revenue

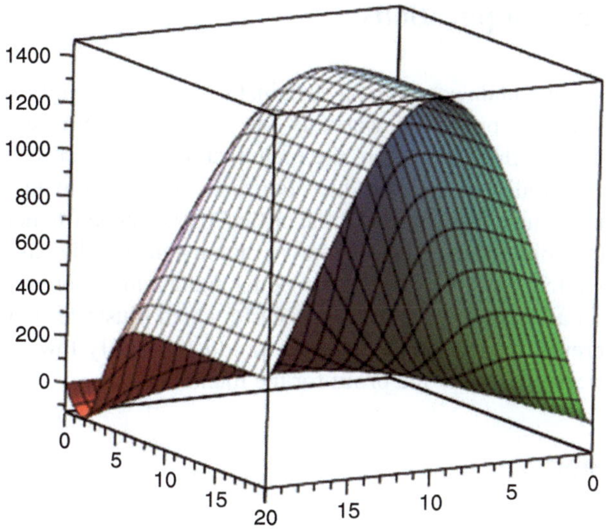

Fig. 1 Concave behavior of profit

Fig. 2 Optimal total maintenance cost

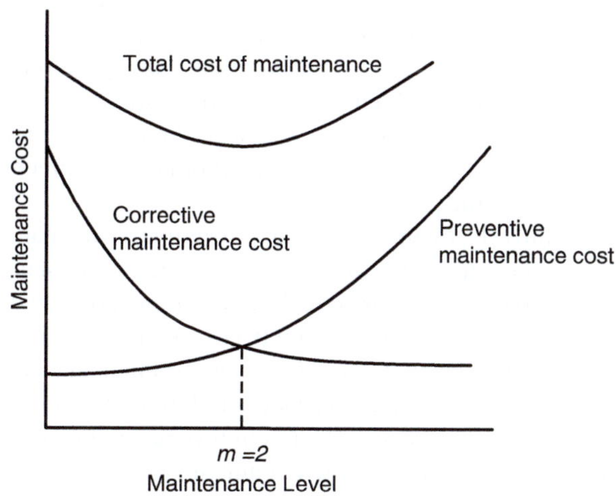

due to repairing the failed products. Further, the concavity of the profit function is established graphically which is shown in Fig. 1.

It can also be deduced that before $m = 2$ the preventive maintenance, i.e., repair cost, is higher as compared to the corrective maintenance cost and beyond maintenance level 2; the preventive maintenance cost is higher. At $m = 2$, the total maintenance cost is minimum as shown in Fig. 2, thereby implying maximum profit.

7 Managerial Implications

For companies to maintain economical sustainability and to strengthen their position in the global market, providing warranty on products is indispensable. Also to cope up with the tough competition, the manufacturing industries should focus on improving the reliability of their product. For this purpose companies should adopt preventive maintenance facility so that the critical parts can be frequently monitored and serviced to avoid failure. The preventive maintenance is the basic maintenance technique that extends useful life of the product and avoids future repair costs. The proposed model addresses the practical scenario which illustrates that by providing time-to-time preventive maintenance of the product not only lowers the product's failure rate but also increases the profit for the manufacturer.

8 Conclusion

This paper developed the warranty analysis model from the manufacturer's perception. The model incorporates the concept of warranty for deciding the optimal sales price and warranty length for a new product diffusion model. We have formulated an optimization problem which determines the maximum profit function of the manufacturer under product warranty and preventive maintenance strategy. The purpose of the problem is to provide managers with the optimal warranty length and per unit selling price along with the optimal preventive maintenance level for the product. The problem also incorporates two-dimensional sales model to estimate the sales of the product.

Intuitively, by implementing preventive maintenance during the warranty period, the manufacturer has to exhibit additional maintenance cost. However, it has been observed that the failure rate decreases tremendously due to preventive maintenance, and therefore, expected warranty cost reduces significantly by providing an optimal maintenance level, especially when the maintenance cost is lower as compared to the loss in revenue due to product failure in warranty period. Thus, the paper infers that the firm can reduce its loss in revenue and increase the product's useful life by providing optimal preventive maintenance policy.

Acknowledgment The research presented in this paper is supported by grants to the first and fourth author from the University of Delhi, R&D Grant No. RC/2015/9677, Delhi, India.

References

1. Glickman TS, Berger PD (1976) Optimal price and protection period decisions for a product under warranty. Manag Sci 22(12):1381–1390
2. Murthy DNP, Blischke WR (2000) Strategic warranty management: a life-cycle approach. Eng Manag IEEE Trans 47(1):40–54

3. Djamaludin I, Murthy DNP, Kim CS (2000) Warranty and preventive maintenance. Int J Reliab Qual Saf Eng 8(02):89–107
4. Aggrawal D, Anand A, Singh O, Singh J (2014) Profit maximization by virtue of price & warranty length optimization. J High Technol Manag Res 25(1):1–8
5. Bass FM (1969) A new product growth model for consumer durables. Manag Sci 15(5): 215–224
6. Singh O, Anand A, Kapur PK, Aggrawal D (2012) Consumer behaviour-based innovation diffusion modelling using stochastic differential equation incorporating change in adoption rate. Int J Technol Mark 7(4):346–360
7. Mansfield E (1961) Technical change and the rate of imitation. Econ J Econ Soc 29(4): 741–766
8. Fourt LA, Woodlock JW (1960) Early prediction of market success for new grocery products. J Mark 25(2):31–38
9. Berke TM, Zaino Jr NA (1991) Warranties: what are they? What do they really cost? In: Proceedings of the reliability and maintainability symposium, Annual, IEEE, pp 326–331
10. Chun YH, Tang K (1995) Determining the optimal warranty price based on the producer's and customers' risk preferences. Eur J Oper Res 85(1):97–110
11. Wu CC, Chou CY, Huang C (2009) Optimal price, warranty length and production rate for free replacement policy in the static demand market. Omega 37(1):29–39
12. Ladany SP, Shore H (2003) Optimal warranty period when sale-price increases with lower specification limit. In: Lenz H-J, Wilrich P-T (eds) Frontiers in statistical quality control, vol 7. Physica-Verlag, Heidelberg, pp 335–345
13. Dekker R (1996) Applications of maintenance optimization models: a review and analysis. Reliab Eng Syst Saf 51(3):229–240
14. Shafiee M, Chukova S (2013) Maintenance models in warranty: a literature review. Eur J Oper Res 229(3):561–572
15. Kapur PK, Singh O, Chanda U, Basirzadeh M (2010) Determining adoption pattern with pricing using two-dimensional innovation diffusion model. J High Technol Manag Res 21(2):136–146
16. Cobb CW, Douglas PH (1928) A theory of production. Am Econ Rev 18(1)
17. Thirez H (2000) OR software LINGO. Eur J Oper Res 124:655–656
18. Chun YH, Lee CS (1992) Optimal replacement policy for a warrantied system with imperfect preventive maintenance operations. Microelectron Reliab 32(6):839–843

Six Sigma Implementation in Cutting Process of Apparel Industry

Reena Nupur, Kanika Gandhi, Anjana Solanki, and P. C. Jha

1 Introduction

Six Sigma is a well-known and accepted concept in the industry for continuously improving the quality of products or processes. The concept explains that the quality of products and processes is the responsibility of all the stakeholders involved in the quality improvement process. Six Sigma is termed as disciplined and data-driven approach for eliminating errors and defects in a process. This approach emphasizes on continuous improvements at all the levels in production process, understanding of customer requirements, business productivity, and financial performance. A term sigma level is a statistical representation of Six Sigma which describes the process performance and quality. Lower sigma level indicates high chances of defects in the process and vice versa. In a project-driven approach, sigma level improves by reducing defects in the process, product, and service. These defects in processes, products, and services can be measured through DPMO (defects per million opportunity). To achieve 6σ, a process must not produce more than 3.4 DPMO. Here, defect is defined as any observation outside the customer specification.

In apparel industry, fashion and quality go hand in hand that deal with a lot of variation and defects in customer demand and manufacturing process.

R. Nupur (✉) • A. Solanki
Department of Applied Mathematics, School of Vocational Studies and Applied Sciences, Gautam Buddha University, Noida, Uttar Pradesh, India
e-mail: reenanupur1981@gmail.com; anjana@gbu.ac.in

K. Gandhi
Bhavan's Usha & Lakshmi Mittal Institute of Management, K. G. Marg, New Delhi, India
e-mail: gandhi.kannika@gmail.com

P.C. Jha
Departmentof Operational Research, University of Delhi, New Delhi, India
e-mail: jhapc@yahoo.com

© Springer Nature Singapore Pte Ltd. 2018

P.K. Kapur et al. (eds.), *Quality, IT and Business Operations*, Springer Proceedings in Business and Economics, DOI 10.1007/978-981-10-5577-5_22

The neck-to-neck competition in the garment business compels manufacturers to produce good quality of apparel in the market to attract customers. Six Sigma methodology is the implementation of a measurement-based strategy that focuses on process improvement and variation reduction, which is accomplished by using DMAIC Six Sigma methodology. The Six Sigma DMAIC process is an improvement system for existing processes falling below specification and looking for incremental improvement. These incremental improvements are done by reducing the present defects in product by using improved process. The total numbers of chances per unit to exhibit a defect are termed as opportunities. Each opportunity must be independent of the other opportunities and must be measurable and observable. The total count of opportunities indicates the complexity of a product or process. After having the knowledge about the complexities in any process, continuous improvement (CI) process is implemented to improve its quality. CI is a policy of constantly introducing small incremental changes in a business in order to improve quality and efficiency. CI can operate at the level of an individual or group, which are specifically brought together to identify potential scope of improvements. Improvements are based on many small changes rather than the radical changes that might arise from research and development. [1] proposes DMAIC Six Sigma approach to improve the processes in healthcare setups to reform and obtain zero error hospitals. [2] discusses in detail about the DMAIC approach of Six Sigma methodology and finds that instead of a structured methodology for a problem-solving, there are problems where it may be ineffective. They emphasized on empirical data usage, statistical tools usage in DMAIC, and lean DMAIC for a better result. [3] implements DMAIC Six Sigma methodology to improve the quality of asbestos roofing at PT BBI. In define phase, SIPOC diagram and Pareto chart; in measure phase, statistical tools like DPMO, sigma level; in analyze phase, fishbone diagram and failure mode and effect analysis (FMEA); in improve phase, DOE and ANOVA; in control phase, some recommendations were given to maintain the improved quality level and decreased DPMO. [4] shows the improvement of information quality; the quality of healthcare would improve to support the patient's satisfaction. The authors in this research have used Six Sigma methodology to improve information quality. The methodology is able to reduce information variance in healthcare, especially in the information that is used in the hospital information system. [5] explains an application of Six Sigma to improve process performance of the critical operational process that leads to better utilization of resources and reduction in variation in a manufacturing enterprise. [6] portrays the improvement in effectiveness of shell and tube heat exchanger in a small-sized furnace manufacturing company by implementing Six Sigma DMAIC methodology to. [7] shows the integration of Six Sigma and quality management system ISO 9000 for the development and continuous improvement of education system in universities. [8] examines the impact of critical success factors (CSFs) of Six Sigma quality program and their impact on performance indicators in some of Lebanese hospitals in Beirut. To achieve the objectives of study, two questionnaires were incorporated and three hypotheses were tested. [9] presents a new framework by integrating different tools and methods like Six Sigma DMAIC, FMEA, TOC,

FC, and swim line diagram to have continuous improvement for product processes and services. This will benefit companies by decreasing their production time and expenditures and increasing product throughput KPI.

In this paper, the concept of quality management tools and continuous improvement is implemented in garment industries to meet the quality specifications of customers. The paper is considering only cutting process of the production process. Four opportunities (distorted pieces, holes/spots, missing parts, raw/frayed edges) associated with the cutting process are identified. DMAIC Six Sigma is applied to identify the defects and improve the quality by reducing the number of defectives by continuous improvement. Pareto chart, cause and effect diagram for all the identified defects, and P-chart are used to have a brainstorming continuous improvement. The concept of DPMO is used to check the nonconformance at every step of the process, and the corresponding sigma level is calculated at every step to have a better interpretation of the research. Thus the aim of the paper is to assure the quality of apparel to the garment manufacturer by improving the sigma level of cutting process by incorporating statistical process control (SPC), Six Sigma tools, DPMO in the phases of DMAIC.

2 Six Sigma Methodology

Six Sigma provides an effective method to focus on customer requirements, through improving the process quality. The current study tries to impart quality to the apparels by controlling the defects during cutting process. To ensure the quality parameters as per buyers' expectation, an inspection is carried out to detect the defects, and then it is controlled to achieve quality specifications. Here, inspection can be defined as the visual examination or review of defects and to check if they meet the required measurements. Detection of defects, determination of causes of defects, reduction of defects, and then proper remedies to these defects can be a help to the apparel industry to achieve quality. The paper tries to increase the sigma level by reducing the DPMO for cutting process of the manufacture of apparel. A continuous improvement process is carried out through DMAIC by imparting many improvement actions to understand and reduce the serious components, which helps to achieve improved quality in the cutting process. The process is divided into four sections, i.e., define phase, measure phase, analyze and improvement phase, and control phase. Here, analyze and improvement phase has been put under the head of "continuous improvement" that discusses the analysis of the present state in the process and explains for the highest defected part of the process. Also, it tells the causes behind the defect. The improvement phase helps in reducing the defects in a specified defect type by using the information provided by analysis phase and the suggestions provided by the concerned departments. The connection between the causes of defects and its improvement process has to be created as both the phases are being processed subsequently.

2.1 Define Phase

Process of converting raw material into finished products is really difficult to maintain quality in the apparel industry. The industry has to suffer a heavy loss if production process of garments is not properly carried out. The production of garments has to go through various processes like cutting process, printing process, stitching process, washing process, and finishing process. The cutting process is the most important phase as once the fabric has been cut, there is hardly any scope to rectify the defects. Therefore, the chances of rejection are also high in such situations. In the highly competitive garment industry, where same quality can be provided by many vendors, a company can sustain only by delivering uncompromised quality product. Therefore, the company cannot afford such type of rejections.

2.2 Measure Phase

To achieve the required sigma level, we first need to know the presence of actual number of defects in the product and process. In TQM, the term DPMO is a quantitative measurement method that computes the total number of defects available in our product and process. By removing these defects, the DPMO value reduces and that leads to improvement in sigma level. Hence, sigma level and DPMO are inversely proportional. Sigma level can easily be calculated from DPMO on Microsoft Excel. In this phase, we collected the data for 20 days for each of the four opportunities as a check sheet (Table 1). DPMO is calculated on the basis of defects observed out of the total pieces produced per day. Mathematically, we can define DPMO as:

$$\text{DPMO} = \frac{\text{Total number of defects dedected}}{\text{No.of units produced} \times \text{opportunity of defects per unit}} \times 1000000$$

where a defect is defined as any part of the product or service that does not meet customer specifications or requirements or causes customer dissatisfaction opportunities which are the total number of chances per unit to exhibit a defect. Each opportunity must be independent of the other opportunities and must be measurable and observable. The total count of opportunities indicates the complexity of a product or process, which is something that can be quantified by a customer. It is a measurable and observable output of a business process. It may manifest itself as physical unit. The data recorded for 20 days for all four defects are collected in a check sheet.

From Table 1, it is obvious that total numbers of 175 products are defective out of 14,020 products. There are four opportunities, namely, distorted pieces, holes/spots, missing parts, and raw/frayed edges. From the above information, DPMO calculated is 3120. Sigma level of 4.23 is obtained from calculated DPMO. DPMO can be

Table 1 Check sheet for cutting process

Defects	Day 1	Day 2	Day 3	Day 4	Day 5	Day 6	Day 7	Day 8	Day 9	Day 10	Day 11	Day 12	Day 13	Day 14	Day 15	Day 16	Day 17	Day 18	Day 19	Day 20	Total
Distorted pieces	4	9	8	0	7	4	5	0	2	4	4	5	4	4	5	0	6	6	4	6	87
Holes/spots	1	4	3	2	1	2	1	2	0	2	1	2	4	3	1	2	0	5	1	4	41
Missing parts	3	3	0	1	4	4	3	1	0	0	3	0	2	0	1	4	0	1	1	1	32
Raw/frayed edges	0	2	1	1	0	0	1	0	0	0	1	0	3	0	1	0	0	1	2	2	15
Total defectives	8	18	12	4	12	10	10	3	2	6	9	7	13	7	8	6	6	13	8	13	175
Total pieces	624	1248	1248	624	1096	1066	419	424	530	534	412	413	997	622	578	428	423	826	631	827	14,020

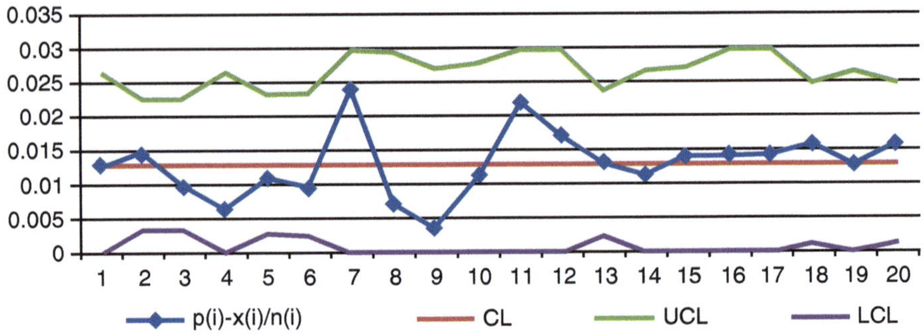

Fig. 1 P-chart showing cutting process

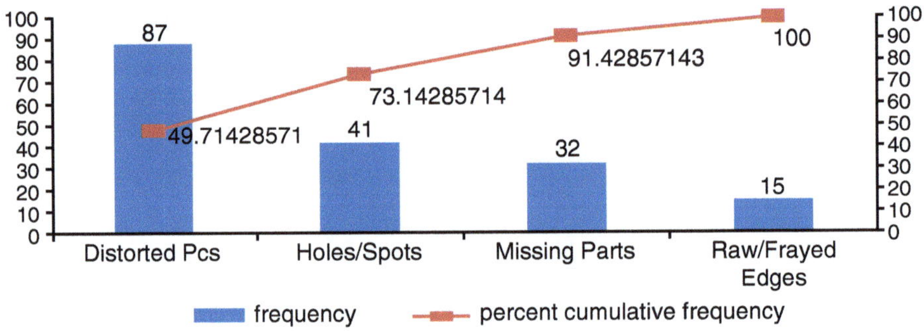

Fig. 2 Pareto chart for the original check sheet

converted into the corresponding sigma level by Six Sigma table or on Microsoft Excel as well. The sigma level of 4.23 says that there are 3120 defects per million.

It is clear from Fig. 1 that production is totally under control, but defectives, p(i), are showing very high fluctuation/variation from their central line. So, it is very much clear that there is further scope of improving the present available quality level. Our aim is to reduce the above calculated DPMO and achieve higher sigma level with continuous improvement using quality improvement tools. DPMO and sigma level are calculated every time after removing the defects one by one with the help of quality improvement tools.

The Pareto chart in Fig. 2 shows that initially distorted pieces contribute maximum error and subsequently other defect types are showing their weight age in the total defects. We start the improvement process with the reduction in defects of distorted pieces, and this reduction is progressed until maximum possible defects are reduced from all the defect types. Thus the aim of the paper is to assure the quality of the apparels to the garment manufacturer by improving the sigma level of cutting process by implementing the Six Sigma.

3 Continuous Improvement Process

The visual appearance, comfort, durability, and fit of a garment can be affected by defects that occur during manufacturing. The defects that occur during the cutting process of apparel production could hardly be removed, hence considered very serious and loss-making. Through the help of cause and effect diagrams, few areas of improvements are identified. These areas are checked and improved through continuous improvement, and the corresponding sigma level calculated is found to be improved. Each and every aspect causing the four defects was considered in the cutting process through quality improvement tools. This helps us to analyze our present problem well. Pareto chart, fishbone diagrams, check sheets, and control charts are among the seven quality improvement tools which are used to analyze our current problem in this paper. Pareto chart is prepared for decision-making and to identify as to which 20% of sources cause 80% of error. Pareto chart helps us determine which category or categories should be the focus of improvement efforts. Another one is cause and effect diagrams which help us to know the critical causes of defects. These critical causes could be improved through the continuous improvement. Hence these are considered as the key tool used in total quality control and Six Sigma. After keen observation of every related causes and the continuous improvement, finally achieved implementations are as follows:

3.1 Improvement Phase: Distorted Pieces

In the first phase of improvement, the paper tries to reduce the defect with the highest percentage of nonconformance, i.e., distorted pieces from Fig. 1. The Pareto chart for the four opportunities is prepared, and it is found that distorted pieces constitute 50% of defects during cutting of cloths. So it is clear that it is urgently required to control this defect of distorted pieces. To know probable causes of distorted pieces, a detailed discussion session is held by the quality team consisting members of different department in the cutting process. Then after getting consent of quality teams on different probable areas, a cause and effect diagram for distorted pieces (Fig. 3) is drawn as shown below.

On multi-voting by the quality team and finding areas of major consent by them, fabrics and pattern are against layout, and improper laying of patterns on fabrics are identified as root causes for the defect of distorted pieces. On checking the above causal factors, the process sigma level increases from 4.23 (DPMO = 3159) to 4.32 (DPMO = 2397) (Table 2). Corrective actions taken on the basis of root causes identified which actually reduced the defects are discussed in control phase for all defect types in this paper. Then to analyze the actual situation, a Pareto chart is prepared as shown in Fig. 4.

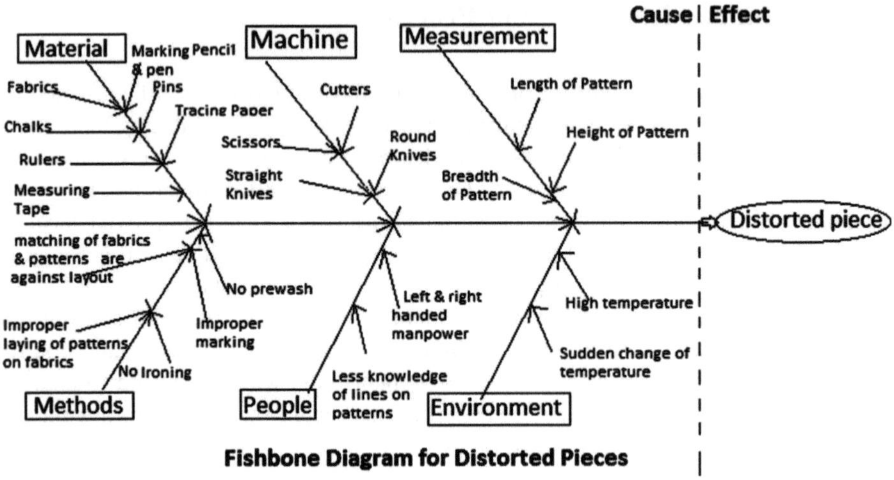

Fig. 3 Fishbone diagram for distorted pieces

3.2 *Improvement Phase: Holes and Spots*

The Pareto chart in Fig. 4 explains the defects of holes/spots that caused the maximum nonconformance to the process. So, now efforts will be directed to reduce this defect. Similarly as above defect type, a fishbone diagram for holes/spots is prepared and shown in Fig. 5.

Detailed discussion on the possible causes followed by the multi-voting by the experts the root causes for holes were identified as making notches and darts; improper pocket positioning whereas improper dyeing and shading were critical causes for spots. On working on these areas well, the improvement table obtained is shown in Table 3.

With the consideration and control of these critical areas, the defects for the holes/spots were reduced from 40 to 19 defects. Sigma level increased from 4.22 (DPMO = 3193) to 4.42 (DPMO = 1758). Then to analyze the actual situation again, a Pareto chart is prepared as shown in Fig. 6.

3.3 *Improvement Phase: Missing Parts*

The Pareto chart in Fig. 6 shows that the defects of missing part cause the maximum nonconformance to the process. Fishbone diagram for the defect of missing part is drawn to work on critical causes of it as shown in Fig. 7.

Table 2 Continuous improvement sheet-Distorted pieces

Defects	Day 1	Day 2	Day 3	Day 4	Day 5	Day 6	Day 7	Day 8	Day 9	Day 10	Day 11	Day 12	Day 13	Day 14	Day 15	Day 16	Day 17	Day 18	Day 19	Day 20	Total
Distorted pcs	0	2	1	1	0	0	1	0	0	1	2	0	2	0	1	0	0	1	2	2	16
Holes/spots	2	3	2	0	2	2	3	0	1	4	3	2	3	2	4	2	4	3	1	4	47
Missing parts	1	3	2	2	0	2	2	2	0	4	2	2	1	0	1	2	0	5	1	4	35
Raw/frayed edges	2	3	0	1	1	0	5	1	0	1	6	0	2	2	1	5	0	1	1	1	33
Total defectives	5	11	5	4	3	4	11	3	1	10	13	4	8	4	7	9	4	10	5	11	132
Total checked	519	1132	987	654	419	581	637	457	713	1095	891	423	713	542	643	634	414	911	643	760	13,769

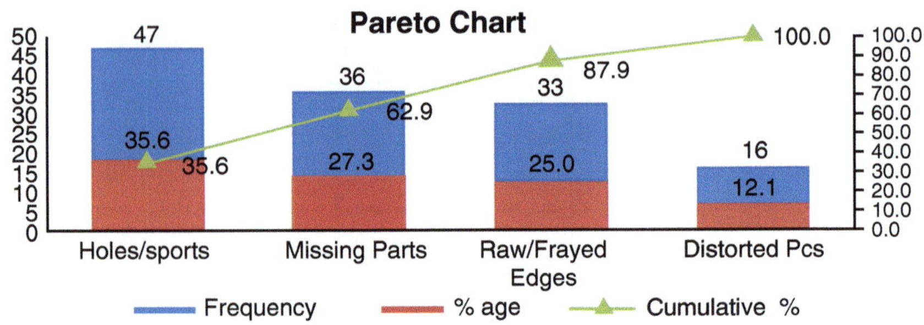

Fig. 4 Pareto chart after reducing distorted pieces

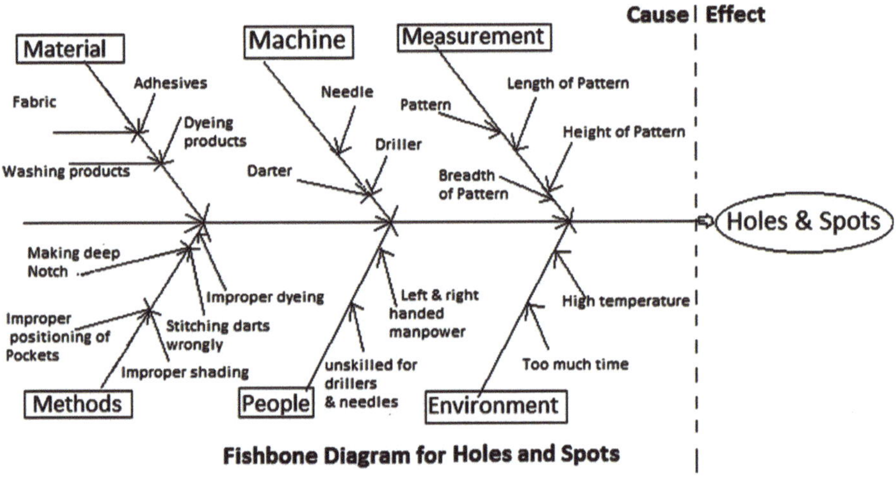

Fig. 5 Fishbone diagram for holes/spots

The critical causes for the defect type are losing patterned fabric; omission of slits, notches, and drills; facing wrong direction on fabrics; missing parts while cutting; and unskilled assembling of fabric parts.

The check sheet after the improvement is as shown in Table 4.

The sigma level is increased from 4.24 (DPMO = 3059) to 4.53 (DPMO = 1231). Pareto chart in Fig. 8 shows the updated state after employing the improvement on missing parts.

3.4 Improvement Phase: Frayed Edges

The Pareto chart in Fig. 8 shows that defects of raw/frayed edges cause the maximum nonconformance to the process. Fishbone diagram for the defect of raw/frayed edges is drawn to work on critical causes as shown in Fig. 9.

Table 3 Continuous improvement sheet – holes/spots

Defects	Day 1	Day 2	Day 3	Day 4	Day 5	Day 6	Day 7	Day 8	Day 9	Day 10	Day 11	Day 12	Day 13	Day 14	Day 15	Day 16	Day 17	Day 18	Day 19	Day 20	Total
Distorted pcs	0	2	2	1	0	0	2	1	0	0	1	3	3	1	0	1	1	2	0	1	21
Holes/spots	0	2	1	1	0	0	4	0	0	1	2	1	3	0	0	0	0	1	1	2	19
Missing parts	1	2	0	1	1	1	7	1	0	0	6	1	1	1	0	3	1	0	1	1	29
Raw/frayed edges	1	2	2	0	1	1	3	0	0	1	2	2	1	1	3	1	2	3	2	1	29
Total defectives	2	8	5	3	2	2	16	2	0	2	11	7	8	3	3	5	4	6	4	5	98
Total checked	212	981	971	637	707	1096	1236	343	423	597	876	664	983	719	289	398	619	753	545	897	13,936

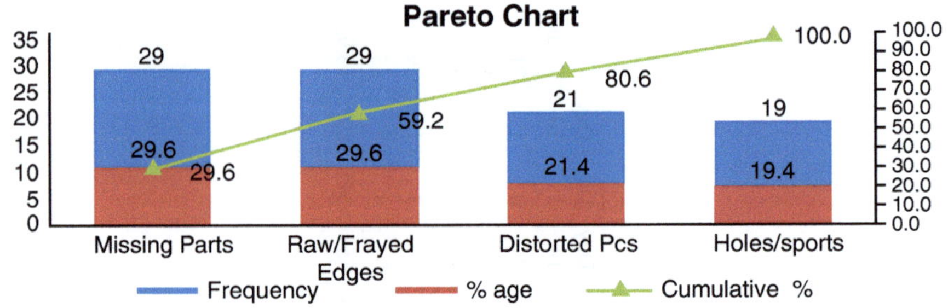

Fig. 6 Pareto chart after reducing the holes/spots

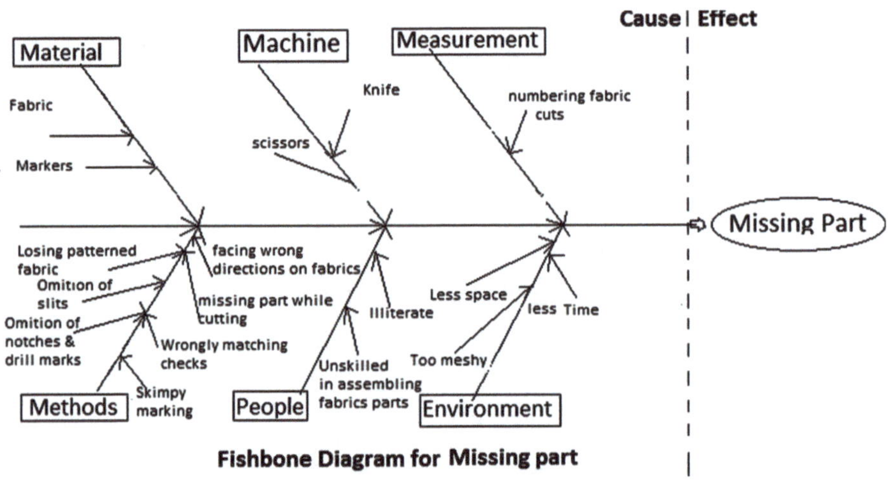

Fig. 7 Fishbone diagram for the missing part

The critical causes for the defect are unsharp scissors, imperfect removal of wrinkles, improper folding of seam lines, and less knowledge. The improvement sheet achieved is shown in Table 5.

The improvement over above identified areas of frayed edges will reduce defects from 19 to 9 defects and sigma level from 4.23 (DPMO = 3123) to 4.63 (DPMO = 878). The final P-chart after reconsidering all the above four defects is shown in the Fig. 10.

The final P-chart (Fig. 10) after reducing the four defects clearly shows that defectives are more cluttered around the central line than the initial P-chart (Fig. 1). It means that the process now is much more stable than the process with which we started. Even after reducing defects up to 878 DPMO from 3159 DPMO, we can see that still there are a lot of defectives which can be controlled and brought near to the central line.

Table 4 Continuous improvement sheet – missing part

Defects	Day 1	Day 2	Day 3	Day 4	Day 5	Day 6	Day 7	Day 8	Day 9	Day 10	Day 11	Day 12	Day 13	Day 14	Day 15	Day 16	Day 17	Day 18	Day 19	Day 20	Total
Distorted pcs	0	2	1	1	1	0	1	1	0	1	3	0	3	0	0	2	0	1	1	2	20
Holes/spots	0	1	1	0	2	1	1	0	0	1	1	0	0	2	1	0	1	1	1	2	16
Missing parts	0	1	0	0	1	0	0	1	0	0	3	0	1	0	0	2	0	0	1	1	11
Raw/frayed edges	1	1	2	0	2	1	1	1	0	1	2	2	1	1	1	0	0	1	2	1	21
Total defectives	1	5	4	1	6	2	3	3	0	3	9	2	5	3	2	4	1	3	5	6	68
Total checked	349	845	942	530	938	769	612	698	738	386	1012	296	1132	763	317	975	412	673	549	875	13,812

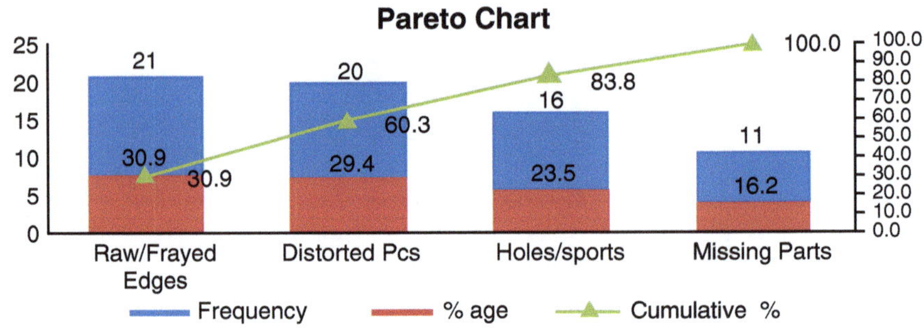

Fig. 8 Pareto chart after reducing missing part

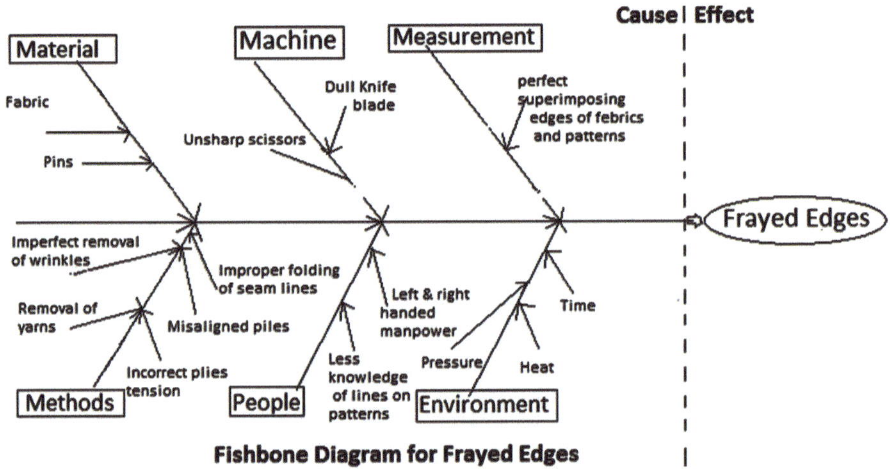

Fig. 9 Fishbone diagram for frayed edges

4 Control Phase

After improving all the phases of cutting process while apparel production, a Pareto chart with current state of process has been shown in Fig. 11.

The above Pareto chart clearly shows that on reducing raw/frayed edges, the scope to decrease the defects in distorted pieces has emerged. Replicating the above steps, again the cause and effect diagram for distorted pieces can be made at this level and then its improvement sheet is prepared. Further, Pareto can be used based on the above obtained improvement table, and the next most serious defect can be improved. The process will go on repeating till the desired sigma level is achieved. The numerical description derived from the above process is as follows:

- Distorted pieces are reduced to 82%, i.e., from 88 defects to 16 defects by exactly following the marker lines when patterns are properly laid on fabrics; fabric spread should have proper tension and free with wrinkles.

Table 5 Continuous improvement sheet – frayed edges

Defects	Day 1	Day 2	Day 3	Day 4	Day 5	Day 6	Day 7	Day 8	Day 9	Day 10	Day 11	Day 12	Day 13	Day 14	Day 15	Day 16	Day 17	Day 18	Day 19	Day 20	Total
Distorted pcs	1	1	0	0	1	0	1	1	0	1	0	1	1	1	1	2	0	1	1	1	15
Holes/spots	0	0	0	0	0	0	1	0	1	0	0	0	1	0	1	0	1	1	0	1	7
Missing parts	0	0	1	0	1	0	0	0	1	1	1	0	1	1	0	1	1	0	1	2	12
Raw/frayed edges	1	1	0	1	1	1	0	0	0	0	1	0	0	0	0	2	0	0	0	1	9
Total defectives	2	2	1	1	3	1	2	1	2	2	2	1	3	2	2	5	2	2	2	5	43
Total checked	624	519	212	349	419	1132	530	423	891	763	549	317	622	713	597	875	707	581	578	846	12,247

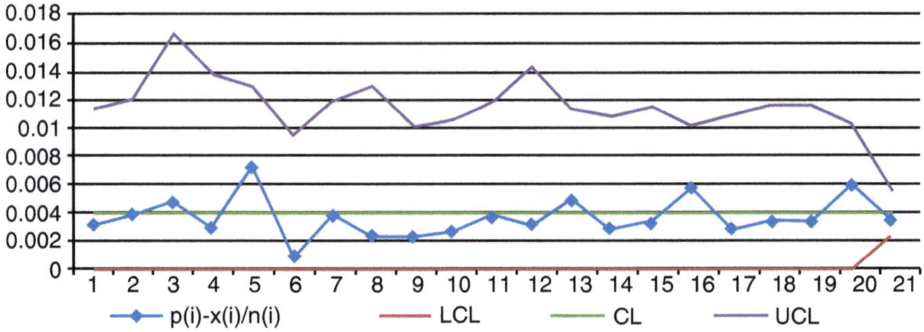

Fig. 10 Final P-chart after reducing all the four defects

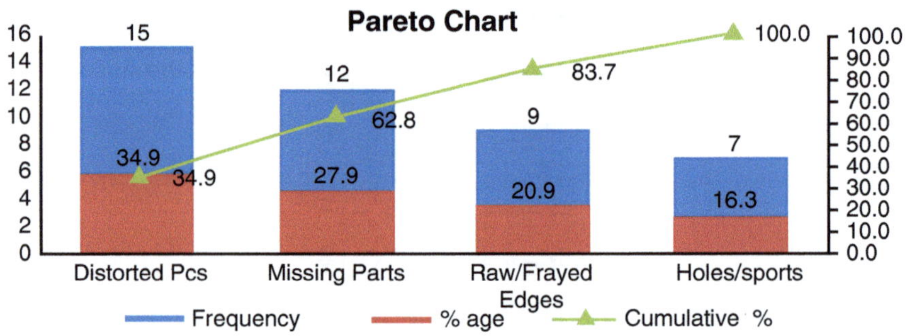

Fig. 11 Pareto chart after reducing raw/frayed edges

- A total of 19 defects (52.5%) of holes/spots could be reduced out of 40 defects by avoiding use of broken needle, using drill machine perpendicular to fabric spread, proper cleaning of machine, ensuring even bleaching, even heat setting according to fabric (especially in case of synthetic), and proper dosing of dyes and chemicals during dyeing and shading.
- The defect of missing parts is reduced from 25 to11, i.e., 56% by proper labeling of parts by marker with correct numbers; double check pattern direction before cutting.
- Even spread of fabric with proper tension, use of sharp knife according to fabric and the proper speed of use of the particular knife, and wrinkle-free plies reduced the defects of frayed edges. Checking these reduces 9 defects out of 19 defects.
- Finally, on reducing approximately 52.6% of defects of the four opportunities, the sigma level is increased from 4.23 to 4.63. It means that a defect of 3123 per million is reduced to 878 defects per million.
- Now, by looking at the two figures (Figs. 1 and 10) of initial stage P-chart and final stage P-chart, it is clear that in final P-chart (Fig. 10), the process has become more stable than before, but few deviations of defectives from central line are still there. This gives us an idea of further improvement. This improvement is continued till we achieve the desired level of the sigma level.

5 Conclusion

Six Sigma quality methodologies have definitely set a benchmark for excellence. It has been done by simply raising the standards of quality through implementation of improvement actions and continuous improvement. The sole objective is to successfully meet desired standards of customer's expectations. With the ever-increasing demand of quality products, quality has become the prime concern for organizations to survive in the global competition and for profitability. It combines elements of powerful disciplines such as statistical quality control that focuses on reducing variation to very low levels for better results. It is validated by the practitioners in industry and academics that there should be Six Sigma between the mean and the specification limits for carrying not more than 3.4 defects per every million in any process. In the present research, the number of defects at each stage in the production process is observed. The plotted results in P-chart (Fig. 1) show the defects under control limits. However, deviation of defectives from the central limit is observed by the concerned department. The department provided ideas of improvement actions to carry out defect reduction process. It was advised that the process still has scope for further improvement and defects reduction. Hence, in the improvement phase, a number of actions have been implemented to reduce the defects. The continuous reduction of causes for defects in the cutting process by using implementation of Six Sigma DMAIC method helped in improving the sigma level.

References

1. Rohini R, Mallikarjun J (2011) Six sigma: improving the quality of operation theatre. Procedia-Social Behav Sci 25:273–280
2. de Mast J Joran Lokkerbol, An analysis of the six sigma DMAIC method from the perspective of problem solving Int J Prod Econ, 139 604–614, 2012.
3. Jonny Z, Christyanti J (2012) Improving the quality of asbestos roofing at PT BBI using six sigma methodology. Procedia-Social Behav Sci 65:306–312
4. Ratnaningtyasa DD, Surendrob K (2013) Information quality improvement model on hospital information system using six sigma. Procedia technol 9:1166–1172
5. Islam M, Khan AM, Khan MR (2013) Minimization of reworks in quality and productivity improvement in the apparel industry. Int J Eng Appl Sci 1(4), ISSN 2305-8269
6. Srinivasan K, Muthu S, Devadasan SR, Sugumaran C (2014) Enhancing effectiveness of shell and tube heat exchanger through six sigma DMAIC phases. Procedia Eng 97:2064–2071
7. Petru PA, Roxana S (2014) Integrating six sigma with quality management systems for the development and continuous improvement of higher education institutions. Procedia-Social Behav Sci 143:643–648
8. Sabry A (2014) Factors critical to the success of six-sigma quality program and their influence on performance indicators in some of Lebanese hospitals. Arab Econ Bus J 9:93–114
9. Jevgeni SE, Roman Z (2015) Framework for continuous improvement of production processes and product throughput. Procedia Eng. 100:511–519

Forecasting of Major World Stock Exchanges Using Rule-Based Forward and Backward Chaining Expert Systems

Sachin Kamley, Shailesh Jaloree, and R. S. Thakur

1 Introduction

In these days, share price forecasting has become the hottest topic in financial time series forecasting and always remains in the limelight due to the volatile states of the economy. This is due to the fact that various business analysts, investors, and researchers have paid attention to the future estimation of stock prices. For the last three decades, AI has emerged as a most powerful area, and it supports various disciplines like natural language processing (NLP), artificial neural network (ANN), robotics, computer vision, expert system, etc. However, the expert system has come forward as most prominent AI techniques and had shown its usefulness in various areas like agriculture, medicine, astrology, engineering, and many more [1].

An expert system is called knowledge-based system that uses knowledge to solve problems, and knowledge must be encoded in some style of rules, facts, procedures, objects, relations, etc. However, the knowledge might be collected from various sources, for instance, books, magazines, newspaper, and reputed journals and sometimes by domain experts [2]. After acquiring the knowledge, it is encoded in a computer where some training and experiences are applied in a specific field.

S. Kamley (✉)
Department of Computer Applications, S.A.T.I., Vidisha, Madhya Pradesh, India
e-mail: skamley@gmail.com

S. Jaloree
Department of Applied Mathematics and Computer Science, S.A.T.I., Vidisha, Madhya Pradesh, India
e-mail: shailesh_jaloree@rediffmail.com

R.S. Thakur
Department of Computer Applications, M.A.N.I.T., Bhopal, Madhya Pradesh, India
e-mail: ramthakur2000@yahoo.com

© Springer Nature Singapore Pte Ltd. 2018

P.K. Kapur et al. (eds.), *Quality, IT and Business Operations*, Springer Proceedings in Business and Economics, DOI 10.1007/978-981-10-5577-5_23

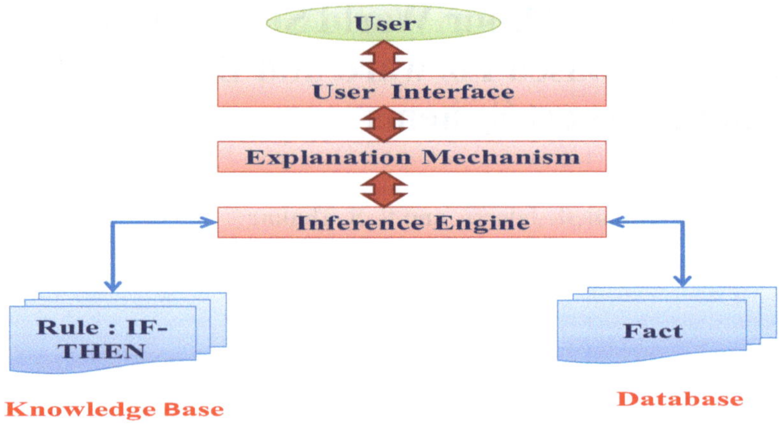

Fig. 1 Anatomy of an expert system [2]

However, the encoded knowledge stored in a appropriate form that must be tested and verified throughout the life of the system [3]. There are various techniques used for symbolic representation of knowledge like formal logic, frame, scripts, semantic net, and so on. Figure 1 shows the anatomy of an expert system.

Figure 1 states the basic anatomy of an expert system. The inference engine is the most powerful component of the system that is responsible for carrying out the reasoning process where the expert system reaches at a particular solution. However, it is also called the brain of the system where it is responsible to link the rules with a knowledge base, whereas facts are provided by the database. Moreover, the knowledge engineer is the person who knows the process of designing an expert system, and explanation component provides a detail reasoning process of the system [4].

Throughout the previous few years, the stock market has shown its worthiness in the fastest-growing economy, but accurate share price prediction is still challenging and a difficult process. Previously, various algorithms have developed for numerical predication, but very few authors have attempted symbolic representation of stock market data. Some important concerns for this research study are:

1. Various fundamental and technical variables still as some important macroeconomic factors are also being thought of to construct the security market knowledge base.
2. This study employed a comparative study between forward and backward chaining methods.
3. This research study can give a foundation for novice share users and researchers by providing a transparent image of expert system tools and techniques.
4. The final focus of this research study is to choose an acceptable reasoning strategy for the stock exchanges as well as improving and extending the inference power of stock market expert system.

Table 1 Brief description of literature review

Contributor	Input variables	Approach	Performance
Zarandi et al. (2012) [5]	Fuzzy linguistic	Fuzzy rule based	Outstanding and lower-risk decisions
Patel (2012) [6]	Fundamental and technical	Fusion models like SVR-RF, SVRANN, etc.	Higher accuracy with low error rate
Mohamed (2014) [7]	Fundamental and technical	Forward chaining- and backward chaining-based inference techniques	Better prediction accuracy than other
Markic et al. (2009) [8]	Corporate performance indicators	EXFIN using production rule	Outstanding and lower-risk decisions
Kamley et al. (2015) [9]	Fundamental factors	Comparative study of forward chaining- and backward chaining-based inference techniques	Better performance of backward chaining but low decision-making
Kamley et al. (2016) [10]	Macroeconomic indicators	Comparative study of forward chaining- and backward chaining-based inference techniques with 50 production rules	Excellent performance of backward chaining and higher decision-making
Chang and Liu (2008) [11]	Fundamental and technical	TSK fuzzy rule	Outstanding with lower-risk decisions
Velumoni and Rao (2016) [12]	Technical variables	Expert system using prospect theory	Low error rate

2 Literature Review

Table 1 briefly describes some significant researcher works over the years.

3 Knowledge Base Design

A good quality of expert system always depends on its knowledge base, but the quality of knowledge base depends on the knowledge that is fed by the knowledge engineer [13]. In an expert system, a knowledge base is maintained employing a set of variables and rule list. Table 2 shows a sample of stock market vocabulary.

Stock market domain is very large and behavior is unpredictable. However, it consists of numerous technical rules; thus, illustration and mapping of all these rules are impossible. During this study, 100 rules are considered for stock market knowledge base [14, 15].

Table 2 Sample of stock market vocabulary

Variable and description
OP, open price; LP, low price; HP, high price; CP, close price; BSE, Bombay Stock Exchange; Hang Seng, China Stock Exchange; VOL, volume; GDP, gross domestic product; EPS, earnings per share; DIVD, dividend; USDP, US dollar prices; OLP, oil prices; FDI, foreign direct investment; INFL, inflation; INTRST, interest rate; PE ratio, price-earnings ratio; STK, stock market; IMP, import

4 Proposed Methodologies

4.1 Forward Chaining

Forward chaining is a data-driven methodology that begins by applying the rules in a forward direction. It is a top-down procedure, i.e., recursively applying the rules over data to generate more data. However, the top most rules are executed each time, and always rule adds a new fact in the database when fired. Any rule can be executed only once. The procedure is stopped once when no further rules can be fired [1, 2, 10, 16]. In forward chaining method, IF part (LHS), i.e., antecedent, is specified first, and the THEN part, i.e., consequent, is specified later. The structure of forward chaining rule is given below.

IF

X

AND

Y

THEN

Z

Figure 2 depicts a flowchart of forward chaining inference procedure.

Figure 2 states that the procedure starts by comparing the premises of each rule (knowledge base) against the fact base for discovering the new applicable rule. In this situation, sometimes conflict resolution arises, i.e., if more than one rule is applied at the same time. At this time, rules are selected based on the priority basis, i.e., ignores those rules whose conclusion is already known and chooses one of the remaining rules. At last, procedure terminates, if no more rules are applicable.

4.2 Backward Chaining

In contrast, backward chaining is called the goal-driven reasoning process, where procedure begins by applying the rules in a reverse direction, i.e., starts from the

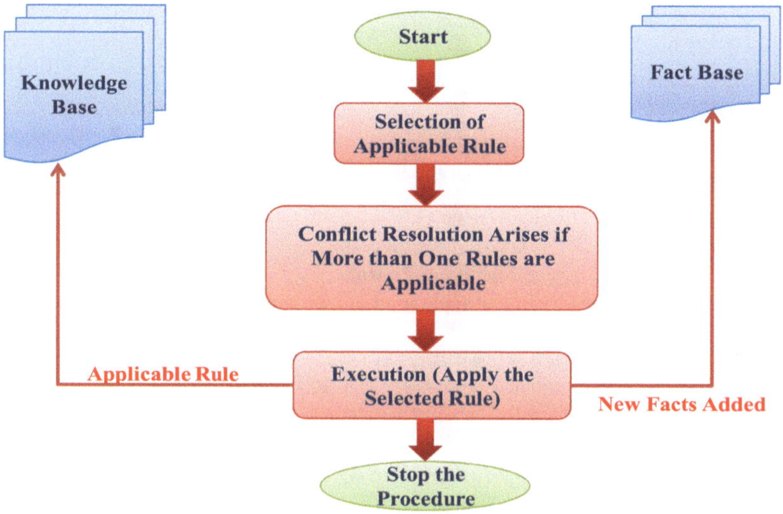

Fig. 2 Flowchart of forward reasoning inference procedure

goal and tries to find more evidences in order to satisfy the goal or hypothesis [2, 4, 10, 16]. In other words, backward chaining procedure works from the goal back to facts. The structure of backward chaining rule is given below.

Z

IF

X

AND

Y

Figure 3 depicts a flowchart of backward chaining inference procedure.

Figure 3 states that procedure starts with a specified goal. Firstly, goal is matched with the fact base; if the goal is found, then the return is true (proved); otherwise, the return is false. After that goal is matched with the conclusion of a rule with a fact base, then the return is true, i.e., proved. Otherwise, try to match the goal with the conclusion of a rule one by one and also check to see if its premises are provable; if yes, then the return is true (proved). In this process, a new goal is pushed on the stack each time, and the process is repeated until the rules are matched with the goal stack and no more rule is left in the knowledge base. Figure 3 is clearly stated that if no rules matched with goal stack, then the procedure reruns false, i.e., can't prove.

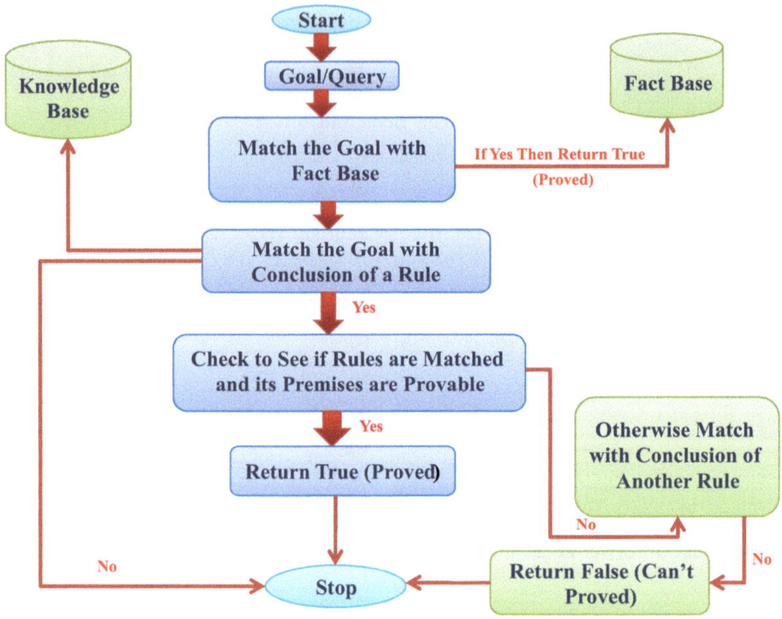

Fig. 3 Flowchart of backward chaining inference procedure

5 Experimental Results

List processing (LISP) is the general purpose, and symbol-oriented language has been getting used extensively for the knowledge representation task. In this study, common LISP 3.0 editor is employed to design the expert system shell, and all the applicable rules are inputted into stock knowledge base in the rule-based format [17]. An inference engine component is accountable for searching all the applicable rules in the knowledge base and updated the fact base once execution of rules is done. Figure 4 shows generation of rule base (knowledge base) from LISP environment.

Now users interact with the system and raised various queries from the system. The questions asked by the system are "what are the impacts of falling US Dollar prices," "what are the impacts of BSE stock exchanges of rising open and high prices of NASDAQ and Hang Sang index," "what are the impacts on all indexes of falling inflation rate and interest rate," "what are the impacts of rising of GDP," and "what are the impacts of rising of FDI on all indices" [10, 18–20]. Therefore, the expert system tries to answer these queries one by one, and the inference engine procedure starts searching all the applicable rules within the knowledge base, and the respective conclusion is drawn. Finally, fact base is updated with the new knowledge. The following command is used to execute the forward reasoning inference procedure:

```
Common Lisp-[Lisp Worksheet]
;; Corman Lisp 3.01 (Patch level 0)
;; Copyright © Corman Technologies Inc. All rights reserved.
;; Unlicensed version, evaluation period expires in 24 days.
'(setq *rule-list* '(
(R1 IF (is rise (USDP) THEN (is BSE up))
(R2 IF (is fall (IMP ECOMY) THEN (is STK down))
(R3 IF (is rise (INTRST INFL) THEN (is STK down))
(R4 IF (is rise (OP LP) and (is rise CP) THEN (is buy shares of all indices))
(R5 IF (is rise (VOL OP ) THEN ((is sell shares )and (is earn profit)))
(R6 IF (is fall (USDP) THEN (is rise share prices))
(R7 IF (is rise (USDP ) THEN (is fall share prices))
(R8 IF (is rise (OLP) THEN (is fall share prices))
(R9 IF (is fall (INFL USDP) THEN (is fall share prices))
(R10 IF (is rise EPS OP) THEN (is rise share prices))
(R11 IF (is fall (IMP EXP) THEN (is fall share prices))
......................
...........................
))
```

Fig. 4 Generation of sample rule base from LISP environment

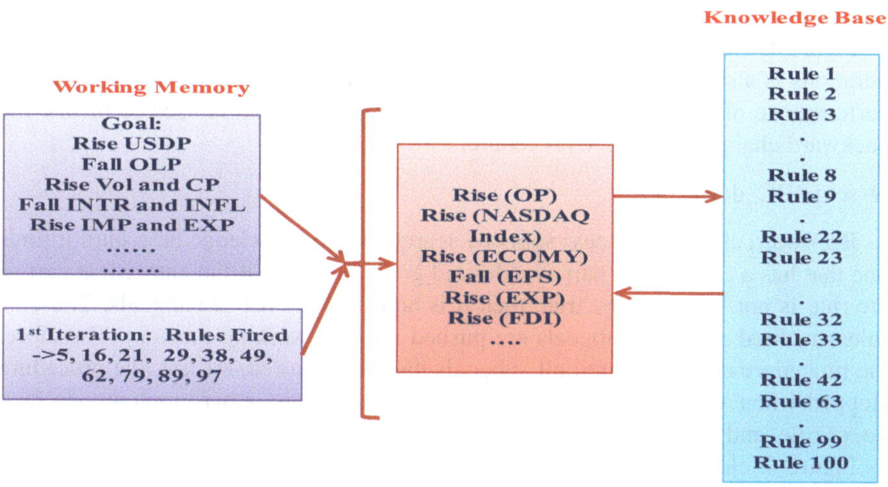

Fig. 5 Production system environment after rule invocation

➔(setq *FC_Infer*T).

Next, the inference process searching for rules proves the specified goal in the knowledge base. As a result, ten rules are fired and new information is added in the working memory. Figure 5 shows production system environment after rule invocation.

However, inference process is repeated until the goal does not become true and knowledge base has no more rules to be fired. Thus, all goals are to be proven. Figure 6 shows a bar graph among fact, no. of rule, and no. of iterations.

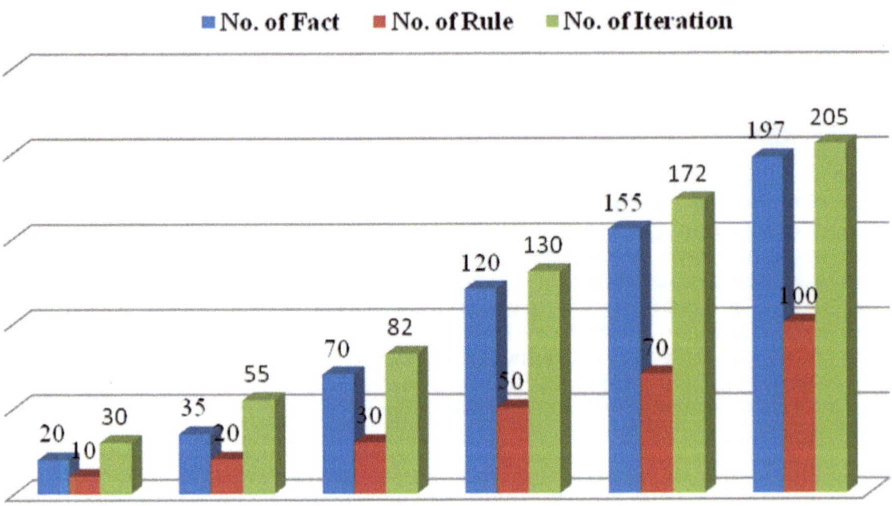

Fig. 6 A bar graph among fact, no. of rule, and no. of iterations

Figure 6 clearly states that when the no. of rules increases, then the no. of iterations is also increasing in the same manner. Thus, the procedure slows up the performance of inference engines. The following command is used to invoke the backward chaining inference procedure:

→(setq *BC_debug*T).

Backward chaining process starts by searching the knowledge base until it finds one that has a consequent part as a desired goal. However, if the antecedent part of the rule is not known to be true, then it is added to the list of subgoals. Thus, 12 rules are fired and new subgoals are pushed into the working memory. Therefore, the procedure continues until all subgoals have been tested. At last, the procedure stopped when all the goals are attempted. Figure 7 shows a bar graph among fact, no. of rule, and no. of iterations.

Figure 7 clearly states that when the no. of rules increases, then the no. of iterations also increases, but very less as compared to forward reasoning approach. Table 3 shows forecasted performance for stock market expert system.

Table 3 depicts that forecasting performance of the stock market provided an opportunity to stock users to invest money in the market and earn profit regarding decision.

6 Conclusion and Future Scopes

In this study, forward chaining- and backward chaining-based expert system inference approaches are implemented. Four noteworthy major world stock exchange data and various technical and macroeconomic indicators are considered for mod-

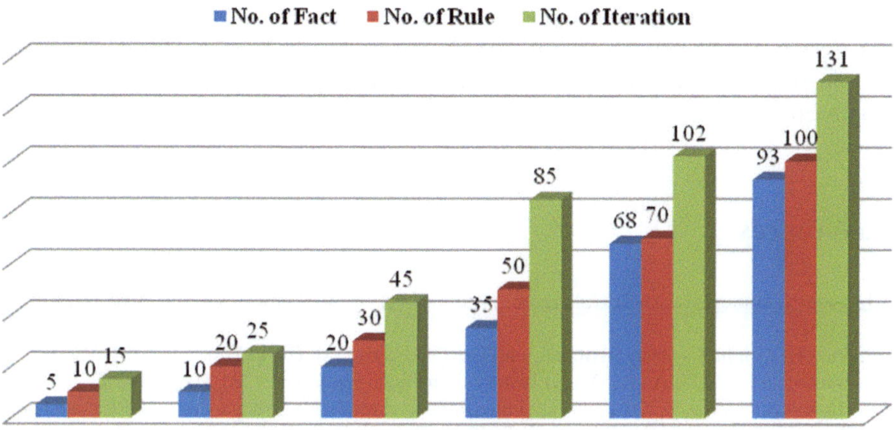

Fig. 7 A bar graph among fact, no. of rule, and no. of iterations

Table 3 Forecasting performance of stock market expert system

S. No.	Fact	Inference result	Investment decision
1	Rise GDP and INFL	Rise Hang Seng index	Sell
2	Fall OLP and GDP	BSE down	Buy
3	Fall EPS	STK down	Buy
4	Fall OP and CP	STK down	Buy
5	Fall INFL and INTR	STK up	Sell
6	Fall GDP	Fall BSE index	Buy
7	Rise UNEMPR	Fall BSE index	Buy
8	Rise FDI	Rise BSE index	Sell
9	Fall OP and CP	Fall BSE index	Buy
10	Fall LP and VOL	Fall BSE index	Buy
11	Rise USDP and INFL	Fall Hang Sang index	Hold
12	Fall GDP and FDI	Fall Nikkei 225 index	Hold
13	Rise USDP	Fall BSE index	Buy
14	Rise OP and HP	STK up and rising share prices	Sell
15	Rise OLP	STK down	Buy
16	Rise Nikkei 225	BSE index up	Sell
17	Fall dividend and EPS	Fall BSE index	Hold
18	Fall OP and HP	Fall Hang Seng index	Buy
19	Fall FDI and PE ratio	Fall BSE and Nasdaq	Buy
20	Rise PE ratio and Fall OP	STK up	Sell

eling the stock knowledge base. Based on the results, it is observed that backward reasoning strategy has outstanding forecasting performance as compared to forward chaining strategy. The sole drawback of forward chaining strategy is that it takes a long time to infer the results, while backward chaining strategy searches the specified goal very quickly. Therefore, this study will be helpful for stock users and brokers to know more about share market conditions and also provide guidance for stock users. Sometimes the rule enhancement process decreases the performance of expert systems. In the near future, fuzzy-based expert system or object-oriented approach will be adopted for higher performance.

References

1. Legg S, Hutter M (2007) Collection of definitions of intelligence. J Front Artif Intell Appl 157:17–24
2. Rich E, Knight K (1991) Artificial intelligence, 3rd edn. McGraw- Hill
3. Buchanan B (2005) A (very) brief history of artificial intelligence. AI Mag 26(4):53–60
4. Russell S, Norvig P (2009) Artificial intelligence: a modern approach, 3rd edn. Prentice Hall
5. Zarandi FM, Neda MH, Bastani S (2012) A fuzzy rule based expert system for evaluating intellectual capital. Adv Fuzzy Syst 12:1–11. doi:10.1155/2012/823052
6. Patel J, Shah S, Thakkar P, Kotecha K (2015) Predicting stock market index using fusion of machine learning techniques. Expert Syst Appl 42(4):2162–2172
7. Mohamed T, Gayar NE, Atiya AF (2014) Forward and backward forecasting ensembles for the estimation of time series missing data. Artif Neural Netw Pattern Recognit (ANNPR), LNAI 8774:93–104
8. Markic B, Tomic D, Pavlovic I (16–21 October 2009) an expert system approach in stock selection attractive for Investment. 13th international research/expert conference, "Trends in the development of machinery and associated technology", Hammamet, pp 297–300
9. Kamley S, Jaloree S, Saxena K, Thakur RS (2015) Forward chaining and backward chaining: rule based expert system approaches for share forecasting and knowledge representation. Presented for 7th International Conference on Quality, Reliability, Infocom Technology and Business Operations (ICQRIT). Sponsored by Springer. University of Delhi, pp 28–30
10. Kamley S, Jaloree S, Thakur RS (2016) Performance comparison between forward and backward chaining rule based expert system approaches over global stock exchanges. Int J Comput Sci Inf Secur (IJCSIS) 14(3):74–81
11. Chang PC, Liu CH (2008) A TSK type fuzzy rule based system for stock price prediction. Expert Syst Appl 34(1):135–144
12. Velumoni D, Rau SS (2016) Cognitive Intelligence based expert system for predicting stock markets using prospect theory. Indian J Sci Technol 9(10):1–6
13. Yunusoglu GC, Selim H (2013) A fuzzy rule based expert system for stock evaluation and portfolio construction: an application to Istanbul stock exchange. Expert Syst Appl 40(3): 908–920
14. Online Stock Market Dataset Available on Yahoo Finance Site. http://www.yahoofinance.com. Accessed 2 Jan 2016
15. Online Macroeconomic Variables Data Available on Site. http://www.indexmondi.com. Accessed 5 Oct 2016
16. Aqil Burney SM, Mahmood N (July2006) A brief history of mathematical logic and applications of logic in CS/IT, J Sci 34(1):61–75
17. Common Lisp Compiler Downloaded from http://www.commonlisp.com. Accessed 25 Oct 2015
18. Gryglewicz S (2011) A theory of corporate financial decisions with liquidity and solvency concerns. J Financ Econ 99(2):365–384
19. Brown and Jennings (1989) On technical analysis. Rev Financ Stud 2(4):527–551
20. Das AP (2008) Security analysis and portfolio management, 3rd edn. I.K. International Publication

Performance Enhancement of AODV Routing Protocol Using ANFIS Technique

Vivek Sharma, Bashir Alam, and M. N. Doja

1 Introduction

The mobile ad hoc network comprises of mobile devices that are connected to each other without wires. Each node in MANET is mobile in nature, so they have dynamical topologies. Such type of network possesses a variety of limitations like no base stations, limited power of each node, limited bandwidth, dynamically changing topologies, etc. These factors lead to primary challenge in routing the information from the sender node to receiver node, network optimization, and link scheduling algorithm. In static network, the shortest path from the source to destination node is usually the optimal route, but in MANETS it is not a well-defined problem due to its dynamically changing topologies. However, to select an optimal routing path and to deliver a message in a MANET, factors such as variable link quality, bandwidth consumption, multiuser interference, power expanded, and topological changes are to be considered. The MANET's routing protocol should be adaptive in nature to consider these effects while taking routing decision.

In the last decade, researchers have been particularly active in optimizing the performance of MANET routing protocols by embedding soft computing techniques to conventional routing protocols [1–5]. But these protocols are prone to attacks due to MANET's environment characteristics. Therefore, to increase the security and efficiency of routing protocol, many researchers propose the utility of fuzzy logic system [6]. Author in [7] calculated the trust value of a node using fuzzy logic system, and if the node has a trusted value greater than threshold, then only it is to be considered. Authors [8] developed a system based on genetic algorithm to detect the intrusion by analyzing the behavior of each node. Authors in [9] proposed routing protocol that enhances the packet forwarding capability of a node based on

V. Sharma (✉) • B. Alam • M.N. Doja
Department of Computer Engineering, Jamia Millia Islamia, New Delhi 110025, India
e-mail: Vivek2015@gmail.com; balam2@jmi.ac.in; mdoja@jmi.ac.in

© Springer Nature Singapore Pte Ltd. 2018

P.K. Kapur et al. (eds.), *Quality, IT and Business Operations*, Springer Proceedings in Business and Economics, DOI 10.1007/978-981-10-5577-5_24

the fuzzy logic system. To solve the transportation system problem of mobile ad hoc networks [10], have calculated the cost of each link and select the optimal path based on calculated cost.

This paper shows the results when delay is also taken as a parameter along with hop count to determine whether to forward a packet to a specific node or not by embedding ANFIS technique to traditional AODV routing protocol. Many nodes establish communication through other nodes by broadcasting its own packet. This may create congestion in the network and affect its performance. So, it become important for the routing protocol to calculate the delay between the nodes so that the congestion can be reduced and that also improves the packet delivery ratio.

The remaining paper is as follows. Section 2 presents adaptive network-based fuzzy inference system for advanced AODV. Section 3 contains the proposed protocol AAODV. In Sect. 4, its simulation model is presented further, and its results are presented and analyzed in Sect. 5. Finally, in Sect. 6, conclusions are discussed.

2 ANFIS for AAODV

ANFIS is called as adaptive network-based fuzzy inference system introduced by J.S.R. Jang. The five layers in an ANFIS system are used in this paper and are shown in Fig. 1. A hybrid learning mode is used to implement different node functions to learn and tune parameters in a FIS. The Sugeno model will be used in this paper that considered its advantage over high-dimension problem.

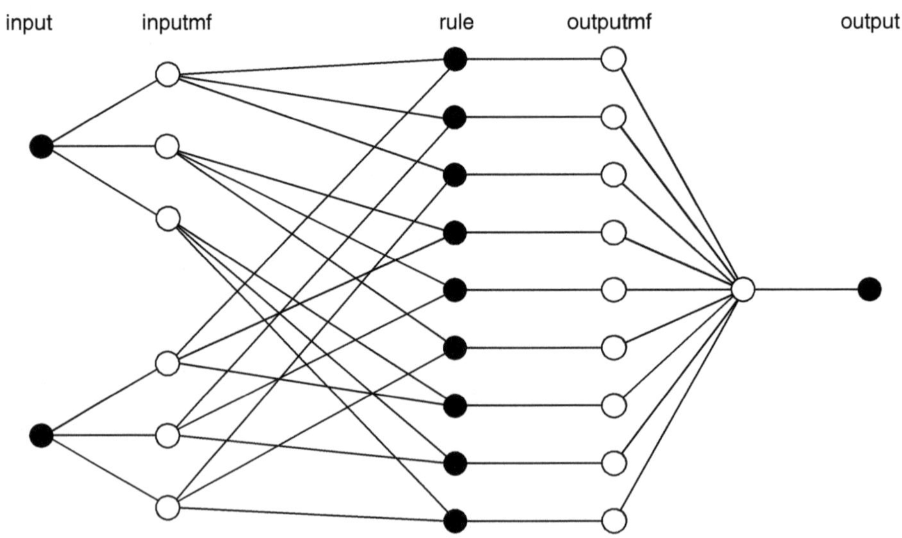

Fig. 1 Architecture of ANFIS

Table 1 Rule value for ANFIS

Rule	Rule values	Output value
Rule 1	If hop count is low and delay is low	Very low
Rule 2	If hop count is low and delay is medium	Low
Rule 3	If hop count is low and delay is high	Medium
Rule 4	If hop count is medium and delay is low	Low
Rule 5	If hop count is medium and delay is medium	Medium
Rule 6	If hop count is medium and delay is high	High
Rule 7	If hop count is high and delay is low	Medium
Rule 8	If hop count is high and delay is medium	High
Rule 9	If hop count is high and delay is high	Very high

Layer 1 generates the membership function for input variable hop count and delay with low, medium, and high as the variable linguistic. The range for input variable hop count is from 0 to 8 nodes and for delay is from 0 to 2. For output, variable linguistic used are low, very low, medium, high, and very high.

In Layer 2, rules are originated as shown in Table 1.

In Layer 3, each node executes normalization of the rules from the preceding layer.

In Layer 4, outputs of the rule are calculated.

In Layer 5, all incoming signals are calculated.

3 Proposed ANFIS AODV (AAODV)

During the routing process, the path information variable is required to take the optimized and effective decision. To counter the uncertain information, an effective mechanism is to be required that can handle the information effectively. For this purpose, we used the ANFIS system.

AODV routing protocol uses its routing table to store and get information for its routing process. It has RREQ packet that store the required information parameter like hop count and delay. The value of each node is taken from this packet and sent to the ANFIS system as input. It processes the system and gives back an output and updates the entry of reverse route source of RREQ message. If the new output is having the value less than the stored one, then the spreading of RREQ message is done until it reaches to its destination. This scheme not only reduces the routing time but also reduced memory overhead required to store all routes until it reaches the destination.

Table 2 Parameters used in simulation environment

Parameter	Values
Routing protocol	AODV, AAODV
No. of nodes	20
Area	700 * 700
Channel capacity	54 Mbps
Traffic type	CBR
Mobility model	Random way point
Simulation time	50, 100, 150, 200, 250

4 Simulation Model

The environment variables that are used in the present simulation are shown in Table 2. The network is modelled with 20 mobile nodes that are placed randomly within $700 \times 700\ \text{m}^2$ area. The channel capacity is 54 Mbps. A random way point model was used where a speed of each node lies in between 0 and 10 m/s. Finally, a traffic generator is also used for constant bit rate (CBR), and the final load of packet is of size 512 bytes. At any particular time, only eight nodes can communicate. The network parameters are simulated at time 50, 100, 150, 200, and 250 seconds by the NS2.35 simulator; the system is implemented by C++ and TCL language. For the implementation of ANFIS, we used MATLAB.

5 Simulation Results

The following parameters are calculated at different simulation times to determine the network performance:

- Packet delivery ratio (PDR): It is the ratio of the sum of all the packets that have been received to the sum of all the packets that have been sent.

$$\text{PDR} = \frac{\sum \text{Number of packet receive}}{\sum \text{Number of packet send}} \tag{1}$$

- End-to-end delay: It is defined as the mean time taken by a packet starting from its source to reach its destination in the network.

$$\text{End to End Delay} = \frac{\sum (\text{arrive time} - \text{send time})}{\sum \text{Number of connections}} \tag{2}$$

From the simulated results, i.e., Figs. 2 and 3, it is clearly observed that the PDR has increased and the average end-to-end delay is improved for 20 nodes, respectively. Therefore, the use of ANFIS with AODV provides much better results than AODV.

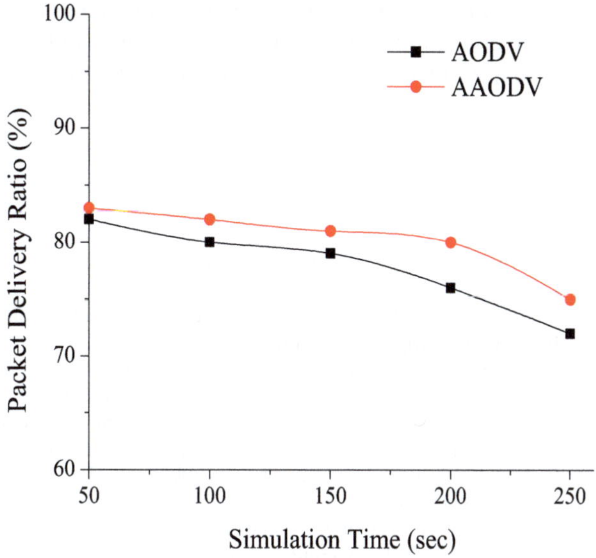

Fig. 2 Packet delivery ratio vs. simulation time

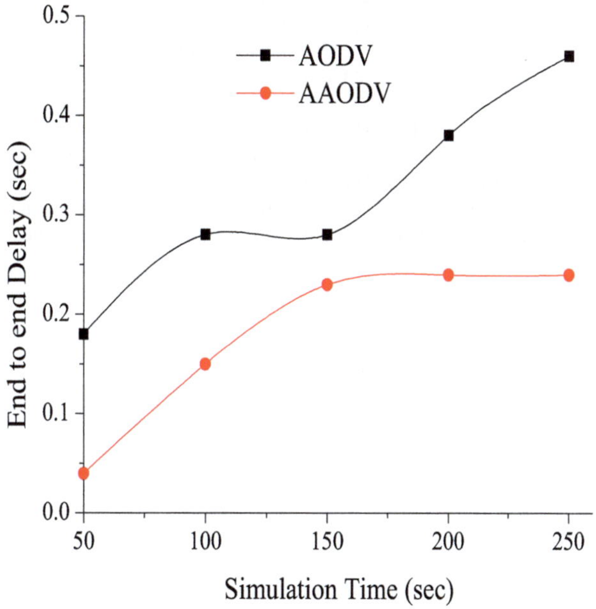

Fig. 3 End-to-end delay vs. simulation time

6 Conclusions

This paper proposes the improvement in standard AODV routing protocol with the use of delay parameter along with hop count parameter while taking the routing decision. The present algorithm embedded ANFIS to AODV routing protocol. The simulation of 20 nodes with simulation time ranging from 50 to 250 s is presented and also compared on the basis of parameters like PDR and throughput. The results show that AAODV performs better than traditional routing protocol.

References

1. Murthy CSR, Manoj BS (2004) Ad Hoc wireless networks: architecture and protocols. Pearson Ltd
2. Sharma V, Alam B (2012) Unicaste routing protocols in mobile Ad Hoc networks: a survey. Int J Comput Appl USA 51:148–153
3. Zhang X, Qian ZH, Guo Y-Q, Wang X (2014) An efficient hop count routing protocol for wireless ad hoc networks. Int J Autom Comput 11(1):93–99
4. Sirisala N, Shoba Bindu C (2015) Uncertain rule based fuzzy logic QoS trust model in MANETs. International conference on advanced computing and communications, pp 55–60
5. Perkins C, Belding-Royer E, Das S (2003) Ad Hoc On-Demand Distance vector (AODV) Routing, RFC 3561
6. Ross TJ (1995) Fuzzy logic with engineering application. McGraw Hills, New York
7. Leo Manickam JM, Shanmugavel S (2007) Fuzzy based trusted Ad Hoc on-demand distance vector routing protocol for MANET. 3rd international conference on wireless and mobile computing, networking and communications
8. Sujatha KS; Dharmar V, Bhuvaneswaran RS (2012) Design of genetic algorithm based IDS for MANET. International Conference on Recent Trends in Information Technology (ICRTIT), pp 28–33
9. Li Y, Weihua H (2010) Optimization strategy for mobile Ad Hoc network based on AODV routing protocol. 6th international conference on Wireless Communications Networking and Mobile Computing (WiCOM), pp 1–4
10. Zuo J, Xin Ng S, Hanzo L (2010) Fuzzy logic aided dynamic source routing in cross-layer operation assisted Ad Hoc networks. 72nd IEEE conference on vehicular technology, pp 1–5

Preservation of QoS and Energy Consumption-Based Performance Metrics Routing Protocols in Wireless Sensor Networks

Ram Bhushan Agnihotri, Nitin Pandey, and Shekhar Verma

1 Introduction

A wireless sensor network is controlled and restrained by the restriction of data bandwidth, and high error rate of the wireless channel limits the quantity of data that can move through over such channel and the communication consistency; the restricted processing command of sensor nodes limited by their physical scope; and limited size and potential of the sensors for group of data and the value of collected data [1]. These limitations are further emphasized by limited liveliness of the batteries [2]. By minimizing the redundant information for transferring the relevant information only that is possible on low bandwidth error leads toward maximizing the processing of sensor nodes; in addition least processing within the network is required for processing power and memory [3]. However, the small cost of processing (per bit cost) as compared to communication entails minimum possible data transfer in situ with maximization of processing [2]. The amount of limited processing potential and memory controls the processing of simple operations in network. This has provided for gathering and processing simple data in wireless sensor networks practical only.

Each sensor node is illustrated by low consistency, limited memory and computational capability, limited battery power, and low precision. The cumulative effect enhances the accuracy and consistency of data, in the presence of huge numbers such sensor nodes deployment over a region. Due to huge redundancy as large number of sensors, this accuracy and consistency affects and senses the similar

R.B. Agnihotri (✉) • N. Pandey
Amity Institute of Information Technology, Amity University, Sec 125, Noida, 201301, India
e-mail: rbagnihotri@amity.edu; npandey@amity.edu

S. Verma
Indian Institute of Information Technology, Allahabad, 211012, India
e-mail: sverma@iiita.ac.in

© Springer Nature Singapore Pte Ltd. 2018
P.K. Kapur et al. (eds.), *Quality, IT and Business Operations*, Springer Proceedings in Business and Economics, DOI 10.1007/978-981-10-5577-5_25

parameters at the similar place and time. If sent to a sink from the sensor nodes, this surge of redundant data stretch collected by sensors over space and time would drain the bounded energy of sensors. It is also quite familiar that energy requirement of communication is very high as compared to computation; hence, it is crucial that only useful data be extracted from the sensed data and communicated to the sink. This would conserve the energy of the sensor nodes to make longer the network lifetime [4–9].

It can be analyzed that divergence in the nature of the network of sensor nodes and each sensor node raises some challenges that require to be addressed [5, 6]. In this matter, the first difficulty occurs due to the low consistency of a sensor node. The deployment of sensors in a different environment, a changeable and unfamiliar kind of values, is sensed. It is, therefore, exceptionally complex to differentiate between normal and abnormal sensor value readings.

To recognize an inconsistent node is vital in the sensor network but quite difficult. The second difficulty occurs at the network level. As declared in advance, sensors have restricted by memory, processing capacity, and battery life limitation. The communication protocols and processing methods must be able to adjust to failures of each node, adapt to changing environment, ensure sensor data integrity, and mind useful information by aggregating spatiotemporal data with communication among the reliable sensor nodes and limited resources.

The third issue is related to resolve the abnormality and unwanted information in the sensed data, and lastly the final problem is to handle the query requirements from the sink by sending information at a desired resolution. The mobility-based applications are quite challenging in WSNs such as monitoring of health and wildlife and search and rescue operations that demand a reliable routing protocols to improve the limitations of mobile sensor nodes. GPS-enabled network also helps to assign the location of node which further reduces the delay in switching from one cluster of one to another. It also helps to improve the packet delivery ratio in the case of data transmission for existing routing protocols. Even though the routing is applied only on the nodes, the routing on the basis of cluster heads can be quite useful in the context of fault tolerance. It will be counted for the cluster heads only to analyze the packet loss in data transmission and reduce the impact of individual mobile nodes [2].

There are so many challenges in wireless sensor network such as power control [7, 8], MAC, routing, and transport. To choose the power levels of transmissions in wireless networks, it is defined on the basis of some constraints of power levels. Power level influences range and also affects the routes. Power levels determine interference. Conceptualization problem occurs for power control in three layers. First problem, quality of reception occurs in physical layer; then network layer faces problem of impact on routing and finally transport layer struck due to higher power impacts congestion. Some of the global impact [9] also happens in power control such as connectivity, which is global property, and clustering which is regional-based property. A routing process using source node with the mobile node is shown in Fig. 1. It shows the overall impact of routing over the hybrid network.

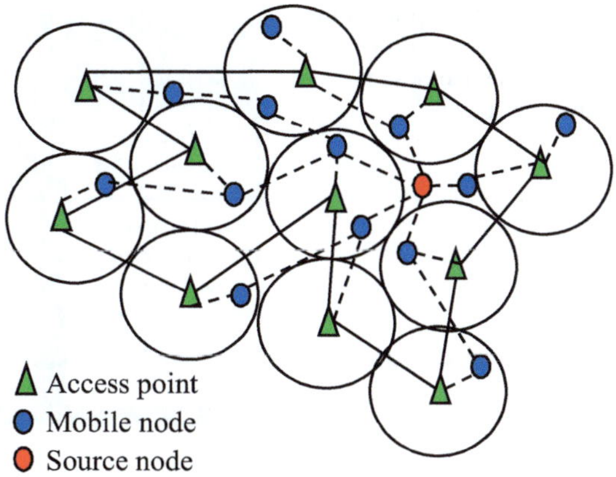

Fig. 1 Block diagram of the hybrid network model

The rest of the paper is structured as follows. Section 2 illustrates the related work. The analysis of the key challenges in WSN for performance metrics in the routing protocols is described in Sect. 3. Experimental analysis is done by simulating two different schemes in Sect. 4. Finally, conclusion section is stated in Sect. 5.

2 Related Work

In 2010, a distributed coding-aware routing (DCAR) protocol is proposed to overcome the network region limitations for detecting possible chances of network coding and also the exploration of the existing routes between a specified source and destination [1]. Figure 1 shows a model for the routing using source node which utilized the access point for connecting the next cluster and mobile nodes. Source nodes are used to utilize the resources in better way. The more availability of source node can also increase the better QoS throughout.

The [10, 11] "next generation architecture of wireless sensor network for communication applications" aims at integrating the communications between sensor nodes by means of supportive diversity and the multi-hop routing where the sensor nodes sense the data and transmitted to its neighbor nodes. The *d* is the vertical and horizontal distance between each sensor node.

Three major layers are used in the wireless networking that play a major role in the WSNs and the existing challenges in WSNs. The performance metrics in routing protocols for challenges in WSN are shown in Fig. 2. The descriptions related to the layers are shown below with the key characteristics.

1. Physical layer optimized for multi-hop wireless networking. Physical layer restructuring has significance in terms of moving toward route and forward activities in physical layer, where routing is defined as to conclude which set

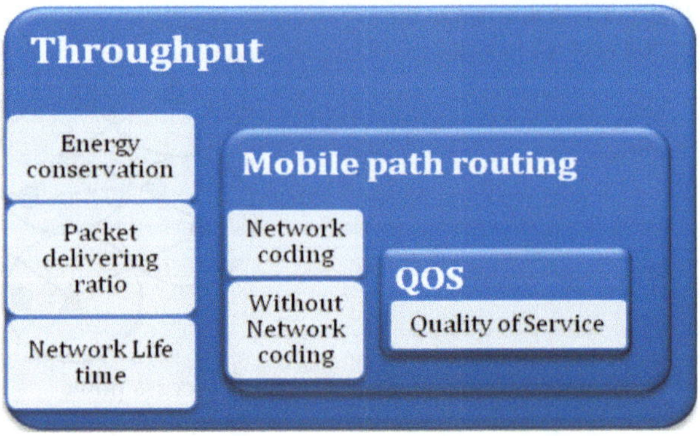

Fig. 2 Block diagram of the challenges in WSN for performance metrics in routing protocols

of nodes send the packet from starting node to destination and forwarding is based onto the transport along this chosen path. The new physical layer has three structures "relay," "transmit," and "receive." Switching process is done at physical layer itself.

2. Access to medium in network for entire (as opposed to single hop) path-centric hops. Atomic unit of operation is equal to multiple hops. Medium access control is path-oriented. Packet does not have to recomplete at every hop. Harness unused resources to increase capacity of path. Cooperative diversity is defined as "Nodes simultaneously retransmit the same packet on different channels, and sets of frequencies are to be distributed over combined receivers. Reduced processing and removal of recontending at every hop will reduce latency. Cooperative transport increases capacity. Path diversity increases path robustness.
3. Cooperative transport of packets. Harness unused resources to increase capacity of path. Concept of "cooperative diversity" depends on nodes simultaneously retransmitting the same packet on different frequencies/channels to be diversely combined at receivers [12].

In the sensing environment, the concerned parameters for sensing node and their related range are to be known and also need to be initialized, before the sensors are deployed in the sensing environment. The time signal function is having three attributes which are being done. If parameters are less, then it shows errors. Then program reads the files and runs in window size.

3 Key Challenges in WSN for Performance Metrics Routing Protocols

The detailed description related to the performance metrics in routing protocols that help to resolve the key challenges in WSNs is described in detail in Tables 1, 2, 3. The key factors related to the performance metric throughput and the relevant improvement in energy conservation, network lifetime, and packet delivering are

Table 1 Performance metric throughput in routing protocol

Performance metric in routing protocol	Key factor	Improvement
Throughput	Cross-layer operation model	The model utilizes a mechanism to restrict the neighbor discovery for broadcasting the dynamic routes only [3]
	A random linear network coding approach toward the code and data packets. The next hop link status and also the quantity of current packets on the receiving end which are sent by the upstream node together help to determine it.	Even though an extended propagation delay and high bit error rate of space information networks [4]
Energy consumption	Cross-layer operation model	In route discovery process, the routing operations embedded the location of the mobile nodes. To adjust the transmission range of the node, it is further utilized by the MAC layer for controlling the transmission power. It helps to decrease the energy consumption of the node(s) which is possible due to the minimized power utilization of the network interface. [3]
	Two different kinds of strategies such as flooding delaying and hop penalty are utilized by energy-efficient survivable routing protocol (ESRP).	The simulation outcomes demonstrate that the energy utilization of traditional routing protocol is less efficient than ESRP. At same time, the number of energy depletion nodes is reduced for load-balancing and the first-energy exhaustion node is also postponed in the appearance. [7]
	In wireless sensors, the maximum energy is consumed by the radio frequency (RF) modules. To determine the routing paths and also preserving the quality of service in routing protocols, routing metrics are significant. In the metric, link quality measurement is maintained for sending packets using the RF module.	To find the distributions for the fittest one, the statistical analysis was tested on link quality indication (LQI) and received signal strength indication (RSSI). Statistical tests on the collected data show a significant correlation between RSSI and short distance in such scenarios. It makes RSSI a routing protocol link quality metric that could be used in devices with limited energy. [8]

described in details with [3, 7, 13] which are given in Table 1. In Table 2, the specific metrics are required for link quality indication (LQI) and received signal strength indication (RSSI), which is key factor related to the quality of service, and the relevant improvement in routing protocol in reference of sending packets is

Table 2 Performance metric quality of service in routing protocol

Performance metric in routing protocol	Key factor	Improvement
Quality of service	Two important metrics are used for the link quality such as link quality indication (LQI) and received signal strength indication (RSSI).	To sustain the measurement of link quality using the RF module, an efficient metrics are to be needed for sending the packets [8].

Table 3 Performance metric multipath routing using network coding in routing protocol

Performance metric in routing protocol	Key factor	Improvement
Multipath routing using network coding	Adding redundant packets	On reducing the delay of packet transmission, the consistency of network transmission can be increased.
	A random linear network coding approach toward the code and data packets. The next hop link status and also the quantity of current packets on the receiving end which are sent by the upstream node together help to determine it.	Even though an extended propagation delay and high bit error rate of space information networks [13]
	The performance of the wireless system has been improved effectively using network coding.	To identify wormholes in rigorous manner and also the correctness, a centralized algorithm is required. Distributed detection algorithm against wormhole in wireless network coding systems (DAWN) is proposed for this distributed wireless network which adaptively detects the wormholes and decide the direction movement of the innovative packets [14].
Multipath routing using without network coding	The stiffness of attacks in hypothetical model, due to mobile nodes of multipath routing protocols	Resiliency against blocking and node isolation-type attacks [9]

illustrated [6]. In Table 3, the key factor related to the performance metric multipath routing using network coding in routing protocol, and the relevant improvement in routing protocol is described in details [9, 13, 14] . The performance metrics in routing protocols for key challenges in WSN are shown in Fig. 2 as detailed block diagram.

Wireless mobile ad hoc network-based routing protocols, such as Ad hoc On-Demand Distance Vector Routing (AODV) [15], Dynamic MANET On-demand (DYMO) [16], and Optimized Link State Routing (OLSR) [17], come in the category of specific Engineering Task Force (IETF) related to Internet. WSNs are to be used by these protocols [18] which keep similar characteristics in the context of wireless. In the category of multi-hop wireless ad hoc networks, Distributed On-demand Multi-Optional (DOMO) protocol is a single-path routing protocol. DOMO is the advanced version of AODV.

4 Simulation Analysis in Routing Protocol with QoS and Energy Consumption-Based Constraints

In the simulation results, QoS-related routing protocol based on data packet redundancy elimination (RP-DPRE) and Distributed On-demand Multi-Optional routing (DOMO) algorithms are shown which plays a key role for improved QoS and less energy consumption in WSNs.

Figure 3 depicts QoS throughput is higher in RP-DPRE in comparison to the others without mobility condition. Initially QoS throughput is quite low (around 50 kbps), due to less number of source nodes. Thereafter, throughput suddenly rises from 50 to 120 kbps, as number source nodes increase from two to six, respectively. Later, this throughput becomes constant onward the use of eight source nodes for transmission.

Figure 3 depicts that initially QoS throughput is almost similar in mentioned different routing protocol with mobility and non- mobility conditions on the availability of small source nodes. RP-DPRE without mobility maintains higher throughput in comparison to RP-DPRE with mobility and others also.

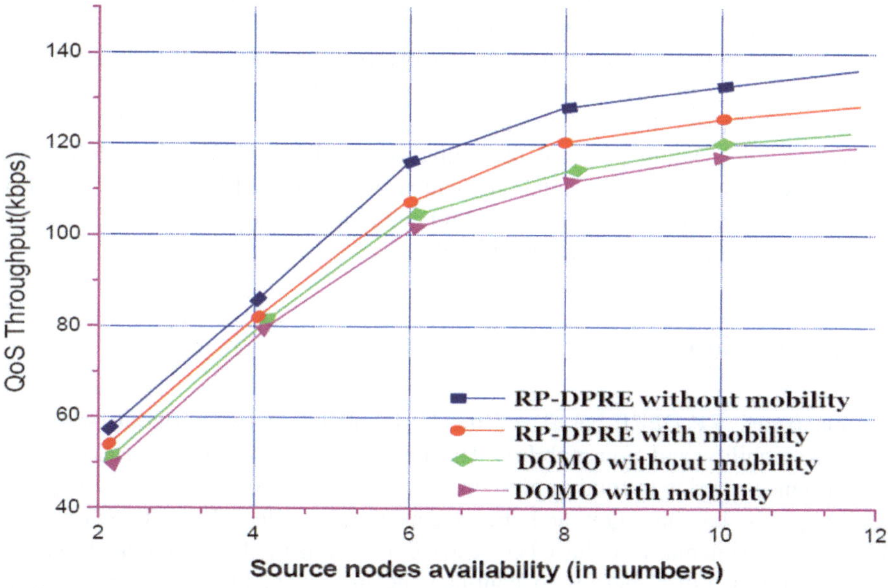

Fig. 3 QoS throughput versus source nodes availability

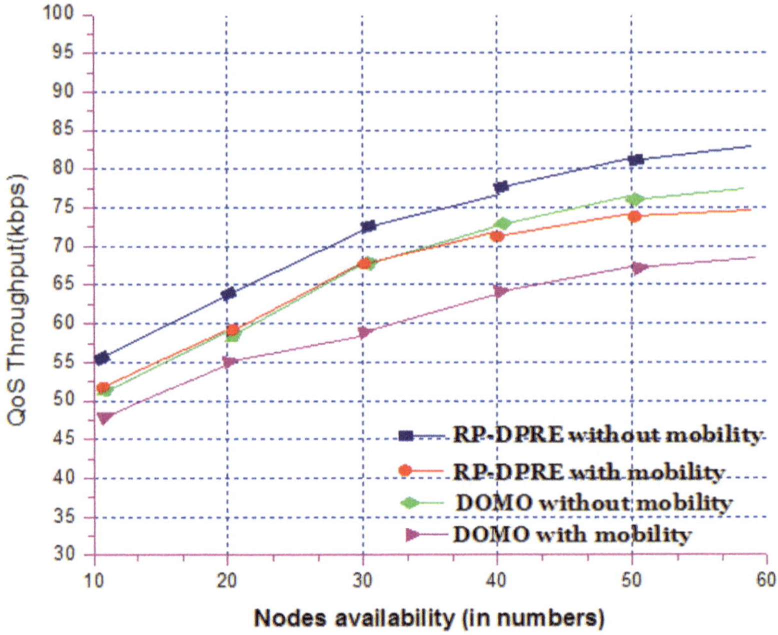

Fig. 4 QOS throughput versus nodes availability

As the node availability increases, the RP-DPRE without mobility keeps sustaining the throughput. DOMO without mobility and RP-DPRE with mobility also achieve almost equal throughput. On the basis of this experimental analysis, RP-DPRE without mobility shows the improved results in comparison to the RP-DPRE with mobility. In this approach, a combination of metrics is approached to improve the results in better way to reduce the energy consumption as the elimination of redundancy level and also the improvement in throughput.

Figure 4 depicts that QoS throughput is always higher in RP-DPRE without mobility-based routing protocol in comparison to RP-DPRE with mobility and others also. As the nodes availability increases, the RP-DPRE without mobility keeps sustaining the throughput. DOMO without mobility and RP-DPRE with mobility also achieve almost equal throughput.

5 Conclusion

In this paper, RP-DPRE-based scheme is simulated in two different scenarios such as mobility and non-mobility. The presented survey on different kinds of performance metrics in routing protocols for WSNs is also considered to improve throughput by handling QoS and energy consumption effect. Even though various kinds of experimental work have been carried out in this field, the simulated analysis helps to check other metrics to continue the improvement in routing protocols processing. Next, we significantly mentioned some key factors, which are essential to do research for improving the routing protocols in WSNs. In the comparison with DOMO with mobility and non-mobility situations, RP-DPRE with mobility even performs better. The combination of other metrics can also improve the results in terms of packet loss and also the improvement in throughput.

References

1. Le J, Lui JCS, Chiu D-M (2010) DCAR: distributed coding-aware routing in wireless networks. IEEE Trans Mob Comput 9(4):596–608
2. Lakhotia J, Kumar R (2014) Fault tolerant and mobility aware routing protocol for mobile wireless sensor network. In: International Conference on Control, Instrumentation, Communication and Computational Technologies (ICCICCT), pp 584–590
3. Al-Jemeli M, Hussin FA (February 2015) An energy efficient cross-layer network operation model for IEEE 802.15.4-based mobile wireless sensor networks. IEEE Sensors J 15(2):684–692
4. Zhao L, Hong X, Liang Q (2004) Energy-efficient self-organization for wireless sensor networks: a fully distributed approach. Proc 47th Ann IEEE Global Telecomm Conf 5(11):2728–2732
5. Lindsey S, Raghavendra C, Sivalingam KM (2002) Data gathering algorithms in sensor networks using energy metrics. IEEE Trans Parallel Distrib Syst 13(8):924–935
6. Bokareva T, Bulusu N, Jha S (2006) Learning sensor data characteristics in unknown environments. In Proceedings of the 1st International Workshop on Advances in Sensor Networks (IWASN 2006). San Jose
7. Yinpeng Y, Yuhuai P, Yejun L, Lei G, Meng S (2014) Survivable routing protocol for green wireless mesh networks based on energy efficiency. Network technology and applications. China Communications, pp 117–124
8. Entezami F, Tunnicliffe M, Politis C (2014) Statistical analysis on wireless sensor network link-quality metrics. IEEE Vehicular Technol Mag. doi:101109/MVT2014.2333693, 18
9. Qi D, Virendra M, Upadhyaya S, Sanzgiri A (2014) Minimum cost blocking problem in multipath wireless routing protocols. IEEE Trans Comput 63(7):1765–1777
10. Culpepper BJ, Dung L, Moh M (2004) Design and analysis of hybrid indirect transmissions (HIT) for data gathering in wireless micro sensor networks. ACM SIGMOBILE Mobile Comput Comm Rev 8:61–83
11. Reznik L, Pless GV, Karim TA (2005) Signal change detection in sensor networks with artificial neural network structure. In IEEE international conference on Computational Intelligence for Homeland Security and Personal Sa fety (CIHSPS), pp 44–51, Orlando
12. Lindsey S, Raghavendra C (2002) PEGASIS: power-efficient gathering in sensor information systems. Proc IEEE Aerospace Conf 3:1125–1130
13. Geng Y, Congjiu Z, Xiaoyu L, Chi Z, Lina W, Yansong L (2014) Research of multi-path routing based on network coding in space information networks. Chin J Aeronaut 27(3):663–669
14. Ji S, Chen T, Zhong S (2015) Wormhole attack detection algorithms in wireless network coding systems. IEEE Trans Mob Comput 14(3):660–674
15. Perkins CE, Royer EM (1999) Ad-hoc on-demand distance vector routing. In: Proceedings WMCSA'99 Second IEEE Workshop on mobile computing systems and applications, 1999, pp 90–100
16. Ratliff S, Dowdell J, Perkins C (2013) Dynamic MANET on-demand (AODVv2) routing
17. Clausen T, Jacquet P, Adjih C, Laouiti A, Minet P, Muhlethaler P et al (2003) Optimized link state routing protocol (OLSR)
18. Heo J, Hong J, Cho Y (2009) EARQ: energy aware routing for realtime and reliable communication in wireless industrial sensor networks. IEEE Trans Indus Inf 5:3–11

Reliability Analysis for Upgraded Software with Updates

Adarsh Anand, Subhrata Das, Deepti Aggrawal, and P. K. Kapur

1 Introduction

In today's globalized informative society, a software system is used in diverse fields and has gained lots of importance in our daily life. It acts like a catalyst for the rapid growth of government and commercial organization. Software not only exists in a variety of products but can also be treated as a very important asset for economic growth [10, 15]. Software has always been targeted by non-state actors in order to achieve illegal profits. To avoid these consequences, software engineers have to secure their product from the burglars starting from its general availability. Security is one of the significant and important attribute of software. In the development of software, many applications are outsourced, and thus it lacks strong integration in terms of software security. There is a growing need to address software security measures across SDLC. In the current global world, software security needs to be addressed holistically and methodically in the same way as attention is given to quality and safety. There are many instances in operational phase of the software in which software security is bypassed leading to the exposure of vulnerability. The developing firms have been carrying out various noteworthy efforts to decrease the number of loop holes, perk up resistance from threats, and shield the reliability of

A. Anand (✉) • S. Das
Department of Operational Research, University of Delhi, Delhi, 110007, India
e-mail: adarsh.anand86@gmail.com; shus.das@gmail.com

D. Aggrawal
University School of Management and Entrepreneurship, Delhi Technical University, East Delhi Campus, Delhi, 110095, India
e-mail: deepti.aggrawal@gmail.com

P.K. Kapur
Amity Center for Interdisciplinary Research, Amity University, Noida, UP, India
e-mail: pkkapur1@gmail.com

© Springer Nature Singapore Pte Ltd. 2018

P.K. Kapur et al. (eds.), *Quality, IT and Business Operations*, Springer Proceedings in Business and Economics, DOI 10.1007/978-981-10-5577-5_26

the merchandise. Software security is fundamental in order to suppress the alarming complexities. One of the effective methods to handle security of software is to provide patches after release of the software.

Nowadays, in the growing importance of technocrat world, the firms have found themselves difficult to sustain in the competitive market. Thus, survival is the utmost concern of the firms for which they adopt the strategy of upgrade. The process of upgradation is focused on adding new features and defects fixations, etc. to an application in the form of installer or additions [3]. Feature enhancement may sometime result into increased fault content due to increased portion of code, and thus firms have to debug them in succeeding versions. The market forces of demand and supply for highly reliable software products lead to successive releases. Several real-life examples can be quoted to supplement the idea of multi-upgradation such as the family of Microsoft Windows, versions of MS Office, and most recent versions of Android software in smartphones [2].

Earlier modeling in multi-upgradation focused on the effect of fault from all its preceding releases for determining the eventual faults debugged in any specific release [11]. Later on, the concept of multi-upgradation has been further extended to account for faults from just previous release [19]. There are many situations in which the testing team may not able to able to remove faults perfectly resulting into imperfect fault removal phenomenon. To model such scenario, Kapur et al. [12] have extended the concept of fault removal process under the just previous release case to cater the effect of imperfect debugging. To study the counter-effect of randomness in software, a multi-upgradation software reliability model based on stochastic differential equation is proposed by Singh et al. [17]. The classification of faults has also been incorporated by some researchers to model the case of just previous release in multi-upgradation. Kapur et al. [13] worked on fault severity modeling, i.e., simple and hard fault modeling. Some researchers have worked on modeling the phenomena of imperfect debugging and stochastic differential equation into severity of faults, to model more realistic scenarios [1, 18]. Recent study by Garmabaki et al. [8] modeled a multi-release model in which focus was given to model the bugs that might be reported from operational phase. In recent past, Anand et al. [2] proposed a generalized multi-upgradation model for faults due to undetected bugs, feature enhancement, and bugs that might be encountered in operational phase, where functional forms of several distributions were considered. Further, Anand et al. [3] modeled a fault removal phenomenon taking into consideration both the faults from the testing of new releases and the reported faults from the operational phase of the preceding releases based on categorization of simple and hard faults during testing and operational phase.

With the escalating need of highly reliable software product among consumers, developers have started to incorporate the concept of providing update to users for fixing the bugs which might get invoked when software is in its operational phase. Software updates have become an integral part and a difficult challenge for software maintenance process. With new versions of software being released on frequent basis, updates in the form of patches have to be scheduled by developers on regular intervals. In this highly malicious environment, security breaches seem to be more common, and due to these threats, more and more firms are becoming possible victim of attacks. In this regard, software developer needs to be on constant watch. Software internal testing and the feedback in terms of bug report given by the users are considered to be important factors while scheduling a patch.

The software that we are using on our systems is in a regular need of attention. Often, most of the software enterprises relocate the resources for a time-bound release of their offering to stay on top of the market. Indeed, most of the firms explicitly incorporate the provision of fixing the vulnerabilities that create security weakness. They scheduled regular release of patches to mitigate the risk due to the available loopholes. With the availability of multigeneration software, the challenge of patch deployment for each version is also becoming a difficult task. As software firms keep on working rigorously to increase their net profit, new upgraded versions of software are often released in the market. With the availability of multiple versions of software, each of which provides varying level of features and scheduling patches for each version requires variable support and maintenance information. Patch is a small program of software which is designed to update or modify existing computer program to fix the error or improve functionality. Patches are commonly referred as bug fixes. They help in protecting the software from unwanted and unpredicted operations which can result in failure. Thus, software patching acts as a useful technique in enhancing the reliability and performance of the software. There are some examples to supplement the assertion of successive releases under patching: monthly security maintenance releases from Samsung Mobile include patches that are developed by Google and Samsung [16]. Recently summary report was given by Microsoft; it was mentioned that the firm provides security update for Windows Kernel (3,171,910), .NET Framework (3170048), etc., in which most severe vulnerabilities could allow security feature bypass and could cause information disclosure [14].

Several researchers have proposed ideas to account for the concept of modeling updates. An economic model for releasing a faulty product early and providing updates later on has been proposed by Arora et al. [5]. Similar to the concept of software debugging, Jiang and Sarkar [9] have formulated a release policy based on the notion of testing in operational phase and suggested the policy under economic implication with patching considered. Another study of cost implication was proposed by Das et al. [6] in which the release decision of software comprises the joint effort of tester and user in order to provide patches. A mathematical model to model the number of bugs removed on inclusion of patching is given by Deepika et al. [7]. They emphasized on modeling the pre- and post-release fault detection under exponential fault detection phenomenon. Also, Anand et al. [4] proposed an economic cost analysis model considering the cost of faults removed in testing phase by user, cost of debugging after release of the product, and market opportunity cost.

In this paper, we have modeled the effect of updates on successive release of software in which the faults of present release comprise faults that might get encountered in testing phase and the leftover faults from testing phase that were debugged in operational phase by both user and tester. The proposed methodology considers rigorous testing in both testing phase and operational phase in successive releases of the product. Here, a new concept to study both upgrade and update in one combinational form has been employed. The outline of the article is arranged as follows: Sect. 2 corresponds to the set of notations and Sect. 3 illustrates our assertion mathematically. Section 4 provides the validity of the proposed methodology numerically. Conclusion is demonstrated in Sect. 5 followed by acknowledgment at last.

2 Notation

$m_i(t)$: Mean value function of fault detection/correction process ($i = 1, 2, 3, 4$)
$F_i(t)$: Probability distributions function for fault removal process ($i = 1, 2, 3, 4$)
a_i: Faults of *i*th version of the software
b_i: Constant rate of fault detection and correction before release of software ($i = 1, 2, 3, 4$)
b_{i1}: Fault correction rate function after release of software ($i = 1, 2, 3, 4$)
τ_i: Release time point ($i = 1, 2, 3, 4$)
$\otimes$: Stieltjes convolution

3 Modeling Framework

Well-established notion in today's competing environment calls for recurrent feature enhancement. Nowadays, many similar products of different firms compete in the market to attain large market share and to have brand value. Normal trend as seen in contemporary globalized world is that firms do not release their complete offering in one go but keep on launching newer versions so as to maintain its position in the market. The step by step procedure of frequent releases with some add-ons is extremely beneficial in context of:

1. Timely penetration into the market which would have been delayed if the firm might have thought of launching complete product in one instance.
2. Successive releases bring product under public eyes as soon as its newer version is released.
3. It increases time duration of product existence in the market.
4. Firms become capable of sustaining competition.
5. Further, judicious use of resources while developing successive release can increase its overall revenue that can be fetched.

Even now, the software is developed in multiple releases where the newer one is acquainted with more advanced functionalities. The successive releases of software can be developed by refining the available functionality and feedback, adding new functions; it can be a combination of both or enhancing the reliability of the software [2]. There can be some time points at which firms want to meet the requirement of users speedily, so they come up with upgrading the versions frequently. Due to hastily coming up with newer versions, some issues might not have been resolved properly, because this firm can update their software in the later hours by providing patch and thereby reduce any problematic issue in near future. Updates are generally software suit or some application that are developed with the notion of fixing loopholes or failure. In certain situations, reports are received from customer ends about improper functionalities, some failure in its usage, or some cases of breeching, which is then the duty of testers to rectify these issues. Overall, the life span of software increases due to patching. Thus, update and upgrade together form a profitable combo for developer; on the whole, it increases the revenue.

Due to complexities associated with the code, all faults cannot be detached in the first version of the software; thus, it might be the case that some bugs remain in the code and might be carried away to its succeeding versions. In order to mathematically model the above concept of upgrade inculcating the concept of update, we divided the total planning horizon in two stages for overall testing of the product, one being the stage of in-house testing in which the major role of the tester is to debug the faults before the software is released in field for usage; while in the succeeding stage, the role of both tester and the users of the software comes into picture leading to prolonged testing even after the release of software for usage. Considering 'τ_i' as the release time of ith version of the software and under the concept of patching, total testing time is divided in two intervals $[0,\ \tau_i]$ and $(\tau_i,\ T]$. The mathematical expression of total faults in the first release of software assuming τ_1 to be its release time is given as follows:

$$m_1(t) = a_1.F_1(\tau_1) + a_1.[1 - F_1(\tau_1)].F_1 \otimes G_1(t - \tau_1) \tag{1}$$

where $F_1(\cdot)$ is the cumulative distribution function for fault removal phenomenon before the release of the software, while $(F_1 \otimes G_1)(t - \tau_1)$ denotes the joint distribution fault removal phenomenon to account for the role of both user and tester, when software is in its operational phase [6].

In order to attract more customers, a company adds some new functionality to the existing software system. Thus, adding some new functionality to the software leads to change in the code. These new specifications in the code lead to increase in the fault content. Now the testing team starts testing the upgraded system, both in testing and for some duration of time in operational phase. Thus, the fault removal phenomenon for second release can be modeled as follows:

$$\begin{aligned} m_2(t) &= a_2F_2(\tau_2) + [a_1 - m_1(t_1)].F_2(\tau_2) \\ &\quad + [a_2 + a_1 - m_1(t_1)][1 - F_2(\tau_2)].F_2 \otimes G_2(t - \tau_2) \\ &= [a_2 + a_1 - m_1(t_1)].F_2(\tau_2) \\ &\quad + [a_2 + a_1 - m_1(t_1)][1 - F_2(\tau_2)].F_2 \otimes G_2(t - \tau_2) \end{aligned} \tag{2}$$

$$m_2(t) = a_2^*.F_2(\tau_2) + a_2^*.[1 - F_2(\tau_2)].F_2 \otimes G_2(t - \tau_2) \tag{3}$$

where $a_2^* = [a_2 + a_1 - m_1(t_1)]$ shows that the eventual fault content for the second version which gets increased due to the remaining faults of its preceding release and faults generated due to new add-ons and $m_1(t_1) = a_1 . F_1(\tau_1) + a_1 . [1 - F_1(\tau_1)] . F_1 \otimes G_1(t_1 - \tau_1)$ denotes the number of bugs which were removed by time 't_1' for first release. The first component in Eq. (3) shows the total number of faults removed in the second release before the product is made available in the market. The second component comprises of remaining number of faults from testing phase which lie latent in the second version of the software, which can be debugged by joint effort of testers and users in the operational phase. The patch can be provided to the users for overcoming faults which occurred while the software is in usage.

There can be a situation in which the firm might be required to add new functionalities in the software for the second time in which new portion of code is added to the existing one, and once again the testing process begins which is further followed by testing in operational phase. Let the time for introduction of the third release be 'τ_3' and (τ_3, t_3) be the interval till prolonged testing of the software is carried out, and also, let users who might encounter some faults in there usage report about them to firms for further removal. The mathematical expression for this fault removal can be given as

$$\begin{aligned} m_3(t) &= a_3 F_3(\tau_3) + \left[a_2^* - m_2(t_2)\right].F_3(\tau_3) \\ &\quad + \left[a_3 + \left[a_2^* - m_2(t_2)\right]\right].[1 - F_3(\tau_3)].F_3 \otimes G_3(t - \tau_3) \\ &= \left[a_3 + \left[a_2^* - m_2(t_2)\right]\right].F_3(\tau_3) \\ &\quad + \left[a_3 + \left[a_2^* - m_2(t_2)\right]\right].[1 - F_3(\tau_3)].F_3 \otimes G_3(t - \tau_3) \end{aligned} \tag{4}$$

$$m_3(t) = a_3^*.F_3(\tau_3) + a_3^*.[1 - F_3(\tau_3)].F_3 \otimes G_3(t - \tau_3) \tag{5}$$

where $a_3^* = \left[a_3 + \left[a_2^* - m_2(t_2)\right]\right]$ demonstrates the actual number of faults for the third release which gets increased due to remaining faults of its previous releases and faults generated due to new features and $m_2(t_2) = a_2^*.F_2(\tau_2) + a_2^*.[1 - F_2(\tau_2)].F_2 \otimes G_2(t_2 - \tau_2)$ indicates the count of faults which were removed by time 't_2' for the second release. The first component in Eq. (5) explains the total number of faults removed in the third version before the product is released. The second component consists of leftover bugs from testing phase which lie dormant in the third version of the software, which can be debugged by combined effort of testers and users in the operational phase.

Similarly, we can write the mathematical expression for the fourth release as follows:

$$\begin{aligned} m_4(t) &= a_4.F_4(\tau_4) + \left[a_3^* - m_3(t_3)\right].F_4(\tau_4) \\ &\quad + \left[a_4 + \left[a_3^* - m_3(t_3)\right]\right].[1 - F_4(\tau_4)].F_4 \otimes G_4(t - \tau_4) \\ &= \left[a_4 + \left[a_3^* - m_3(t_3)\right]\right].F_4(\tau_4) \\ &\quad + \left[a_4 + \left[a_3^* - m_3(t_3)\right]\right].[1 - F_4(\tau_4)].F_4 \otimes G_4(t - \tau_4) \end{aligned} \tag{6}$$

$$m_4(t) = a_4^*.F_4(\tau_4) + a_4^*.[1 - F_4(\tau_4)].F_4 \otimes G_4(t - \tau_4) \quad (7)$$

where $a_4^* = [a_4 + [a_3^* - m_3(t_3)]]$.

The above fault removal phenomenon can be extended to n-versions of the software given as follows:

$$m_n(t) = a_n^*.F_n(\tau_n) + a_n^*.[1 - F_n(\tau_n)].F_n \otimes G_n(t - \tau_n) \quad (8)$$

where $a_n^* = [a_n + a_{n-1}^* - m_{n-1}(t_{n-1})]$.

To facilitate the analysis of the above proposed methodology of upgrade inculcating update in each version of the software, we have considered the case of four versions of the software. The distribution function to account for removal phenomenon is assumed to be exponential in nature before the release of the software. In this period $(\tau_i, \ t_i]$, both the tester and user are jointly participating in fault removal phenomena with different rates of debugging. Taking into consideration $F_i(t) \sim$ exponential and $G_i(t) \sim$ constant distribution in nature, we get

$$(F_i \otimes G_i)(t) = \left(1 - e^{-b_i t}\right) \quad (9)$$

The above Eqs. (1), (3), (5), and (7) considering the conjoint effect of testers and users in operational phase can be rewritten as

$$m_1(t) = a_1\left(1 - e^{-b_1\tau_1}\right) + a_1.\left(e^{-b_1\tau_1}\right).\left(1 - e^{-b_{11}(t-\tau_1)}\right) \quad (10)$$

$$m_2(t) = a_2^*\left(1 - e^{-b_2\tau_2}\right) + a_2^*.\left(e^{-b_2\tau_2}\right).\left(1 - e^{-b_{21}(t-\tau_2)}\right) \quad (11)$$

$$m_3(t) = a_3^*\left(1 - e^{-b_3\tau_3}\right) + a_3^*.\left(e^{-b_3\tau_3}\right).\left(1 - e^{-b_{31}(t-\tau_3)}\right) \quad (12)$$

$$m_4(t) = a_4^*\left(1 - e^{-b_4\tau_4}\right) + a_4^*.\left(e^{-b_4\tau_4}\right).\left(1 - e^{-b_{41}(t-\tau_4)}\right) \quad (13)$$

4 Data Analysis

To validate the aforesaid concept, we have used a real software failure data [21] which comprises of four releases. We use least squares method, one of the most significant methods to estimate model parameters in the software reliability field. For parameter estimation, we apply statistical package SAS software for analysis purpose [20]. Tables 1 and 2 illustrate the parameter estimates and comparison criteria of each release of the proposed methodology. The performance analysis of the proposed model is measured by the four common criteria SSE, MSE, RMSE, and R^2, i.e., the coefficient of determination.

Table 1 Parameter estimates

Parameters	Release 1	Release 2	Release 3	Release 4
a	103.599	120.85	62.32	44.3713
b_i	0.10466	0.0935	0.09481	0.05225
b_{i1}	0.23964	0.2433	0.37657	0.14233

Table 2 Comparison criteria

Criteria	Release 1	Release 2	Release 3	Release 4
SSE	66.9007	267.5	44.4631	14.8315
MSE	3.9353	16.72	4.4463	0.927
Root MSE	1.9838	4.089	2.1086	0.9628
R^2	0.9959	0.9893	0.9913	0.9958

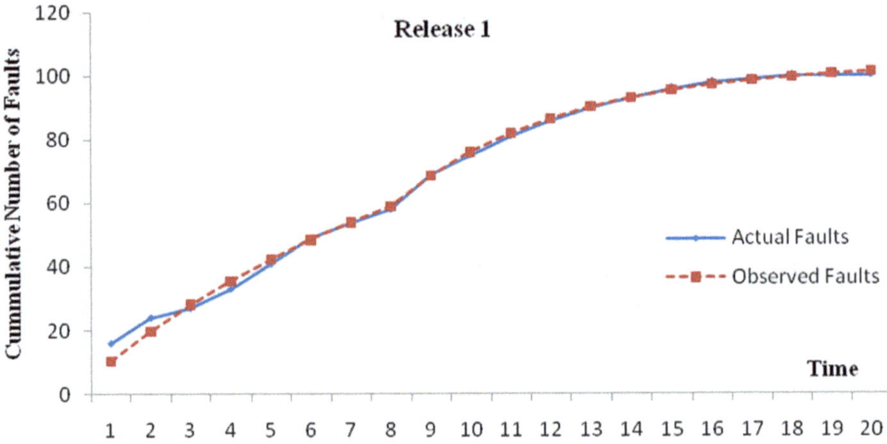

Fig. 1 Goodness of fit curve for release 1

Both Tables 1 and 2 show the good illustration of better predictive power of the proposed approach. From Table 2, it can be observed that the value of SSE, MSE, and root MSE of the fourth release is quite lesser as compared to other releases. Thus, release 4 gives better fit. Figures 1, 2, 3, and 4 have the pictorial representation of multi-upgradation modeling inculcating the prolonged testing methodology.

5 Conclusion

Efficient and reliable software is crucial for firms to provide its customer satisfaction. The ultimate goal of firms is not only to have competitive advantage over rivals but also to acquaint its offering with higher level of reliability, for which they inculcate the notion of prolonged testing, i.e., the concept of patching. The proposed approach has added a new dimension in the field of software engineering to model a realistic scenario. In this paper, a methodical approach to model the significant

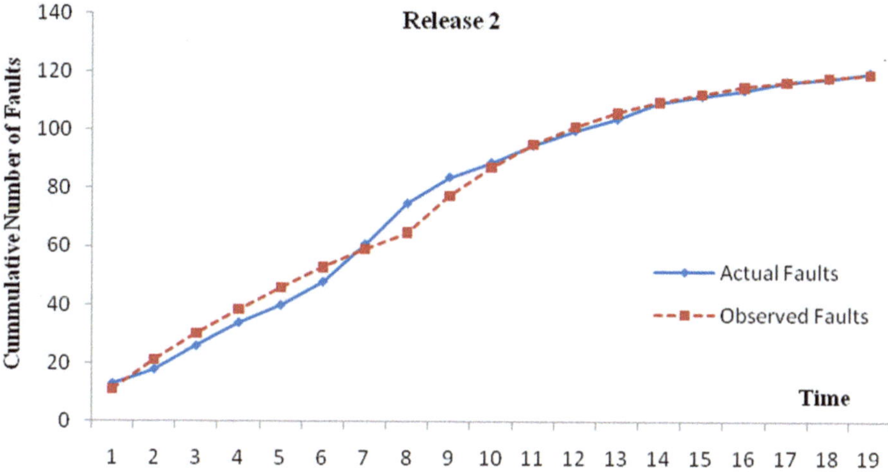

Fig. 2 Goodness of fit curve for release 2

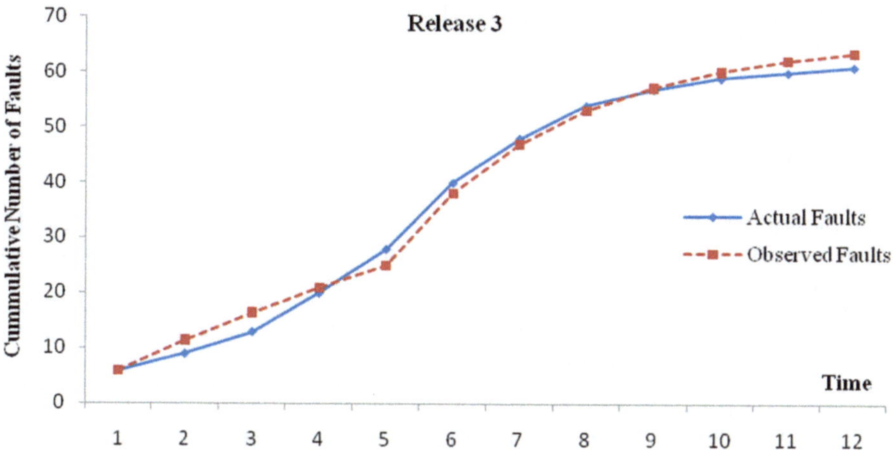

Fig. 3 Goodness of fit curve for release 3

idea of updates and upgrades has been incorporated. Updates mean to provide a security band for the product, while upgrade leads to adding new add-ons in the existing product to uphold their position. Further to supplement the proposition, we analyzed it on failure data, and the result obtained shows that coming with updates and upgrades enhances the predictability of failures.

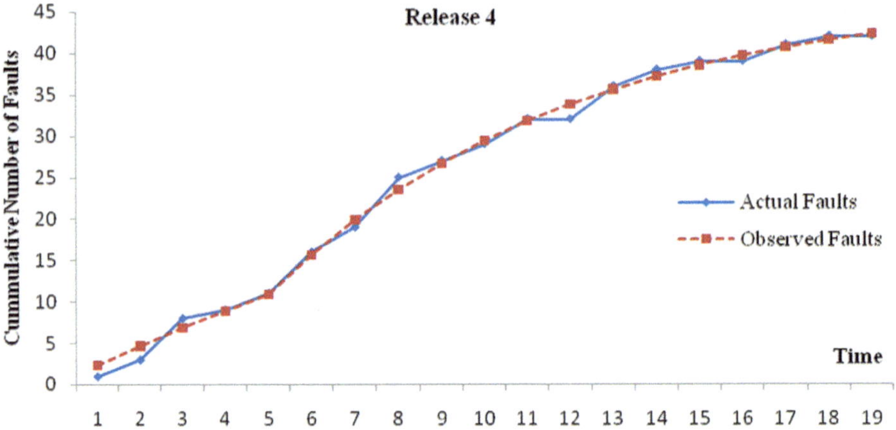

Fig. 4 Goodness of fit curve for release 4

Acknowledgment The research work presented in this paper is supported by grants to the first author and third author from the University of Delhi R&D Grant No. RC/2015/9677, Delhi, India.

References

1. Aggarwal AG, Kapur PK, Garmabaki AS (2011) Imperfect debugging software reliability growth model for multiple releases. In: Proceedings of the 5th national conference on computing for nation development-INDIACOM, New Delhi
2. Anand A, Singh A, Kapur PK, Das S (2014) Modeling conjoint effect of faults testified from operational phase for successive software releases. In: Proceedings of the 5th international conference on life cycle engineering and management (ICDQM), pp 83–94
3. Anand A, Singh O, Das S (2015) Fault severity based multi up-gradation modeling considering testing and operational profile. Int J Comput Appl (0975–8887) 124(4):9–15
4. Anand A, Agarwal M, Tamura Y, Yamada S (2016) Economic Impact of software patching and optimal release scheduling. Qual Reliab Eng Int 33(1):149–157. doi:10.1002/qre.1997
5. Arora A, Caulkins JP, Telang R (2006) Sell first, fix later: impact of patching on software quality. Manag Sci 52(3):465–471
6. Das S, Anand A, Singh O, Singh J (2015) Influence of patching on optimal planning for software release & testing time. CDQM- An Int J 18(4):81–92
7. Deepika AA, Singh N, Dutt P (2016) Software reliability modeling based on in-house and field testing. CDQM- An Int J 19(1):74–84
8. Garmabaki AHS, Aggarwal AG, Kapur PK, Yadavali VSS (2014) The Impact of bugs reported from operational phase on successive software releases. Int J Prod Qual Manag 14(4):423–440
9. Jiang S, Sarkar S (2003) Optimal software release time with patching considered. In: Proceedings 13th annual workshop information technologies and systems, Seattle, pp 61–66
10. Kapur PK, Pham H, Gupta A, Jha PC (2011) Software reliability assessment with OR application. Springer, Berlin
11. Kapur PK, Tandon A, Kaur G (2010) Multi up-gradations software reliability model. ICRESH, pp 468–474

12. Kapur PK, Singh O, Garmabaki A, Singh J (2010) Multi up-gradation software reliability growth model with imperfect debugging. Int J Syst Assur Eng Manag 1(4):299–306
13. Kapur PK, Anand A, Singh O (2011) Modeling successive software up-gradations with faults of different severity.In: Proceedings of the 5th national conference on computing for nation development, ISSN 0973–7529 ISBN 978–93–80544-00-7,
14. Microsoft Security Bulletin Summary for July 2016. https://technet.microsoft.com/en-us/library/security/ms16-jul.aspx. 27 July 2016
15. Pham H (2006) System software reliability. Springer, London
16. Samsung Android Security Updates http://security.samsungmobile.com/smrupdate.html, 27 July, 2016
17. Singh O, Kapur PK, Anand A, Singh J (2009) Stochastic differential equation based modeling for multiple generations of software. In: Proceedings of fourth International Conference on Quality, Reliability and Infocom Technology (ICQRIT), Trends and Future Directions, Narosa Publications, pp 122–131
18. Singh O, Kapur PK, Anand A (2011) A stochastic formulation of successive software releases with fault severity. In: Industrial engineering and engineering management, pp 136–140
19. Singh O, Kapur PK, Khatri SK, Singh JNP (2012) Software reliability growth modeling for successive releases. In: Proceeding of 4th International Conference on Quality, Reliability and Infocom Technology (ICQRIT), pp 77–87
20. SAS Institute Inc. (2004) SAS/ETS user's guide version 9.1. Cary, NC: SAS Institute Inc.
21. Wood A (1996) Predicting software reliability. IEEE Comput 11:69–77

Quantitative Software Process Improvement Program Using Lean Methodology

Mahesh Kuruba and Prasad Chitimalla

1 Introduction

Quantifying the process improvements has always been challenging. The software process improvement programs are predominantly designed and performed with reference to frameworks like CMMI®, ISO, ITIL, etc. Typically, such process improvement programs take 1 $\sim$ 3 years depending on the size and complexity of the organization structure. Such programs find it difficult to demonstrate quantitative benefits until the journey is complete. At times, due to the absence of quantitative data, demonstrating benefits even after completion of process improvement program is challenging. The challenges faced in process improvement programs are predominantly commitment from senior management and change management of the practitioners [1]. Due to the nature of the conventional approach, the practitioners start experiencing improvements and realizing the benefits much later in the program. This affects the motivation of the practitioners in the program and the benefits that it can provide them. Considering the magnitude of the process improvements happening globally, there is a little evidence of the quantified results [2]. This is due to lack of methods by which organizations quantify the potential for improvement and demonstrate the benefits. These characteristics of program design inhibit process improvement programs in the organization, due to its inability to garner management and practitioners' support. Hence, it's essential to design and adopt a quantitative-based approach for process improvement.

Lean methodology adopted from manufacturing organizations offers promising results with little or no data. Experience has shown that even in the absence of system collected data, the data collected through practitioners during lean assessments can quantify the potential benefits due to process improvement. Also, the traditional

M. Kuruba (✉) • P. Chitimalla
Global Consulting Practice, Tata Consultancy Services, Pune, India
e-mail: m.kuruba@tcs.com

© Springer Nature Singapore Pte Ltd. 2018

P.K. Kapur et al. (eds.), *Quality, IT and Business Operations*, Springer Proceedings in Business and Economics, DOI 10.1007/978-981-10-5577-5_27

improvement frameworks like SCAMPI do not quantify the opportunity due to improvement for a specific finding. In contrast, lean methodology indicates the potential for improvement for a given solution, which makes it easy for management and practitioners alike to prioritize and implement such solutions.

In this paper, we present our experience of demonstrating benefits of process improvements quantitatively using lean methodology. The experience is from a process improvement program in software maintenance and support environment. We discuss typical organization challenges in Sect. 3; the factors that are considered while selecting the methodology for process improvement are discussed in Sect. 4; and the process improvement methodology is being discussed in Sect. 5 and the results achieved across multiple organizations in Sect. 6; comparison of SCAMPI and Lean methodology based assessments in Sect. 7; tailoring the lean based assessment methodology in Sect. 8 and follows our conclusions in Sect. 9.

2 Literature Review

Lean methodology has been traditionally looked at minimizing waste, delivering value to customer, and maximizing flow. Jonsson [3] has reviewed perspectives about lean principles in software development by various authors. Petersen and Wohlin [4] proposed a novel approach to blend quality improvement paradigm and lean software development practices. On the other hand, organizations have adopted lean thinking for software development projects [5–8]. Middleton and Joyce [5] based on their data collection over a period of 12 months reported results in improvement of lead time to deliver software by 37%, consistency of delivery rose by 47%, and defects reported by customers fell by 24%. Staats and Upton [6] have shared the experiences of an Indian IT company adopting lean. Pernstal et al. [9] have identified gaps in application of lean to large-scale software development. However, what would be interesting to see is a business case for senior management to go for lean. In addition to applying lean principles, the management needs a business case for process improvement programs. Though there are ROI models [10] to demonstrate the benefits at the program level, it doesn't provide the potential benefits that the process improvement program can bring in and the corresponding benefit from each improvement area. Raffo and Kellner [11] have modeled and simulated software processes to evaluate quantitatively.

Hence, there is a need for a process improvement methodology that (a) estimates the potential benefit due to process improvement program and (b) estimates the potential benefit due to a specific improvement area or a process change (c) determining the benefit after implementing the solution. The benefit could be cycle time reduction or cost reduction of a process. In this paper, we share our experience of implementing such methodology using lean principles.

3 Organization Challenges

Some challenges the organization was facing to provide efficient application maintenance and support services to their customers are as follows:

- No standard processes for maintenance and support activities
- Unable to scale up to meet IT requirements of customer
- Unable to meet service quality requirements

The organization didn't have much visibility into the process/service performance; they did not have an organization-level standard process to provide services. The organization was struggling to identify the quality metrics for their processes while the customer was expecting quality of services, and there were no service levels agreed between them. Since the organization didn't have visibility on their current process performance, they could not define service levels.

The process improvements initiated internally were not always providing the results to the organization. The organization needed a structured method to improve processes. At this point of time, the organization engaged us to help them in improving their processes and the program went on for 1.5 year. Subsequently, our customer went on to implement this methodology for other units in the organization.

As a first step (as shown in Fig. 1), program planning workshop was conducted along with the organization key stakeholders to understand various challenges the organization is facing and determine appropriate methodology for process improvement suitable to the organization. It involveda series of discussions with senior managers and middle-level managers to understand the organization's challenges. Based on the discussions, the program objectives are defined ensuring IT strategy that is aligned to business strategy. The key objectives identified for the program are defining organization-level processes and defining quality metrics and process performance.

The various organization challenges for process improvement program identified are:

(a) No incentives: The benefit that the organization will receive due to process improvement, in terms of productivity improvement, could mean reduction in cost of application maintenance and reduced revenue. This is a typical challenge faced by organizations at initial maturity levels.
(b) Lack of structured and proven approach: Practitioners in the organization tried implementing process improvements in their respective divisions, but largely they were not successful due to unawareness of any structured and proven

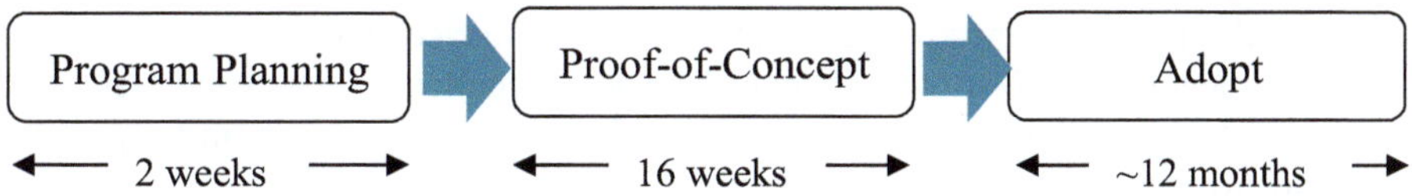

Fig. 1 Process improvement program phases

approach for process improvement. They may not be aware of the various factors that need to be considered during process improvement activities.

(c) Lack of acceptance from practitioners: The unsuccessful process improvement programs in the past resulted in uncertainty among the practitioners. They were not certain if such improvement programs will be beneficial and thus made them skeptical about the process improvement programs.
(d) Assumed negative impact on day-to-day operations: There were apprehensions if the process improvement activities need a lot of effort which will impact their day-to-day operations.

Considering such cultural aspects, challenges, and apprehensions of the stakeholders, a bottom-up approach with focus on quantitative benefits, a 16-week lean-based process improvement cycle, was identified as the most appropriate approach for the organization.

4 Guiding Principle for Process Improvement

Considering the above challenges and organization context, the following guiding principles were considered while designing the program:

(a) Customer-focused improvements: Considering the challenges the organization is facing in providing required services to the customer, it is vital to demonstrate the improvements resulting in improved customer experience due to process improvements.
(b) Proof-of-concept of the methodology: To demonstrate that the proposed lean methodology would actually help the organization in their process improvement, conducting proof-of-concept becomes a critical step in the process improvement journey. The objective of proof-of-concept is to demonstrate the methodology, obtain acceptance from practitioners, and make necessary changes to the methodology if required. Based on proof-of-concept results, the organization can proceed with adoption of the methodology in the rest of the organization as shown in Fig. 1.
(c) Efficient and effective processes: The organizations which intend to speed up their processes in meeting customer requirements need to apply lean to make the processes efficient and effective.
(d) Focus on benefit realization: The benefit realization will enable the management to have a business case for process improvement.
(e) Target quick wins: To instill confidence among the practitioners and the management, early benefits would demonstrate that the methodology will be useful for process improvement and help in gaining acceptance.
(f) Incremental organizational change: The main challenge associated with this kind of programs is acceptance from practitioners and management for process improvement. To obtain acceptance:

- The program planning along with the organization will assist in understanding their ground challenges and design the program with an appropriate methodology.
- Lean methodology will help in engaging practitioners from planning, identifying pain points, performing causal analysis, and designing and implementing appropriate solutions.
- Lean methodology will enable the organization with an incremental improvement and change.

(g) Bottom-up approach: In consensus based organizations it is critical to obtain acceptance at various levels. This calls for improving processes at the division level and then roll up to standardize at the organization level.

(h) Leverage internal best practices and best-in-class processes: It is recommended to leverage the current best practices of the organization and enhance them with the best-in-class processes observed globally.

(i) Change agents: Change agents from the organization are critical in such process improvement activities, as they understand the organization dynamics and will be in a position to steer the decisions or changes accordingly.

(j) Quantitative improvement methodology: In organizations where the work culture is detail oriented and inclined toward data, lean methodology which is primarily a data-driven methodology will be appropriate for process improvement. In some organizations, the data may not be available for analysis, and the organization will be concerned how they will be able to succeed in the absence of data. The level of details captured during the value stream map workshop will quickly show the organization the way they work, the bottlenecks they have in the processes, and the process performance for the various requests that they receive.

5 Process Improvement Cycle

The lean-based process improvement cycle is a 16-week process organized in four phases, namely, initiate, assess, develop, and deploy, as shown in Fig. 2. The key activities performed during these four phases include identifying problems, identifying causes, brainstorming for solutions, designing solutions, implementing solutions, and demonstrating benefits.

Initiate Phase The scope in terms of processes and organization units is defined. Planning for the workshops and preliminary data collection is performed during this phase.

Assess Phase During this phase, the processes are analyzed to identify the problems and estimate the potential for process improvement. To identify problems/pain points in the process, a detailed value stream map (VSM) is prepared along with the practitioners. The value stream map captures the as-is detailed activities performed

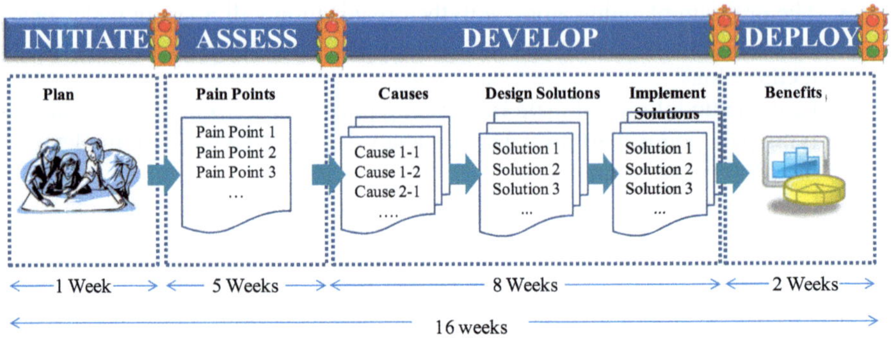

Fig. 2 Process improvement cycle

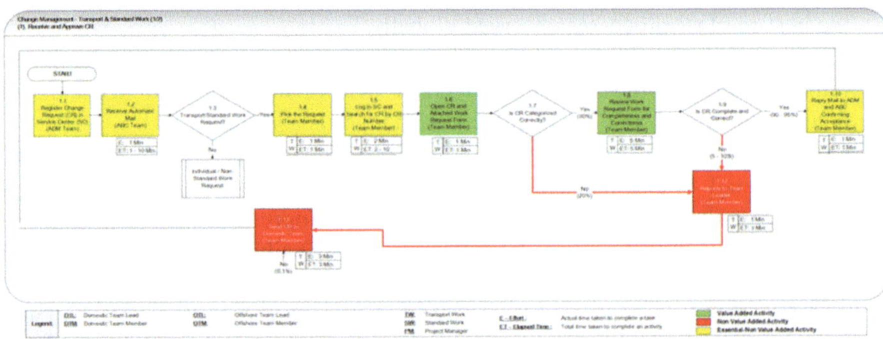

Fig. 3 Value stream map of a process – sample

by the team as shown in Fig. 3. It will also capture the effort, elapsed time, volume of requests, and percentage of times a specific path in the VSM is followed. This data captured during the assessment will be used to quantify the inefficiencies in the process and derive estimated potential benefits. The process performance in terms of effort and elapsed time for each process is determined as shown in Fig. 4. The summary of performance for all the processes is shown in Fig. 5. The assessment report, which provides a summary view of key process efficiency metrics like value-added ratio (VAR), first time right (FTR), and VA/NVA effort distribution per instance for all processes, is shown in Fig. 6.

Develop Phase The causes are identified jointly with the division members using a five-why technique, and then solutions for the causes are identified through structured brainstorming. Estimate potential benefit by implementing each solution. Cost/effort to implement these solutions is also estimated to perform cost-benefit analysis. The identified solutions are prioritized, selected, and implemented. Based on the solutions implemented, an improved to-be VSM process is defined by practitioners. The performance metrics like VAR and FTR are computed post improvements. This will demonstrate the extent to which the improvement has

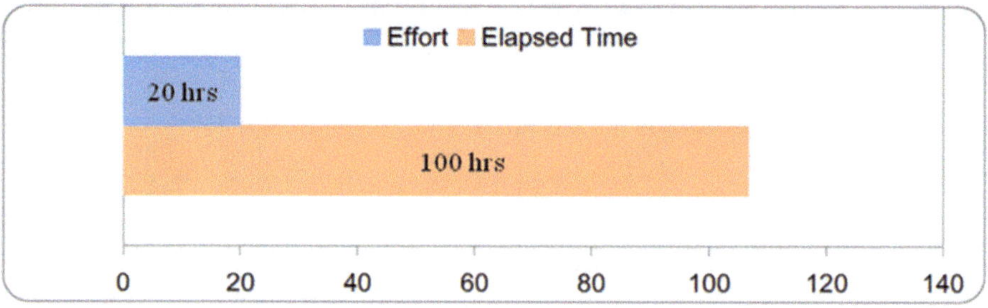

Fig. 4 Process unit cost and performance for a service – sample

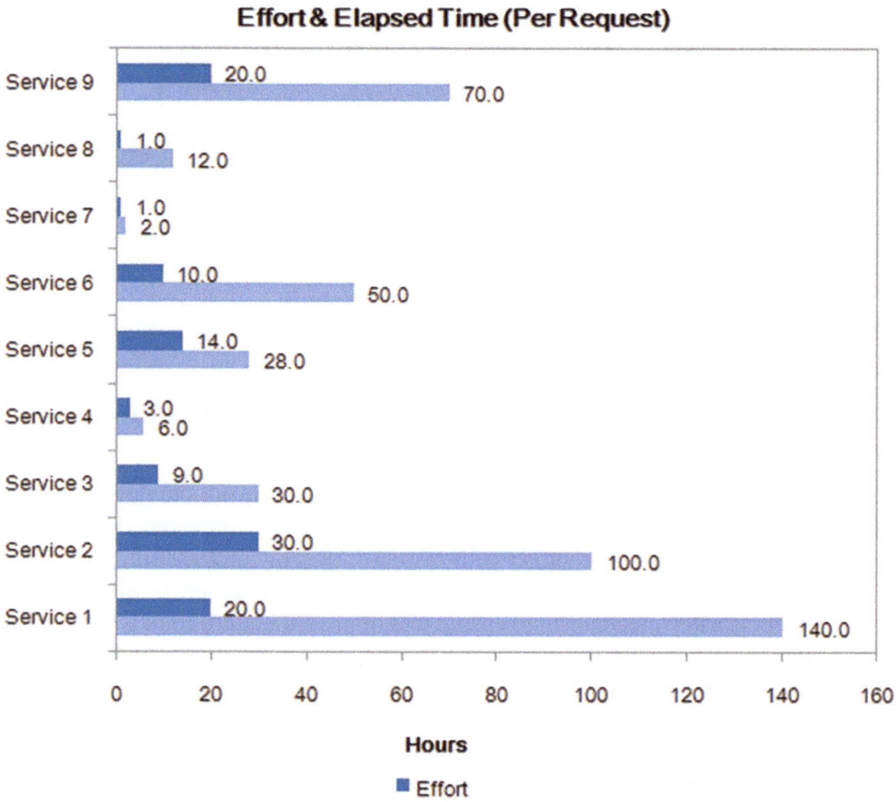

Fig. 5 Process unit cost and performance for multiple services – sample

happened. Also the reduced effort/elapsed time per request is visualized. The summary of problems, causes, solutions, and their prioritization is shown in Fig. 7. Thus, the improvement that each improvement has brought in and the benefit realized is tracked. Some of the quick-win solutions are implemented in a short cycle time to demonstrate the benefits. Typically quick wins are implemented during

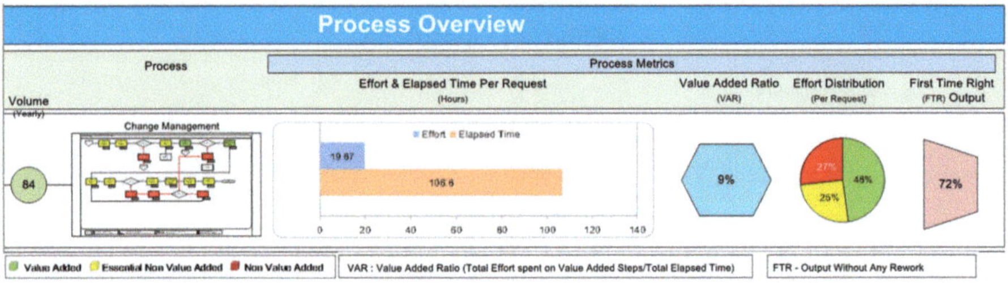

Fig. 6 Overview of quality and performance metrics for a department (assessment output – sample)

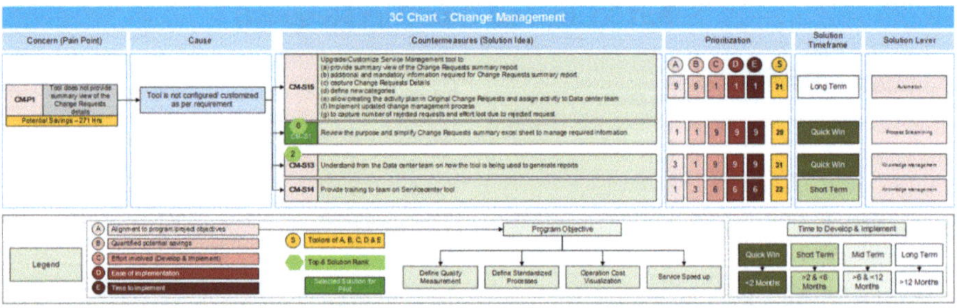

Fig. 7 Overview of solutions for process improvement (improvement output – sample)

8 weeks of develop phase. During this process, the process champions from the team and the practitioners also get trained on the methodology.

Deploy Phase As not all solutions can be implemented during the develop phase itself, an implementation plan for the remaining solutions is prepared. The same is handed over to process champions with responsibility of implementing, tracking, and realizing benefits. Governance structure with representation from the senior management is a key mechanism to ensure that improvement program is successful. The governance team periodically reviews the progress of the improvement program with the focus on benefits realized and reported. To encourage process champions for their active participation in the process improvement journey and spread the culture of continuous improvement, rewards and recognition program is set up.

Piloting the lean-based process improvement methodology enabled change in the organization by engaging the division members in all the phases of the 16-week cycle, to instill confidence in the division members that processes can be improved in a shorter time frame. The structured methodology ensured that the division members can be successful in improving their processes, which otherwise was not possible from their past experience.

6 Quantitative Results

Based on the experience of implementing lean in several organizations, the typical benefits that are demonstrated using the 16-week process improvement cycle are shown in Table 1:

(a) The assessments indicate about 13% ~61% with an average of 27% potential savings due to process improvement.
(b) The realized benefits from process improvements are in the range of 15% ~23% with an average benefit of 19%.
(c) Typical value-added ratio ranges from 5% ~47% with an average value of 14.7%.
(d) The first time right capability averages at 46%, with as minimum as 7% and maximum being 92%.
(e) The effort per cycle has a typical distribution of VA, NVA, and ENVA activity categories as 45%, 22%, 33%, respectively.
(f) Provides visibility in terms of process performance as shown in Figs. 4 and 5.
(g) The organization will start understanding and implementing the methodology in a short span of 16 weeks, resulting in cultural change in the organization.

Table 1 Results from various process improvement programs

Benefits			
	Lowest (%)	Highest (%)	Average (%)
Lean metrics			
Potential improvement opportunities identified through lean assessment	13	61	27
Benefits delivered through lean implementation	15	23	19
Cycle time reduction	25	61	46
Productivity improvement	6	69	23
Service effectiveness improvement	18	27	24
Defect reduction	12	94	58
Capacity improvement	22	25	24
Service efficiency improvement	8	32	21
Assessment metrics			
Value-added ratio (VAR)	5	47	14.7
First time right (FTR)	7	92	46
Average effort distribution (VA, NVA, ENVA) (45%, 22%, 33%)			
Value-added activity (VA) effort distribution	19	58	45
Non-value-added activity (NVA) effort distribution	9	42	22
Essential non-value-added activity (ENVA) effort distribution	18	50	33

The values vary based on the organization, the processes, and the scope of the process itself. Based on the initial process improvement cycle experience, the methodology was further refined to suit the organization constraints during the "adopt" phase. Few projects are currently in their "adopt" phase.

7 Comparison with SCAMPI Assessments

Some of the key differences between SCAMPI and Lean assessments are as follows (Table 2):

8 Tailoring to Organization Needs

Based on the organization scope (like organization, processes), business objectives, and constraints, the lean assessments and improvement cycle need to be tailored in terms of approach, selection of lean tools, and duration of the lean improvement cycle. The following are some of the tailoring guidelines (this is not an exhaustive list):

- Based on the business objectives, for instance, cost reduction, cycle time reduction, relevant lean tools, and metrics need to be selected, measured, and monitored across the program.
- The duration of the 16-week cycle may need to be increased if the practitioners are not available or if the solutions selected cannot be implemented in a short cycle. Organization needs to refrain from increasing the duration beyond 22 weeks as it might reduce the interest of the practitioners and may lead to loss of "window of opportunity."

Table 2 Comparison of SCAMPI and lead assessment

SCAMPI assessment	Lean assessment
A quick organizational level appraisal method based on few sample projects	A detail-oriented method covering end-to-end details of the process
Doesn't provide quantitative view for improvement	It's a quantitative benefit visualization method
Performed against the CMMI® framework	Not specific to any framework. However a framework can be used to understand the practices and plug the gaps from the respective processes
Predominantly leverages the defined processes and validates with the practitioners	Focusses on processes being practiced on the ground than those defined/documented. Required process details are collected through discussions with the practitioners
Limited interaction with process practitioners	High level of interaction with process practitioners to leverage their experiential data

- Customer should be part of the lean assessments. In case of their nonavailability or inaccessibility, the customer facing the role in the organization like sales people can be part of the assessments and improvements.
- Lean assessments may be applied to a subset of the organization. However, this will limit the extent to which the organization can achieve benefits.
- Lean methodology is domain agnostic and can be applied to both business and IT processes (development, maintenance, support, and infrastructure).

9 Conclusion

The solution of quantitative-based process improvement enables the organization in:

(a) Visualizing the scope for process improvement
(b) Understanding the causes for process issues
(c) Identifying relevant solutions for the issues
(d) Quantifying the potential benefit by implementing solutions
(e) Implementing the solutions (quick wins)
(f) Realizing the quantitative benefits
(g) Creating a cultural change in the organization

Based on the successful implementation of the lean methodology, the organization could reap some quick benefits in a short span of time. Also, acceptance of the methodology by both the managers and practitioners enabled the organization to proceed with the program. However, to achieve the total potential savings due to process improvement, the organizations need to continue implementing all the solutions identified. Governance mechanism was established to ensure that the benefits achieved are monitored and reported.

References

1. Kandt R (2003) Ten steps to successful software process improvement. In: Proceedings of 27th annual international computer software and applications conference, Hong Kong, China
2. DACS ROI Dashboard, http://www.dtic.mil/ndia/2004cmmi/CMMIT4WedPM/McGibbonDACSROIDashboard.pdf
3. Jonsson H (2012) Lean software development: a systematic review. IDT Mini-conference on interesting results in computer science and engineering, Sweden
4. Petersen K, Wohlin C (2010) Software process improvement through the lean measurement (SPI-LEAM) method. J Syst Softw 83(7):1275–1287
5. Middleton P, Joyce D (2012) Lean software management: BBC worldwide case study. IEEE Trans Eng Manag 59(1):20–32
6. Staats B, Upton D (2009) Lean principles, learning, and software production: evidence from Indian software services, Harvard Business School Technology & Operations Mgt. Unit Working Paper No. 08–001

7. Widman J, Hua SY, Ross SC (2010) Applying Lean principles in software development process – A case study. Issues Inf Sys 11(1)
8. Bocock L, Martin A (2011) There's something about lean - a case study. In: Agile conference, August 2011, pp 10–19
9. Pernstål J, Gorschek T, Feldt R (2013) The lean gap: a review of lean approaches to large-scale software systems development. J Syst Softw 86(11):2797–2821
10. Rico DF Software Process Improvement (SPI): modeling return on investment. http://www.cmcrossroads.com/sites/default/files/article/file/2013/XDD3704filelistfilename1_0.pdf
11. Raffo D, Kellner M (1999) Modeling software processes quantitatively and evaluating the performance of process alternatives .Wiley, IEEE Computer Society Press

Selection of Optimal Software Reliability Growth Models: A Fuzzy DEA Ranking Approach

Vijay Kumar, V. B. Singh, Ashish Garg, and Gaurav Kumar

1 Introduction

The software reliability valuation is essential to estimate and forecast the trust ability and working of a software system. Nowadays, software plays an important role in our daily lives. Therefore, software companies are concerned with consumer's requirement, quality of the product, and timely delivery of product. Also, the increasing use of software system in critical application requires that the reliability of this system must be so high that the system performance does not stance hazard to human life. Software reliability is the most essential feature to analyze the performance of software, and it is also very difficult to find out whether the software being delivered is consistent or not. The users or purchaser feedback, i.e., trouble news, system aging, and complaints or compliments, indicates the reliability of any software product. In order to estimate the reliability, cost, and release time of software, the software reliability growth models (SRGMs) play a vital role. Superlatively, SRGMs give an idea of symbolizing the development process and permit software engineers to estimate about the excepted future trust ability of software which is under process. Such methods permit managers to precisely assign time, money, and human resources to a project, in critical business

V. Kumar (✉)
Department of Mathematics, Amity School of Engineering & Technology, New Delhi, India
e-mail: vijay_parashar@yahoo.com

V.B. Singh
Delhi College of Arts & Commerce, University of Delhi, Delhi, India
e-mail: vbsinghdcacdu@gmail.com

A. Garg • G. Kumar
Department of Electronics & Communication Engineering, Amity School of Engineering & Technology, New Delhi, India
e-mail: aashish.garg2013@gmail.com; gk19952610@gmail.com

© Springer Nature Singapore Pte Ltd. 2018

P.K. Kapur et al. (eds.), *Quality, IT and Business Operations*, Springer Proceedings in Business and Economics, DOI 10.1007/978-981-10-5577-5_28

application, and judge when a portion of software has reached a point. At present, numbers of SRGMs increase rapidly every year, but in spite of this, there is no model that is applicable in all cases. In general, the existing SRGMs are data set dependent; hence, it is not possible to recognize in advance about any model which has to be used in a specified case. Therefore, the ranking of SRGMs is a need for the software mangers, i.e., which SRGM to be used for a specific data set. We present here a fuzzy data envelopment analysis (DEA) ranking approach for assortment and ranking of different nonhomogeneous Poisson process (NHPP)-based SRGMs. Fuzzy DEA is a tool to compare the performance of a set of activities or organizations under uncertain environment. It is a very effective method to evaluate the relative efficiency of decision-making units (DMUs), since the data of production processes cannot be precisely measured in some cases; the uncertain theory has played an important role in DEA. Further, this paper is organized as follows: Sect. 2 provides the literature about the SRGMs. In Sect. 3, symbols and notations have been described. Fuzzy DEA, credibility, and the CCR model have been explained thoroughly in Sect. 4. Data collection and parameter estimation have been described in Sect. 5. Methodology to apply genetic algorithm has been described in Sect. 6. Furthermore, result and discussion are described in Sect. 7. Conclusion drawn is described in Sect. 8.

2 Literature Review

A lot of SRGMs have been proposed by the researchers and scientists in the last five decades. Goel and Okumoto [1] are the first to describe a nonhomogeneous Poisson process model. Yamada et al. [2] proposed NHPP-based SRGM that can represent the exponential and S-shaped growth curve based on the parameter values. Also, this model plays an important role in software failure data. Abdel-Ghaly [3] showed that various technique of model selection result in various models being chosen. Khoshgoftaar and Woodcock [4] proposed a technique to choose a reliability model in the middle of several alternatives by means of the log-likelihood function. The goodness-of-fit (GOF) model selection criterions are applied by Lyu and Nikora [5]. Logistic growth model simply fits the cumulative number of detected faults at a given time and also predicts the economic population growth which has to be applied for the estimation of software reliability growth [6]. Musa et al. [7] developed a model considering only testing time as the criterion in the software reliability growth process. Pham Zhang IFD [8] proposes a software reliability growth model based on a nonhomogeneous Poisson process (NHPP) that includes a logistic-exponential testing coverage perform with imperfect debugging. In recent time, Kapur et al. [9] proposed a model using optimal control theory to study the behavior of allocation of resources. Kumar et al. [10] and Kumar and Sahni [11] used optimal control theory to optimize the allocation of limited resources and used genetic algorithm to determine the optimum value of the testing efforts for a single-release software model.

Apart from the models discussed above, many SRGMs have been developed for the different set of assumptions in the last five decades. In the direction of ranking SRGMs, little amount of work has been done by researchers. Sharma et al. [12] first proposed a distance-based approach (DBA) to rank SRGMs. Sharma et al. [7] define a distance called Euclidean composite distance (CD) between each different SRGM to the best possible state and provide the ranking to the different SRGM based on the values of composite distance. The minimum value of composite distance defines the most efficient SRGM.

3 Notations and Symbols

In this paper, symbols and variables means:

$x_i = (x_{i1}, x_{i2}, \ldots, x_{ip})$: Input vector of DMU_i, $i = 1,2,\ldots,$n
$x_0 = x_i = (x_{01}, x_{02}, \ldots, x_{0p})$: Input vector of the target DMU_0
$y_i = (y_{i1}, y_{i2}, \ldots, y_{iq})$: Output vector of DMU_i, $i = 1,2,\ldots,$n
$y_0 = (y_{01}, y_{02}, \ldots, y_{0q})$: Output vector of the target DMU_0
$\widetilde{x}_i = \left(\widetilde{x}_{i1}, \widetilde{x}_{i2}, \ldots, \widetilde{x}_{ip}\right)$: Fuzzy input vector of DMU_i
$\widetilde{y}_i = \left(\widetilde{y}_{i1}, \widetilde{y}_{i2}, \ldots, \widetilde{y}_{iq}\right)$: Fuzzy output vector of DMU_i
$u \in R^{m \times 1}$: Vector of input weights
$v \in R^{m \times 1}$: Vector of output weights
y_i^{r4}: Value of output 2 corresponding to the fourth row
y_i^{r3}: Value of *output* 1 corresponding to the third row
x_i^{r2}: Value of input 2 corresponding to the second row
x_i^{r1}: Value of input 1 corresponding to the first row
$i = 1, 2, 3, 4, 5$ for various SRGMs selection
L: The level of credibility contains in information from source S
(1-L): Represents the complement credibility assigned to the alternation data source D

4 Methodology

Data envelopment analysis (DEA), a technique developed by Charnes et al. [13], can be used to evaluate and compare organizational units that use many different inputs to produce various outputs over a specified period of time, but there is no concept of uncertainty and approximate reasoning in DEA models.

4.1 CCR Model

The criteria for determining whether a DMU is efficient or inefficient can be easily determined by formulation of a linear programming problem, which can be easily solved by applying genetic algorithm (GA), and thus the relative efficiency for each DMU can be calculated. A complete DEA solves a linear program, one for each of the DMU. The mathematical form of a CCR model is described as follows:

$$h_k = \text{Max} \sum_{r=1}^{s} u_k y_{rk}$$

Subject to:

$$\sum_{i=1}^{m} v_i x_{ij} - \sum_{r=1}^{s} u_{ry} y_{rj} \geq 0, \qquad j = 1 \ldots n,$$

$$\sum_{i=1}^{m} v_i x_{ik} = 1,$$

$$u_r \geq 0 \text{ for } r = 1, \ldots, s,$$

$$y_r \geq 0 \text{ for } r = 1, \ldots m. \tag{1}$$

4.2 DEA Model

The most often used DEA model is CCR model. In this model, the efficiency of entity evaluated is obtained as a ratio of the weighted output to the weighted input subject to the condition that the ratio for every entity is not larger than one. Mathematically, it is described as follows:

$$\max \theta = \left\{ v^T y_0 / u^T x_0 \right\}$$

Subject to:

$$v^T y_j \leq u^T x_j$$

$$u \geq 0;$$

$$v \geq 0; \tag{2}$$

Definition
Efficiency: DMU_0 is best if $\theta^* = 1$, where θ^* is the optimal value of (2)

4.3 *Fuzzy DEA Ranking Approach*

Although today there exist several techniques for evaluating the relative efficiencies of a set of models, DEA has been extensively used for evaluating the performances and thus efficiency. Basically, DEA is a mathematical programming model applied to observational data for a set of DMUs to calculate the respective efficiencies. The basic DEA results group the DMUs into two sets, those that are efficient and those that are inefficient. Apart from the advantages of DEA, some limitations should also be considered. We required a precise measurement of input and output parameters of a DMU, in order to effectively apply DEA, but for many practical applications, it is difficult to measure them accurately. Instead the respective input and output data can be given as a fuzzy variable.

Let $\tilde{x}_i$ and $\tilde{y}_i$ be the fuzzy input and output vector. As we know that the fuzzy constraints $v^T\tilde{y}_j \le u^T\tilde{x}_j$ do not define a deterministic feasible set, we employ the idea of chance constraint programming (CCP) which was developed by Charnes et al. [13]. A confidence level (1-α) at which it is most required that the fuzzy constraints hold may be provided. Hence we have obtained the constraints as follows:

$$Cr\left\{v^T\tilde{y}_j \le u^T\tilde{x}_j\right\} \ge 1-\alpha, \quad j = 1, 2, \ldots .n \tag{3}$$

where, Liu and Liu [14] presented a credibility measure $Cr\{A\}$ to express the chance that fuzzy event A occurs. Liu [16] provide the details about credibility measure in his work. The main aim of this model is to maximize the ratio of $v^T\tilde{y}/u^T\tilde{x}$. The greater the value is, the more efficient the DMU. When this ratio comes out to be unity, then there is 100% efficiency in the given DMU. Considering the chance constraints defined by Eq. (3), the fuzzy DEA model can be written as follows:

$$\max \theta = Cr\left\{v^T\tilde{y}_0/u^T\tilde{x}_0\right\} \ge 1$$

Subject to:

$$Cr\left\{v^T\tilde{y}_j \le u^T\tilde{x}_j\right\} \ge 1-\alpha \quad j = 1, 2, \ldots .n$$

$$u \ge 0; \quad v \ge 0; \tag{4}$$

in which $\alpha \in (0, 0.5)$.

The greatest value of objective function is the most efficient *DMU* that is ranked. If there exists at least one DMU with the optimal value, $\theta * \geq \alpha$. Therefore, we have the following definition:

Definition 2 (α-Efficiency) *DMU* is α-efficient if $\theta * \geq \alpha$, where $\theta *$ is an optimal value of Eq. (4).

The problem defined by Eq. (4) is equivalent to following linear programming model:

$$\max \theta = v^T y_0^{r_4} - u^T x_0^{r_1}$$

Subject to:

$$2v^T \left(y_0^{r_4} - y_0^{r_3}\right) + 2u^T \left(x_0^{r_2} - x_0^{r_1}\right) = 1;$$

$$v^T \left[(1 - 2\alpha)\, y_j^{r_4} + 2\alpha y_j^{r_3}\right] - u^T \left[(1 - 2\alpha)\, x_j^{r_1} - 2\alpha x_j^{r_2}\right] \leq 0; \quad j = 1, 2, 3 \ldots.., n;$$

$$u \geq 0; \quad v \geq 0; \tag{5}$$

5 Data Collection

The purpose of this study is to examine the fitness of the fuzzy DEA approach so that a widespread ranking of the different SRGMs could be made combining various attributes toward SRGMs for a data set. In this paper, we have considered five SRGMs to demonstrate the applicability of the technique discussed in Sect. 4. The paper included NHPP-based SRGMs, namely,

- Generalized Goel model [6]
- Logistic growth model [6]
- Modified Duane model [6]
- Yamada imperfect debugging model [2]
- Pham Zhang IFD model [8]

The mean value function and intensity function of these SRGMs are summarized in Table 1.

In software reliability prediction, parameter estimation is the most important. SRGMs are useful only if estimation of its parameters is possible. We have used nonlinear regression technique in statistical package for social sciences (SPSS) software for estimation of parameters. The data set is obtained from Tandem Computers [16] which is described in Table 2.

The estimated parameters for the above SRGMs using SPSS software are described in Table 3.

Table 1 Summary of the software reliability growth models

Model name	Mean value function $m(t)$	Intensity function $\lambda(t)$
Generalized Goel [3]	$a\left(1-e^{-bt^c}\right)$	$abct^{c-1}e^{-bt^c}$
Logistic growth [3]	$a/(1-ke^{-bt})$	$\frac{abke^{-bt}}{(1+ke^{-bt})^2}$
Modified Duane [3]	$a[1-(b/(b+t))^c]$	$acb^c(b+t)^{1-c}$
Yamada imperfect debugging model [11]	$\frac{ab\left(e^{at}-e^{-bt}\right)}{(a+b)}$	$\frac{ab\left(ae^{at}+e^{-bt}\right)}{a+b}$
Pham Zhang IFD [5]	$a-e^{-bt}\left(1+(b+d)t+dt^2\right)$	$ae^{-bt}[bt(b-d)+d(b^2t^2-1)]$

Table 2 Tandem Computers software failures (data set) [17]

Weeks	CPU hours	Defects found	Weeks	CPU hours	Defects found
1	519	16	11	6539	81
2	968	7	12	7083	86
3	1430	9	13	7487	90
4	1893	33	14	7846	93
5	790	41	15	8250	96
6	3058	49	16	8564	98
7	3625	54	17	896	99
8	445	58	18	9102	100
9	5218	69	19	961	100
10	586	75	20	10,000	100

Table 3 Estimation results for SRGMs

Model name	Input parameters
Generalized Goel [3]	$a = 68.154$, $b = 7.821\times10^{-3}$
Logistic growth [3]	$a = 106.818$, $b = 0.889$
Modified Duane [3]	$a = 207.8$, $b = 64.05$
Yamada imperfect debugging model [11]	$a = 52.47$, $b = 0.224$
Pham Zhang IFD [5]	$a = 83.64$, $b = 0.350$

Firstly, for the application of fuzzy DEA approach to the SRGM, we have to determine the input and output parameters. The values a and b as shown in the Table 3 are used as the input parameters for various SRGMs. The mean value function $m(t)$ and reliability function

$$R(t) = e^{-(m(t+\Delta t)-m(t))}$$

are used as output parameters. To write the input and output parameters in the form of fuzzy inputs and outputs, firstly, we have calculated the lower limits and upper limits of the input and output parameters. For example, if a parameter has any value p, then its lower limit for the 1% of original value will be $(p - 0.01*p)$, and for the upper limit, we have added the 1% value of original value to the original value, i.e., $(p + 0.01*p)$. The values of lower limit and upper limit of inputs and outputs, respectively, are shown in Table 5. Further, we have formulated an LPP for each SRGM using Eq. (5) with the value of credibility measure $(\alpha) = 0.4$. Further, we have used genetic algorithm (GA) to obtain efficiency for each SRGM after the formation of respective LPP.

6 Genetic Algorithm

Genetic algorithm is a very useful tool for evaluating the relative efficiencies of given SRGMs. It was developed by John Holland who invented it in the early 1970s. It provides a random search algorithm [17], which is simulated on the law of natural selections and the genetic information recombination with the population. GA suggests the endurance of the fittest between individuals over consecutive generation. It also sustains the population of n chromosomes connected with fitness value. The various steps required for evaluating the efficiency are described below:

(a) *Selection*: It is used to generate next generation of population by choosing the solutions from available populations. It is done in order to direct the search to promising areas within the search space. Selection is based on the value of fitness function.
(b) *Crossover*: It is used to generate new child (offspring) by exchanging the genetic code of obtained solutions. Crossover rate is defined by taking the probability of parent solution being crossed over. Ideally, the range of crossover rate lies between 0.6 and 0.9.
(c) *Mutation*: It is used to maintain the total number of genetic characteristics in the genetic makeup of population. The mutation probability mutates the new offspring of the entire in chromosome. Mutation rates are used to stop the best solutions from being effected.

The value of parameters considered for optimization is summarized in Table 4. Further, we have defined the fitness function for the problem stated in Eq. (5). The fitness function is given by:

$$\text{Fitness function} = v1 \times y_0^{r^4} + v2 \times y_0^{r^4} - u1 \times x_0^{r^1} - u2 \times x_0^{r^1} \quad (6)$$

To find the optimal value of efficiency by GA, we have used MATLAB software (R2010a) with VC++ (6.0) compiler, and we have chosen the following appropriate parameters to apply genetic algorithm, which are shown in Table 4.

Optimum efficiency for each SRGM can be evaluated by using the above fitness function. Ideally there is no optimum method of fixing these values. But usually, in practice a small size of population is being combined with a high crossover

Table 4 Values of parameters of GA

Parameters	Values
Size of population	100
Probability of crossover	0.90
Probability of mutation	0.01
Convergence condition of upper bound	1000
Difference between two consecutive iterations	0.00001

Table 5 Illustrative example

Model name	Generalized Goel	Logistic growth	Modified Duane	Yamada imperfect debugging model	Pham Zhang IFD
Input 1	(67.86846, 68.554, 69.6954)	(107.80867, 107.818, 107.8933)	(206.811, 208.92, 10.989)	(51.609, 52.17, 52.6457)	(83.52036, 84.364, 85.20764)
Input 2	(0.0078546, 0.007934, 0.0080134)	(0.8733, 0.89, 0.9067)	(63.4095, 64.05, 64.6905)	(0.21186, 0.214, 0.21614)	(0.34056, 0.344, 0.34744)
Output 1	(2.041773548, 2.0669756, 2.08863795)	(104.6558106, 104.6683, 104.6767425)	(43.100145, 43.5355, 43.970855)	(0.832194, 0.84060, 0.849006)	(82.82934, 83.6660, 84.5086)
Output 2	(0.946370421, 0.95592969, 0.96548898)	(0.478697873, 0.483533, 0.488368533)	(0.1866645, 0.18855, 0.1904355)	(0.93191445, 0.941397, 0.950740977)	(0.83813895, 0.846605, 0.85507105)

Table 6 Results of efficiency with $\alpha = 0.4$

SRGM modal name	Efficiency
Generalized Goel	0.40139495594
Logistic growth	0.705878493786138
Modified Duane	0.641000941708439
Yamada imperfect debugging model	0.40139887668857
Pham Zhang IFD	0.401457346259342

probability to obtain an optimal solution. The input and output parameters for the five SRGMs are given in Table 5, and efficiency obtained for the respective SRGMs is described in Table 6.

7 Result and Discussion

Fuzzy DEA is a very realistic way of evaluating the relative efficiencies and providing ranking to the various SRGMs. The input parameters a and b are obtained from SPSS software, and output parameters are obtained by solving the mean value function and reliability function for respective SRGMs. By using the value of input and output parameters, a LPP for each SRGM is developed. After making an LPP, the resulting LPP is implemented in genetic algorithm for obtaining the efficiency

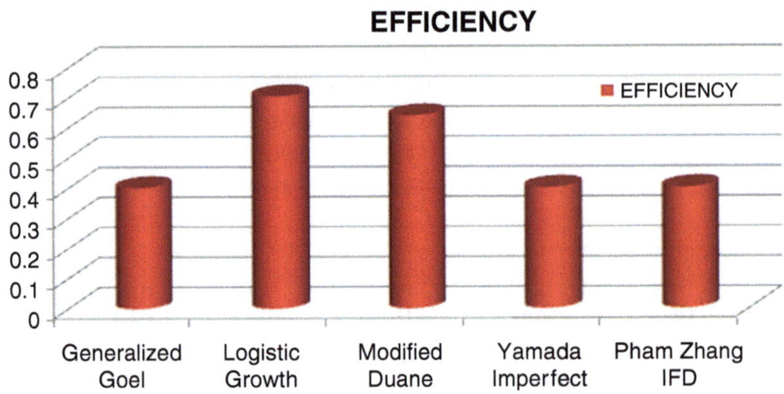

Fig. 1 Efficiency of each SRGM

for the respective SRGM. The efficiency for the different model is described in Table 6 with $\alpha = 0.4$. The results in Table 6 clearly show that the logistic growth model is ranked at number 1 based on the above analysis and is followed by modified Duane, Pham Zhang, Yamada imperfect debugging model, and generalized Goel. The plot between SRGMs and their respective efficiencies has been shown in Fig. 1.

8 Conclusion and Future Scope

In this paper, we formulated a new approach to rank various SRGMs with fuzzy inputs and outputs based on credibility measure and using linear programming model (LPP). The proposed method is suitable for ranking SRGMs based on efficiency which is obtained by using genetic algorithm. Fuzzy efficiency means to gauge the reliability of its crisp equivalent. The comparison analysis of the proposed SRGMs is made by two common criteria, namely, mean value function and reliability function. Based on efficiency, logistic growth model is the most efficient model among the five SRGMs. In the future, the work can be extended by considering different set of SRGMs and taking more input and output variables.

References

1. Goel AL, Okumoto K (1979) Time-dependent error-detection rate model for software reliability and other performance measures. IEEE Trans Reliab 28(3):206–211
2. Yamada S, Tokuno K, Osaki S (1992) Imperfect debugging models with fault introduction rate for software reliability assessment. Int J Syst Sci 23(12):2241–2252
3. Abdel-Ghaly PYC, Littlewood B (1986, September) Evaluation of competing software reliability predictions. IEEE Trans Softw Eng SE-12(12):950–967

4. Khoshgoftaar TM, Woodcock T (1991) Software reliability model se- lection: a case study. In: Proceedings of the 2nd international symposium on software. Reliability engineering. Austin, IEEE Computer Society Press, pp 183–191
5. Lyu M, Nikora A (1992) CASREA-A computer-aided software reliability estimation tool. In: proceedings of the fifth international workshop on computer-aided software engineering, Montreal, CA pp 84–95
6. Huang CY, Lyu MR, Kuo SY (2003, March) A unified scheme of some non-homogenous Poisson process models for software reliability estimation. IEEE Trans Softw Eng 29(3):81–89
7. Musa JD, Okumoto K (1983) A logarithmic poisson execution time model for software reliability measurement. In: Conference proceedings 7th inter- national conference on software engineering, pp 230–237
8. Pham H (2006) System software reliability. Springer, London
9. Kapur PK, Pham H, Chanda U, Kumar V (2013) Optimal allocation of testing effort during testing and debugging phases: a control theoretic approach. Int J Syst Sci 44(9):1639–1650
10. Kumar V, Khatri SK, Dua H, Sharma M, Mathur P (2014) An assessment of testing cost with effort dependent FDP and FCP under learning effect: a genetic algorithm approach. Int J Reliab Qual Saf Eng 21(6):1450027
11. Kumar V, Sahni R (2015) An effort allocation model considering different budgetary constraint on fault detection process and fault correction process. Decis Sci Lett 5(1):143–156
12. Sharma K, Garg R, Nagpal CK, Garg RK (2010) Selection of optimal software reliability growth models using a distance based approach. IEEE Trans Reliab 59(2):266–276
13. Charnes A, Cooper WW, Rhodes E (1978) Measuring the efficiency of decision making units. Eur J Oper Res, Article PDF (2218 K) View Record in Scopus Citing articles (8254), 2:429–444
14. Liu B, Liu YK (2002) Expected value of fuzzy variable and fuzzy expected value models. IEEE Trans Fuzzy Syst 10(4):445–450
15. Liu B (2004) Uncertainty theory: an introduction to its axiomatic foundations
16. Wood A (1996) Predicting software reliability. IEEE Comput 29(11):69–77
17. Goldberg DE (1989) Genetic algorithms in search of optimization and machine learning. Addison-Wesley, Reading

Significance of Parallel Computation over Serial Computation Using OpenMP, MPI, and CUDA

Shubhangi Rastogi and Hira Zaheer

1 Introduction

There are massive jumps in technology which leads to expansion of existing system to make it fast and more efficient. Parallel computation provides a fast system with high price tags too [1–3]. A parallel algorithm is cost optimal if its execution time is faster than the running time of the fastest known sequential algorithm for the same problem. Generally programming has been evolved for serial computation. There are two ways of implementing a parallel algorithm: one, to modify the existing sequential algorithm and, second, to implement a new algorithm especially for parallel execution. If the sequential algorithm consists of some independent groups of instructions, then it can easily be modified to run in parallel otherwise; one has to implement a new algorithm. Today's computing scenarios consist of multiple nodes connected by a network. Every computer or node consists of multi-core processors and many core accelerators like graphics processing units (GPUs). Inside a system, OpenMP framework can be used to implement parallelism, between different nodes, MPI can be used, and to make use of GPUs, CUDA is the framework.

S. Rastogi
Ajay Kumar Garg Engineering College, Ghaziabad, India
e-mail: Shubhangi05.rastogi@gmail.com

H. Zaheer (✉)
Indian Institute of Technology, Roorkee, India
e-mail: hirazaheeriitr@gmail.com

© Springer Nature Singapore Pte Ltd. 2018

P.K. Kapur et al. (eds.), *Quality, IT and Business Operations*, Springer Proceedings in Business and Economics, DOI 10.1007/978-981-10-5577-5_29

2 Necessity of Parallel Computing

An algorithm's efficiency depends on two factors: time complexity and space complexity. Space is manageable now but time is the prominent issue nowadays. The major objective of parallel computation is to make high-speed systems. If a work is done by a solitude processor, then it will execute steps one by one in a serial fashion, but if few processors work simultaneously to achieve the same objective, then obviously it will produce the output in very less time.

2.1 Speedup Factor

If some processors are working simultaneously, then the speedup achieved may be calculated as:

$$S(p) = \frac{\text{Execution time using one processor (best sequential algorithm)}}{\text{Execution time using a multiprocessor with } p \text{ processors}} = \frac{t_s}{t_p}$$

Where

t_s = execution time on a single processor.
t_p = execution time on a multiprocessor using "p" processors.
$S(p)$ = speedup achieved by using multiprocessor.

Everything has a limit; so Amdahl has given a law to define the limit up to which speedup can be gained, and this law is known as Amdahl's law. According to which:

$$S(p) = \frac{t_s}{ft_s + \frac{(1-f)t_s}{p}} = \frac{1}{f + \frac{(1-f)}{p}} \qquad \lim_{p\to\infty} S(p) = \frac{1}{f}$$

By increasing the number of processors p and make it to infinite, the whole quantity $(1-f)/p$ will become zero. So on further increasing the processors, there is no increase in speedup (Fig. 1).

In any parallel computation, each individual processor is working serially. So there is a serial portion always. In a parallel computation, there are always a serial portion and a parallel portion.

So, if "f" is the serial portion, then $(1 - f)$ will be the parallel portion and their respective execution times would be ft_s and $t_s(1 - f)$/p. So if 90% of the system is parallel, then speedup will be 20 times. But in actual there are factors like synchronization, load balancing, etc. involved so there is decrease in speedup after a point in reality (Fig. 2).

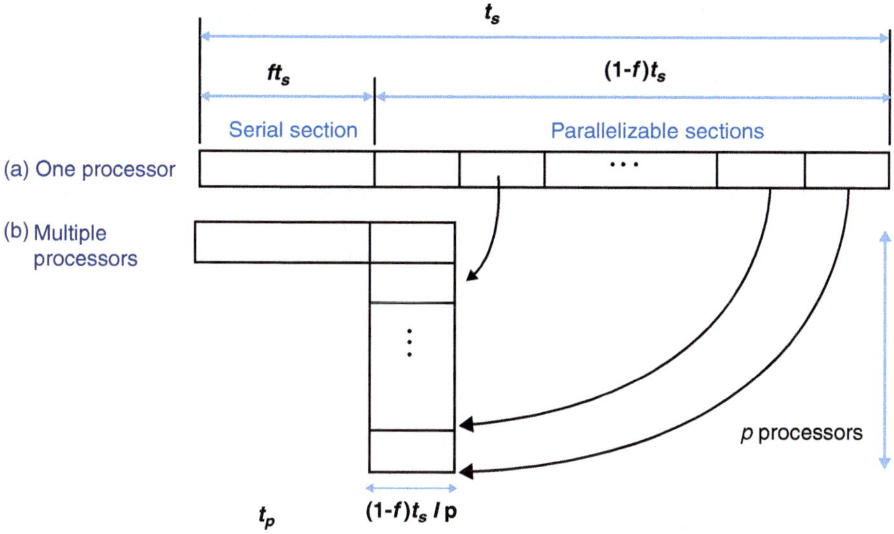

Fig. 1 The graph of Speedup versus number of processors

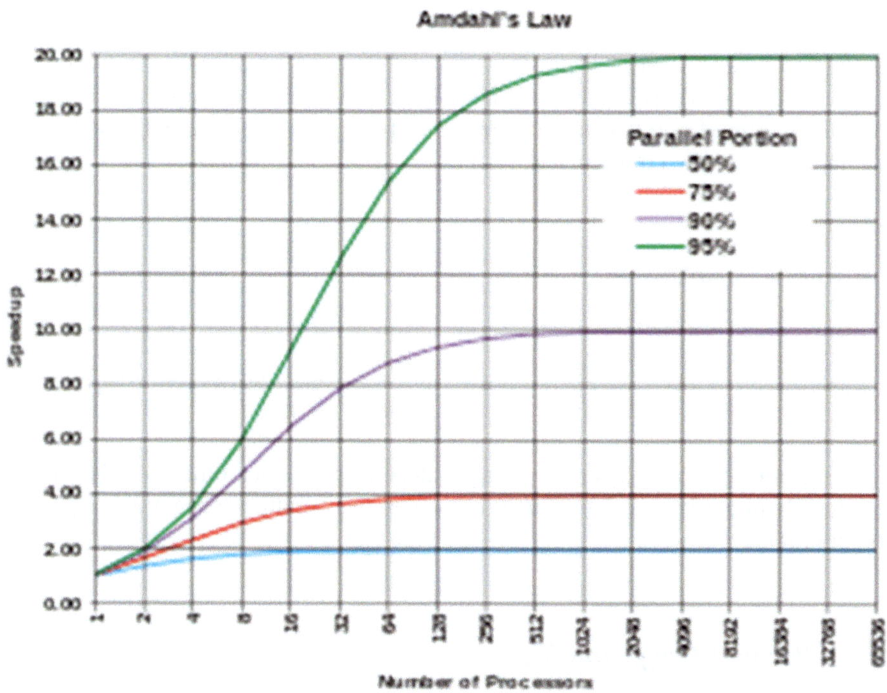

Fig. 2 Speedup vs number of processors

3 Parallel Programming vs Serial Programming

3.1 Serial Programming

If there are n instructions in a program and input of ith instruction is dependent on the output of (*i*-1)th instruction, so instructions are executed one after another in a sequence. Below an example is given which a part of C program to calculate the summation of all the elements of an array A of size N:

```
void main()
{
 int i, N=1000;
 double A[N], B[N], C[N];
 for (i=0; i<N; i++)
  {
  A[i] = B[i] + C[i]
  }
}
```

In this program every instruction is dependent on its previous instruction. It performs N operations one by one. Now let's take another example of calculating the summation of corresponding elements of two given arrays A and B and finally storing the result in a third array C:

```
For(i=0;i<10;i++)
{
C[i]=A[i]+B[i];
Printf("Resultant array %d" C[i]);
}
```

But in this program, instructions are independent of each other. So it's futile to execute them one by one. This forces us to introduce another concept that is parallel programming.

3.2 Parallel Programming

In this paper, three parallel programming tools OpenMP, MPI, and CUDA are introduced. Each is explained in detailed.

3.2.1 OpenMP

OpenMP stands for open multiprocessing and is based on shared memory architecture which has multithreaded capacity. It is based on fork-join method, i.e., parallelizing the code with the help of master thread which forks some required number of threads and assigns some task accordingly. So, all threads can work independently at the same time. Finally join the results of all the individual thread and get the final output [4, 5] (Fig. 3).

The main components of OpenMP are the constructs used for the creation of threads, workload balancing, synchronization, user-level routines, and environmental routines (Fig. 4).

Now consider our first example in OpenMP.

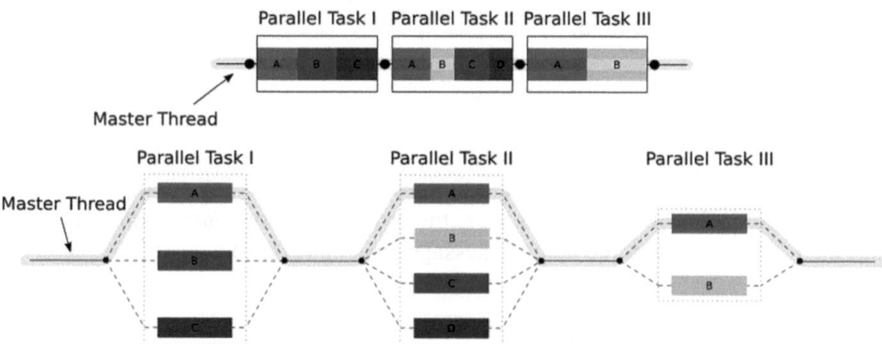

Fig. 3 Parallelization using multiple threads

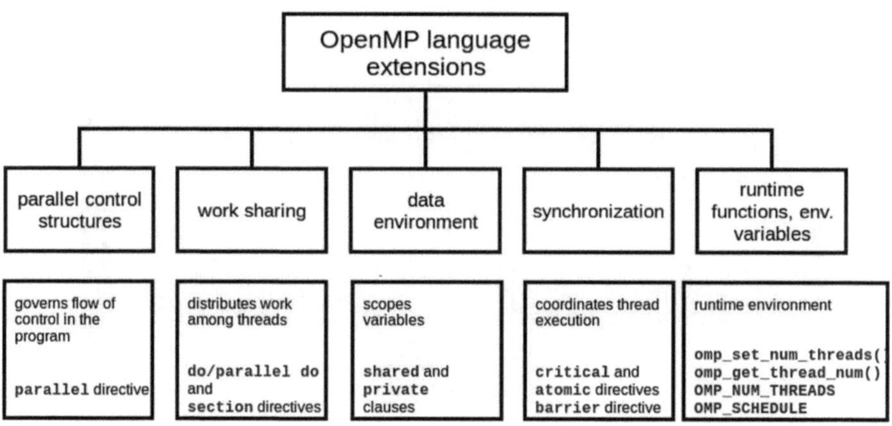

Fig. 4 OpenMP language extensions

```
#include "omp.h"
void main()
{
 int i, N=1000;
 double A[N], B[N], C[N];
#pragma omp parallel for
 for (i=0; i<N; i++) {
  A[i] = B[i] + C[i];
 }
}
```

So by simply adding *#pragma omp parallel for* above the for loop, we can parallelize the loop. If there is N number of processors available, then we will obtain the output in one unit of time.

3.2.2 MPI

MPI stands for message passing interface which works in inter-process manner. MPI is based on distributed system. Here the processes work concurrently and share intermediate results with the help of message passing (Fig. 5).

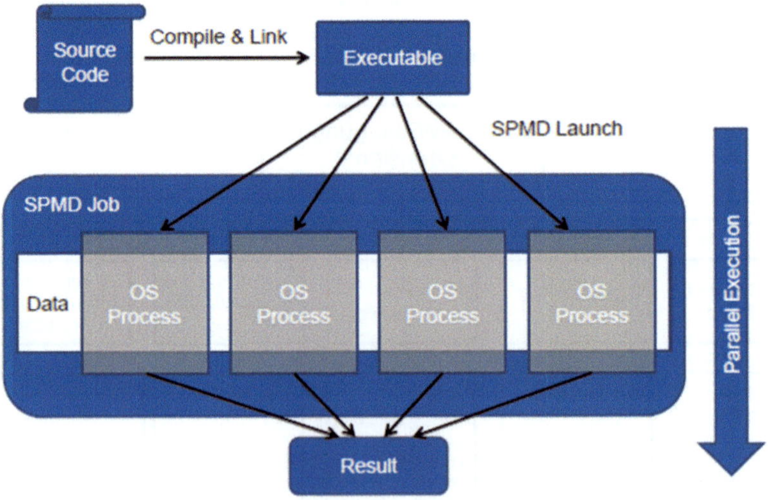

Fig. 5 Multiple instances of a single program working on multiple (parts of) data

```
#include <stdio.h>
#include <mpi.h>
int main(int argc, char *argv[])
{
 int rank, size;
 MPI_Init(&argc, &argv);
 MPI_Comm_rank(MPI_COMM_WORLD, &rank);
 MPI_Comm_size(MPI_COMM_WORLD, &size);
 printf("Hello world! I am %d of %d\n", rank, size);
 MPI_Finalize();
 return(0);
}
```

This is a simple code in MPI.

MPI_Init: Before starting any parallel work, MPI needs to initialize arguments such as processes, input to program, etc.

MPI_Comm_rank: Communicator rank is used to determine the rank of calling process. (Each process is assigned a unique ID, i.e., rank.)

MPI_Comm_size: Communicator size to determine the size of the topology.

MPI_Finalize: To terminate all the communications.

The message exchange among processes is done by using send and receive calls which is the part of MPI programming [6, 7]. As we can see MPI is not as simple as OpenMP, but there are some more problems in MPI like deadlock if code is not written carefully.

3.2.3 CUDA

NVIDIA introduced a parallel computing architecture CUDA which stands for compute unified device architecture. It is the computing engine in NVIDIA graphics processing units or GPUs that is accessible to software developers with the help of industry standard programming languages.

GPU is more apt for data parallel computation. As it can be seen in Fig. 6, there are multiple arithmetic logic units (ALUs); it means same program is executed on many data components in parallel. So it has high ratio of arithmetic operations to memory operations. Although memory latency is high here, it is overcome by high-speed calculations. Its programming needs to be understood from the zero level, so it is being included in the paper (Fig. 7).

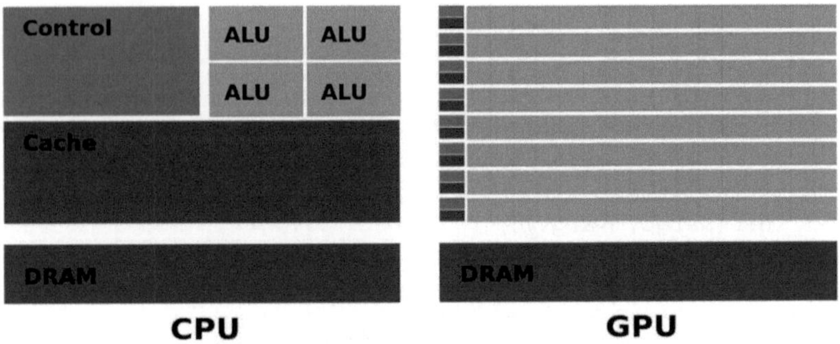

Fig. 6 Architectures of CPU and GPU

Fig. 7 Processing flow of CUDA

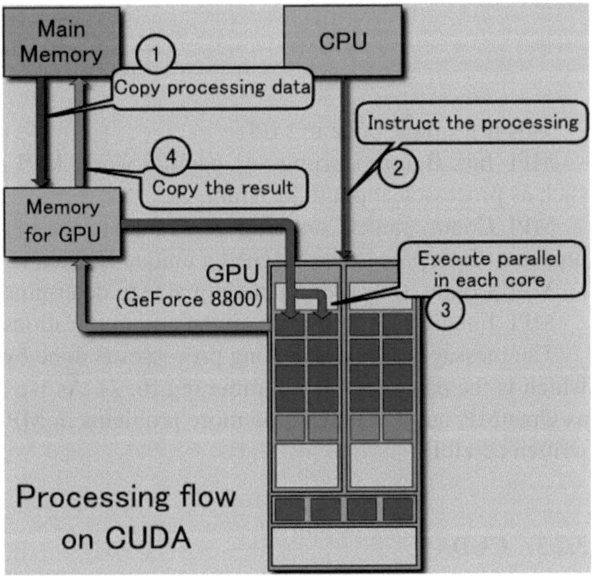

3.3 *Hybrid Approach*

In any parallel program, there is a sequential portion also. Parallel programming is based on fork-and-join method. These three parallel approaches can be combined together in one single program. OpenMP is used within a process like by using some number of threads to run simultaneously, while MPI is the parallelism obtained between the processes. CUDA makes use of GPU (graphics processing unit) in which multiple processing units are used for parallelism. So, all three in one programs can be used.

4 Conclusion

There is a need of fast systems in today's world. So parallel computation is an important concept which speeds up the system and reduces the overall time. This paper describes the calculation of speedup factor also using Amdahl's law. This paper covers the serial and parallel programming. There are three frameworks used for parallel programming, OpenMP, MPI, and CUDA. OpenMP involves parallelism at thread level. MPI uses multiple cores to work in parallel, while CUDA uses multiple accelerating units, i.e., GPUs. Their programming is very simple and this paper discussed some programming concepts also. There are hybrid systems which use all the three frameworks in a single program which in turn massively reduces the overall time [7–9]. So, it can be said that parallel computation takes less time in comparison to serial computation. In future, the program times can be calculated to analyze the speedup obtained in different parallel programs over serial programs.

References

1. El-Rewini H, Abd-El-Barr M (2005) Advanced computer architecture and parallel processing, vol 42. Wiley, Hoboken
2. Zhang Y, Sarkar T (2009) Chapter 1: introduction. In: Parallel solution of integral equation-based EM problems in the frequency domain, 1, Wiley-IEEE Press, Hoboken
3. Winterbottom N. Parallel processing and database architectures. In: Design and application of parallel digital processors, 1988. International Specialist Seminar on the, vol, no, pp 48–52, 11–15 April 1988
4. Alonso P, Cortina R, Martínez-Zaldívar FJ, Ranilla J (2009) Neville elimination on multi- and many-core systems: OpenMP, MPI and CUDA. J Supercomput, in press, doi:10.1007/s11227-009-0360-z. SpringerLink Online Date: Nov. 18, 2009
5. Yang CT, Huang C-L, Lin CF (2011) Hybrid CUDA, OpenMP and MPI parallel programming multicore GPU clusters. Comput Phys Commun 182(1). Elsevier
6. Bodin F, Bihan S (2009) Heterogeneous multicore parallel programming for graphics processing units. J Sci Program 17(4):325–336. doi:10.3233/SPR-2009-02
7. Atasoy NA, Sen B, Selcuk B (2012) Using gauss – Jordan elimination method with CUDA for linear circuit equation systems. Volume 1. INSODE Elsevier
8. Adams NM, Kirby SPJ, Harris P, Clegg DB (1996) A review of parallel processing for statistical computation. Stat Comput Springer 6(1)
9. Crockett TW (1997) An introduction to parallel rendering. Parallel Comput 23(7)

Software Release and Patching Time with Warranty Using Change Point

Chetna Choudhary, P. K. Kapur, A. K. Shrivastava, and Sunil K. Khatri

1 Introduction

With the enormous use of information technology, computer systems are widely used to control safety-critical systems and day-to-day entities. Increased demand of high-quality software is the necessity in these application areas. To determine reliability of the system, the software reliability must be evaluated with utmost care. Developing reliable software is the main concern among the software industry. Proper scheduling of processes, limitation of resources, unrealistic requirements, etc. had a great impact on software reliability. Many softwares nowadays are interdependent among the modules; it is very hard to develop the consistent software. It is particularly tough as soon as there is interdependence, and it is rather difficult to identify whether there is reliable software being delivered. Once the software is delivered, one way to major its reliability is through customer feedback like problem being reported, system outages, complaints or compliments, etc. However, by then it is too late; software vendors need to know whether their products are reliable before they are shipped to customers. Software reliability models attempt to provide that information.

C. Choudhary (✉)
Amity School of Engineering & Technology, Amity University, Noida, UP, India
e-mail: chetna.choudhary13@gmail.com

P.K. Kapur • A.K. Shrivastava
Amity Center for Interdisciplinary Research, Amity University, Noida, UP, India
e-mail: pkkapur1@gmail.com; kavinash1987@gmail.com

S.K. Khatri
Amity Institute of Information Technology, Amity University, Noida, UP, India
e-mail: sunilkkhatri@gmail.com

© Springer Nature Singapore Pte Ltd. 2018

P.K. Kapur et al. (eds.), *Quality, IT and Business Operations*, Springer Proceedings in Business and Economics, DOI 10.1007/978-981-10-5577-5_30

Various researchers are working on software reliability growth models to predict the reliability under different sets of parameters. These software reliability growth models generally follow non-homogenous Poisson process (NHPP) with consideration that a constant fault detection rate (FDR) and a fault detection process depend only on the residual fault content [3]. With consideration to specific environment, software programs are executed and improved with detection and correction of bugs/errors. Generally SRGMs take into consideration that, in fault detection process, the faults which occurred due to each failure are independent and random with time according to the distribution given by Musa et al. [12]. Factors such as running environment, testing strategy, defect density, and resource allocation can affect the failure distribution. To cover more faults within the shorter period of time, testing teams keep including new approaches and techniques not yet been used, or in consultation with some experts, we need to preform software risk analysis. Hence, the fault detection rate may not be continuously smooth and can be changed at some time moment τ known as change point. Foremost, Zhao [8] used the change point in software and hardware reliability. Then Huang et al. [13] included the change point to calculate the software reliability growth modeling with testing effort functions. Kapur et al. proposed the multiple change points in software reliability growth modeling for fielded software [13]. Looking into the importance of the change point models in the reliability estimation, an SRGM based on stochastic differential equations incorporating change point concept has been proposed by Kapur et al. [16].

Software requires testing to overcome the faults which in turn increase reliability. However which results in consumption of resources like work force, processing hours etc. which acquire more cost to the organization. However, in testing phase if bugs are not eliminated, then users will face the consequences of faults in the operational time which will increase the debugging cost of the software compared to the testing phase [2]. The delayed testing not only results in increase in reliability of software but also the development cost of software. This will further delay the software from releasing which cause thrashing in terms of market opportunity. On the other hand, insufficient testing intended for fault confiscation shortens the cost of software improvement but increase the hazard of more faults during operational phase. Hence it is significant to determine the optimal time to release the software from the point of developer.

Therefore, in order to minimize the associated risk, the organization needs proper scheduling of different prerelease and post-release phases with due consideration to the change point. From the many decades, researchers are working on software release time problem [1–3, 12, 19]. Yamada [15] anticipated a software cost model under warranty to determine the optimal time to release software. Dohi [4] projected release policies with due consideration to debugging time lag. Yamada and Osaki [5] proposed problem related to release time on the basis of deprecated cost and optimized reliability objective. Kapur and Garg presented the release policies based on testing effort by means of inclination toward intensity of failure [6]. They furthermore incorporated the concept of cost of penalty while considering software release time problem [7]. Huang and Huang and Lyu anticipated release time policies by considering the consequence of testing effort overheads [8]. Yun and Bai projected the software release time problems presuming software life cycle to be random [9]. With competitive market surroundings, a need to deliver a better product has increased. This need is captured by the users in terms of what we can call a software warranty, which acts as an indicator to judge its reliability with presumption that a longer warranty period indicates higher reliability [1]. Warranty is an assurance between the seller and buyer that a product is error-free and will work as stated, and if defect occurs, it's a vendor's job to rectify those defects [2]. The duration of warranty is decided by the developer. In warranty phase rectifying a bug then reporting a fault is negligible for the user. Pham and Zhang included service contract and hazard cost to the conventional cost function [10]. Kapur [11] refined a multi-objective maximization problem to determine release time. Kapur [11] formulated release policy for the exponential change point SRGM.

Though it is a difficut job for a testing team to develop a software which is completely free of errors. But in order to ensure that users face least number of errors during operation phases, today software industry continue to test even after release it. As software releases, it is noted that either the rate of failure decreases or else it is invariable, depending on whether there is augmentation consistency of software throughout the operational phase or not (see Figs. 1 and 2).

Figure 1 illustrates graph for reliability of software for the case when no testing is done after release. Hence there is no reliability growth. On the other hand, in Fig. 2 we see that there is growth in reliability after a certain time interval due to

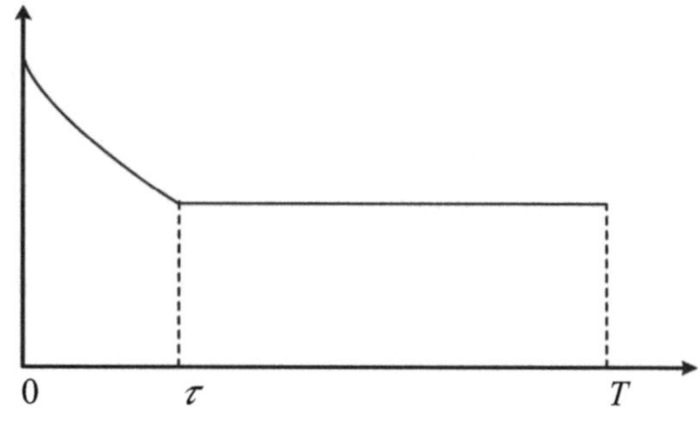

Fig. 1 Growth of software reliability without patching

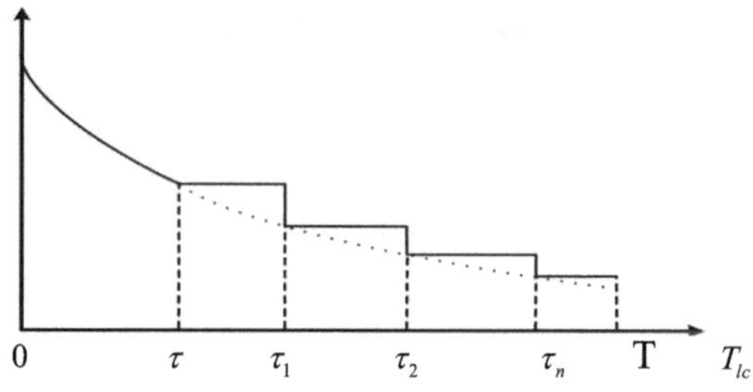

Fig. 2 Growth of software reliability including patching

release of patches/updates after release. This time interval is the difference between two patching times after release, and the update can also be termed as fix, an instruction set that is used to repair flaws which successfully deceives the testing side during prerelease testing phase of software. Well-timed delivery for updates is able to serve intention and will stipulate the premature release of patch results in indecent fixation of failures, and delayed updates would lead to more failures during operational phase and will incur more charge to the development side. Cavusoglu [20] considered the patch release for security and executive tribulations from the insight of both the seller and the organization. They formed a game theoretic model to gain knowledge of balancing the update, and cost management takes place between the merchant and organization. Consequently, Okamura [24] and Luo [21] considered a bug discovery process which is nonhomogeneous and examines both sporadic and asporadic update management models. Lately Dev [3] developed a cost model for finding optimal release policies for security patch management. Arora [22] has shown the transposition between releasing buggy software and saving in patching the buggy software later through a cost model. They have also worked on showing the disadvantage of releasing the software before time and updating later on. Jiang [17] estimated scheduling strategies of software where they have shown the advantage of early software release with continued testing after the release. Kapur [24] outlined a comprehensive framework to determine the optimum time to release and to stop testing software along with the benefit of early release and continuing testing even after release.

Based on the above literature, we see that no research was conducted to consider the effect of change point in finding the optimal time to release the software and updating the software under warranty. This gap analysis has led us to develop a model that can offer a solution to the above problem. In the proposed work, a generalized outline is provided to develop a cost model for determining the optimum time to release and to update software under warranty using change point; therefore, the total expected cost can be minimized. Remaining sections are structured as follows: In Sect. 2, we will talk about model formation specifying necessary presumptions essential for the model. Section 3 will give you an idea about the developed cost model using change point. In Sect. 4, numerical illustration is catered using a real-life data set to authenticate the projected work. Section 5 will discuss the final conclusion.

2 Framework of Modeling

This part of the section pays more attention on the structuring outlines for the proposed work. The following are the notations and presumptions worn in structuring the proposed work.

2.1 Notations

$m(t)$	Expected number of faults that are to be removed by time t or mean value of fault
A	Number of faults initially lying dormant in the software
b1	Rate of fault detection before change point
b2	Rate of fault detection after change point
$F(t)$	Fault distribution function
τ_c	Change point
τ	Release time of software
τ_p	Release time of patch
W	Length of warranty
T_{lc}	Life cycle of the software
C_1	Testing cost per unit time
C_2	Market opportunity cost
C_3	Fault detection/removal cost by tester before releasing the software
C_4	Fault detection/removal cost communicated by means of the user after release of software in warranty
C_5	Fault detection/removal cost communicated by means of the consumer after patch releasing in warranty period
C_6	Fault removal cost communicated by means of the user after warranty

2.2 Assumptions

1. Phenomenon for fault removal pursues NHPP whole time of the software life cycle.
2. Every time a fault is experienced, an instantaneous debugging attempt takes place to discover the cause of failure in order to confiscate it.
3. Number of faults left in the software equally affects the failure rate.
4. Fault fixation process is ideal and no fresh faults are being introduced during debugging process.
5. Process of fault detection autonomously and identically circulated.
6. Number of bug's latent in the software is unchanging.
7. Life cycle of software is finite.
8. Patch cost is negligible.
9. Opportunity cost is pretended to be invariably growing, twofold continually differentiable convex function of r. As per approximated outcome of the learning, the results to a great extent accelerated through definite functional outline of market opportunity cost; hence, we are taking into consideration the form given by Jiang and Sarkar (2011).

2.3 Devising the Model

Through hazard rate we formulate the mean value function for the augmented amount of faults eliminated which is shown as

$$\frac{\mathrm{d}m(t)}{\mathrm{d}t} = \left(\frac{f(t)}{1 - F(t)}\right)(a - m(t)) \tag{1}$$

where $\frac{f(t)}{1-F(t)}$ is the failure detection rate.

On solving the above equation by considering the initial condition as $m(0) = 0$, we get

$$m(t) = \mathrm{a} \cdot F(t) \tag{2}$$

2.4 Model Framework

In the proposed framework, we have subdivided the software life cycle $[0, T_{\mathrm{lc}}]$ into five subparts, i.e., testing phase of software before change point phase $[0, \tau_c]$, software testing phase after change point $[\tau_c, \tau]$, releasing of patch $[\tau, \tau_p]$, phase of warranty $[\tau, \tau + w]$, and software operational phase $[\tau + \mathrm{w}, T_{\mathrm{lc}}]$. Software time frame can be depicted as in Fig. 3. Also it is to be mentioned that each subdivision of fault detection process follows NHPP phenomenon.

First Phase [0, $\boldsymbol{\tau_c}$*]* During the phase, testers are detecting or removing the bugs. The number of fault detected is given by

$$m\left(\tau_c\right) = a \cdot F\left(\tau_c\right) \tag{3}$$

where $F_1(\tau_{\mathrm{c}})$ rate of fault detection using these bugs is detached commencing from software before change point in $[0, \tau_{\mathrm{c}}]$.

Second Phase [$\boldsymbol{\tau_c, \tau}$*]* For this interval change point has been considered as software and is ready to be released by the developer. The developers at some moment change the testing strategies to find more number of bugs in product. The number of bugs detached in this phase is shown as

$$m\left(\tau - \tau_c\right) = a \cdot \left(1 - F_1\left(\tau_c\right)\right) \cdot \left(1 - \frac{1 - F_2\left(\tau\right)}{1 - F_2\left(\tau_c\right)}\right) \tag{4}$$

Fig. 3 Life cycle of software

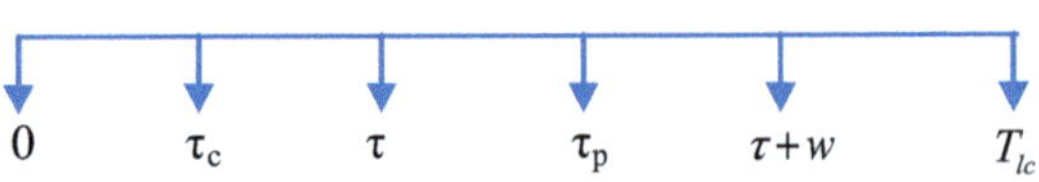

where $a \cdot (1 - F_1(\tau_c))$ is the number of bugs removed after change point. $\left(1 - \frac{(1-F_2(\tau))}{1-F_2(\tau_c)}\right)$ is the fault removal rate after change point.

Third Phase [$\boldsymbol{\tau}, \boldsymbol{\tau}_p$] During this phase update is prepared in reference to the bugs detected by users. At τ_p, a fix/update will be provided to fasten the number of faults reported by users. The total number of faults detected in this phase is given below as

$$\begin{aligned} m\left(\left(\tau + \tau_p\right) - \tau\right) &= (a - m(\tau)) \cdot F_3\left(\left(\tau + \tau_p\right) - \tau\right) \\ &= \left(1 - \left[1 - \frac{1-F_1(\tau_c) \cdot (1-F_2(\tau))}{1-F_2(\tau_c)}\right]\right) \cdot F_3\left(\left(\tau + \tau_p\right) - \tau\right) \end{aligned} \tag{5}$$

In the above equation, $a\left(1 - \left[1 - \frac{(1-F_1(\tau_c))(1-F_2(\tau))}{1-F_2(\tau_c)}\right]\right)$ represents the outstanding bugs which were left hidden during the first and second phases. $F_3((\tau + \tau_\text{p}) - \tau)$ is the rate for fault detection/removal by the user in $[\tau, \tau_\text{p}]$.

Fourth Phase [$\boldsymbol{\tau}_p, \boldsymbol{\tau} + w$] In warranty phase the user faces failure because of the faults which were not corrected in the first patch. The amount of faults detected/removed by end of this phase is summarized as

$$\begin{aligned} m\left((\tau + w) - \left(\tau + \tau_p\right)\right) = a \cdot &\left(1 - \left[1 - \frac{(1-F_1(\tau_c))(1-F_2(\tau))}{1-F_2(\tau_c)}\right]\right) \\ &\cdot \left(1 - F_3\left(\tau + \tau_p\right) - (\tau)\right) \\ &\cdot F_4\left((\tau + w) - \left(\tau + \tau_p\right)\right) \end{aligned} \tag{6}$$

In the above equation, $a \cdot \left(1 - \left[1 - \frac{(1-F_1(\tau_c))(1-F_2(\tau))}{1-F_2(\tau_c)}\right]\right) \cdot \left(1 - F_3\left(\tau + \tau_\text{p}\right) - (\tau)\right)$ represent the outstanding faults remained hidden during the second and third intervals. $F_4((\tau + w) - (\tau + \tau_\text{p}))$ is the rate of fault removal during the interval $[\tau_\text{p}, \tau + w]$.

Fifth Phase [$\boldsymbol{\tau} + w, T_{lc}$] During the post warranty interval, the amount of bugs discovered is summarized

$$\begin{aligned} m\left(T_{lc} - (\tau + w)\right) = a &\left(1 - \left[1 - \frac{(1-F_1(\tau_c))(1-F_2(\tau))}{1-F_2(\tau_c)}\right]\right) \\ &\left(1 - F_3\left(\tau + \tau_\text{p}\right) - \tau\right)\right) \left(1 - F_4\left((\tau + w) - \left(\tau + \tau_p\right)\right)\right) \\ &F_5\left(T_\text{lc} - (\tau + w)\right) \end{aligned} \tag{7}$$

In this interval, the faults, those that are left undiscovered during last intervals, can cause failure, and the same will be communicated by the user for elimination. The amount of faults left is substituted as

$$a \cdot \left(1 - \left[1 - \frac{(1-F_1(\tau_c))\cdot(1-F_2(\tau))}{1-F_2(\tau_c)}\right]\right)$$
$$\left(1 - F_3\left(\tau + \tau_p\right) - \tau\right)\right)\left(1 - F_4\left((\tau + w) - \left(\tau + \tau_p\right)\right)\right).$$

$F_5(T_{lc} - (\tau + w))$ expresses the rate of fault removal during the interval $[\tau + w, T_{lc}]$.
Now, the cost model will be formulated in the next section to calculate the total cost consumed for the fault removal as per the Eqs. (3, 4, 5, 6, 7) phase wise.

3 Cost Model

Through this section we put emphasis on finding the amount of cost incurred for "a" initial number of faults detected throughout the software life cycle. In this we will discuss briefly cost related to testing, market opportunity, and amount spent in removing the faults per phase.

Cost of Testing The amount incurred while testing the software till release via testing side is defined as cost of testing. The cost for this phase is measured by doing product of testing cost with the time duration for which the testing is conducted. Let c_1 denote the testing cost per unit time, and τ is the release time of the software as per the proposed model. So the sum of testing cost is represented by

$$c_1 \cdot \tau \tag{8}$$

Market Opportunity Cost This can be termed as the cost incurred due to postponement in release of the software. Market opportunity cost is inversely proportional to the release of the software. More time taken for the release will increase the opportunity cost. This can be represented by the below equation, which is same as provided by Jiang and Sarkar [17]:

$$c_2 \cdot \tau^2 \tag{9}$$

Software Release Cost Before Change Point For the interval $[0, \tau_c]$, testers work separately for detecting the bugs in the software. This amount, i.e., the fault removed by testers per unit time is expressed asc_3. As obtained using Eq. (3), the number of faults detected/removed by the testing team before change point is represented by $m(\tau_c)$. Hence the cost can be formulated in the interval $[0, \tau_c]$as

$$Cost_{0,\tau_c} = c_3 \cdot m\left(\tau_c\right) \tag{10}$$

Software Release Cost After Change Point In the given interval $[\tau_c, \tau]$, the tester is functioning independently for detection process. Therefore, the cost of removing the bugs by testers is expressed asc_4. Now, the cost incurred for the interval is deliberated by multiplying cost with the faults detected. The total cost of removing the faults is specified by

$$Cost_{\tau,\tau_c} = c_4 \cdot m(\tau - \tau_c) \tag{11}$$

Patch Release Cost Because of the increased market competition, organizations cannot risk in delaying software release. Hence software is released with some faults still present in phase of testing. Because of this the users are provided warranty on the software to fix the faults with updates/patch by the developer during operational phase. Service contract is assurance provided to users about software that it will continue to perform efficiently during operational phase; otherwise, the organization will provide solution in the form of updates for the failure-causing faults, or they will replace the product. On facing failure within software, the users report the problem to the organization for fixation. The organization has to again test the software which is subject to additional charge to the firm. Using Eq. (5), the total amount of bugs is expressed by $m((\tau + \tau_p) - \tau)$ communicated by the user and solved by the tester. Suppose c_5 represents the fault removal, cost per unit time and the cost incurred for the interval are shown as

$$cost_{\tau,\tau_p} = c_5 \cdot m\left(\left(\tau + \tau_p\right) - \tau\right) \tag{12}$$

Software Under Warranty Cost With the patch release, the user still may find some faults during the interval $[\tau_p, \tau + w]$ because of the faults remaining in the patch phase. Due to warranty of the software, firms need to confiscate the failure-causing faults. By using Eq. (6), the amount of fault detached subsequent to update release and is denoted by $m((\tau + \mathrm{w}) - (\tau + \tau_p))$. Suppose c_6 represents the removal cost per unit fault, then the detection cost can be expressed as

$$cost_{\tau_p,\tau+w} = c_6 \cdot m\left((\tau + w) - \left(\tau + \tau_p\right)\right) \tag{13}$$

Post Warranty Cost As the software warranty gets over, the organization does not provide any support to any kind of failures that occurred, and sometimes they make agreement with the user for removing the faults all the way through the software life cycle. This cost has been incorporated while modeling the cost function. During $[\tau + w, T_{lc}]$, failures are encountered only by the users due to number of faults remained unobserved post warranty period. Using Eq. (7), we can obtain the total number of faults. Let $c_7$denote the removal cost of faults so the total cost incurred is given by

$$Cost_{\tau+w,T_{lc}} = c_7 \cdot m\left(T_{lc} - (\tau + w)\right) \tag{14}$$

Total Cost This can be summed as all the cost used throughout the life cycle of software. By adding together the entire cost as per Eqs. (8, 9, 10, 11, 12, 13, 14), the total cost function can be expressed as

$$totalcost = c_1 \cdot \tau + c_2 \cdot \tau^2 + Cost_{0,\tau_c} + Cost_{\tau,\tau_c} + Cost_{\tau,\tau_p} + Cost_{\tau_p,\tau+w} + Cost_{\tau+w,T_{lc}} \quad (15)$$

In the subsequent section, validation of the projected model has been performed through a numerical illustration.

4 Numerical Illustration

On the basis of supposition that the fault is identical and is autonomously distributed, it is clear that the rate of fault removal follows exponential distribution given as $F_i(t) = 1 - e^{-b_i \cdot t}$ where b_i denotes rate of fault detection in "i" interval. The parameters are estimated using Woods [18] data set. As per the data, the software is being tested for 20 weeks to detect 100 bugs for the given duration. The change point analyzer is used to find the change point of the data sets which is taken to be eighth week. The parameter estimation is done using the SPSS on the basis of least square method on the data set that gives 'a' $= 356.937$ and $b_1 = 0.0419$ and $b_2 = 0.1326$. This depicts that software initially has 357 faults which were detected/removed by the testing team with rate b_1 and b_2. We have assumed the software life cycle $T_{lc} = 100$.Usually rate of fault detection by the user and by tester is dissimilar due to different competence. Suppose $r1$, $r2$, $r3$ denotes fault detection rate ratio for the user with respect to tester in third, fourth, and fifth intervals.

The sum of faults detected/removed in interval $[0, \tau_c]$ represented as

$$m(\tau_c) = a\left(1 - e^{-b_1 \cdot \tau_c}\right)$$

Therefore from Eq. (10) the cost associated with $m(\tau_c)$ number of faults is specified as

$$c_3 \cdot a(1 - \exp(-b_1 \cdot \tau_c)) \quad (16)$$

Using Eq. (11), we get the faults removed in interval $[\tau_c, \tau]$ as $a \cdot (1 - (1 - \exp(-b_1 \cdot \tau_c)))\left(1 - \frac{(1-(1-\exp(-b_2 \cdot \tau)))}{1-(1-\exp(-b_2 \cdot \tau_c))}\right)$ and the cost associated cost with $m(\tau - \tau_c)$ is specified as

$$c_4 \cdot a(1 - (1 - \exp(-b_1 \tau_c)))\left(1 - \frac{(1 - (1 - \exp(-b_2 \tau)))}{1 - (1 - \exp(-b_2 \tau_c))}\right) \quad (17)$$

Amount of faults detached in $[\tau, \tau_p]$ is expressed as

$$m\left(\left(\tau + \tau_p\right) - \tau\right) = a\left(1 - \frac{(1-(1-\exp(-b_1 \cdot \tau_c))) \cdot (1-(1-\exp(-b_2 \cdot \tau)))}{1-(1-\exp(-b_2 \cdot \tau_c))}\right) \cdot \left(1 - \exp\left(-b_2 \cdot r_1 \cdot \left(\tau + \tau_p - \tau\right)\right)\right)$$

and associated cost associated with $m((\tau+\tau_{\mathrm{p}})-\tau)$ number of faults is represented by

$$c_5\cdot a\cdot\left(1-\frac{(1-(1-\exp(-b_1\cdot\tau_c)))\cdot(1-(1-\exp(-b_2\cdot\tau)))}{1-(1-\exp(-b_2\cdot\tau_c))}\right)\cdot\left(1-\exp\left(-b_2\cdot r_1\cdot\left(\tau+\tau_{\mathrm{p}}-\tau\right)\right)\right) \tag{18}$$

Amount of faults detached in $[\tau_{\mathrm{p}},\tau+w]$ is represented by

$$\begin{aligned}&m\left((\tau+\mathrm{w})-\left(\tau+\tau_p\right)\right)\\&=a\cdot\left(1-\frac{(1-(1-\exp(-b_1\cdot\tau_c)))\cdot(1-(1-\exp(-b_2\cdot\tau)))}{1-(1-\exp(-b_2\cdot\tau_c))}\right)\\&\quad\cdot\left(1-\left(1-\exp\left(-b_2\cdot r_1\cdot\left(\tau+\tau_p-\tau\right)\right)\right)\right)\\&\quad\cdot\left(1-\exp\left(-b_2\cdot r_2\cdot\left((\tau+w)-\left(\tau+\tau_p\right)\right)\right)\right)\end{aligned}$$

and the associated cost associated with $m((\tau+w)-(\tau+\tau_P))$given number of faults is given by

$$\begin{aligned}&c_6\cdot a\cdot\left(1-\frac{(1-(1-\exp(-b_1\cdot\tau_c)))\cdot(1-(1-\exp(-b_2\cdot\tau)))}{1-(1-\exp(-b_2\cdot\tau_c))}\right)\\&\cdot\left(1-\left(1-\exp\left(-b_2\cdot r_1\cdot\left(\tau+\tau_{\mathrm{p}}-\tau\right)\right)\right)\right)\\&\cdot\left(1-\exp\left(-b_2\cdot r_2\cdot\left((\tau+w)-\left(\tau+\tau_{\mathrm{p}}\right)\right)\right)\right)\end{aligned} \tag{19}$$

Also the amount of faults removed in $[\tau+w,T_{\mathrm{lc}}]$ is given by

$$\begin{aligned}m\left(T_{\mathrm{lc}}-(\tau+w)\right)&=a\cdot\left(1-\frac{(1-(1-\exp(-b_1\cdot\tau_c)))\cdot(1-(1-\exp(-b_2\cdot\tau)))}{1-(1-\exp(-b_2\cdot\tau_c))}\right)\\&\quad\cdot\left(1-\left(1-\exp\left(-b_2\cdot r_1\cdot\left(\tau+\tau_{\mathrm{p}}-\tau\right)\right)\right)\right)\\&\quad\cdot\left(1-\left(1-\exp\left(-b_2\cdot r_2\cdot\left((\tau+w)-\left(\tau+\tau_{\mathrm{p}}\right)\right)\right)\right)\right)\end{aligned}$$

and the cost associated with $m(T_{\mathrm{lc}}-\tau+\mathrm{w})$ given amount of faults is shown as

$$\begin{aligned}&c_7\cdot a\cdot\left(1-\frac{(1-(1-\exp(-b_1\cdot\tau_c)))\cdot(1-(1-\exp(-b_2\cdot\tau)))}{1-(1-\exp(-b_2\cdot\tau_c))}\right)\\&\cdot\left(1-\left(1-\exp\left(-b_2\cdot r_1\cdot\left(\tau+\tau_p-\tau\right)\right)\right)\right)\\&\cdot\left(1-\left(1-\exp\left(-b_2\cdot r_2\cdot\left((\tau+w)-\left(\tau+\tau_p\right)\right)\right)\right)\right)\end{aligned} \tag{20}$$

By taking phase wise testing cost consequently, we obtain the total cost with single patching specified by

$$\begin{aligned}
totalcost &= c_1 \cdot \tau + c_2 \cdot \tau^2 + c_3 \cdot a\Big(1 - \exp\left(-b_1 \cdot \tau_c\right) \\
&+ c_4 \cdot a\left(1 - \left(1 - \exp\left(-b_1 \tau_c\right)\right)\right)\left(1 - \tfrac{(1-(1-\exp(-b_2 \tau)))}{1-(1-\exp(-b_2 \tau_c))}\right) \\
&+ c_5 \cdot a \cdot \left(1 - \tfrac{(1-(1-\exp(-b_1 \cdot \tau_c)))\cdot(1-(1-\exp(-b_2 \cdot \tau)))}{1-(1-\exp(-b_2 \cdot \tau_c))}\right) \\
&\quad \left(1 - \exp\left(-b_2 \cdot r_1 \cdot \left(\tau + \tau_p - \tau\right)\right)\right) \\
&+ c_6 \cdot a \cdot \left(1 - \tfrac{(1-(1-\exp(-b_1 \cdot \tau_c)))\cdot(1-(1-\exp(-b_2 \cdot \tau)))}{1-(1-\exp(-b_2 \cdot \tau_c))}\right) \\
&\cdot \left(1 - \left(1 - \exp\left(-b_2 \cdot r_1 \cdot \left(\tau + \tau_{\mathrm{p}} - \tau\right)\right)\right)\right) \\
&\cdot \left(1 - \exp\left(-b_2 \cdot r_2 \cdot \left(\left(\tau + w\right) - \left(\tau + \tau_{\mathrm{p}}\right)\right)\right)\right) \\
&+ c_7 a \cdot \left(1 - \tfrac{(1-(1-\exp(-b_1 \cdot \tau_c)))\cdot(1-(1-\exp(-b_2 \cdot \tau)))}{1-(1-\exp(-b_2 \cdot \tau_c))}\right) \\
&\cdot \left(1 - \left(1 - \exp\left(-b_2 \cdot r_1 \cdot \left(\tau + \tau_p - \tau\right)\right)\right)\right) \\
&\cdot \left(1 - \left(1 - \exp\left(-b_2 \cdot r_2 \cdot \left(\left(\tau + w\right) - \left(\tau + \tau_p\right)\right)\right)\right)\right) \\
&\cdot \left(1 - \exp\left(-b_2 \cdot r_3 \cdot \left(T_{\mathrm{lc}} - \left(\tau + w\right)\right)\right)\right)
\end{aligned} \tag{21}$$

now defining the cost parameter values before doing the optimization of the above given equation and providing all cost values as $c_1 = 30$, $c_2 = 5$, $c_3 = 20$, $c_4 = 30$, $c_5 = 55$, $c_6 = 55$, and $c_7 = 135$. Also let us suppose that the organization generally provides the user with warranty of 6 months, i.e., w = 24 (weeks) on the software and fault detection rate ratios r1 = 0.5, r2 = 0.5, and r3 = 0.6. It should be noted that values obtained are purely through experience. In general cost of per unit fault during a given phase is directly relative to the ratio of assets consumed to the removed fault within that period. It is to be noted that post warranty period of software is much longer as compared to any other phases of software life cycle; hence it consumes high amount of resources. Also the number of faults removed during this period is low as the maximum number of faults is debugged during testing and warranty phase. Hence cost per unit fault removal during post warranty period is highest. Based on the similar arguments, magnitude of cost per unit fault removal in other phases can be described. Now based on the above assumptions, we will describe the phase-wise fault detection with the associated cost. This is to inform that the service contract and update time be measured since time of release τ, because after release only updates and warranty can be provided to the customer. By substituting the cost values in Eq. (21) and performing optimization using MAPLE for the total cost function, we acquire optimal outcome as software release time $\tau^* = 21.19$ weeks; optimal time to release the patch $\tau_p^* = 1$ and the optimized cost value as 14385.69. The below given table (Table 1) summarizes the fault detection by the tester as well as user with the help of over mentioned values phase wise.

Table 1 Descriptions of faults phase wise

Phase	MVF	Number of removed faults
Software time to release before change point $[0, \tau_c]$	$m(\tau_c)$	101.83 (102 approx.)
Software release time before change point $[\tau_c, \tau]$	$m(\tau - \tau_c)$	210.7753 (211 approx.)
First patch release time $[\tau, \tau_p]$	$m((\tau + \tau_p) - \tau)$	2.8446 (3 approx.)
Warranty time $[\ \tau_p, \tau + w]$	$m((\tau + w) - (\tau + \tau_p))$	32.4592 (32 approx.)
Operational time $[\tau + w, T_{lc}]$	$m(T_{lc} - \tau + w)$	8.90770

5 Conclusion

Through this paper we anticipated a comprehensive cost model to find the optimum time for release and for patch of software sold beneath warranty using change point subsequently to reduce the total anticipated software expenditure. In the existing effort, we have carefully calculated the case of distinct patching; however, from the similar lines of anticipated model, we can broaden our work for multiple patches. In the future we are capable to expand our model by taking into consideration the budget and consistency constraint on the optimum time to release and to patch. This model provides information to the developers that how much more testing is required to test the software after changing the testing strategies after a moment of time τ, and due to the repetitive nature of updates, developers can release the subsequent patches less costly. As more failure reports are sent by the user or the tester, it will provide more opportunities for the developer to detect and further confiscate the corresponding fault. When the next update is released and executed, software failure rate will be reduced accordingly.

References

1. Kapur PK, Khatri SK, Singh O, Shrivastava AK (2014) When to stop testing under warranty using SRGM with change point. In the IEEE Xplore conference proceeding of International Conference on IT in Business, Industry & Govt., CSIBIG held during March 8–9, 2014 at Sri Aurobindo Institute of Technology Indore Ujjain Highway Indore, pp 200–205
2. Kapur PK, Pham H, Gupta A, Jha PC Software reliability assessment with OR applications (Springer Series in Reliability Engineering)
3. Goel AL (1985) Software reliability models: assumptions, limitations and applicability. IEEE Trans Softw Eng SE-ll:1411–1423
4. Dohi T, Kaio N, Osaki S (1997) Optimal software release policies with debugging time lag. Int J Reliability, Quality and Safety Engineering 04(03)
5. Yamada S, Osaki S (1987) Optimal software release policies with simultaneous cost and reliability requirements. Eur J Oper Res 31:46–51
6. Kapur PK, Garg RB (1991) Optimal software release policies for software systems with testing effort. Int J Syst Sci 22(9):1563–1571

7. Kapur PK, Garg RB (1989) Cost-reliability optimum release policies for software system under penalty cost. Int J Syst Sci 20:2547–2562
8. Zhao M (1993) Change-point problems in software and reliability. Commun Stat Theory Methods 22(3):757–768
9. Yun WY, Bai DS (1990) Optimum software release policy with random life cycle. IEEE Trans Reliab 39(2):338–353
10. Pham H, Zhang X (1999) A software cost model with warranty and risk costs. IEEE Trans Comp 48(1):71–75
11. Kapur PK, Agarwal S, Garg RB (1994) Bi-criterion release policy for exponential software reliability growth models. Rech Oper/Oper Res 28:165–180
12. Musa JD, Iannino A, Okumoto K (1987) Software reliability: measurement, prediction, applications. Mc Graw Hill, New York
13. Kapur PK,Singh VB, Anand S (2007) Software reliability growth model of fielded software based on multiple change-point concept using a power function of testing time. In: Kapur PK, Verma AK (eds) Quality reliability and infocom technology. MacMillan India Ltd., New Delhi, pp 171–178
14. Kapur PK, Garg RB, Aggarwal AG, Tandon A (2009) General framework for change-point problem in software reliability and related release time problem. In proceedings of ICQRIT
15. Yamada S (1994) Optimal release problems with warranty period based on a software maintenance cost model. Trans IPS Jpn 35(9):2197–2202
16. KapurPK, SinghVB, Sameer A (2007) Effect of Change-Point on Software Reliability Growth Models Using Stochastic Differential Equations. Published in the proceedings of 3rd International Conference on Reliability and Safety Engineering. Misra RB, Naikan VNA, Chaturvedi SK Goyal NK (eds), (INCRESE-2007), Udaipur, pp 320–333
17. Jiang Z, Sarkar S, Jacob VS (2012) Post-release testing and software release policy for enterprise-level systems. Inf Syst Res 23(3), Part 1 of 2, 635–657
18. Wood A (1996) Predicting software reliability. IEEE Comput 29:69–77
19. Ompal S, Kapur PK, Shrivastava AK, Kumar V (2015) Release time problem with multiple constraints. Published in Int J Syst Assur Eng Manag 6(1):83–91
20. Cavusoglu H, Cavusoglu H, Zhang J (2008) Security patch management: share the burden or share the damage? INFORMS 54:657–670
21. Luo C, Okamura H, Dohi T (2015) Optimal planning for open source software updates. Proc IMechE Part O. doi:10.1177/1748006x15586507
22. Arora A, Caulkins JP, Telang R (2006) Research note: sell first, fix later: impact of patching on software quality. Manag Sci 52(3):465–471
23. Okamura H, Tokuzane M, Dohi T (2009) Optimal security patch release timing under non-homogeneous vulnerability-discovery processes. Proceedings of the 20th International Symposium on Software Reliability Engineering (ISSRE'09), Mysuru, pp 120–128
24. Kapur PK, Shrivastava AK (2015) When to release and stop testing of a software: a new insight, published in Conference Proceedings of International Conference on Reliability, Infocom Technology and Optimization (Trends and Future Directions), held during October 2–4, 2015 at Amity University, Uttar Pradesh, pp 1–6

Two-Dimensional Framework to Optimize Release Time and Warranty

Nitin Sachdeva, P. K. Kapur, and Ompal Singh

1 Introduction

Software developers can easily achieve the target reliability of their system by evaluating its reliability during development, especially during the testing phase. In order to quantify this reliability, test cases are executed as inputs randomly selected from some known distributions. Then a software reliability growth model is applied to predict total testing resources employed in terms of efforts required to meet the targeted reliability requirements. However, such testing procedures are directed toward finding defects/bugs in a more random approach.

Nowadays businesses are striving to employ disruptive innovations due to the fast pace/short life of any product in the market. Developers are therefore under great pressure to deliver more complex software with increasingly aggressive schedules and with limited resources. Testers are therefore expected to validate the reliability and quality of software even faster and with fewer resources at their disposal. One of the challenging questions to be answered by the development team has always been how long they should continue to test. What level of testing resources would be enough to meet the desired level of reliability? Undoubtedly, both factors, technical quality/reliability and economic market opportunity, are key determinants of testing/release time.

Despite the tremendous success of information technology in the last two to three decades, developers are consistently looking for effective ways to manage their testing resources specifically testing time and effort. Predominantly, several studies

N. Sachdeva (✉) • O. Singh
Department of Operational Research, University of Delhi, Delhi, India
e-mail: nitin.sach@gmail.com; drompalsingh@live.com

P.K. Kapur
Amity Center for Interdisciplinary Research, Amity University, Noida, UP, India
e-mail: pkkapur1@gmail.com

© Springer Nature Singapore Pte Ltd. 2018

P.K. Kapur et al. (eds.), *Quality, IT and Business Operations*, Springer Proceedings in Business and Economics, DOI 10.1007/978-981-10-5577-5_31

in the past have laid special emphasis on considering release time as testing stop time [3, 13] which is clearly summarized in the works of Kapur et al. [13]. Perpetually the thrust in these studies has been toward assessing the cost-benefit analysis. Till the time continued testing is justified by predetermined budget, reliability can be improved. Software reliability is, of course, one important criterion based on which testing stops; however, the underlying assumption of corresponding release time and testing stop time has always dominated in the literature.

Nowadays, with intense competition and client pressure to release the software as soon as possible, testers are expected to validate the quality of software even faster and with even fewer resources. Several researchers in the past have tried to evaluate both the optimal testing time and testing effort to be employed [5, 18, 21, 30].

Several testing effort-based software reliability growth models (SRGMs) have been proposed by researchers in the past couple of decades [9, 13, 23, 25, 26, 33, 34, 41]. Testing effort-based SRGM was first introduced by Yamada et al. [41]. Using the logistic testing effort function, Huang and Kuo [8] proposed another SRGM. Kapur et al. proposed [15] an SRGM with a testing effort-based learning process. Further, a unified framework for modeling testing effort-based SRGM was also proposed by Kapur et al. [16]. Under the assumption of imperfect debugging, Ahmad et al. [1] proposed an S-shaped SRGM with testing effort. Inoue and Yamada [10] suggested a lognormal testing effort-based SRGM. Peng et al. [32] also proposed a testing effort-based SRGM with detection and correction as two stages in an imperfect debugging environment. Later on in the same year, Zhang et al. [42] proposed a unified framework for the similar SRGM in imperfect debugging environment. Further working on the same lines, Li et al. [24] converted the testing effort functions to S-shaped learning function and proposed yet another SRGM with imperfect debugging. Lately, researchers are working on proposing testing effort-based SRGMs for multiple versions of software Singh et al. [37].

Recently Sachdeva et al. [36] proposed a generalized framework for a software testing model with separate release and testing time by optimizing overall testing cost including the market opportunity cost related to delayed release in the market. The proposed study tends to work under the similar environment wherein testing is believed to be carried on into the post-release phase or warranty phase of the software in order to minimize total testing resources, i.e., testing time and efforts.

In software reliability testing literature, the emphasis has been on optimizing release time by assuming optimal release time to coincide with testing stop time Kapur et al. [13]. However, it is quite possible to relax this assumption and adjust these two time points at a notable difference in the life cycle of the software. In other words, testers can continue testing the software post its release. The idea of optimizing both the release and testing stop time is of immense importance to the developers and in turn affects the company's bottom line. Too early a release results in severe risk of software failures in the field, while a delayed release proposition leads to increase production cost and loss of market opportunity. Hence, in this paper, we try to handle this complex decision for product managers and propose a two-dimensional cost model to optimize both the release and testing stop time.

Optimal software release policies with a debugging time lag, proposed by Dohi et al. [4], is one of the pioneering works in the software release policy area, while release time problems with cost minimization and reliability maximization objectives have been highlighted in the works of Yamada and Osaki [40]. Kapur and Garg [18] proposed testing effort-based release policies and also introduced the concept of penalty cost in modeling release time problem of software (1989). Huang [6], Huang and Lyu [5], and Huang et al. [7] discussed release policies considering the effect of testing effort expenditure. Later, Kapur et al. [12] developed release time problem to minimize the cost of software having a random life cycle, subject to achieving a desired level of intensity. Further, Kapur et al. [12] developed a software cost model incorporating the cost of dependent faults along with independent faults. Pham and Zhang [35] incorporated warranty and risk cost to the traditional cost function, and further Kapur et al. [17, 20] proposed a unique model to incorporate warranty as a measure of customer's expectation and satisfaction. Kapur et al. [21, 22, 30, 31] proposed a price and warranty length optimization models to emphasize on the impact of warranty length on the release of the software. The idea to incorporate separate release and testing stop time in our modeling framework along with optimizing warranty presents a completely unique and effective way for developers to minimize their overall testing cost. Thus this paper is written in the spirit to evaluate these times separately for the developers.

Additionally, a discrete time-based diffusion model to identify the release of a new product using MCDM approaches [31] was proposed by Sachdeva et al. [27] to overcome the biasedness present in continuous time data. Later on, some researchers worked on bi-criterion release time problem; Kapur et al. [19] developed a multi-objective optimization problem for determination of release time. In our present study, we have considered a two-dimensional cost framework based on testing time and testing efforts employed during various phases of software testing. We make use of classical Cobb-Douglas [2] production function to explain the multidimensionality in our study.

Kapur et al. [13, 17, 20, 21, 30] and Singh et al. [29] have proposed several multi-upgradation models to determine optimal release time of the next release of software. By optimizing overall software testing cost, Kapur et al. [17] proposed a cost model for successive software release. In our current study, we bring forward the concept of separate release and testing stop time, and our results show how quick release in the market can help organization in saving a lot on testing cost and capitalizing on the market opportunity and with continued testing how overall reliability of the software can be enhanced.

The major focus of this paper is to build a robust as well as cost-effective strategy for finding optimal release policy of large-scale custom-based information systems. Our major objective is to develop a release policy that can help quickly release software to the client in order to minimize opportunity loss risk while optimizing overall testing cost. We intend to consider testing cost along with market opportunity cost. The key contributions of our work include presenting a generalized framework for modeling of post-release testing to optimize release time and testing effort. We intend to address several specific questions related to optimal release policy.

First, by considering the trade-offs between testing time and market opportunity, how can we obtain the optimal release time? We provide a generalized framework to model this scenario and an illustrative understanding of the same.

Second, we pioneer in presenting how the two dimensions of release time and testing effort can be simultaneously optimized in order to minimize cost to achieve the desired level of reliability at the release time. The model proposed in this paper incorporates the effect of these two vital factors simultaneously using Cobb-Douglas production function which is explained in Sect. 3.1.

Third, in the literature related to cost modeling for single and multiple releases of software [13, 14, 33, 34] it's assumed that testing and operational environment phases are governed by same debugging distribution functions, which is quite contrary to today's testing practices followed by practitioners. Accordingly, we consider a more realistic scenario that debugging process followed by the testers, pre- and post-release of the software, takes place differently, and therefore, we propose a generalized cost framework with different distribution functions for different phases of a software testing cycle in our paper.

Lastly, we compare the no post-release testing policy with the one where testing continues post-release to showcase how the latter policy outperforms the former in terms of lowering overall testing cost by 38%, reducing release time almost by 50%, and delaying testing stop time considerably thereby increasing overall testing effort during the pre- and post-release phase by 35%.

Section 2 discusses the cost model considered for addressing the main objective of this study aside from those mentioned above. Various costs included in the proposed model are explained in this section with notations. In Sect. 3, cost model with two cases is proposed: (1) *release policy with no post-release testing but warranty provided and* (2) *release policy with post-release testing and warranty provided.* Section 4 explains policy implications with two proposed cost optimization models subjected to reliability and budgetary constraints. In Sect. 5, the numerical illustration of the proposed models using Tandem Computers data set is presented [39] along with the sensitivity analysis in Sect. 6. Section 7 provides the conclusion and future scope of the research undertaken.

2 Cost Model Development

In order to determine the optimal release time and optimal warranty length corresponding to testing stop time of software, we make use of the four major costs identified from the literature and are presented in Table 1.

After the functional forms of each of these proposed costs are obtained, optimal testing stop time and release time of the software can be obtained by minimizing the total of all the four costs. In the following sections, we formulate the proposed cost model using the *notations* mentioned below. All other costs of testing and debugging process which is involved in the testing process are considered to be negligible in this study.

Table 1 Four Major Costs

Cost	Explanation	References
Per Unit Testing Cost	Refers to testing planning, test case generation, test case execution and analysis of testing cost. Also, cost of CPU hours consumed during testing is included in Per Unit Testing Cost	Kapur et al. [13]
Debugging Cost During Testing	Refers to cost associated with bug removal process during testing	Okumoto and Goel [28]
Debugging Cost Due to Software Failure in the Field	Refers to the debugging cost of those bugs associated with faults identified in the operational phase of software. It also included loss of revenue due to system downtime, liability cost, & customer's dissatisfaction cost	Kapur et al. [8]; Dalal and Mallows [3]
Market Opportunity Cost	Refers to the market-related cost or the opportunity loss cost due to late entry into the market	Dalal and Mallows [3]; Singpurwalla and Wilson [38])

Notations

a: *Expected bugs to be eventually detected*
b: *Failure detection/correction rate by tester's testing*
b': *Failure detection rate by user's usage*
$m(T_{\text{lc}})$: *Number of faults removed during the lifecycle of the software*
$m(\tau)$: *Expected number of faults removed by employing* τ *testing resources*
$F(\tau)$: *Distribution functions for fault correction*
s: *Testing Time of the software*
s_w: *Testing stops time of the software*
^{s}lc: *Software life cycle*
w: *Warranty length*
τ: *Testing resources*
τ_w: *Testing resources employed during the warranty phase*
$^{\tau}\text{lc}$: *Testing resources employed till the end of life cycle post warranty phase*
^{u}: *Testing Effort*
α: *Output elasticity of testing time*

3 Proposed Cost Models

Assumptions The proposed cost models are based on the failure observation/fault removal phenomenon as modeled by nonhomogeneous Poisson process (NHPP) Kapur et al. [13]. The entire fault detection/correction process follows typical NHPP criterions like the independence of bug detection process, the occurrence of failures due to remaining number of faults in the system, detection/correction to be an instantaneous process, perfect debugging, and a finite number of faults to be removed in the finite life cycle of the software.

We also consider that the market opportunity cost is monotonically increasing twice with the continuously differentiable convex function of T as suggested by Jiang et al. [11].

Kapur et al. [13] proposed a unified approach to modeling the fault detection/correction process using the hazard rate approach in deriving the mean value function of a cumulative number of faults removed. Based on this, we have

$$\frac{\mathrm{d}m(t)}{\mathrm{d}t} = \frac{f(t)}{1 - F(t)}(a - m(t)) \tag{1}$$

where, $h(t) = \frac{f(t)}{1-F(t)}$ is the time dependent fault correction rate per remaining faults. In solving the above equation, we obtain

$$m(t) = a.F(t) \tag{2}$$

3.1 Modeling of Two-Dimensional SRGM

In literature, most of the SRGMs are developed as time-based models, whereas in reality testing time alone can never provide in-depth understanding of the software reliability. Therefore, in this paper, we make use of testing time and testing effort as the two dimensions for determining software reliability. This two-dimensional model is based on the two-dimensional software reliability model proposed by Inoue and Yamada [10]. Here, we consider testing time and testing effort as the two factors contributing combined together as testing resources employed by the developer. Therefore, in the proposed two-dimensional SRGM, a number of faults removed are considered to be dependent upon the testing resources τ, which is defined as

$$t \cong s^{\alpha} u^{1-\alpha} \tag{3}$$

$$0 \leq \alpha \leq 1$$

where

τ: testing resources
s: testing time
u: testing effort
α: output elasticity of testing time

Let $\{N(s,u), s \geq 0, u \geq 0\}$ be a two-dimensional stochastic process representing the cumulative number of software failures by time "s" and testing effort "u." A two-dimensional NHPP with a mean value function m (s, u) is formulated as

$$\Pr(N(s,u) = n) = \frac{(m(s,u))^n}{n!}\exp(-m(s,u)) \tag{4}$$

The mean value function with respect to testing resources is given as follows:

$$m(\tau) = aF(\tau) \tag{5}$$

The cumulative number of faults removed $m(\tau)$ is dependent on testing resources τ·τ is a two-dimensional variable, with testing time and testing effort as its dimensions. The two-dimensional models are useful as they can show the effect of two aspects of a variable in which the result is dependent. Therefore, (2) can be rewritten as follows:

$$m(\tau) = aF\left(s^{\alpha}u^{1-\alpha}\right) \tag{6}$$

Measuring the cost related to testing is an important step in releasing the basic version of software. The basic cost consists of three parts: per unit testing cost during the testing phase, debugging cost during the testing phase, and debugging cost during operational phase or cost of debugging in the field. The costs of debugging in testing phase include all expenses that arise when a bug is identified and removed. The debugging cost in operational phase contains all expenses that occurred when a reported bug is identified and removed during operational phase including also the indirect liability and loss of goodwill cost.

$$\begin{aligned} C_{\text{total}} &= C_{\text{per unit testing cost}} + C_{\text{debugging cost intesting phase}} + C_{\text{debugging cost inoperational phase}} \\ C_{\text{total}} &= c_1.\tau + c_2.m(\tau) + c_3.[a - m(\tau)] \end{aligned} \tag{7}$$

Here c_1, c_2, and c_3 denote per unit cost of testing, the cost of debugging during testing phase, and cost of debugging during operational phase, respectively.

When no post-release testing is considered for software and no warranty is provided to the users and testing stops at the time of release, the existing cost function as proposed by Kapur et al. [13] would suffice. However, any bugs reported during the operational phase of the software leads to higher cost of debugging for the developer. The cost of removing a bug during the operational phase will be larger than debugging it during testing as it involves the indirect cost such as the liability cost and the loss of goodwill resulting from customer's dissatisfaction along with the direct cost associated with bug removal in the field.

The proposed cost models try and answer these questions about how to minimize overall testing cost and still try and capitalize market opportunity. Not only the developer is concerned about the reliability of the software to be released, but also the fear of software failure in the field could prove to be a costly affair if desired reliability is not attained during the testing phase of the software.

We proposed two cases here highlighting the key difference with the existing cost model. A release policy with no post-release testing but warranty provided to the customer is proposed with market opportunity cost involved in Case 1. In Case 2, we have tried to model a scenario where developer continues to test its software even after its release. The idea is to capture how the trade-off between capitalizing market opportunity and achieving desired reliability is to be carried out to ensure minimal overall testing cost. During the post-release testing phase, we have also considered that the developer provides a warranty to its client thereby ensuring users of the product quality and developer's confidence in the product.

Case 1: Release policy with no post-release testing but warranty provided
When no patch release is considered but software developer provides a warranty to the user for a finite period of time post release of the software, it acts as an assurance for the quality of testing and reliability of software to the user. Although the testing posts the release of the software stops, developer bears the warranty cost which includes the cost of debugging post-release till a finite time length.

$$\begin{aligned} C_{\text{total}} &= C_{\text{per unit testing cost}} + C_{\text{debugging cost in testing phase}} + C_{\text{debugging cost in warranty phase}} \\ &\quad + C_{\text{debugging cost in operational phase}} + C_{\text{market opportunity cost}} \\ C_{\text{total}} &= c_1.\tau + c_2.m(\tau) + c_3.[m(\tau_w) - m(\tau)] + c_4[a - m(\tau_w)] + c_5.\tau^2 \end{aligned} \tag{8}$$

Here, c_1, c_2, c_3, c_4, and c_5 denote the per unit testing cost, debugging cost in the warranty phase, the debugging cost in the operational phase, and market opportunity cost, respectively. Also, $m(\tau_w)$ denotes the number of faults detected during the phase $s \rightarrow s_w$ wherein warranty is provided by the developer, and hence $m(\tau_w)$ can be written as

$$\begin{aligned} m(\tau_w) &= aF(\tau_w) \\ &= a\ F\left(s_w^{\alpha} u^{1-\alpha}\right) \\ &= a\ F\left((s+w)^{\alpha} u^{1-\alpha}\right) \end{aligned} \tag{9}$$

In case we consider a more realistic environment wherein testing team employs different resource allocations before and after release of the software and also during operational phase post warranty period, then the above cost equation can be rewritten as

$$\begin{aligned} C_{\text{total}} &= c_1.\tau + c_2.a.F_1(\tau) + c_3.a.[1 - F_1(\tau)].F_2(\tau_w - \tau) \\ &\quad + c_4.a.[1 - F_1(\tau)][1 - F_2(\tau_w - \tau)].F_3(\tau_{\text{lc}} - \tau_w) + c_5.\tau^2 \end{aligned} \tag{10}$$

In simple words, testing environment is not the same as the operational environment.

Although software developers would not like to bear the debugging cost post warranty period is over, we still consider this cost to ensure that the software is well supported throughout its life cycle. It is quite possible that the client had agreed in service level agreements (SLA) to pay the maintenance charges post warranty phase, but the debugging cost on the part of developer cannot be ignored.

In case we consider the operational phase to be large enough that all the remaining bugs from the warranty phase can be assumed to be detected during this period, then the fraction of the faults to be removed by testers will be 1, i.e., $F_3(\tau_{lc} - \tau_w) = 1$ for a large value of $\tau_{lc} - \tau_w$.

Here we have to understand the difference between $m(\tau_w) - m(\tau)$ and $a.[1 - F1(\tau)].F_2((\tau w) - \tau))$. The first term denotes the total number of bugs detected during the warranty period $[s, sw]$, when we consider the same testing environment during testing and warranty period.

In the second term, $a.[1 - F1(\tau)].F_2(\tau_w - \tau)$, $a.[1 - F1(\tau)]$ denotes the remaining number of bugs from the testing period of software, from which some of the bugs are detected with proportion $F2(\tau_w - \tau)$ during the warranty period. In such a scenario, testing team operates under a different environment, and $F2(\tau_w - \tau)$ denotes the cumulative probability distribution of bugs detected post testing period of the software.

Case 2: Release policy with post-release testing and warranty provided

When post-release testing is considered, a software system is released before testing stops. The idea to continue testing even after the software is released is to ensure that the reliability of the software could be enhanced and hence minimize the threat of software failure in the field. In this case, we consider that the developer does not only continue to invest their resources in software testing post release but also provide their client with a desired and predecided level of warranty.

The overall objective of providing warranty during the post-release testing period is to overcome client's fear of software failure during this phase. Warranty acts as an assurance to the client that the minimal level of reliability has been attained before releasing the software. It also assures the client that the product is of expected quality although it is released early to them.

The bigger challenge to be addressed in this phase is by the developer as he needs to continue testing; however, with this scenario, the developer would be keen to optimize both the release time and the length of the warranty period, so as to minimize overall testing cost.

$$\begin{aligned} C_{\text{total}} = {} & C_{\text{per unit testing cost}} + C_{\text{debugging cost in pre-release testing phase}} \\ & + C_{\text{debugging cost in warranty phase}} \left(C_{\text{debugging cost in post-release testing phase}}\right) \\ & + C_{\text{debugging cost in operational phase}} + C_{\text{market opportunity cost}} \end{aligned}$$

During this policy, per unit testing cost amounts to

$$C_1(\tau) = c_1.\tau \tag{11}$$

During this policy, the cost associated in the testing period $[0, s]$, denoted by $C_2(\tau)$ is given by,

$$C_2(\tau) = c_2.m(\tau) = c_2 a.F_1(\tau) \tag{12}$$

We also consider the market opportunity cost due to late entry in the market as

$$C_5(\tau) = c_5.\tau^2 \tag{13}$$

In the second period, $[s,\ s_w]$, both testers and users simultaneously work on the software, however, with different failure rates. During this phase, the only fraction of remaining faults from the prerelease phase would be detected either by the tester or by users.

Cost associated with bug detected by users is given by

$$C_{\text{users}}(\tau) = c_3.(a - a.F_1(\tau)).\int_{\tau}^{\tau_w} \overline{F}_2(x).f_3(x-\tau)\,dx \tag{14}$$

Note that during this period, the cost of debugging completely depends on who detects the bug (user or tester) first.

Cost associated with bug detected by testers is given by

$$C_{\text{testers}}(\tau) = c_4(a - a.F_1(\tau)).\int_{\tau}^{\tau_w} \overline{F}_3(x-\tau).f_2(x)dx \tag{15}$$

The second part of the above equation represents cost associated with those faults which are only detected by testers before the users could detect them. That means no extra cost in terms of indirect cost is levied post release if testers continue to detect bugs from the remaining count of bugs.

In the operational phase, only a fraction of the faults remaining from warranty phase is to be detected and are detected by users only. Cost associated with this phase is given as

$$C'_{\text{users}}(\tau) = c_6.(a - a.F_1(\tau)).\left(1 - \left(\int_0^{\tau_w-\tau} \overline{F}_2(x).f_3(x)dx + \int_0^{\tau_w-\tau} \overline{F}_3(x).f_2(x)dx\right)\right) .F_4(\tau_{\text{lc}} - \tau_w) \tag{16}$$

By adding all the abovementioned six costs, we have the expected total cost for the post-release testing policy

$$C_{\text{total}} = C_1(\tau) + C_2(\tau) + C_{\text{users}}(\tau) + C_{\text{testers}}(\tau) + C'_{\text{users}}(\tau) + C_5(\tau)$$

$$C_{\text{total}} = c_1.\tau + c_2.a.F_1(\tau) + c_3.(a - a.F_1(\tau)).\int_{\tau}^{\tau_w} \overline{F}_2(x).f_3(x-\tau)\,dx$$
$$+ c_4(a - a.F_1(\tau)).\int_{\tau}^{\tau_w} \overline{F}_3(x-\tau).f_2(x)dx$$
$$+ c_6.(a - a.F_1(\tau)).\left(1 - \left(\int_0^{\tau_w-\tau} \overline{F}_2(x).f_3(x)dx + \int_0^{\tau_w-\tau} \overline{F}_3(x).f_2(x)dx\right)\right)$$
$$.F_4(\tau_{\text{lc}} - \tau_w) + c_5.\tau^2 \tag{17}$$

where $c_1, c_2, c_3, c_4, c_5,$ & c_6 denote per unit testing cost, debugging cost prerelease of the software, debugging cost during post-release of the software for bugs detected and reported by users, debugging cost during post-release of the software for bugs detected and reported by testers, market opportunity cost, and debugging cost in the operational phase post testing stops, respectively.

4 Policy Implications

In the literature, several optimization models concerning software testing cost have been proposed wherein the primary focus is upon minimizing testing cost subject to reliability and budgetary constraints. Kapur et al. [13], [33, 34]), and Wood [41] have all concluded that the optimal testing time is clearly related to the desired level of reliability predefined by the developer. Across these models testing stop time is considered to coincide with software release time. GO (1980) emphasized on two important catalysts which influence the optimal release time of the software:

Reliability: Optimal testing time should be based on predefined reliability.
Budget: Based on testing budget, optimal testing time is obtained.

Jiang et al. [11] showed that the post-release testing assists in capitalizing market opportunity by releasing the software early to the clients (proposition 2), and at the same time it leads to reduced overall testing cost (proposition 4). The authors in their research work compared various scenarios to validate their propositions, and their analysis shows that in post-release testing both users and testers contribute toward bug detection process, though not directly connected in regard to detecting bugs, but certainly the process, leads to optimal cost and decoupling testing stop time from release time. In the proposed framework, the fault detection process is strongly assumed to follow an exponential distribution.

Through this work, we suggest an alternative generalized framework for software managers to help them achieve an effective trade-off between capitalizing market opportunity and reducing overall testing cost. We observed that when we consider the generalized testing process, the effectiveness of the model to identify both testers and user's effort is heightened. While devising the objective function, we considered the above differences and incorporated them into the analyses.

The general software debugging process as proposed by GO model (1980) is considered to model the cost model.

In order for software developers to minimize their total testing cost over a fixed time horizon and compare the two proposed scenarios of post-release testing and no post-release testing, we have formulated two cost models considering post-release testing and no post-release testing.

4.1 *Without Post-release Testing*

From Eq. (10), we have the total testing cost as

$$\begin{aligned} C_{\text{total}} = {} & c_1.\tau + c_2.a.F_1(\tau) + c_3.a.[1 - F_1(\tau)].F_2(\tau_w - \tau) \\ & + c_4.a.[1 - F_1(\tau)][1 - F_2(\tau_w - \tau)].F_3(\tau_{\text{lc}} - \tau_w) + c_5.\tau^2 \end{aligned}$$

where τ denotes the optimal testing stop and release time, τ_w denotes the optimal testing resources employed during the warranty period, and τ_{lc} refers to the finite time horizon considered as the lifetime of the software. Now we consider that in order to release the software, developers ascertains the release time based on reliability and budgetary constraints, which are stated as $R\left(\frac{x}{\tau}\right) \geq R_0$ and $C(\tau) \leq C_B$. Here, R_0 denotes the predefined reliability value set as per the service level agreements between the developer and client, and $R\left(\frac{x}{\tau}\right)$ denotes the reliability of the system at the time of release, i.e., with testing resource "τ." Similarly, C_B represents the allocated budget, and $C(\tau)$ denotes the total testing cost. These constraints are taken to consider the practical scenario, in which these constraints are imposed by the firm on the testing team to meet the user's requirement of reliable software within a fixed budget to achieve the ultimate goal of high gains.

The optimization problem for this case without post-release testing can now be rewritten as

$$\begin{aligned} \text{Min } C_{\text{total}} = {} & c_1.\tau + c_2.a.F_1(\tau) + c_3.a.[1 - F_1(\tau)].F_2(\tau_w - \tau) \\ & + c_4.a.[1 - F_1(\tau)][1 - F_2(\tau_w - \tau)].F_3(\tau_{\text{lc}} - \tau_w) + c_5.\tau^2 \qquad \text{(P1)} \end{aligned}$$

Subject to :

$$R\left(\frac{x}{\tau}\right) \geq R_0, C_{\text{total}} \leq C_B, \tau \geq 0$$

4.2 With Post-release Testing During Warranty Phase

Based on Eq. (17) for debugging process, the proposed cost model for post-release testing scenario is given as

$$C_{\text{total}} = C_1(\tau) + C_{\text{users}}(\tau) + C_{\text{testers}}(\tau) + C'_{\text{users}}(\tau) + C_5(\tau)$$

Once again subject to the reliability and budgetary constraint, i.e., $R\left(\frac{x}{\tau}\right) \geq R_0$, $C(\tau, \tau_w) \leq C_B$, we can write the optimization problem as

$$\text{Min } C_{\text{total}} : C_1(\tau) + C_{\text{users}}(\tau) + C_{\text{testers}}(\tau) + C'_{\text{users}}(\tau) + C_5(\tau) \quad \text{(P2)}$$

$$\begin{aligned}\text{Min } C_{\text{total}} :\ & c_1.\tau + c_2.a.F_1(\tau) + c_3.(a - a.F_1(\tau)).\int_{\tau}^{\tau_w}\overline{F}_2(x).f_3(x-\tau)\,dx \\ & + c_4(a - a.F_1(\tau)).\int_{\tau}^{\tau_w}\overline{F}_3(x-\tau).f_2(x)dx \\ & + c_6.(a - a.F_1(\tau)).\left(1 - \left(\int_0^{\tau_w-\tau}\overline{F}_2(x).f_3(x)dx + \int_0^{\tau_w-\tau}\overline{F}_3(x).f_2(x)dx\right)\right). \\ & F_4(\tau_{\text{lc}} - \tau_w) + c_5.\tau^2\end{aligned}$$

$$\text{Subject to}: \quad R\left(\frac{x}{\tau}\right) \geq R_0; C_{\text{total}} \leq C_B; \quad \tau, \tau_w \geq 0$$

5 Numerical Illustration

Using Tandem Computers (Woods [39]) data set, the application of the proposed models is presented in this section. We assume that the testing team's bug detection rate remains the same during the testing and operational phase of the software testing process, i.e., $b_1 = b_2 = b$(say). Assuming $b_3 = rb$ is the rate by which users detect the faults, where r is the ratio of user's fault detection rate with respect to tester's testing rate during the phase$[\tau, \tau_w]$. Also b_4 the failure detection rate of users in the last phase, $[s_w, s_{\text{lc}}]$ is given by $b_4 = r'b$ where r$'$ is ratio of user's fault detection rate with respect to tester's testing rate during the phase $[\tau_w, \ \tau_{\text{lc}}]$. Note that $r' \geq r$ due to increased user's base (Sect. 2).

Hence F_1, F_2, F_3 and F_4 in Eq. (17) are given by

$$F_1(\tau) = \left(1 - e^{-b.\tau}\right) = \left(1 - e^{-b.\left(s^{\alpha}u^{1-\alpha}\right)}\right),$$

$$F_2((\tau_w) - \tau) = 1 - e^{-b((\tau_w) - \tau)} = 1 - e^{-b\left((s+w)^{\alpha}u^{1-\alpha} - s^{\alpha}u^{1-\alpha}\right)},$$

$$F_3((\tau_w) - \tau) = 1 - e^{-rb\cdot((\tau_w) - \tau)} = 1 - e^{-rb\left((s+w)^{\alpha}u^{1-\alpha} - s^{\alpha}u^{1-\alpha}\right)}$$

$$F_4(\tau_{\text{lc}} - (\tau_w)) = 1 - e^{-r'b(\tau_{\text{lc}} - (\tau_w))} = 1 - e^{-r'b\left(s_{\text{lc}}{}^{\alpha}u^{1-\alpha} - (s+w)^{\alpha}u^{1-\alpha}\right)}$$

Using the above distribution functions in Eqs. (14), (15), and (16), we can obtain the respective cost of removing faults by both testers and users during the respective phases given by these equations.

Therefore, the total cost function when no post-release testing and no warranty provided is given by

$$\begin{aligned} C^{\tau}{}_{\text{total}} &= c_1.\tau + c_5.\tau^2 + c_2.a.\left(1 - e^{-b.\tau}\right) + c_3.a.\left(e^{-b.(\tau_{\text{lc}}-\tau)}\right) \\ &= c_1\left(s^{\alpha}u^{1-\alpha}\right) + c_5.\left(s^{\alpha}u^{1-\alpha}\right)^2 + c_2.a.\left(1 - e^{-b.s^{\alpha}u^{1-\alpha}}\right) \\ &\quad + c_3.e^{-b.\left(s_{\text{lc}}{}^{\alpha}u_{\text{lc}}{}^{1-\alpha} - s^{\alpha}u^{1-\alpha}\right)} \end{aligned} \tag{18}$$

Total cost function with no post-release testing but warranty is provided can be given by

$$\begin{aligned} C^{\tau}{}_{\text{total}} &= c_1.\tau \;+\; c_2.a.F_1(\tau) \;+\; c_3.a.\left[1 - F_1(\tau)\right].F_2(\tau_w - \tau) \\ &\quad + c_4.a.\left[1 - F_1(\tau)\right]\left[1 - F_2(\tau_w - \tau)\right].F_3(\tau_{\text{lc}} - \tau_w) + c_5.\tau^2 \\ &= c_1\left(s^{\alpha}u^{1-\alpha}\right) + c_2.a.\left(1 - e^{-b.s^{\alpha}u^{1-\alpha}}\right) \\ &\quad + c_3.a.e^{-bs^{\alpha}u^{1-\alpha}}\left(1 - e^{-b\left((s+w)^{\alpha}u^{1-\alpha} - s^{\alpha}u^{1-\alpha}\right)}\right) \\ &\quad + c_4.a.e^{-bs^{\alpha}u^{1-\alpha}}.e^{-b\left((s+w)^{\alpha}u^{1-\alpha} - s^{\alpha}u^{1-\alpha}\right)} \\ &\quad .\left(1 - e^{-b\left(s_{\text{lc}}{}^{\alpha}u^{1-\alpha} - (s+w)^{\alpha}u^{1-\alpha}\right)}\right) + c_5.\left(s^{\alpha}u^{1-\alpha}\right)^2 \end{aligned}$$

Total cost function with post-release testing being considered is given by

$$\begin{aligned} & c_1\tau + c_2.a.\left(1 - e^{-b.\tau}\right) + c_3.a.e^{-b\tau}\left(1 - e^{-b(1+r)(\tau_w-\tau)}\right)\cdot\tfrac{r}{1+r} \\ &\quad + c_4.a.e^{-b\tau}\left(1 - e^{-b(1+r)(\tau_w-\tau)}\right)\cdot\tfrac{1}{1+r} \\ &\quad + c_6.a.e^{-b\tau}.e^{-b(1+r)(\tau_w-\tau)}.\left(1 - e^{-r'b(\tau_{\text{lc}}-\tau_w)}\right) + c_5.\tau^2 \\ &= c_1\left(s^{\alpha}u^{1-\alpha}\right) + c_2.a.\left(1 - e^{-b.s^{\alpha}u^{1-\alpha}}\right) \\ &\quad + c_3.a.e^{-bs^{\alpha}u^{1-\alpha}}\left(1 - e^{-b(1+r)\left(\left((s+w)^{\alpha}u^{1-\alpha} - s^{\alpha}u^{1-\alpha}\right)\right)}\right)\cdot\tfrac{r}{1+r} \\ &\quad + c_4.a.e^{-bs^{\alpha}u^{1-\alpha}}\left(1 - e^{-b(1+r)\left(\left((s+w)^{\alpha}u^{1-\alpha} - s^{\alpha}u^{1-\alpha}\right)\right)}\right)\cdot\tfrac{1}{1+r} \\ &\quad + c_6.a.e^{-bs^{\alpha}u^{1-\alpha}}.e^{-b(1+r)\left(\left((s+w)^{\alpha}u^{1-\alpha} - s^{\alpha}u^{1-\alpha}\right)\right)}.\left(1 - e^{-r'b\left(\left(\tau_{\text{lc}}{}^{\alpha}u^{1-\alpha} - (s+w)^{\alpha}u^{1-\alpha}\right)\right)}\right) \\ &\quad + c_5.\left(s^{\alpha}u^{1-\alpha}\right)^2 \end{aligned} \tag{19}$$

Table 2 Optimal values of release time, testing stop time, testing effort, and total cost

Model	Optimal release time and effort	Optimal testing stop time and effort	Optimal warranty length	Optimal cost
P1	(10.09, 5499.99) (s^*, u^*)	(35.05, 11500.55) ($s_w{}^*$, $u_u{}^*$)	35 weeks	8538.15
P2	(7.05, 3998.58) (s^*, u^*)	(30, 8494.55) ($s_w{}^*$, $u_u{}^*$)	30 weeks	6267.70
No post-release testing and no warranty	(14.93, 8000.04) (s^*, u^*)		–	10115.30

In order to obtain the optimal release and testing stop time, we make use of both the cases of with and without post-release testing using the following parameter values. Suppose $r=0.5$, $r'=0.6$, $\alpha=0.9$, $\tau_{\text{lc}}=200$, $c_1=50$, $c_2=30$, $c_3=100$, $c_4=30$, $c_5=5$, $c_6=100$, $R_0=0.60$, and $c^{\tau}{}_B=50000$. Note that $a=m(\tau_{\text{lc}})$.

Using Eqs. (16) and (17) and the above parameter values in problems (P1) and (P2) and optimizing it by using maple software, we obtain the optimal values of release time, testing stop time, warranty length, and testing effort (Table 2).

We clearly observe that optimal release time when post-release testing is considered is shorter than the release time when post-release testing is not considered (s^* for P2 (7.05) is less than s^* for P1 (10.09)). Another advantage of the proposed strategy of continuing testing after the early release of the software ensures lowering of the warranty period provided (35 weeks in P1, whereas it is 30 weeks in P2). Also the testing effort employed in case of P1 at both the release time (5499.99) and testing stop time (11500.55) in model P1 seems to be on the higher side when compared to release time effort (3998.58) and testing stop time effort (8494.55).

Quite evident if the testers continue testing and the users considerably support them in providing them with bugs detected at their end, the overall testing cost (6267.70) including the market opportunity and other debugging cost seems considerably less as compared to the case P1 (8538.15) and the case of no post-release and no warranty cost (10115.30). It clearly indicates that the proposed framework can help firms reduce overall testing cost and also insure lower market opportunity cost.

Based on the given parameters, we estimated the total number of faults in the software as 130. The number of faults removed during various phases of post-release testing policy is presented in Table 3. In case of post-release testing, system is released early with approximately 35% of the faults lying undetected in the system.

During post-release testing, both testers and users together would detect 38 of the remaining bugs during warranty period. Therefore the users are under a less risky situation when dealing with early release because of post-release testing as the reliability of the system post its release keeps on increasing with time.

Table 3 Phase-wise fault detected using the proposed model

Phase		Number of faults removed
Prerelease testing phase $[O, \tau]$		86.86 (approximately 87)
Post-release testing phase $[\tau, T]$	Total	38.05 (approximately 38)
	Tester	14.35 (approximately 15)
	User	23.70 (approximately 24)
Posttesting stop phase $[T, T_{lc}]$		5.7 (approximately 6)

We understand from Table 1 that early release of the software does possess the risk of more undetected bugs, but not all these bugs would result in software failure in the field. Therefore, tester's testing the software even post its release ensures that these bugs are fixed before they cause any trouble to the final users. Additionally, both testers and users are now involved in the debugging process after the software release, thereby reducing the expected number of undetected bugs once the testing stops. Hence, significantly less number of failures is expected to occur under the proposed model in comparison to the existing debugging cost models.

The overall behavior of the cost function under the two compared policies as shown in Fig. 1 demonstrates the cost-effectiveness of the proposed post-release policy. With increasing testing time, we observe that the overall cost tends to increase, but with the early release of software along with adequate warranty period, the overall cost is kept under control, and an optimal number of bugs are also removed until the testing stops. The proposed model thus helps to evaluate not only the optimal release time, testing stop time which is equivalent to warranty length, but also the optimal testing efforts to be employed.

We now tend to evaluate the sensitivity of the model parameters on optimal release and testing stop time along with optimal testing efforts in the next section.

6 Sensitivity Analysis of the Model Parameters

From the discussion given in the preceding section, it is good to know that the optimal decision-making depends on various parameters that may not be precisely estimated.

The use of sensitivity analysis will help the analyst to understand how changing the parameters of the model will affect the decision outcome. The decision model is then rerun by holding all other parameters constant. We have conducted sensitivity analysis by calculating relative change of optimal release time, testing stop time, and testing effort at both the release and testing stop time based on two very critical parameters, R_0 and α.

The sensitivity of the optimal software release time/effort with respect to a model parameter can be quantified by $\Omega^{s^*}_{p,\theta}$ and $\Omega^{u^*}_{p,\theta}$, which are the relative change of s^* , u^* when θ is changed by $100p$ percent, i.e.,

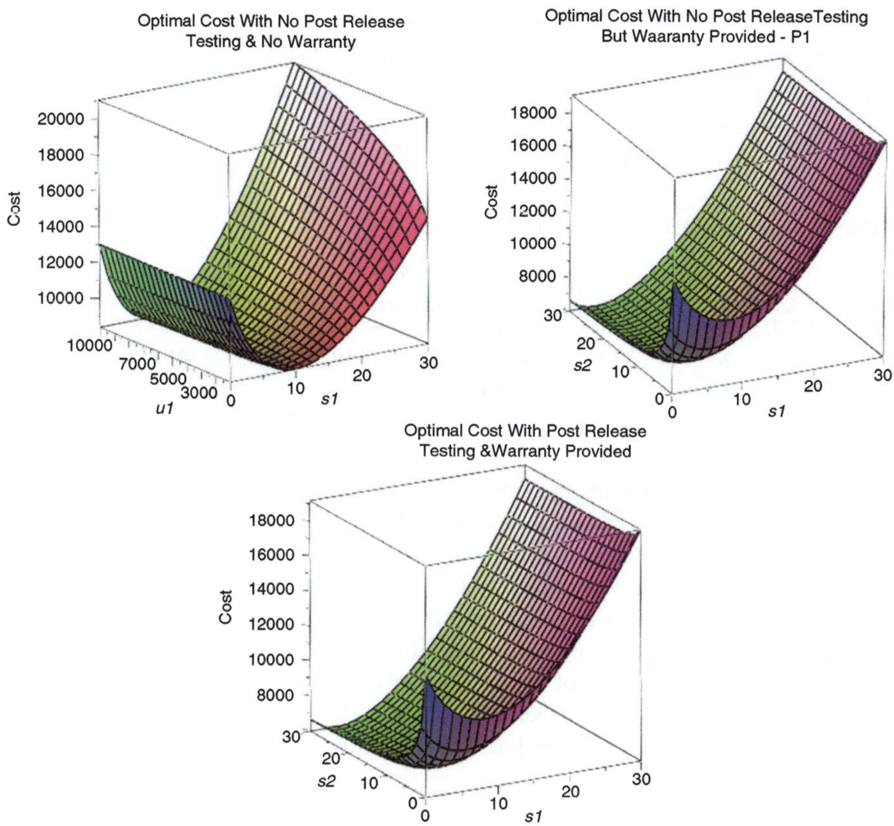

Fig. 1 Optimal cost comparison for existing versus proposed methodology

$$\Omega_{p,\theta}^{s*} = \left| \frac{s^* (\theta + p\theta) - s^* (\theta)}{s^* (\theta)} \right| \tag{20}$$

$$\Omega_{p,\theta}^{u*} = \left| \frac{u^* (\theta + p\theta) - u^* (\theta)}{u^* (\theta)} \right| \tag{21}$$

where $p\theta$ is the amount of change in positive or negative direction and θ is parameter which is selected for the change. Two sub-cases for sensitivity are considered as follows.

6.1 Case 1: Sensitivity Analysis Based on R_0

Here relative change of optimal release time, testing stop time, and respective testing efforts are calculated using Eqs. 20 and 21. The optimal software reliability under the original values of model parameters is $R_0 = 0.7$, and relative change in release

Table 4 Sensitivity analysis results based on change of R_0

$\Omega^{s^*}_{p,\theta}/ R_0$	−14.29%	+14.29%	+28.57%	$\Omega^{u^*}_{p,\theta}/R_0$	−14.29%	+14.29%	+28.57%
Ω_s	−33.33	44.68	123.12	Ω_u	−24.99	25.04	25.04
Ω_{s_w}	−31.67	18.33	33.33	Ω_{u_w}	41.27	−5.82	76.58
Ω_C	−3.9	13.28	48.75				

Table 5 Sensitivity analysis results based on change of α

$\Omega^{s^*}_{p,\theta}/\alpha$	−11.11%	−22.22%	−33.33%	−44.44%	$\Omega^{u^*}_{p,\theta}/\alpha$	−11.11%	−22.22%	−33.33%	−44.44%
Ω_s	−99.95	−99.95	−100	−100	Ω_u	−0.02	−0.33	−0.88	−1.7
Ω_{s_w}	16.67	26.67	33.33	34.33	Ω_{u_w}	17.19	18.97	41.27	59.92
Ω_C	−0.87	−1.49	−2.04	−2.04					

time and effort along with testing stop time and respective effort based on reliability change has been calculated and shown in Table 4.

It can be seen that the sensitivity of $s*$ with respect to R_0 is quite significant, e.g., when R_0 increases/decreases by 14%, and $s*$ increases/decreases by 45/33%, respectively. Further the testing stop time increases/decreases by 18/32% with 14% increase/decrease of R_0. Also the testing effort $u*$ increases/decreases by 25/24% with 14% increase/decrease of R_0, respectively. We also observe that the overall cost increases by almost 49% when R_0 is increased by 29% clearly emphasizing on the impact of attaining a desired level of reliability at the time release leading to quite high overall testing cost.

6.2 *Case 2: Sensitivity Analysis Based on α*

Relative change of optimal release, testing stop time, and respective efforts is calculated by using Eqs. 20 and 21 by changing the value of parameter α; the result is shown in Table 5. The optimal release, testing stop time, and effort to release the software are obtained at $\alpha = 0.9$.

It can be seen that the sensitivity of $s*$ with respect to α is significant, e.g., when α is reduced by 11–44%, $s*$ reduces by 99–100%. Here the testing stop time s_w* gradually increases from 16 to 34% with 11–44% reduction in values of α. Further the testing effort $u*$ does not show major effect by changing α as reduction of less than 1% is observed with 11–44 decrease in the value of α. Quite insignificant reduction in the overall testing cost, 0–2% is observed with decreasing values of α.

7 Conclusion, Managerial Implications, and Future Research Directions

Historically, testing stop time and software release time were considered to be synonymous wherein software developer's sole focus was on providing bug-free highly reliable software. With advancements in IT industry and ever-changing marketing dynamics came profound and impatient clients demanding highly efficient systems in lesser time. This implies early software release and hence more bugs still lying dormant in the system. In order to handle this situation, the concept of patch release became quite useful in which most of the software developers after the release of the software would continue testing the software and release small programs to fix those remaining bugs in the system.

Now, in existing software reliability literature, it has been invariably assumed that testing and operational phases are governed by the same debugging distribution which implies similar technical environment before and after the release of the software. With the current changing scenario, the seldom the company would employ similar testing resources post release of the software.

In this paper, we propose a novel generalized software release and testing stop time policy that helps meet the advancement challenges to a large extent. We have formulated a new cost modeling framework for obtaining optimal release time, and warranty length, testing stop time and at each time point how much testing effort is employed, in a scenario of continued testing post release of the software.

Our findings indicate that although early release of the software is quite detrimental in deciding company's future as there lies a higher risk of software failure in the field, with post-release testing policy, the risk of software failure reduces, and the chances of capitalizing on the market opportunity by early entrants ensures profitability in the long run.

Additionally, these days for software developers, it is quite common to release their product early into the market to capitalize on the opportunity presented by markets and continue to release patches later on to ensure higher software reliability. However, a crucial decision faced by developers is to optimize warranty period post release as debugging during warranty phase is often accompanied by a huge cost.

We have attempted not only to model this scenario with the help of new cost modeling approach wherein we consider both testers and users playing a significant role in optimizing total testing cost to obtain optimal testing time, warranty length, and release time of the software but also incorporated this scenario in the presence of two-dimensional modeling by considering testing effort as the second dimension of time. We made use of famous Cobb-Douglas production function from the economic theory to formulate this two-dimensional testing cost modeling framework.

For practical usage, we provide a numerical illustration for further application insights into the framework. Findings of our research work have important managerial implications for managing large-scale enterprise level software development, especially the testing process. Our results provide for sufficient proofs that when market opportunity cost is significantly high, detaching software release time from the testing stop time, developer can ensure huge cost saving in terms of early software delivery in reducing software failures in the field which indirectly reduces warranty cost and overall testing cost due to shortening of testing stop time. It also provides the developer with an opportunity to capitalize on the market by early entry due to the shortened release time and enough testing post release of the software. This flexibility can even provide firms with an ability to increase their software reliability which indirectly ensures lower client risk post release of the software along with increased customer satisfaction.

Furthermore, sensitivity analysis of important model parameters is carried out. In order for developers to attain higher software reliability, our proposed model provides for efficient trade-offs between optimal release time and testing stop time. It suggests that the optimal release time shrinks, while the testing stop time stretches for attaining higher reliability at testing stop time. However, if higher reliability is demanded at the time of release, then both software release and testing stop times are stretched.

Sensitivity analysis clearly provides intriguing insights regarding testing stop time which indirectly corresponds to warranty length a developer should provide. Depending upon the mutual agreement between developer and client, an optimal release time along with optimal warranty length can be obtained using our proposed framework.

Finally, numerous research directions can be unveiled from the proposed study. Firstly, we assume that the warranty period coincides with the testing stop time, whereas there is always a possibility to provide for warranty during the operational phase which begins at posttesting stop phase of the software. Secondly, for illustration purpose, we made use of exponential distribution to model bug detection process in order to simplify the numerical extent in the framework. This can easily be extended to obtain an analytical solution for various other distributions used in the literature by Kapur et al. [8] and Pham [25]. Finally, we considered only four major costs to model the proposed framework and assumed all other costs to be negligible. It would be interesting to see what happens if we consider say patching cost or may be some other costs in the proposed framework.

References

1. Ahmad N, Khan MGM, Rafi LS (2010) A study of testing effort dependent inflection S-shaped software reliability growth models with imperfect debugging. Int J Qual Reliab Manag 27(1):89–110
2. Cobb CW, Douglas PH (1928) A theory of production. Am Econ Rev 18(Supplement):139–165
3. Dalal SR, Mallows CL (1988) When should one stop testing software? J Am Stat Assoc 83:872–879
4. Dohi T, Kaio N, Osaki S (1997) Optimal software release policies with debugging time lag. Int J Reliab Qual Saf Eng 04(03)

5. Huang CY, Lyu MR (2005) Optimal release time for software systems considering cost, testing-effort, and test efficiency. IEEE Trans Reliab 54(4):583–591
6. Huang CY (2005) Cost-reliability-optimal release policy for software reliability models incorporating improvements in testing efficiency. J Syst Softw 77:139–155
7. Huang CY, Kuo SY, Lyu MR (1999) Optimal software release policy based on cost and reliability with testing efficiency. In: Proceedings of 23rd IEEE annual international computer software and applications conference, Phoenix, pp 468–473
8. Huang CY, Kuo SY (September 2002) Analysis of incorporating logistic testing-effort function into software reliability modeling. IEEE Trans Reliab 51(3):261–270
9. Huang CY, Kuo SK, Lyu MR (2007) An assessment of testing effort dependent software reliability growth models. IEEE Trans Reliab 56(2):198–211
10. Inoue S, Yamada S (2013) Lognormal process software reliability modeling with testing-effort. J Softw Eng Appl 6:8–14
11. Jiang Z, Sarkar S, Jacob VS (2012) Postrelease testing and software release policy for enterprise level systems. Inform Syst Res, Informs 23(3):635–657
12. Kapur PK, Garg RB, Kumar S (1999) Contribution to hardware and software reliability. World Scientific publishing Co. Pvt. Ltd., Singapore
13. Kapur PK, Pham H, Gupta A, Jha PC (2011) Software reliability assessment with OR application. Springer, Berlin
14. Kapur PK, Pham H, Singh JNP, Sachdeva N (2014) When to stop testing multi up-gradations of software based on cost criteria. Int J Syst Sci Oper Logist 1(2):84–93
15. Kapur PK, Goswami DN, Bardhan A (2007) A general software reliability growth model with testing effort dependent learning process. Int J Model Simul 27(4):340–346
16. Kapur PK, Ompal S, Aggarwal AG, Kumar R (2009) Unified framework for developing testing effort dependent software reliability growth models. WSEAS Trans Syst 4(8):521–531
17. Kapur PK, Sachdeva N, Singh JNP (2014) Optimal cost- a criterion to release multiple versions of software. Int J Softw Assur Eng Manage 5(2):174–180
18. Kapur PK, Garg RB (1991) Optimal software release policies for software systems with testing effort. Int J Syst Sc 22(9):1563–1571
19. Kapur PK, Agarwal S, Garg RB (1994) Bi-criterion release policy for exponential software reliability growth models. Recherche Oper/Oper Res 28:165–180
20. Kapur PK, Anand A, Sachdeva N (2014) Profit estimation for a product under warranty: an evaluation based on Customer's expectation and satisfaction. Int J Reliab Qual Software Eng 21(6):1–16
21. Kapur PK, Kumar U, Sachdeva N, Anand A (2012) Optimal price & warranty length. Commun Dependability Qual Manag Int J 15(1):5–13
22. Kapur PK, Sachdeva N (2015) Managing warranty length & price in the presence of customer dissatisfaction. Amity Global Bus Rev Amity University 10(1):64–74
23. Kuo SY, Huang CY, Lyu MR (2001) Framework for modeling software reliability, using various testing-efforts and fault-detection rates. IEEE Trans Reliab 50(3):310–321
24. Li Q, Li H, Lu M (2015) Incorporating S-shaped testing-effort functions into NHPP software reliability model with imperfect debugging. Journal of Syst Eng and Elec 26(1):190–207
25. Lin CT, Huang CY (2008) Enhancing and measuring the predictive capabilities of testing effort dependent software reliability models. J Syst Software 81:1025–1038
26. Musa JD (2004) Software reliability engineering: more reliable software, faster and cheaper
27. Sachdeva N, Singh O, Kapur PK (2015) Optimal launch of a new generation of technology: a multi-attribute approach discrete time diffusion process. Int J Technol Market 10(3):3–23
28. Okumoto K, Goel AL (1980) Optimum release time for software systems based on reliability and cost criteria. J Syst Software 1:315–318
29. Singh O, Kapur PK, Shrivastava AK, Kumar V (2015) Release time problem with multiple constraints. Int J Syst Assur Eng Manage 6 (1):83–91
30. Kapur PK, Pham H, Aggarwal AG, Kaur G (2012) Two-dimensional multi-release software reliability modeling and optimal release planning. IEEE Trans Reliab 61(3):758–768

31. Kapur PK, Khatri SK, Singh O, Sachdeva N (2015) "Multi criteria based optimal launch time of a new product" quality, reliability and infocomm technology and industrial technological management. I.K. International Publishing House Pvt. Ltd., pp 289–306
32. Peng R, Li YF, Zhang WJ, Hu QP (2014) Testing effort dependent software reliability model for imperfect debugging process considering both detection and correction. Reliab Eng Syst Saf 126:37–43
33. Pham H (2006) System software reliability. Springer, London
34. Pham H (2006) System software reliability. Reliability engineering series, Springer
35. Pham H, Zhang X (1999) A software cost model with warranty and risk costs. IEEE Trans Comp 48(1):71–75
36. Sachdeva N, Kapur PK, Singh O (2016) Generalized framework for optimal pre & post release software testing in presence of warranty. Accepted International Journal of Procurement Management
37. Singh J, Singh O, Aggrawal D, Anand A, Singh I (2012) A flexible reliability growth model for various releases of software under the influence of testing resources. J Pure Appl Sci Technol NLSS 2(2):23–35
38. Singpurwalla N, Wilson S (1994) Software reliability modeling. Internat Statist Rev 62(3):289–317
39. Wood A (1996) Predicting software reliability. IEEE Comput 11:69–77
40. Yamada S, Osaki S (1987) Optimal software release policies with simultaneous cost and reliability requirements. Eur J Oper Res 31:46–51
41. Yamada S, Ohtera H, Narihisa H (1986) Software reliability growth models with testing effort. IEEE Trans Reliab 35(1):19–23
42. Zhang C, Cui G, Liu H (2014) A unified and flexible framework of imperfect debugging dependent SRGMs with testing-effort. J Multimed 9(2)

Technological Capabilities Impacting Business Intelligence Success in Organisations

Rekha Mishra and A. K. Saini

1 Introduction

Business intelligence (BI) is tools, technology and a concept that assist businesses in managing large amounts of information to identify and develop new opportunities and enable decision-making. BI is an approach that makes use of technology to assist the organisation and achieve their long-term goals by aligning the business with the organisation strategy. Rud, Olivia [1] gave a wholesome definition for business intelligence that highlights this blend of technological and organisational perspective. They defined BI as a set of theories, methodologies, processes, architectures and technologies that transform raw data into meaningful and useful information for business purposes. BI can handle large amounts of information to help identify and develop new opportunities. Making use of new opportunities and implementing an effective strategy can provide a competitive market advantage and long-term stability.

Though penetration of BI is still low, BI-driven businesses have shown high performance, and therefore it is considered a high priority for many companies across the globe. Incidentally, Gartner's [2] report came out with their research survey that showed that less than 30% of the BI project are successful, implying that the rate of success of BI project is very low. This suggests that existing BI success model fails to serve the purpose, and hence there is need for more research. The purpose of this research is to develop an update instrument to assess how technological capabilities impact success of business intelligence in an organisation, as having the right technological BI capabilities is important for an organisation to realise maximum benefits from its BI investment.

R. Mishra (✉) • A.K. Saini
University School of Management Studies (USMS), Guru Gobind Singh Indraprastha University, New Delhi, India
e-mail: rekhareflection@gmail.com; aksaini1960@gmail.com

© Springer Nature Singapore Pte Ltd. 2018
P.K. Kapur et al. (eds.), *Quality, IT and Business Operations*, Springer Proceedings in Business and Economics, DOI 10.1007/978-981-10-5577-5_32

2 Literature Review

Business intelligence success has been defined in terms of net benefit gained from BI capabilities which have been studied from organisational and technological perspectives, but still organisations fail to achieve BI success. For simplification in this study, we have focused only on understanding and analysing technological capabilities impacting business intelligence success in organisations today. We explored BI research literature spanning over the last 15 years to assess the key technological factors impacting BI success. The following review is an outcome of this analysis.

When it comes to technological factors impacting BI success, a majority of literature have focused on data and content quality. Some recent studies like, for example, by Abdinnour-Helm et al. [3], emphasised on the need for interaction among subsystems, for example, integrating business intelligence and knowledge management to deal with challenges of the modern-day systems. Bill Hostmann [4] work discussed the need to have timely access to the relevant, reliable information to enable the business manager make every day's decisions since access, consistency and quality of information is a must for information system success. Chung and Nunamaker's [5] work explored the role of visual framework to ease information overload and offer practical implications to information system users. Deng and Chi [6] study developed a comprehensive and dynamic view of system to solve problems in organisations in regard to system use by using seven emergent constructs of system use problems in the area of reporting, data, workflow and role authorization, users' lack of knowledge, system error and user–system interaction. Isik [7] research provided a better understanding of BI success by proposing a framework that examines the impact of technological capabilities on BI success, in the presence of different decision environments. McKeen et al. [8] work provided insight on how to identify and implement effective IT capabilities. The result is a five-step framework for enhancing IT capabilities. Mircea et al. [9] paper presents aspects of the mix of BI and cloud computing, putting the stress on the integration of a cloud BI solution within organisations. An ROI indicator was also used to assess the advantage of cloud BI over non-cloud BI. Popovič et al. [10] proposed a model of the relationship between business intelligence systems and information quality and also investigated impact of business intelligence systems' maturity on the quality of content and media quality. Popovič et al. [11, 12] emphasised on understanding how business intelligence system (BIS) dimensions are interrelated and how they affect BIS use. BIS maturity is determinant of information access quality and information content quality. But information content quality is more significant as compared to information access quality. Pretorius and Wijk [13] article proposed that while designing information visualisation techniques, one needs to focus on the data, as new awareness is acquired about end user requirements and simultaneously more new requirements for end user are known with this approach. Ramakrishnan [14] paper explores the factor that governs organisation BI goals and their data collection strategy. It also provides them with a model to support decision-making. Sangar and Iahad [15] study pointed out that many BIS implementations are not successful because they are time-consuming and expensive. They also proposed a framework for critical success factor for BI success based on project implementation life cycle. Watson [16] paper discussed four important BI trends that have a role to play in BI success. These trends were scalability, pervasive BI, operational BI and the BI-based organisation.

The technological capabilities contributing to BI success studied in the researches seem to suggest that they can be expanded to redefine grouping of the constructs used to define BI success model.

3 Research Gap, Theoretical Framework and Hypothesis

What is the impact of technological capability on business intelligence success in organisation today? From the literature review it appears that there is a lack of consensus on how to achieve success with BI, and a consistent model, explaining the BI success has not yet emerged. This suggests that there are gaps in the research to be filled and that researches have perhaps overlooked one or more key constructs for BI success model. Hence a research needs to be undertaken to provide a better understanding of BI success by examining the impact of technological capabilities on BI success in the current scenario.

Following the analysis of prior studies for BI success, an update model in terms of revised constructs for BI success is proposed as shown in Fig. 1. Business intelligence success is the dependent variable, and technological capability is the independent variable. The assumption is that technological capability determines business intelligence success in an organisation. Also since this study explores the role of technological capability only in BI success, all the other capabilities are assumed to be standard for study purpose.

Following the review of latest literature, we tabulated various technological capability constructs attributed to BI success. From these constructs, we identified those constructs which are pointed out as the fundamental for success of BI systems in the researches done till now but perhaps not put together or grouped and explored simultaneously to understand their impact on system success.

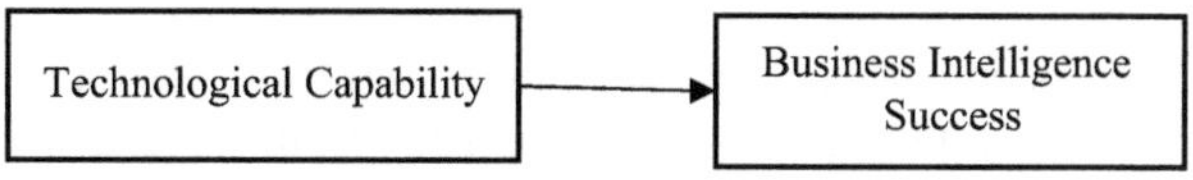

Fig. 1 Theoretical model for BI success (Source: Literature review)

Table 1 Construct: technological capability

Variable	Source
Integration	Abdinnour-Helm et al. [3]
Visual framework	Chung and Nunamaker [5]
Reporting, data, workflow and role authorisation, users' lack of knowledge, system error, user–system interaction	Deng and Chi [6]
User access	Eckerson [17]
Access, consistency, quality of information, intuition involved, interaction with other systems	Hostmann [4]
Data source, data reliability, data type, information system interaction, user access	Isik [7]
IT capability framework	McKeen et al. [8]
Cloud BI	Mircea et al. [9]
Content quality	Popovič et al. [10]
Information access quality	Popovič et al. [11]
Effective visualisation	Pretorius and Wijk [13]
Data collection strategies	Ramakrishnan et al. [14]
Implementation life cycle	Sangar and Iahad [15]
Scalability, pervasive BI, operational BI	Watson [16]

Source: Literature review

Table 2 Construct: business intelligence success

Variable	Sources
Decision support	Isik [7]
EUCS	Hou [18]
System quality, information quality, service quality, use, user satisfaction and net benefits	Petter et al. [19]
Information quality, user-friendly, overall satisfaction	Søilen [20]

Source: Literature review

3.1 Technological Capabilities

Table 1 is an exhaustive list of construct identified from literature review. For this study the technological capability has been defined using Isik [7] scale.

3.2 BI Success

Various constructs have been identified from literature review to operationalise the business intelligence success as shown in Table 2. BI success was defined using the following constructs, decision-making, information timeliness, information precision, user-friendly and overall satisfaction drawn from Søilen [20] scale.

The following hypotheses have been formulated to test impact of technological capability in achieving BI success in organisation.

H1a: There is no relationship between the data type and BI success in an organisation.
H1b: There is no relationship between the data source and BI success in an organisation.
H1c: There is no relationship between the data reliability and BI success in an organisation.
H1d: There is no relationship between the system interaction and BI success in an organisation.
H1e: There is no relationship between the user access and BI success in an organisation.

4 Research Methodology

To find the items contributing to constructs of technological capability and business intelligence success, an adapted scale was developed. To check the reliability of the scale, an online survey was created using Google forms. The survey respondents were end users and solution developer for BI systems from middle and executive management and having experience varying from 5 years to less than 25 years and having different educational backgrounds and hailing from different industries. These respondents were selected from using author's contacts list on LinkedIn. A list of 150 people who fit the above-mentioned criteria was drawn from the contact list. The online questionnaire, shown in Annexure, was emailed to all the people on the list to collect data.

The respondents were asked to indicate their organisation type, work experience, education level, level in management and type of industry. Information was also sought on various technological capabilities they thought impacted BI success in their organisation. Information was also sought on their definition of BI success. A five-point Likert scale-based questionnaire was developed following literature review for carrying out the survey. The literature suggests that five-point scale appears to be less confusing, and to increase response rate also with a five-point scale, it is quite simple for the interviewer to read out the complete list of scale descriptors [21].

5 Finding and Discussion

For confirming validity and reliability of the scale, a pilot study was conducted. To find questionnaire reliability, we measured the internal consistency, which are the most popular methods of estimating reliability. Cronbach's alpha or internal consistency test was used [22]. Mathematically, internal consistency is the average of all possible split-half correlations. Conceptually, it measures how well the items function together, that is, do people respond consistently with their standing on the construct of interest or not.

A total of 35 responses were received and were usable. The response rate for our study was 23% as only 35 respondents out of 150 replied to our survey. When the sample size is small, the Cronbach's alpha coefficients were reliable provided the first eigenvalue of the principal component analysis was greater than 6 Yurdugül [23]. Using this approach, the Cronbach's alpha value was found to be 0.79 which is good since a minimum alpha of 0.6 sufficed for early stage of research.

The sample respondents were mainly from IT/ITeS, banking and retail industry. 80% of the respondents were from private organisation, 17% were from government and 3% self-employed. 60% of the respondents were from middle management, while 40% were from executive management. The demographic profile of the respondents is shown in Table 3.

Business intelligence success was measured using Søilen [20] work. The construct comprised of five items. This is shown in Table 4.

Technological capability was measured using Isik's [7] work, as shown in Table 5.

Table 3 Respondent profile

Demographic characteristics	Percentage
Organisation type	
Private	80
Government	17
Self-employed	3
Work experience	
5 yr. to less than 10 yr	10
10 yr. to less than 15 yr	14
15 yr. to less than 20 yr	14
20 yr. to less than 25 yr	10
Educational level	
Graduate	10
Postgraduate	45
Professional	40
PhD.	5
Level in management	
Executive management	40
Middle management	60
Type of industry	
IT/ITeS	60
Retail	5
Finance	30
Others	5

Source: Primary data analysis

Table 4 Success construct

Construct	Item	Code
BI success	Supports for decision-making	BIS1
	Precise information	BIS2
	Timely information	BIS3
	User-friendly	BIS4
	Overall satisfaction	BIS5

Source: Literature review

Table 5 Technological capability construct

Construct	Item	Code
Data reliability	Internal data reliability	DR1
	Internal data conflict	DR2
	Internal data accuracy	DR3
	Update internal data	DR4
	Reliable external data	DR5
	External data conflicts	DR6
	Accuracy of external data	DR7
	Update external data	DR8
Data source	Accessible internal data source	DS1
	Usable internal data source	DS2
	Comprehensible internal data sources	DS3
	Concise internal data sources	DS4
	Available external data sources	DS5
	Usable external data sources	DS6
	Comprehensible external data sources	DS7
Data type	Accurate quantitative data	DTy1
	Comprehensive quantitative data	DTy2
	Consistent quantitative data	DTy3
	High-quality quantitative data	DTy4
	High-quality qualitative data	DTy5
	Accurate qualitative data	DTy6
	Comprehensive qualitative data	DTy7
	Consistent qualitative data	DTy8
System interaction	Unified view of business data and processes	IS1
	Links among multiple business applications	IS2
	Enterprise information resources information	IS3
	Seamless access to data, applications and systems	IS4
User access	Quality access	UA1
	Access authorisation	UA2

Source: Literature review

Table 6 Descriptive statistics: BI success

Code	Mean	Standard deviation
BIS1	3.9	0.5
BIS2	3.9	0.9
BIS3	3.7	0.7
BIS4	3.2	1.1
BIS5	3.7	0.8

Source: Primary data analysis

Table 7 Descriptive statistics: data reliability

Code	Mean	Standard deviation
DR1	3.5	1.1
DR2	3.3	1.0
DR3	3.7	0.8
DR4	3.9	0.9
DR5	3.5	0.8
DR6	3.4	0.7
DR7	3.4	0.8
DR8	3.2	1.0

Source: Primary data analysis

Likert-type scales are useful when you are measuring latent constructs, that is, characteristics of people such as attitudes, feelings, opinions, etc. Latent constructs are generally thought of as unobservable individual characteristics that are believed to exist and cause variations in behaviour. A single item with this scoring is ordinal, so not a good choice for the kind of statistical analyses that require a numerical or interval-level variable. Likert scale is summing together several Likert-type variables to create a scale that can produce an interval-level variable. Since we have a series of Likert-type questions that when combined describe a trait or attitude, we use means and standard deviations to describe the scale and get an assessment about respondent opinions.

As shown in the Table 6 for BI success which is measured using Likert-type scale, the median for all items is found to be 3.7 which corresponds to overall satisfaction with BI. And rightly so as has been shown in literature, business intelligence success is dependent on various factors which are specific or peculiar to the industry, BI maturity and management level in organisation. Hence BI success cannot be generalised; overall satisfaction seems to be a better measure of BI success.

The remaining constructs are measured using a Likert scale, as shown in Table 7 for data reliability. The mean for items was found to be 3.5 which correspond to internal data reliability and external data reliability and hence implies the better measure for data reliability construct.

As shown in Table 8, for data source, the mean for all the items was found to be 3.1 which correspond to concise internal data source. Hence concise internal data source is a better measure of data source.

As shown in Table 9, for data type, the mean for all the items was found to be 3.4 which correspond to item accurate qualitative data and item consistent qualitative data. Hence these items are the better measures of data type.

Table 8 Descriptive statistics: data source

Code	Mean	Standard deviation
DS1	3.9	0.8
DS2	3.2	0.7
DS3	3.3	1.1
DS4	3.1	1.1
DS5	2.8	1.4
DS6	2.7	1.2
DS7	2.7	1.1

Source: Primary data analysis

Table 9 Descriptive statistics: data type

Code	Mean	Standard deviation
DTY1	3.6	1.0
DTY2	3.6	0.8
DTY3	3.9	0.7
DTY4	3.1	1.0
DTY5	3.0	1.2
DTY6	3.4	0.9
DTY7	3.1	1.0
DTY8	3.4	1.0

Source: Primary data analysis

Table 10 Descriptive statistics: system interaction and user access

Code	Mean	Standard deviation
IS1	3.0	1.1
IS2	3.1	1.4
IS3	2.8	1.2
IS4	2.8	1.3
UA1	3.7	0.9
UA2	4.0	1.0

Source: Primary data analysis

As shown in Table 10, for system interaction the mean for system interaction items was found to be 2.9 which tend towards item unified business data and processes. Hence a better measure of system interaction variable is unifies business data and processes.

Also as shown in Table 10, for user access items, the mean was found to be tending towards 4 which correspond to item satisfaction with user access. Hence a better measure of user access items variable is satisfaction with user access.

As seen from the data in various tables, there is very little variance, with all responses at either 3 or 4 is observed, implying standard OLS regression would not be an appropriate analysis method. More complex techniques, such as logistic regression, would be better. But due to small sample size, it is not feasible to carry out this analysis. So, in our case the data from a survey are analysed with the

Table 11 Spearman correlation coefficient analysis

Dependent variable		BIS3	DS1	DTy3	DR4	UA1	IS2
BIS3	Correlation coefficient	1.000	0.411[a]	0.218	0.355[a]	0.188	0.206
	Sig. (two-tailed)		0.014	0.208	0.037	0.280	0.235

Source: Primary data analysis
[a]Correlation is significant at the 0.05 level (two-tailed)

objective of explaining the association in a dependent variable which is business intelligence success. We wish to find out what other variables might be strongly related to higher or lower scores on the dependent variable. The Likert scale-based survey data was found to be not normally distributed. To measure the association among variables, Spearman correlation coefficient analysis was chosen. The results were shown in Table 11. Data source and data reliability were found to have significant role in overall satisfaction with BI in organisation.

6 Limitation and Further Work

For simplification, this study assumed the role and impact of other capabilities and dimensions to be standard while exploring the role of various technological capabilities in BI success. Another limitation is that respondents were selected from author's contact list only on LinkedIn, and these respondents came from diverse industry and background. Perhaps a larger sample size and from one industry or sector only would have provided a more accurate assessment of the situation.

7 Conclusion

The study provides an insight to all managers at various levels in organisation to better understand the technological capability requirements and best practices for attaining BI success and leverage the benefits it provides. The findings also help clarify the definition of BI success as per the current business requirements across the globe. The pilot survey helped develop a valid and reliable instrument for further research. The finding will serve as road map for software development team developing BI solutions for organisations, and for BI researches it provides a knowledge base from where they can take up further empirical research for analysis, and for project managers and BI solution developers, it serves as a guide to formulate an effective and solution development strategy.

Annexure

Questionnaire

1. Highest education level you have completed?	4. What is your level in the organisation?
Graduate degree	Executive management
Postgraduate degree	Middle management
PhD	Operational management
	Other (please specify) _______
2. What is your organisation type?	5. Which below best describes your industry?
Private	Manufacturing
Government	Finance
Self-employed	IT/ITeS
	Education
3. What is your work experience?	Wholesale and retail trade
5 year to less than 10 year	Transportation
10 year to less than 15 year	Banking
15 year to less than 20 year	Manufacturing
20 year to less than 25 year	Utilities
	Government
	Insurance
	Other (please specify) ______

Please choose the response which best describes your satisfaction with each of the following:

BI success	Strongly dissatisfied	Somewhat dissatisfied	Neither satisfied or dissatisfied	Somewhat satisfied	Strongly satisfied
The BI that you are using supports your decision-making					
The BI that you are using provides precise information you need					
The BI that you are using provides information you need in time					
The BI that you are using is user-friendly					
Overall satisfaction with BI that you are using overall					

Please choose the response that best describes each of the following statements:

Data sources	Strongly disagree	Somewhat disagree	Neither agree nor disagree	Somewhat agree	Strongly agree
The internal data sources used for your BI are readily available					
The internal data sources used for your BI are readily usable					
The internal data sources used for your BI are easy to understand					
The internal data sources used for your BI are concise					
The *external* data sources used for your BI are readily available					
The *external* data sources used for your BI are readily usable					
The *external* data sources used for your BI are easy to understand					

Please choose the response that best describes each of the following statements:

Data types	Strongly disagree	Somewhat disagree	Neither agree nor disagree	Somewhat agree	Strongly agree
Your BI provides accurate quantitative data					
Your BI provides comprehensive quantitative data					
Your BI provides consistent quantitative data					
Your BI provides high-quality quantitative data					
Your BI provides high-quality *qualitative* data					
Your BI provides accurate *qualitative* data					
Your BI provides comprehensive *qualitative* data					
Your BI provides consistent *qualitative* data					

Please choose the response that best describes each of the following statements:

Data reliability	Strongly disagree	Somewhat disagree	Neither agree nor disagree	Somewhat agree	Strongly agree
Internal data collected for your BI is reliable					
There are inconsistencies and conflicts in the internal data for your BI					
Internal data collected for your BI is accurate					
Internal data for your BI is updated regularly					
External data collected for your BI is reliable					
There are inconsistencies and conflicts in the *external* data for your BI					
External data collected for your BI is accurate					
External data for your BI is updated regularly					

Please choose the response that best describes each of the following statements:

User access	Strongly disagree	Somewhat disagree	Neither agree nor disagree	Somewhat agree	Strongly agree
Are you satisfied with the quality of the way you access your BI					
The way you access your BI is fits well to the types of decisions you make using my BI					

Please choose the response that best describes each of the following statements:

Interaction with other systems your BI provides...	Strongly disagree	Somewhat disagree	Neither agree nor disagree	Somewhat agree	Strongly agree
Your BI provides a unified view of business data and processes					
Your BI provides links among multiple business applications					
Your BI provides a comprehensive electronic catalogue of the various enterprise information resources in the organisation					
Your BI provides easy and seamless access to data from other applications and systems					

References

1. Rud OP (2009) Business intelligence success factors: tools for aligning your business in the global economy (Vol. 18). Wiley, Hoboken
2. Gartner (2012) Gartner Says Fewer Than 30 Percent of Business Intelligence initiatives Will Align Analytic Metrics Completely With Enterprise Business Drivers by 2014. Retrieved from http://www.gartner.com/newsroom/id/1891515
3. Abdinnour-Helm S, Lengnick-Hall ML, Lengnick-Hall CA (2003) Pre-implementation attitudes and organizational readiness for implementing an enterprise resource planning system. Eur J Oper Res 146(2):258–273
4. Hostmann B (2007), BI competency Centres: bringing intelligence to the business, business performance management, Gartner, November 2007
5. Chung C, Nunamaker N (2005) A visual framework for knowledge discovery on the web: an empirical study of business intelligence exploration. J Manag Inf Syst/Spring 21(4):57–84
6. Deng X, Chi L (2012) Understanding post adoptive behaviours in information systems use: a longitudinal analysis of system use problems in the business intelligence context. J Manag Inf Syst 29(3):291–326
7. Isik O (2009) Business intelligence success: an empirical evaluation of the role of BI capabilities and organization's decision environment. AMCIS 2009 Doctoral Consortium, 13
8. McKeen JD, Smith HA, Singh S (2005) Developments in Practice XVI: A Framework for Enhancing IT Capabilities. Commun Assoc Inf Syst 15:Article 36
9. Mircea M, Ghilic-Micu B, Stoica M (2011) Combining business intelligence with cloud computing to delivery agility in actual economy. J Econ Comput Econo Cybernetics Stud 45(1):39–54
10. Popovič A, Coelho PS, Jaklič J (2009) The impact of business intelligence system maturity on information quality. Inf Res 14(4)

11. Popovič A, Hackney R, Coelho PS, Jaklič J (2012) Towards business intelligence systems success: effects of maturity and culture on analytical decision making. Decis Support Syst 54(1):729–739
12. Popovič A, Turk T, Jaklič J (2010) Conceptual model of business value of business intelligence systems. Manag J Contemp Manag Issues 15(1):5–30
13. Pretorius A, Wijk J (2009) What does the user want to see? What do the data want to be? Inf Vis 8:153–166
14. Ramakrishnan T, Jones MC, Sidorova A (2012) Factors influencing business intelligence (BI) data collection strategies: an empirical investigation. Decis Support Syst 52(2):486–496
15. Sangar A, Iahad N (2013) Critical factors that affect the success of Business Intelligence Systems (BIS) implementation in an organization. Int J Sci Technol Res 2(2)
16. Watson HJ (2009) Tutorial: business intelligence – past, present, and future. Commun Assoc Inf Syst 25, Article 39
17. Eckerson (2003) Smart companies in the 21st century use business intelligence (BI) solutions, The Data Warehousing Institute (TDWI), 2003
18. Hou C (2012) Examining the effect of user satisfaction on system usage and individual performance with business intelligence systems: an empirical study of Taiwan's electronics industry. Int J Inf Manag 32(6):560–573
19. Petter S, DeLone W, McLean E (2008) Measuring information systems success: models, dimensions, measures, and interrelationships. Eur J Inf Syst 17:236–263
20. Søilen KS (2012) An evaluation of business intelligence software systems in SMEs–a case study. J Intell Stud Bus 2(2)
21. Dawes JG (2008) Do data characteristics change according to the number of scale points used? An experiment using 5 point, 7 point and 10 point scales. Int J Mark Res 51(1)
22. Nunnally JC, Bernstein IH (1994) Validity. Psychometr Theory. 99–132
23. Yurdugül H (2008) Minimum sample size for Cronbach's coefficient alpha: a Monte-Carlo study. Hacettepe Üniversitesi Eğitim Fakültesi Dergisi 35(35)

Testing Time and Effort-Based Successive Release Modeling of a Software in the Presence of Imperfect Debugging

Vijay Kumar, P. K. Kapur, Ramita Sahni, and A. K. Shrivastava

1 Introduction

In the present connected world, software has become pervasive in every way of life; in our culture and in our daily routine. There has been an incomparable growth in the availability and accessibility of technology among mankind. Due to the rising dependence on software systems, it becomes a tedious task for the software firms to keep up the pace with the demands flowing from every corner of the country. Software systems are being quickly updated with latest features keeping in mind the requirements of the users. The most beneficial way to survive in this revolutionized competitive environment is to develop the software in different phases and to release the software in multiple releases. Launching software in various releases offers quite a lot of advantages to the firms and as well as to the customers. On the other hand, this process is highly complicated as there is a risk of new release containing a bug, thereby increasing the fault content in the software. Hence, at times upgrades of software cause the program to break down.

Since human beings are majorly involved in the debugging process, the possibility of imperfection cannot be denied in reality. Testing, the most critical phase,

V. Kumar (✉)
Department of Mathematics, Amity School of Engineering and Technology,
New Delhi, 110061, India
e-mail: vijay_parashar@yahoo.com

P.K. Kapur • A.K. Shrivastava
Amity Center for Interdisciplinary Research, Amity University, Noida, UP, India
e-mail: pkkapur1@gmail.com; kavinash1987@gmail.com

R. Sahni
Department of Applied Mathematics, Amity Institute of Applied Sciences,
Amity University, Noida, India
e-mail: smiles_ramita@yahoo.co.in

© Springer Nature Singapore Pte Ltd. 2018

P.K. Kapur et al. (eds.), *Quality, IT and Business Operations*, Springer Proceedings in Business and Economics, DOI 10.1007/978-981-10-5577-5_33

focuses primarily on detection and removal of dormant faults in the software. Testing team puts best efforts and adopts well-planned strategies to ensure error-free software. But, no software can be made 100% free from bugs. During the fault correction process, some faults remain undetected in the testing process, some detected faults are not repaired completely, and some new faults are added during the removal of detected faults. The above situations cause imperfect debugging in the software. Another situation may arise, in which an error gets replaced by another. This situation causes error generation in the software. Error generation in software is more hazardous than imperfect debugging process as in former the fault content increases and in latter the fault content remains unaltered. Imperfect debugging and error generation results in the software are due to insufficient knowledge of the testing team.

A two-dimensional two-stage model for fault detection and fault correction process under the effect of two types of imperfect debugging has been proposed in this paper. Cobb-Douglas Production Function has been incorporated to portray the testing resources and time for the multiple software releases. The faults detected in the operational phase of the previous release or left incomplete are incorporated in the next release. The paper is structured as follows: In Sect. 2, review of literature is depicted. In Sect. 3, the assumptions and notations for the proposed modeling are depicted. Section 4 briefs about the unification scheme for two-stage SRGM (Software Reliability Growth Model) under imperfect debugging. This section further describes the two-dimensional two-stage modeling of the proposed models. The data validation of the proposed models and the parameter estimation results along with comparison criteria are discussed in Sect. 5, and finally the conclusion is presented in Sect. 6.

2 Literature Review

In the past years, software reliability modeling has gained a lot of importance. Several researchers and scientists have developed, studied, and applied many SRGMs to measure and predict the reliability of the software. A remarkable attempt was made in this field by Goel and Okumoto [1]. They developed a model with failure detection rate based on a nonhomogeneous Poisson process (NHPP). Schneidewind [2] was another pioneer who developed a model with time-dependent failure rate for single release. Lately, abundant research has been carried out by taking into consideration only a single release of the software and by assuming that development activity is based only on one factor, either testing time or testing effort. Goel and Okumoto [1], Musa et al. [3], and Ohba [4] created models for single release of the software considering only testing time as the criterion in the software reliability growth process. Researchers like Yamada et al. [5] and Huang and Kuo [6] considered testing effort as an alternative approach in the software reliability growth process. In recent times, Kapur et al. [7–9] and Kumar and Sahni [10] adopted a new technique using optimal control theory to study the behavior of allocation of resources for a single release software model. Kumar et al. [11, 12] used optimal control theory to optimize the allocation of limited resources and used genetic algorithm to determine the optimum value of the testing efforts for a single release software model.

In this technology-driven society, there is a strong need of upgrading the software and releasing software in different versions. A multi-upgradation model was proposed by Kapur et al. [13]. They considered testing phase to be a single-stage process in a perfect debugging environment. No time lag between detection and correction process was assumed. They also considered the cumulative faults in a release are dependent on all the earlier releases. However, in practical situation, there exists a time gap between the fault detection and fault correction process. Schneidewind [2] initiated the concept of modeling fault correction process as a two-stage process with a constant delay in time between detection and correction process. Further, Xie et al. [14] described modeling of fault correction process with delayed detection process with a deterministic delay. All detection-correction modeling discussed in literature is based on single release of the software. In our paper, the testing process is considered to be a two-stage process of fault detection and fault correction process in a two-dimensional multi-release framework.

In the last five decades, many SRGMs were originated in a perfect debugging environment. In reality, this debugging process is imperfect in nature. The idea of imperfect debugging was originated by Goel [15]. The model proposed by Ohba and Chou [16] is based on error generation applied to Goel and Okumoto [1] model with an exponential fault detection rate. This model is referred as imperfect debugging model. Peng et al. [17] proposed a framework considering imperfect debugging phenomenon to develop testing effort-dependent fault detection and fault correction process models. Kapur et al. [18] developed two general frameworks for deriving several software reliability growth models based on NHPP in the presence of error generation and imperfect debugging. Pachauri et al. [19] described SRGM with Weibull testing effort function in imperfect debugging environment with constant and time-varying fault detection rates, respectively. Singh et al. [20] developed a two-stage fault detection and correction model under the effect of two types of imperfect debugging for multiple releases of the software. Kapur et al. [21] proposed a multi-upgradation model in imperfect debugging environment. Wang et al. [22] considered a log logistic distribution fault content function in an imperfect debugging model. Kapur et al. [23] presented a two-stage detection-/correction-based model with testing effort incorporating imperfect debugging in multi upgradation of software. They assumed exponential distribution as detection process and logistic distribution function as correction process.

Earlier, one-dimensional model was developed with respect to either testing time or testing effort, but nothing fruitful was done by taking together the combined effect of testing time and resources. However, in the most recent time, SRGMs are integrating the effect of both the time of testing and resources on the reliability of the software. Ishii and Dohi [24] tested the dependence of testing time as a testing effort. Inoue and Yamada [25, 26] used two-dimensional Weibull-type SRGM to study reliability assessment. Recently Kapur et al. [27] presented a two-dimensional mathematical structure for multiple release model considering the remaining fault content of the previous release as well as the new faults introduced. Further they assumed that the fault eradication process is governed both by testing time and testing resources. For this they have used well-known Cobb-Douglas function to examine the behavior of testing time and resources consumption. A two-dimensional model was also proposed by Inoue et al. [28].

3 Assumptions and Notations for the Proposed Modeling

3.1 Assumptions

1. Using an NHPP, the fault detection and fault removal process are modeled.
2. Software failure is caused during execution due to the remaining faults present in the software.
3. A time lag is observed between detection and correction process.
4. The faults remaining in the software affect the failure rate equally.
5. On observation of software failure, an immediate correction process begins, and the following may occur:

   (a) Fault amount is decreased by 1 with the probability p.
   (b) Fault amount remains the same with probability $1 - p$.

   This assumption depicts the effect of imperfect debugging.

6. With a constant probability, new faults are produced during fault correction process in the software irrespective of the fact that faults are removed successfully or not.

   The effect of error generation is depicted by assumption 6.

3.2 Notations

$m_i(\tau)$: Expected number of faults removed by time τ for ith release ($i = 1$ to 4)
$f(\tau), g(\tau)$:Probability density function
$F(\tau), G(\tau)$: Distribution function for fault detection and fault correction process
τ_{i-1}: Time for ith release ($i = 2$ to 4)
a_i:Initial fault content for ith release ($i = 1$ to 4)
a: Total initial fault content
$\otimes$: Stieltjes convolution
$*$:Convolution operator
b_i: Fault detection rate for ith release ($i = 1$ to 4)
β, b: Constant

p_i: Probability of perfect debugging for ith release ($i = 1$ to 4)
α_i: Error generation rate for ith release ($i = 1$ to 4)
s: Testing time
u: Resources
$m(s, u)$:Cumulative number of faults removed by time s and with the usage of resources u
p= Probability of perfect debugging
α= Error generation rate

4 Modeling Development

4.1 Unification Modeling in Software Reliability

The one-stage unified modeling is proposed by Kapur et al. [21] and is described as

$$m(\tau) = aF(\tau) \tag{1}$$

where $F(\tau)$ is the failure time distribution.

Equation (1) can be written as

$$\frac{dm(\tau)}{d\tau} = \frac{f(\tau)}{1 - F(\tau)}[a - m(\tau)] \tag{2}$$

where $\frac{f(\tau)}{1-F(\tau)}$ is one-stage failure detection/correction rate.

$[a - m(\tau)]$ denotes the expected number of faults remaining in the software at time τ.

A generalized framework for unified modeling approach for two stages which can be described as Musa [3] to incorporate the effect of fault detection process given by the distribution function F(τ) and fault correction process given by the distribution function $G(\tau)$ can be written on the same lines of Eq. (1) as follows:

$$m(\tau) = a(F \otimes G)(\tau) \tag{3}$$

Hence,

$$\frac{dm(\tau)}{d\tau} = a(f^{*}g)(\tau) \tag{4}$$

We can rewrite (4) as

$$\frac{dm(\tau)}{d\tau} = \frac{(f^{*}g)(\tau)}{[1 - (F \otimes G)(\tau)]}[a - m(\tau)] \tag{5}$$

Table 1 Mean value functions

$F(\tau)$	$G(\tau)$	$m(\tau)$
$T \sim \exp(b)$	$T \sim \gamma(\alpha_1, \beta_1)$	$\frac{a}{1-\alpha}\left[1-\left(1-\Gamma(\tau,\alpha_1,\beta_1)+\frac{e^{-b\tau}}{(1-b\beta_1)^{\alpha_1}}\Gamma\left(\tau,\alpha_1,\frac{\beta_1}{1-b\beta_1}\right)\right)^{p(1-\alpha)}\right]$
$T \sim \exp(b)$	$T \sim \log(b,\beta)$	$\frac{a}{1-\alpha}\left[1-\left(1-\left(\left(1-e^{-b\tau}\right)+(1+\beta)e^{-b\tau}.\ln\left(\frac{(1+\beta)e^{-b\tau}}{1+\beta e^{-b\tau}}\right)\right)\right)^{p(1-\alpha)}\right]$

$$\text{i.e., } \frac{dm(\tau)}{d\tau} = r(\tau)\left[a - m(\tau)\right] \tag{6}$$

where $r(\tau) = \frac{(f^*g)(\tau)}{[1-(F \otimes G)(\tau)]}$ which is fault detection rate/fault correction rate.

Now combining the concepts of imperfect debugging and error generation in the proposed modeling, we have the following differential equation:

$$\frac{dm(\tau)}{d\tau} = \frac{(f^*g)(\tau)}{[1-(F \otimes G)(\tau)]}p\left[a + \alpha m(\tau) - m(\tau)\right] \tag{7}$$

Solving (7) with the initial condition $m(0=0)$, we get the following solution:

$$m(\tau) = \frac{a}{1-\alpha}\left[1 - (1-(F \otimes G)(\tau))^{p(1-\alpha)}\right] \tag{8}$$

In the following Table 1, mean value functions for existing SRGMs corresponding to different $F(\tau)$ and $G(\tau)$ as proposed by Singh et al. [20] have been given.

4.2 Multi Upgradation Modeling

1. *Modeling for Release 1*

Let us assume at time $\tau = \tau_1$ the first version of the software is released. The mathematical equation describing the cumulative number of faults removed in the first release is given by

$$m_1(\tau) = a_1^*(F_1 \otimes G_1)(\tau), \qquad 0 \le \tau \le \tau_1 \tag{9}$$

where $a_1^* = \frac{a_1}{1-\alpha_1}$

and $(F_1 \otimes G_1)(\tau) = \left[1 - (1-(F \otimes G)(\tau))^{p_1(1-\alpha_1)}\right]$.

2. *Modeling for Release 2*

During proper planning for second release, the firm comes across the following aspects:

1. The elimination of the faults left from release 1
2. Newly produced faults due to functional enhancement of the software to fulfill the essential demands of the users

Keeping these two aspects in mind, the modeling is done in the following way:

$a_1^* \left(1 - \left(F_1 \otimes G_1\right)(\tau_1)\right)$ represents the undetected faults remaining in the first release or the faults introduced due to error generation while testing process.
a_2^*represents the fault content of the second release.
$(F_2 \otimes G_2)(\tau - \tau_1)$ represents the fraction of faults removed during the testing of second release.

Therefore, the mathematical equation depicting the cumulative number of faults removed in second release is given by

$$m_2(\tau) = a_2^* \left(F_2 \otimes G_2\right)(\tau - \tau_1) + a_1^* \left[1 - \left(F_1 \otimes G_1\right)(\tau_1)\right] \left(F_2 \otimes G_2\right)(\tau - \tau_1)$$
$$m_2(\tau) = \left[a_2^* + a_1^* \left[\left[1 - \left(F_1 \otimes G_1\right)(\tau_1)\right]\right] \left(F_2 \otimes G_2\right)(\tau - \tau_1);\tau_1 \le \tau \le \tau_2 \quad (10)$$

where $a_2^* = \frac{a_2}{1-\alpha_2}$
and $\left(F_2 \otimes G_2\right)(\tau) = \left[1 - \left(1 - \left(F \otimes G\right)(\tau - \tau_1)\right)^{p_2(1-\alpha_2)}\right]$.

3. *Modeling for Release 3 and 4*

Based on similar lines of second release, the mathematical equation for release 3 and release 4 is given as follows:

$$m_3(\tau) = \left[a_3^* + a_2^* \left[1 - \left(F_2 \otimes G_2\right)(\tau_2 - \tau_1)\right]\right] \left(F_3 \otimes G_3\right)(\tau - \tau_2);\ \tau_2 \le \tau \le \tau_3 \quad (11)$$

where $a_3^* = \frac{a_3}{1-\alpha_3}$
and $\left(F_3 \otimes G_3\right)(\tau) = \left[1 - \left(1 - \left(F \otimes G\right)(\tau - \tau_2)\right)^{p_3(1-\alpha_3)}\right]$.
Similarly,

$$m_4(\tau) = \left[a_4^* + a_3^* \left[1 - \left(F_3 \otimes G_3\right)(\tau_3 - \tau_2)\right]\right] \left(F_4 \otimes G_4\right)(\tau - \tau_3);\tau_3 \le \tau \le \tau_4 \quad (12)$$

where $a_4^* = \frac{a_4}{1-\alpha_4}$
and $\left(F_4 \otimes G_4\right)(\tau) = \left[1 - \left(1 - \left(F \otimes G\right)(\tau - \tau_3)\right)^{p_4(1-\alpha_4)}\right]$.

In the above equations by taking $\alpha_i = 0$ and $p_i = 1$ (i = 1 to 4), we get the model proposed by Kapur et al. [27]. Also by using different $F(\tau)$ *and* $G(\tau)$, we can derive the existing models in the literature, as well as we can develop some new SRGMs.

4.3 Mean Value Function for Proposed SRGM Models

We have the following proposed two-stage two-dimensional models assuming $\tau = s^x u^{1-x}$(Cobb-Douglas form) where s= time and u= usage of resources and $0 \leq x \leq 1$.

SRGM-1 Here, the fault detection process is represented by an exponential function, and the fault correction process is represented by a gamma function where b is the fault detection rate per remaining fault:

$$F(s,u) = \exp\left(b, s^x u^{1-x}\right); G(s,u) = \gamma\left(\alpha_1, \beta_1, s^x u^{1-x}\right)$$

Thus, mean value function for proposed SRGM-1 is:

$$m(s,u) = \frac{a}{1-\alpha}\left[1 - \left(1 - \Gamma(s,u,\alpha_1,\beta_1) + \frac{e^{-bs^x u^{1-x}}}{(1-b\beta_1)^{\alpha_1}} \Gamma\left(s,u,\alpha_1,\frac{\beta_1}{1-b\beta_1}\right)\right)^{p(1-\alpha)}\right] \quad (13)$$

SRGM-2 Here, the fault detection process follows exponential distribution, and the fault correction process follows logistic distribution:

$$F(s,u) = \exp\left(b, s^x u^{1-x}\right); G(s,u) = \log\left(b, \beta, s^x u^{1-x}\right)$$

Thus, mean value function for proposed SRGM-2 is:

$$m(s,u) = \frac{a}{1-\alpha}\left[1 - \left(1 - \left(\left(1 - e^{-bs^x u^{1-x}}\right) + (1+\beta)\, e^{-bs^x u^{1-x}} . \ln\left(\frac{(1+\beta)\, e^{-bs^x u^{1-x}}}{1+\beta e^{-bs^x u^{1-x}}}\right)\right)\right)\right] \quad (14)$$

5 Model Validation and Parameter Estimation

In software reliability prediction, parameter estimation is the most important. SRGMs are useful only if estimation of its parameters is possible. The proposed models are validated using four release data set collected from Tandem Computers [29]. We have used nonlinear regression technique in Statistical Package for Social Sciences (SPSS) software for estimation of parameters. Goodness of fit criteria is used to compare and analyze the fitness of the model. With the exception in release 1, the fault content left out in the previous release is considered in the next release.

5.1 Criteria of Comparison

1. MSE*:* The mean square error is defined as

$$\mathrm{MSE} = \sum_{i=1}^{k} \frac{\left(m\left(t_i\right) - y_i\right)^2}{k}$$

where y_i denotes the observed values of the parameters and $m(t_i)$ denotes the estimated values of the parameters and k denotes the number of observations. Lower the MSE, less will be the fitting error; thus, goodness of fit will be better.

2. R^2*:* This coefficient is defined as the ratio of the sum of squares resulting from the trend model to that from the constant model subtracted from 1:

$$R^2 = 1 - \frac{\text{residual SS}}{\text{corrected SS}}$$

It ranges in value from 0 to 1. The larger the value of R^2, the better the model explains the variation in the data.

3. Bias: Prediction error ($\mathrm{PE_i}$) is the difference between the observed and the estimated number of faults. Bias is evaluated by taking the average of the $\mathrm{PE_i}$. The lower the value of bias, the better will be the goodness of fit.
4. Variation: The standard deviation of prediction error is known as variation:

$$\text{Variation} = \sqrt{\left(\frac{1}{N-1}\right)\sum\left(\mathrm{PE_i} - \mathrm{Bias}\right)^2}$$

The higher the value of variation, the poor is the goodness of fit and vice versa.

5. RMSPE: It is a measure of closeness with which a model predicts the observation:

$$\mathrm{RMSPE} = \sqrt{\left(\mathrm{Bias}^2 + \mathrm{Variation}^2\right)}$$

The lesser its value, the better is the goodness of fit.

5.2 Results and Discussions

Initially for each SRGM, the data set of first release is considered, and using SPSS, estimation of parameters is done. The predicted values thus obtained from estimation process help in estimating the faults left in release 1. With these leftover faults of release 1, together with the data set of release 2, parameters in second release are evaluated. Similar process is carried for third and fourth release. The estimation results for various parameters in proposed SRGM-1 and SRGM-2 are depicted in Table 2.

The following Table 3 depicts the comparison criteria for proposed SRGM-1 and SRGM-2 for successive four releases of the software.

From the above comparison results, it is inferred that for release 1 and 2, the value of R^2 for SRGM-1 is higher than that of SRGM-2. Also, values of MSE, bias, variation, and RMSPE for SRGM-1 are lower than that of SRGM-2. Thus, for

Table 2 Estimation results for SRGM-1 and SRGM-2

SRGM-1	Release 1	Release 2	Release 3	Release 4
a	125	128	64	44
b	0.098	0.147	0.3	0.16
α	0.093	0.05	0.05	0.084
α_1	0.085	0.6	1.32	0.88
p	0.86	0.8	0.952	0.88
β_1	0.045	0.104	0.28	0.14
x	0.32	0.32	0.33	0.46
SRGM-2	**Release 1**	**Release 2**	**Release 3**	**Release 4**
a	102	123	65	45
b	0.38	0.50	0.86	0.37
α	0.16	0.106	0.097	0.023
β	1.82	5.138	37	1.8
p	0.69	0.64	0.97	0.61
x	0.47	0.40	0.45	0.53

Table 3 Comparison results for SRGM-1 and SRGM-2

SRGM-1	Release 1	Release 2	Release 3	Release 4
MSE	6.08633	7.439442	4.74215	1.082816
R^2	0.993	0.994	0.989	0.994
Bias	−0.053	−0.53263	−0.44833	−0.05
Variation	2.530555	2.748324	2.225754	1.067864
RMSPE	2.53111	2.799461	2.270458	1.069034
SRGM-2	**Release 1**	**Release 2**	**Release 3**	**Release 4**
MSE	13.82091	8.198125	3.522244	0.955186
R^2	0.983	0.994	0.992	0.995
Bias	−0.757	−0.54211	−0.55162	−0.07716
Variation	3.73431	2.888488	1.873632	1.000983
RMSPE	3.810271	2.93892	1.953148	1.003953

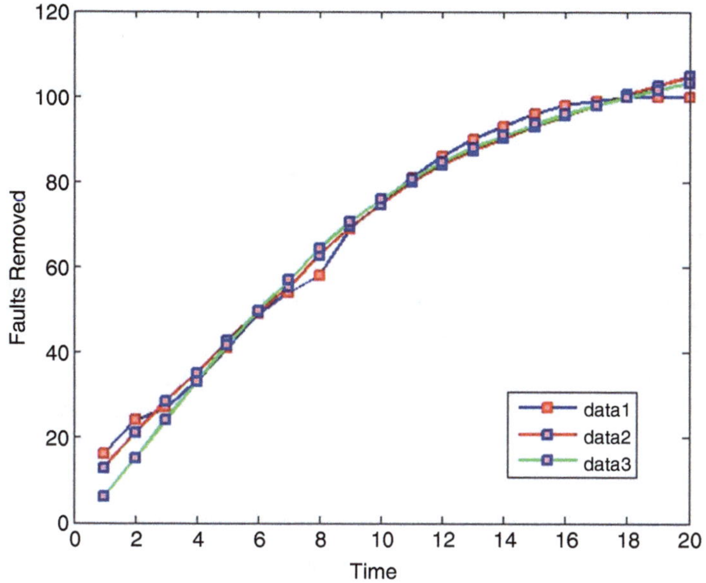

Fig. 1 Goodness of fit curve for release 1

release 1 and 2, SRGM-1 is giving the best goodness of fit. For release 3 and 4, the value of R^2 for SRGM-2 is higher than that of SRGM-1. The values of MSE, variation, and RMSPE for SRGM-2 are lower than that of SRGM-1. But, value of bias for SRGM-2 is higher than that for SRGM-1. Hence, bias is not a good measure for comparison of data set for release 3 and 4. Thus, we can conclude that SRGM-2 gives best fit to the observed for release 3 and 4 according to other four criteria of comparison. Figures 1–4 represent the goodness of fit between the estimated and the actual values of SRGM-1 and SRGM-2 for four releases. In the below figures, data 1 represents the actual faults removed for each release; data 2 represents the estimated faults removed in SRGM-1 for each release; data 3 represents the estimated faults removed in SRGM-2 for each release.

6 Conclusion

In this paper, testing time and effort-dependent multi-release modeling under the effect of two types of imperfect debugging has been developed. In the proposed modeling, we have considered testing phase to be two-stage fault detection and fault correction process. Here, SRGMs are proposed from existing SRGMs using well-known Cobb-Douglas function. The leftover faults in the previous release are also incorporated in addition to the new fault content. Validity of the proposed SRGMs has been tested on Tandem Computer data set. Estimated parameters for SRGM-1

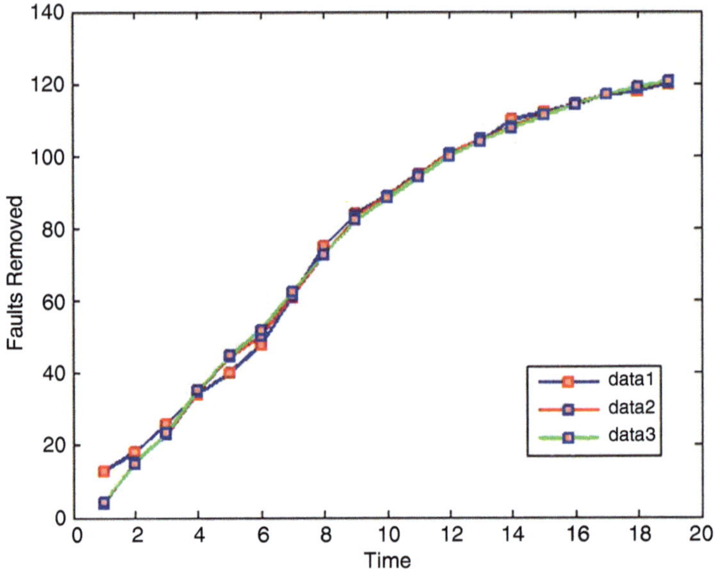

Fig. 2 Goodness of fit curve for release 2

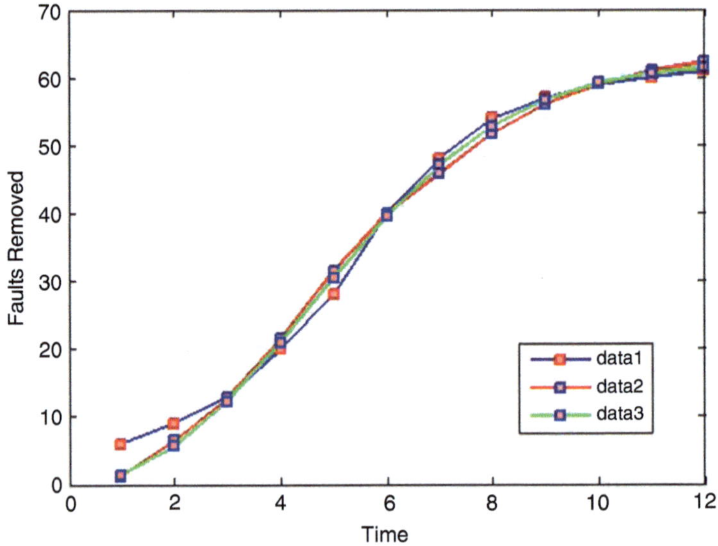

Fig. 3 Goodness of fit curve for release 3

and SRGM-2 in each release are given in the paper. In addition, the comparison analysis of the proposed SRGMs is made by five common criteria, namely, bias, MSE, R^2, variation, and RMSPE. It is seen that SRGM-1 gives best goodness of

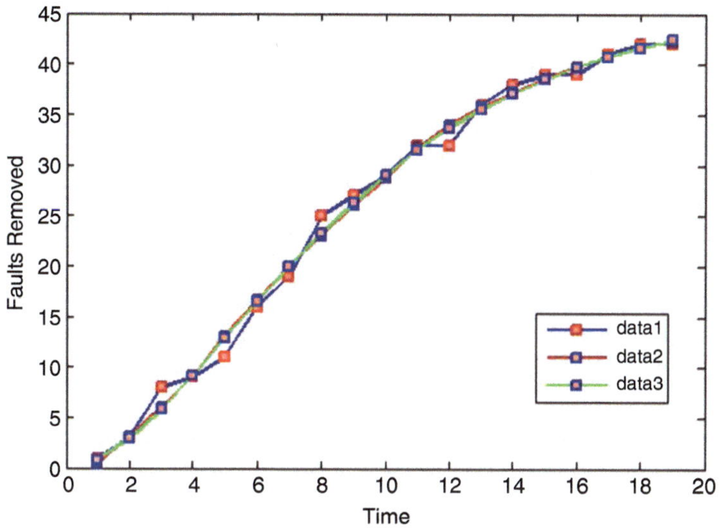

Fig. 4 Goodness of fit curve for release 4

fit for release 1 and 2 and SRGM-2 gives best goodness of fit for release 3 and 4. The graphs are also illustrated to depict the good fit results of both SRGMs for all releases. In the future, models can also be extended by incorporating the cost function or by considering the impact of number of test cases executed and testing coverage in fault correction process.

References

1. Goel AL, Okumoto K (1979) Time dependent error detection rate model for software reliability and other performance measures. IEEE Trans Reliab R-28(3):206–211
2. Schneidewind NF (1975) Analysis of error process in computer software. Sigplan Notices 10(6):337–346
3. Musa JD, Iannino A, Okumoto K (1987) Software reliability:measurement, prediction, applications. McGraw Hill, New York, pp 267–276
4. Ohba M (1984) Software reliability analysis models. IBM J Res Dev 28(4):428–443
5. Yamada S, Hishitani J, Osaki S (1993) Software-reliability growth with a Weibull test-effort. IEEE Trans. Reliab 42(1):100–106
6. Huang CY, Kuo SY (2002) Analysis of incorporating logistic testing-effort function into software reliability modeling. IEEE Trans Reliab 51(3):261–270
7. Kapur PK, Pham H, Chanda U, Kumar V (2013) Optimal allocation of testing effort during testing and debugging phases: a control theoretic approach. Int J Syst Sci 44(9):1639–1650
8. Kapur PK, Kumar V (2011) Testing resource allocation for fault detection and correction processes under dynamic environment. In the proceedings of national conference on Computing for Nation Development (INDIACOM-2011,New Delhi), pp 331–336
9. Kapur PK, Chanda U, Kumar V (2010) Dynamic allocation of testing effort when testing and debugging are done concurrently, communication in dependability and quality management. Int J Serbia 13(3):14–28

10. Kumar V, Sahni R (2015) An effort allocation model considering different budgetary constraint on fault detection process and fault correction process. Decis Sci Lett 5(1):143–156
11. Kumar V, Khatri SK, Dua H, Sharma M, Mathur P (2014) An assessment of testing cost with effort dependent FDP and FCP under learning effect: a genetic algorithm approach. Int J Reliab Qual Saf Eng 21(6):145002
12. Kumar V, Kapur PK, Taneja N, Sahni R (2015) On allocation of resources during testing phase incorporating flexible software reliability growth model with testing effort under dynamic environment. Int J Oper Res (In press)
13. Kapur PK, Tandon A, Kaur G (2010) Multi up-gradation software reliability model. 2nd International Conference on Reliability, Safety and Hazard (ICRESH-2010), pp 468–474
14. Xie M, Hu QP, Wu YP, Ng SH (2006) A study of the modeling and analysis of software fault-detection and fault-correction process. Qual Reliab Eng Int 23:459–470
15. Goel AL (1985) Software reliability models: assumptions, limitations and applicability. IEEE Trans Softw Eng SE-11:1411–1423
16. Ohba M, Chou XM (1989) Does imperfect debugging effect software reliability growth. In: Proceedings of 11th international conference of software engineering, pp 237–244
17. Peng R, Li YF, Zhang WJ, Hu QP (2014) Testing effort dependent software reliability model for imperfect debugging process considering both detection and correction. Reliab Eng Syst Saf 126:37–43
18. Kapur PK, Pham H, Anand S, Yadav K (2011) A unified approach for developing reliability growth models in the presence of imperfect debugging and error generation. IEEE Trans Reliab 60(1)
19. Pachauri B, Kumar A, Dhar J (2014) Software reliability growth modeling with dynamic faults and release time optimization using GA and MAUT. Int J Appl Math Comput 242:500–509
20. Singh O, Kapur PK, Shrivastava AK, Das L (2014) A unified approach for successive release of a software under two types of imperfect debugging. In: Proceedings of 3rd international conference on reliability, infocom technologies and optimization, pp 275–280
21. Kapur PK, Garmabaki AHS, Singh J (2011) Multi up-gradation software reliability model with imperfect debugging, international congress on productivity, quality, reliability, optimization and modelling. ICPQROM, New Delhi
22. Wang J, Wu Z, Shu Y, Zhang Z (2014) An imperfect software debugging model considering log logistic distribution fault content function. Int J Syst Softw 100:167–181
23. Kapur PK, Singh O, Shrivastava AK, Singh JNP (2015) A software up-gradation model with testing effort and two types of imperfect debugging. IEEE Xplore proceedings of international conference on futuristic trends in computational analysis and knowledge management, pp 613–618
24. Ishii T, Dohi T (2006) Two-dimensional software reliability models and their application. Proceedings of 12th Pacific Rim international symposium dependable computing, pp 3–10
25. Inoue S, Yamada S (2008) Two-dimensional software reliability assessment with testing-coverage. Second international conference on secure system integration and reliability improvement
26. Inoue S, Yamada S (2009) Two-dimensional software reliability measurement technologies. In the proceedings of IEEE, IEEM
27. P. K. Kapur, H. Pham, A. G. Aggarwal and Gurjeet Kaur 2012 Two dimensional multi-release software reliability modeling and optimal release planning IEEE Trans Reliab 61(3):758–768
28. Inoue S, Fukuma K, Yamada S (2010) Two-dimensional change- point modeling for software reliability assessment. Int J Reliab 17(6):531–542
29. Wood A (1996) Predicting software reliability. IEEE Comput 29(11):69–77

The Role of Educational ERP in the Isomorphic Development of the Newly Started Higher Education Institutions

Said Eid Younes

1 Introduction

The positive trend in the growth of the number of higher education institutions listed in the International Handbook of Universities (Published by the International Association of Universities) is evident. For instance, the number of listed higher education institution in the years 2010, 2012, and 2015 increased from over 9200 in 2010 to over 15,000 in 2012 reaching over 18,000 in 2015.

The last two decades witnessed large changes not only in the number of students in higher education institutions (HEIs), or students enrolled in HEIs [14], but also in the configuration and structure of HEIs [3, 8]. HEIs are nowadays more autonomous, competitive, and accountable. These changes were in response to social, economic, and cultural shifts in the surrounding environment [11]. HEIs are working in globalized market, where the number of international students is increasing in HEIs [20].

The newly started HEIs function in multifaceted work challenges: complying with the regulatory licensing requirements, gaining credibility right from the start, offering novel and innovative academic programs, agile following of the labor market demands and trends, and attracting enough number of students to cross the break-even economic threshold. Taking into consideration that new HEIs share the pool of students with legacy HEIs, the earlier mentioned challenges are magnified.

Legacy HEIs evolved over decades, some over centuries, through which work experience and reputation were established. New HEIs have to convey their

S.E. Younes (✉)
Department of Information Systems, College of Economics, Management and Information Systems, University of Nizwa, Nizwa, Sultanate of Oman
e-mail: syoues@unizwa.edu.om

© Springer Nature Singapore Pte Ltd. 2018

P.K. Kapur et al. (eds.), *Quality, IT and Business Operations*, Springer Proceedings in Business and Economics, DOI 10.1007/978-981-10-5577-5_34

credibility using various techniques: collaborating or affiliating with one of the reputable legacy HEIs, offering highly specialized academic program expected to close a gape in the labor market, or offering lenient admission requirements.

Collaboration, affiliation, and accreditation by reputable professional bodies (as quality assurance bodies) are some of the widely used practices by new HEIs. Such arrangements reflect the intention of the newly established HEIs to gain acceptance, credibility, and therefore the legacy of the well-established HEIs. The aforementioned arrangements are better explained and understood in the light of the institutional theory.

In the following section, a brief introduction to the institutional isomorphism theory in the context of HEIs will be presented. The role of the enterprise resource planning (ERP) systems in institutional isomorphism will be established in section III. Section IV presents the techniques used by new HEIs to establish institutional isomorphism using educational ERP.

2 The Institutional Isomorphism Theory

According to the institutional theory, institutions interact with the surrounding environment to receive support and legitimacy through the elaboration of the rules and requirements which the organizations must confirm to [7, 9, 15, 18]. The application of such rules and the adherence to the organizational requirements grant the organization acceptance from the surrounding environment, legitimacy, and increased survival capabilities. This process is called the institutional isomorphism.

The three processes shaping the institutional isomorphism are explained as follows [6, 7, 17, 19]:

- Coercive isomorphism: Organizational conformity is achieved through the use of power and sanctions and occurs by organizational adherence to laws and rules established by institutional actors. In the context of HEIs, the coercive isomorphism is achieved by confirming to the governmental or licensing bodies' rules and regulations.
- Mimetic isomorphism: It is the tendency to remain similar to peer organizations in order to cope with the organizational environment. Copied plans or actions reduce risks and uncertainty in the organization and increase predictability. In HEI context, the mimetic isomorphism is manifested through copying the experience of the peer institutions either indirectly or through affiliation and cooperation.
- Normative isomorphism: It is related to the professional standards in the organizational field. In the HEI context, the normative isomorphism is connected to beliefs and practices created by national and international quality assurance systems and professional certification bodies [16].

The isomorphic characteristics of the HEIs have been extensively studied in literature [5, 10, 16].

The fourth type of isomorphism is suggested for the technology-dependent organizations, the technical isomorphism [4]. Technical isomorphism results from embeddedness of organizational procedures and business rules within technology [2]. This fourth dimension of isomorphism plays an important role in HEIs as these institutions are increasingly becoming technology dependent through the use of information systems (IS).

It has become well established now that HEIs are increasingly shifting to integrated software solutions provided by ERP systems as replacement and upgrading of the legacy IS used by the HEIs [1]. The next paragraph presents the role of ERP as a tool of the technical isomorphism in particular and as a binding agent to the institutional isomorphism.

3 ERP as a Tool of Technical Isomorphism

The proliferation use of ERPs in organizations, to cope with the fast pace of changes in the business environment and to replace the legacy IS, has become evident [12]. The HEIs are no exception to this fact, as many HEIs worldwide are adopting the ERP solutions [1].

Reference [2] presents a case study of the application of ERP system (SAP) at Serviceco, a Dutch utilities firm, where the well-established business rules and organizational procedures embedded in SAP ERP pressured the organization toward conformance to SAP standards. As the use of ERP leads to the enhancement of the interorganizational similarities in operational processes, ERP-induced isomorphism was recognized in the organization.

This case study presented a case of achieving institutional isomorphism through the adaptation of ERP solution. SAP ERP helped in achieving technical isomorphism in the organization which in turn contributed to institutional isomorphism.

To better understand how ERP can be the tool to implement the technical isomorphism and to augment the organizational endeavor to achieve the institutional isomorphism, we have to look again into the main trends of the institutional thinking [17]. Theories classify the institutional thinking in two main trends: macro-level and microlevel.

The macro-level thinking describes the effect of external and environmental factors as the main determinant to institutional behavior. Coercive, mimetic, and normative isomorphism processes are the mechanisms used by organizations to align themselves with the environment.

The microlevel thinking describes the technology institutionalization process. Reference [21] defined three stages of technology institutionalization:

- Habitualization stage: shared social meanings are produced.
- Objectivation stage: facts become independent as a reality and sharing the experience of this reality with others.
- Sedimentation stage: a routine behavior of objectified facts is practiced.

Contrary to the macro-level thinking, the microlevel thinking is limited to factors within the boundaries of the organization, and the stages listed above describe how the technology is institutionalized in the organization.

Accordingly, technology isomorphism can be explained by the microlevel three stages of the institutional thinking [4].

The adaptation of an ERP solution in an organization will initiate the technology institutionalization process as an internal process. Proper deployment and effective use of ERP in organizations take ERP users through the three stages of the microlevel isomorphism and bind various external isomorphism (macro-level) processes leading to institutional isomorphism.

4 Institutionalization Isomorphism in the Newly Started HEIs

The starting of a new HEI is a complex, challenging, and noble endeavor. New HEIs are expected to perform like legacy HEIs, assure to their stakeholders quality education, and meet the critical startup challenges (financial, operational, acceptance by academic community, building reputation, etc.).

At the startup, new HEIs are subject to the following types of organizational pressures:

- Statutory licensing bodies' pressure: failing to comply with the statutory regulations and rules may lead to the denial of licensing.
- Peer competitors' pressure: inadequate competition strategies to cope with the competitor's pressure may jeopardize the operational viability of the HEI.
- Recognition pressure: represented by the strife of the new HEIs to gain legitimacy and acceptance from the academic community.
- Operational pressure: frequent staff change, part-time faculty members dependence, and different organizational orientations of the new staff which may lead to difficulties in unified implementation of policies and procedures [10].

The above-listed isomorphic pressures are not mutual exclusive processes and could be interdependent [17]. The new HEIs respond to these pressures by adopting institutional isomorphism processes.

Table 1 summaries the HEIs possible isomorphic responses.

New HEIs have no option to respond or not to the coercive pressure when it is related to the sheer existence of the organization, as in the case of licensure requirements.

On the other hand, mimetic isomorphism was given higher emphasis in HEIs [5]. Reference [13] placed it to be the most pertinent account of isomorphism change in Australian universities.

Table 1 HEI's possible isomorphic responses

Pressure	Isomorphic pressure	Isomorphic response
Statutory licensing	Coercive	Comply with rules and regulations
Peer competitors	Coercive	Implement strategies to diffuse the competitors pressure
Recognition pressure	Mimetic/normative	Adopt the strategies, policies, and practices followed by legacy HEIs
		Fulfill the requirements of the professional accreditation bodies
Operational pressure	Normative	Ensure unified implementation of the adopted policies and procedures

Accordingly, mimetic isomorphism is an important process through which new HEI can benchmark with well-performing organizations (optimal or near to optimal performance), reduce uncertainty, and increase predictability.

Normative isomorphism is the real challenge of the new HEIs. Albeit, new HEIs will start functioning under complete set of policies and procedures (which could be part of the licensure or professional bodies accreditation requirements), yet the orientation diversity of the stakeholders (academic and administrative staff and students) of the new HEIs makes achieving normative isomorphism follow typical learning curve track. This difficulty is due to the nature of normative isomorphism which is mostly concerned with the moral and pragmatic aspect of legitimacy [19].

The use of IS in organizations is viewed as the result of normative influence [17]. ERP provides unified learning environment for the stakeholders, guides the learning process, and ascertains that HEI is achieving its role properly. The new generations of educational ERPs facilitate planning, managing, and monitoring of all functional applications of teaching and research [1]. Therefore, it is expected to support and accelerate various operations at different levels of isomorphism processes.

Referring to paragraph II, new HEIs adaptation of ERP introduces technical isomorphism which in turn binds the three isomorphic processes followed in the HEI.

At the microlevel isomorphism, ERP supports the three stages of the technical isomorphism as follows:

- Habitualization stage: organizational functions and operations are performed only through ERP. Deviation from the standard procedures embedded in ERP is discouraged. This stage may witness resistance to change or adaptation.
- Objectivation stage: successful habitualization stage leads to increased dependence on the ERP. Core of pilot experienced users disseminates best practices of ERP functions.
- Sedimentation stage: using ERP is part of the daily routine in the organization. Even new members joining the organization takes ERP functions for granted.

Table 2 The role of ERP in supporting the various stages of institutional isomorphism at the macro level

Isomorphic process	Educational ERP support
Coercive	Most of the licensing requirements state that HEI should operate and run its functions and operations through ERP
Mimetic	Acquiring educational ERP used by other HEIs (well-performing HEIs) helps the new HEI to have operational features similar to the well-established HEIs, reduce the risks, and increase predictability
Normative	Teaching and research policies and procedures which are embedded in the educational ERP unify the learning and accumulation of experience processes within HEI

The microlevel stages constitute the technical isomorphism which induces isomorphism at the internal level within the organization (HEI) and supports the isomorphism processes at the macro-level. Table 2 summarizes the role of ERP in supporting the various stages of institutional isomorphism at the macro-level.

5 Conclusion

The number of students joining HEIs is projected to increase in upcoming decade and thus the number of new HEIs joining the family of existing HEIs.

New HEIs face the challenges of legitimacy, support, and recognition from the surrounding environment. Further, new HEIs face internal challenges concerning creating homogeneous work culture, cultivating best practices, and achieving HEIs goals and objectives.

The proper adaptation of educational ERP in the new HEI helps the organization to achieve institutional isomorphic features and reach the optimal institutional norms to gain acceptance, legitimacy, and recognition of the surrounding environment.

References

1. Abugabah A, Sanzogni L, Alfarraj O (2015) Evaluating the impact of ERP systems in higher education. Int J Inf Learn Technol 32(1):45–64
2. Batenburg R, Benders J, Blonk VDH (2008) Technical isomorphism at work: ERP-embedded similarity-enhancing mechanisms. Ind Syst Data Syst 108(1):60–69
3. Beliklie I, Kogan M (2007) Organization and governance of universities, higher Education Policy 20(4):477–493
4. Benders J, Batenburg R, Van der Blonk H (2006) Sticking to standards. Technical and other isomorphic pressure in deploying ERP=systems. Inf Manag 43(1):194–203
5. Croucher G, Woelert P (2015) Institutional isomorphism and the creation of the unified national system of higher education in Australia: an empirical analysis. J Hig Educ:1–15. doi:10.1007/s10734-015-9914-6

6. Currie WL (2011) Institutional theory of information technology. In: The Oxford handbook of management information systems: critical perspectives and new directions. Oxford University Press, Oxford, pp 137–173
7. DiMaggio PJ, Powel WW (1991) Iron cage revisited: institutional isomorphism and collective rationality in organizational fields. Am Sociol Rev 48:147–160
8. Enders J, Boer H, Weyer E (2013) Regulatory autonomy and performance: the reform of higher education re-visited. High Educ 65(1):5–23
9. Gates GS (1997) Isomorphism, homogeneity and rationalization in universities retrenchment. Rev High Educ 20(3):253–275
10. Hodson P, Connolly M, Younes S (2008) Institutionalization in a newly created private university. Qual Assur Educ 16(2):141–147
11. Jarvis D (2014) Policy Transfer, neo-liberalism or coercive institutional isomorphism? Explaining the emergence of regulatory regime of quality assurance in the Hong Kong higher education sector. Policy Soc Assoc Elsevier 33:237–252
12. Kallunki JP, Laitinen EK, Silvola H (2011) Impact of enterprise resource planning systems on management control systems and firm performance. Int J Account Inf Syst 12(1):20–39
13. Marginson S, Considine M (2000) The enterprise university: power, governance and reinvention in Australia. Cambridge press, Cambridge
14. Maslen G (2012) Worldwide student numbers forecast to double by 2025. University World News, Issue No:209
15. Meyer J, Rowan B (1977) Institutionalized organizations formal structures as myth and ceremony. Am J Sociol 83:340–363
16. Mizikaci F (2011) Isomorphic and diverse features of Turkish private higher education, program for research on private higher education. Educational Administration & Policy Studies, University of Albany, State of New York
17. Pishdad, A., Haidar, A., Koronios, A. (2012) Technology and organizational evolution: an institutional perspective. J Innov Bus Best Pract 2012 Article ID 655615
18. Scott WR (1994) Institutional analysis: variance and process theory approaches. In: Scott WR, Meyer JW (eds) Institutional environments and organizations : structured complexity and individualism. Sage, Thousand Oaks, pp 81–99
19. Scott WR (2001) Institutional and organizations, 2nd edn. Sage, Thousand Oaks
20. UNESCO 2015. http://www.uis.unesco.org/Education/Pages/international-student-flow-viz.aspx
21. Zucker L (1977) The role of institutionalization in cultural persistence. Am Sociol Rev 4(25):726–743

When to Start Remanufacturing Using Adopter Categorization

Nitin Sachdeva, P. K. Kapur, and Ompal Singh

1 Introduction

With over several decades of research, one of the most important questions in the mind of any manager is how to manage my product life cycle? Product life cycle management involves several successive strategies employed by the manager as a product goes through its life cycle. Specifically, in manufacturing, it typically starts with raw material extraction, followed by component production, assembly, testing, and distribution to customers for the use and final disposal at the product end-of-life. Due to the rapid development of technology and product development processes, this product life cycle has reduced significantly. On the other side, strong customer preferences and his volatile nature have sharply accelerated product diffusion dynamics.

Shortened life cycle coupled with customer's volatile preferences has led to increasing amounts of products and components being abandoned, including electrical and electronic products, household appliances, automobiles, and even industrial machinery, to name a few. Many world-class companies have realized that reverse logistics practices, combined with source reduction processes, can be used to gain competitive advantage and at the same time achieve sustainable development [28]. Manufacturers now view the "reverse logistics system" as a revenue opportunity instead of a cost-minimization approach and are managing their reverse supply chains as a source of value [14, 15].

N. Sachdeva (✉) • O. Singh
Department of Operational Research, University of Delhi, Delhi, India
e-mail: nitin.sach@gmail.com; drompalsingh@live.com

P.K. Kapur
Amity Center for Interdisciplinary Research, Amity University, Noida, UP, India
e-mail: pkkapur1@gmail.com

© Springer Nature Singapore Pte Ltd. 2018
P.K. Kapur et al. (eds.), *Quality, IT and Business Operations*, Springer Proceedings in Business and Economics, DOI 10.1007/978-981-10-5577-5_35

To resolve this challenge of how to manage these end-of-life products, an Extended Producer Responsibility (EPR) approach has emerged as a leading environmental policy, whereby a producer's responsibility for a product is extended to the postconsumer stage of the product life cycle [26]. The ultimate goal of EPR is to minimize the total environmental effects associated with the overall product life cycle and, in particular, to reduce the quantity of waste and environmental pollution from end-of-life product disposal. EPR legislation is also expanding worldwide to the USA, Japan, South Korea, and other countries. The products of most concerns under EPR legislation span a wide range of categories, including automobiles, large and small household appliances, and telecommunications and IT equipment. On the contrary, it has been quite significantly observed that several of these cast-off products are well within their life span and contain substantial amounts of residual value. A lot of the components of such products can be readily reused for manufacturing products with original functionality. Successful business cases such as Xerox remanufacturing used photocopiers and toner cartridges [19] and Mercedes-Benz and Ford collecting and disassembling their end-of-life vehicles to recover valuable components as spare parts for both consumers and commercial customers [33] are some of the well-established companies following the practice of closed-loop product life cycle. However, it is evident that the current efforts toward implementing environmentally superior end-of-life strategies to form a closed-loop product life cycle are far from widespread.

Here, one of the most important concerns is that we cannot solely rely on legislation to pressurize manufacturing units to consider taking the product end-of-life responsibility. A lot depends on essentially the engineers and designers as how they tend to resolve this key sustainability challenge by designing techniques which are futuristic and sustainable. Thereby, it's the need of the hour for manufacturing companies to design products keeping in mind the economic consequences of product end-of-life. Some of the challenges these companies face and we have tried to highlight through our research include the following: How can the economic benefits of component reuse be modeled in a remanufacturing scenario? And how optimal time to start remanufacturing, under the product diffusion dynamics in the market, affects component reuse in remanufacturing? And what is the volume of reused components available for remanufacturing?

The paper has twofold objectives. First, the main aim of this paper is to obtain optimal profit under product remanufacturing using the profit maximization framework proposed by Sachdeva et al. [31]. Second, we attempt to study the optimal time for OEMs to start remanufacturing by considering adopter categorization, in the presence of product diffusion dynamics. We make use of the famous Bass diffusion model [4] to quantify the returned quantity and obtain the adopter categorization, which helps to gain meaningful managerial insights. Through the numerical illustration, we demonstrate the effect adopter categorization and product diffusion dynamics on quantity returned for remanufacturing.

As for the beginning, we present an overview of the literature in the area of remanufacturing, closed-loop supply chain, product diffusion, and adopter categorization in Sect. 2. In Sect. 3, we present the basic cost model for such a production system. Subsequently, an optimization model based on profit maximization is proposed in Sect. 4 wherein various costs of production, inventory, remanufacturing, acquisition, and disposal are being considered. In Sect. 5 we present the numerical illustration to provide managers with simple but powerful insights to help them apply our proposed methodology in their specific problem instances. Section 6 concludes this paper with a brief description of our proposed optimization framework in the presence of remanufacturing under adopter categorization and summary of our findings.

2 Background and Literature Review

With the product end-of-life management gaining a lot of attention in the past few decades and legislation making original equipment manufacturers (OEM) responsible for managing end-of-life product, several academicians and researchers have shown a keen interest in this area. End-of-life are those product returns whose reuse is not feasible, and they can only be of value by their material recycling or possibly reusing their components for products with lower specifications, such as phones make use of expensive silver extracted from old phone's motherboard. Apple launched its high-tech robot Liam which is capable of deconstructing their iPhones and removing important components for reuse.

Quite a significant number of review papers have addressed various characteristics of end-of-life cycle management, like from closed-loop supply chain [9, 11, 18, 32], marketing of remanufactured products [2, 8], reverse logistics [6, 7, 10], economics of remanufacturing [3, 13, 20], etc.

The research area of remanufacturing goods for sustainable and environment-friendly growth has seen unprecedented growth in the last decade or so. Ever-rising consumer expectations and demands for good quality product at reasonable price prompt manufacturers to innovate newer ways of manufacturing products. Remanufacturing provides for enough opportunities to OEMs to strengthen the bottom line by huge cost savings rising out of lesser component procurement toward the end-of-life cycle of the product. However, the biggest challenge for OEMs is to understand the product diffusion dynamics and how it impacts the overall quantity available to them for reuse. In this paper, an attempt to obtain an optimal time to start remanufacturing is studied by making use of adopter categorization proposed by Mahajan and Muller [21].

Optimal time to start remanufacturing is the latest time when OEM should begin remanufacturing so that beyond the time where optimal profit is obtained, the remaining potential in the market can be met by the remanufactured products alone, and there exist no need to procure new components for manufacturing of these products. In other words quantity of remanufactured products available to OEM, if remanufacturing begins at optimal τ, is sufficient to meet the remaining market demand for that product beyond T^* (Fig. 1).

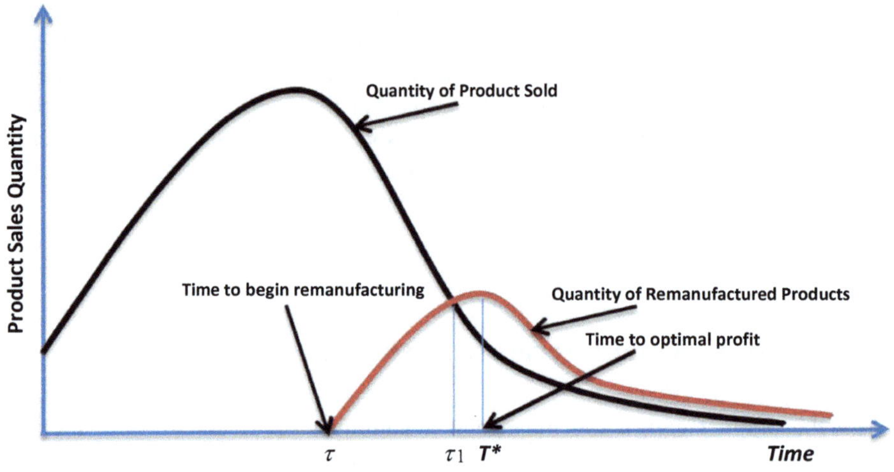

Fig. 1 Diffusion of both new and remanufactured products

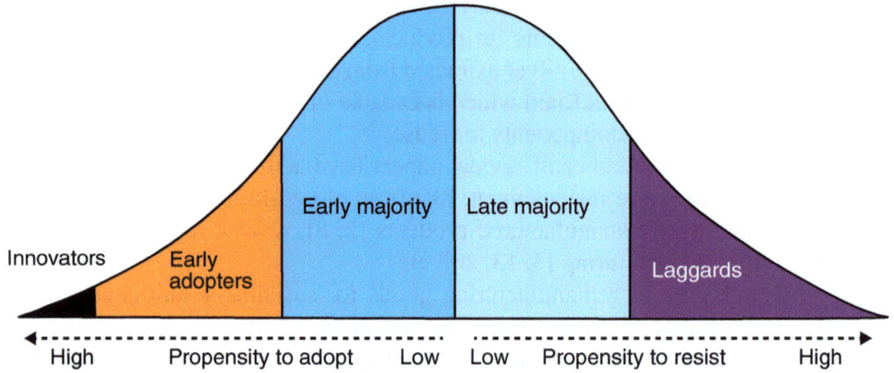

Fig. 2 Roger's adopter categorization

Rogers [30] classic book, the *Diffusion of Innovations*, states that the adoption process for any innovation follows normal curve with respect to time. This means that adoption rate of an innovation is relatively slower during the early stages of adoption until it stabilized itself. Later on, more customers begin to demand the product and subsequently the product growth increases. Once again toward the end of the product life cycle, growth decelerates.

Rogers [30] proposed the categorization of adopters of any innovation as innovators (2.5%), early adopters (13.5%), early majority (34%), late majority (34%), and laggards (16%) (Fig. 2).

Later, Mahajan and Muller (1992) also proposed the adopter categorization using Bass model [4] (Table 1).

Table 1 Adopter categorization based on time

Adopter category	Time interval covered	Expression for the time interval	Expression for the adopter category size
Innovators	Initiation of the diffusion process		P
Early adopters	Up to T1	$\frac{1}{p+q}\ln\left\{2+\sqrt{3}\frac{p}{q}\right\}$	$\frac{1}{2}\left(1-\frac{p}{q}\right)-\frac{1}{\sqrt{12}}\left(1+\frac{p}{q}\right)-p$
Early majority	T1 to T*	$\frac{1}{p+q}\ln\left\{2+\sqrt{3}\right\}$	$\frac{1}{\sqrt{12}}\left(1+\frac{p}{q}\right)$
Late majority	T* to T2	$\frac{1}{p+q}\ln\left\{2+\sqrt{3}\right\}$	$\frac{1}{\sqrt{12}}\left(1+\frac{p}{q}\right)$
Laggards	Beyond T2		$\frac{1}{2}\left(1+\frac{p}{q}\right)-\frac{1}{\sqrt{12}}\left(1+\frac{p}{q}\right)$

In this paper, the focus is to obtain optimal time to start remanufacturing by making use of the analytical logic underlying the classical adopter categorization approach proposed by Rogers. Since varied products are adopted at a different rate in the market with significantly different time of adoption, it is quite certain that manufacturer faces the challenge as to when to start asking consumers for returning the goods, for remanufacturing purpose. Mahajan and Muller (1990) in one of their pioneering work compared the adopter categories generated by the classical approach (Rogers [30]) with Bass [4] diffusion model. Bass model helps in categorizing these adopters as innovators and imitators and suggests that at any given time, both innovators and imitators adopt the product, however, at a different rate because of their difference in getting influenced either through advertising or word of mouth.

Here in this paper, authors have focused on making use of this adopter categorization and time to adopt as the basis to obtain an optimal time to start remanufacturing. Using the five categories of adopters as suggested by Rogers, and their timings of adoption as time to begin remanufacturing, we optimize the overall profit of the manufacturer based on quantity returned and cost saving attached with the entire process of remanufacturing.

3 The Proposed Model

In this study, we consider remanufacturing as a process that involves acquiring back of used products from consumers, disassembling, testing, refurbishing, replacing, reassembling, and finally testing again. We assume that only a fraction of returned products can be used for remanufacturing. As in practice, products have multiple components, out of which some of them can be remanufactured and to varied degrees. In such circumstances, we assume only a single remanufacturing level α simply because most of the products that exist in the market today have one or may be two main components that are most valuable, and they are major drivers of remanufacturing.

The manufacturer's sole goal is to maximize the total profit of introducing the product with or without the use of reusable components, calculated over the life cycle of the product. This can be done by determining, first, the various costs involved in the production, inventory, reverse logistics, refurbishing, and disposal and, second, the diffusion dynamics of the product.

To develop this model, we introduce assumptions concerning diffusion process of the product over its life cycle, quantity reused, reverse logistic cost components along with other cost components of the entire production cycle in Sect. 3.1 and further formulate the optimization problem in Sect. 4.

3.1 Assumptions

The general forecasting of returned goods would involve analyzing the demand for the end-of-life products, the supply chain of existing product, and several cost components involved in the reverse logistics process. To simplify the analysis, we consider only one component to be of value to the manufacturer, which, in future research work can easily be extended to any number of components. The fraction of the total volume of the component under considerations is assumed to pass the quality control test post its reprocessing and is then kept as inventory for newer product reassembling. We do assume that remanufactured products comprise of reusable components from the inventory along with other newer components in order to meet the customer demand of standard level of quality and warranty as desired out of new products with no reusable components.

We also assume that the remaining quantity of components that is not fit for reuse will be disposed of, and the product disposal costs are to be incurred by the manufacturer under extended producer responsibility.

3.2 Product Diffusion and Quantity Returned Dynamics Using Adopter Categorization

In practical scenarios product demand in the market is neither infinite nor stationary, rather in most usual cases it follows a finite life cycle in the market with a dynamic diffusion pattern. The earlier theory of adoption and diffusion of new products has been profoundly presented by Rogers [30] wherein the thrust was upon studying the various stages of product life cycle: introduction, growth, maturity, and decline (Fig. 3). Thus the innovation diffusion models depict the dynamics of new product sales in the market over the finite time horizon.

Bass model [4] is considered to be as a milestone in the innovation diffusion literature. Impact of mass communication and word of mouth on the innovation adoption in any social system was very effectively mapped by this famous model.

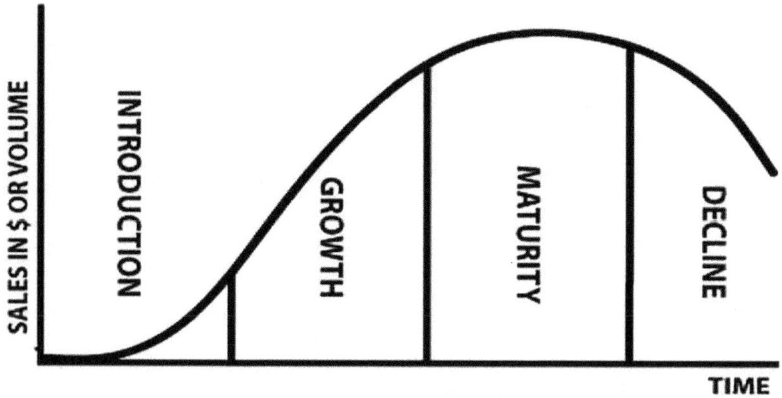

Fig. 3 Roger's product life cycle

Some of the other diffusion models, proposed by Fourt and Woodlock [12, 16, 17], Sachdeva et al. [31], Singh et al. [27], Norton and Bass [25], and Wilson and Norton [34], underlie the Bass model and quite effectively modeled the innovation diffusion process, but Bass model still remains unmatched.

Based on the Bass model, potential adopters are categorized on the basis of how they are affected through the two different channels of communication. One group primarily gets affected by the mass media when making a purchase decision, while the other group known as the imitators gets affected by word-of-mouth communication, when buying an innovation. Mathematically the Bass model is expressed as follows:

$$\frac{dN(t)}{d(t)} = \left[p + q\frac{N(t)}{m}\right].[m - N(t)]$$

Using the initial condition $N(0) = 0$, Bass solved the equation as:

$$N(t) = m\frac{1 - e^{-(p+q)t}}{1 + \left(\frac{q}{p}\right)e^{-(p+q)t}} \tag{1}$$

Since each product attracts fewer buyers during the early phase and with time, this customer adoption rate tends to increase before it finally reaches its pinnacle and begins to lose its attractiveness. Accordingly, it is reasonable to assume that the customer's adoption behavior process follows a hump-shaped curve. Therefore, the first derivative of eq. (1) represents the abovementioned phenomenon significantly well.

$$N'(t) = m\frac{\left(\frac{(p+q)^2}{p}\right)e^{-(p+q)t}}{\left(1+\left(\frac{q}{p}\right)e^{-(p+q)t}\right)^2} \tag{2}$$

Interesting $N'(t)$ reaches its maximum value $S(t^*) = m\left(\frac{p+q}{4q}\right)^2$ at:

$$t^* = \frac{1}{(p+q)}\ln\left(\frac{q}{p}\right) \tag{3}$$

3.3 Reverse Channel: Quantity Returned

The focus of manufacturer in utilizing reused components for the manufacturing of new products is to obtain that subset of components returned that are usable when the product is returned, for example, the inner casing of a used tire can be easily reused in the manufacturing of new tire. One of the important reasons why not all collected products can be used for remanufacturing is the limited durability of its reusable components. We assume that when the product reaches its end of the cycle, the reverse logistics process fed into the supply chain by the manufacturer plays a vital role. During the specific time, τ of the product life cycle, a consumer consumes the product and then thinks of returning it through the reverse channel offered by the manufacturer. During this process, we assume for our modeling framework that the product spends very little to no time as delay in the reverse channel.

With the proposed model, we assume only a fraction of existing adopters at any time period return their product, which is dependent upon the return rate and cost of acquisition of the product. Further, only a part of the returned products would finally result in the quantity available for reuse for manufacturing which is dependent upon the yield rate [22, 23].

In the literature, relationship between the return rate and cost of acquisition is studied either as a linear model [24] or as customer surplus model [1, 29]. We consider that the manufacturer collects a certain percentage r of all the current product adopters $N(t-\tau)$ after they reach their end of useful lives with their current adopters. The idea of $r\%$ adopters returning the product is based on the simple fact that not all current users of the product would be willing to return their product; customers always have the freedom to sell off their goods to a third party involved in collection of used products; or the manufacturer's cost of acquiring from certain regions could be relatively higher than expected.

Post recovery of the used products, it goes through various phases of disassembly, cleaning, refurbishing, and inspection, which results in identifying δ percent of goods that can be used in remanufacturing process. This remanufacturing yield δ is restricted by technical constraint of durability and market constraint of demand.

The remaining part of the collected and inspected product is declared not fit for remanufacturing and is considered as a disposal cost to the company.

Here, we consider that the potential quantity available for remanufacturing products and new product sales together governs the product diffusion dynamics after a time τ in the product life cycle as depicted in Fig. 1. We assume that after the product diffuses in the market at some time point τ, the product would be then made available to the manufacturer for remanufacturing purpose. This time point includes the time duration of the consumer using the product and the time it takes to bring the product back to the manufacturer through the reverse logistics process.

We consider that the diffusion dynamics based on Fig. 1 and Eq. 1 of the components to be remanufactured also follows similar hump shape but with a different level of magnitude and time span depending on the return r and yield δ rates.

The total volume of quantity to be reused can then be written as:

$$Q_{\text{remanuf}} = \alpha.N(t-\tau) \tag{4}$$

where $\alpha = r.\delta$ is the effective return rate finally used in remanufacturing.

It is quite evident from Eq. (4) that the available remanufactured products will follow a similar diffusion pattern as that of new product sales $N(t)$. It is quite interesting to observe that there exists time point τ_1 further to which the remanufactured products would be able to cater to the remaining market demand, and not all the returned end-of-life products need to be reused for remanufacturing as the market demand diminishes as shown in Fig. 1. Therefore, the company can save a lot by not buying any further raw materials and consume all the returned quantity in order to meet the remaining customer demand. We expand on this in the next subsection on cost components.

3.4 Cost Components

The proposed cost structure contains four important but different groups of operations – manufacturing, reverse logistics, remanufacturing, and disposal for formulating our economic model of production system considering the use of both new and reused components for manufacturing. The process of remanufacturing begins only when the products that have reached their end-of-life from customer's perspectives are returned to the manufacturer using the reverse logistics process and are then fed into the main production line one's reprocessed (Fig. 4).

Remanufacturing acts as a closed-loop supply chain and hence can help manufacturers save an enormous amount of capital to be invested in raw materials and also save in terms of large product disposal cost. Further, the manufacturer promotes sustainable development in the economy overall by reusing the components through remanufacturing.

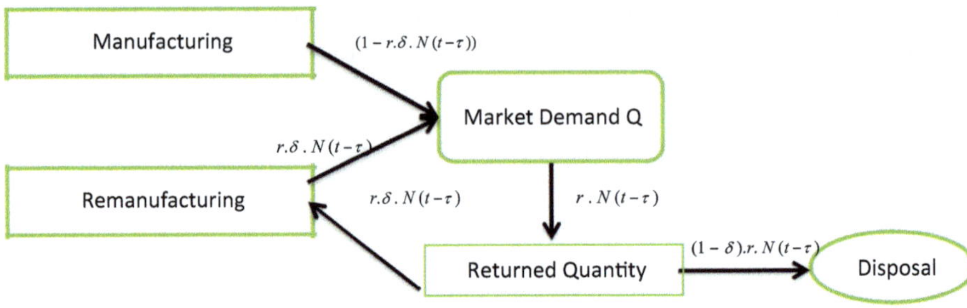

Fig. 4 Product flow during remanufacturing

In order to frame the simplified version of the proposed model, we consider only one component at the end-of-life product that is to be reused in remanufacturing. As highlighted earlier too, most of the products that exist in the market today have one or may be two main components that are most valuable, and they are major drivers of remanufacturing. Our proposed model can be easily generalized to multiple components of the product too. We assume that for this component, only a small proportion of the total quantity returned to the manufacturer would be made available after inspection phase, and the remaining components are for disposal or recycling. We also assume that these reusable components are merged with the other required new components for manufacturing fresh products, and they form part of the inventory in place of existing components, which would have been there in the absence of remanufacturing scenario.

Here we consider formulating the cost structure by focusing on the acquisition, reversal, reprocessing, and disposal phase of the production cycle under a remanufacturing environment. We assume the following four cost components in our proposed structure: unit production cost $P_{Q\text{manuf}}$, unit remanufacturing Cost C_{remanuf}; unit acquisition cost C_{acq}, and unit disposal Cost C_{disposal}. $P_{Q\text{remanuf},N}$ includes cost of manufacturing new products with reusable components along with the inventory carrying cost of existing and reusable components; C_{remanuf} includes reverse logistics cost (cost of transportation mainly), refurbishing cost, and the cost of inspection; C_{acq} includes cost of acquiring the product back from the consumer; and C_{disposal} includes cost of disposing/recycling one component not suitable for remanufacturing.

4 The Optimization Model

In the absence of remanufacturing process, the total production cost of any manufacturer includes simply the $P_{Q\text{manuf}}$ cost of producing Q_{manuf} units and inventory carrying cost $I_{Q\text{manuf},\,N\text{adoption}}$ for $(Q_{\text{manuf}} - N_{\text{adoption}})$units of inventory, where N_{adoption} denotes the number of units adopted in the market and the fixed cost F_{cost}.

In the combined manufacturing and remanufacturing environment, the total cost of producing Q units with a mix of new and reusable components is given as:

$$\begin{aligned}\psi_{\text{prodcost}} &= F_{\text{cost}} + \left[P_{Q_{\text{manuf}}}.Q_{N,t} - P'_{Q_{\text{remanuf}}}.Q_{\text{remanuf}}\right] \\ &\quad + \left[I_{Q,N_{\text{adoption}}}.\left[Q_{N,t} - N(t)\right] - I_{Q,N_{\text{adoption}}}.\alpha.N\left(t-\tau\right)\right] \\ &\quad + C_{\text{remanuf}} + C_{\text{disposal}} + C_{\text{acq}}\end{aligned} \tag{5}$$

where Q denotes total quantity produced by time t, $P'_{Q_{\text{remanuf}}}$ the cost of producing Q_{remanuf} units; $I_{Q,N_{\text{adoption}}}$ represents inventory carrying cost for $(Q - N_{\text{adoption}})$ units of inventory; $I_{Q_{\text{remanuf}},N(t-\tau)}$represents inventory carrying cost for $\alpha N(t-\tau)$ units of inventory used in remanufacturing; C_{remanuf} denotes cost of reverse logistics, refurbishing, and inspection; and C_{disposal} represents cost of disposal or recycling of components unfit for remanufacturing.

Here, $\left[P_{Q_{\text{manuf}}}.Q_{N,t} - P'_{Q_{\text{remanuf}}}.Q_{\text{remanuf}}\right]$ denotes the reduction in production cost based on our assumption that remanufactured products would comprise of reusable components from the inventory. Thereby saving the procurement cost of newer components and reducing the overall production cost by a quantity equivalent to Q_{remanuf}. Additionally, we also assume that these reusable components are merged with the other required new components for manufacturing fresh products, and they form part of the inventory in place of existing components, which would have been there in the absence of remanufacturing scenario. Hence manufacturers realized a decrease in inventory cost which is given by the term $\left[I_{Q,N_{\text{adoption}}}.\left[Q_{N,t} - N(t)\right] - I_{Q,N_{\text{adoption}}}.\alpha.N\left(t-\tau\right)\right]$.

Further, the remaining quantity of components that is not fit for reuse will be disposed of, and the product disposal costs are to be incurred by the manufacturer under extended producer responsibility.

The proposed economic function defined in this paper consists of revenue function $R_{r_s,N_{\text{adoption}}}$, where r_s is the price of the product and the production cost function ψ_{prodcost} as discussed earlier. The proposed profit function ϕ_{profit} is then given by (6). The following two subsections provide detailed discussions of this function:

$$\phi_{\text{profit}} = R_{r_s,N_{\text{adoption}}} - \psi_{\text{prodcost}} \tag{6}$$

4.1 Revenue Function

Remanufacturing is often considered to generate economic and energy benefits to the manufacturer. Most real remanufacturing systems are constrained by both market demand and limited effective yield rate. In order to maximize the total profit by maximizing cost savings due to remanufacturing, we consider the following revenue function at any time t in the product life cycle:

$$R_{r_S,N(t)} = r_S.N(t) \tag{7}$$

where $N(t)$ clearly indicates the demand of the product at time t and r_S denotes the price of the product in the market.

4.2 Production Cost

Let Q denotes the quantity produced by the manufacturer and is dependent on the demand of the product, expressed by a number of units sold, $N(t)$, and on the product's stage in its life cycle denoted by the time t. The relationship between these factors is expressed by the Cobb and Douglas [5] type production function:

$$Q_{N,t} = \theta.N^{\eta}.t^{1-\eta} \tag{8}$$

where θ and η are positive parameters.

Let $F_{\cos t}, P_{Q_{\text{manuf}}},$ *and* $I_{Q_{\text{manuf}},N_{\text{adoption}}}$ denotes the fixed setup cost, per unit production cost and per unit inventory carrying cost, respectively. The total cost comprising of fixed and the variable component at any time t and demand $N(t)$ without remanufacturing is given as:

$$\begin{aligned} \psi_{\text{prodcost}} &= F_{\text{cost}} + P_{Q_{\text{manuf}}} + I_{Q,N_{\text{adoption}}} \\ \psi_{\text{prodcost}} &= \left[F_{\text{cost}} + P_{Q_{\text{manuf}}}.Q_{N,t} + I_{Q,N_{\text{adoption}}}.\left[Q_{N,t} - N(t)\right]\right] \end{aligned} \tag{9}$$

Let $F_{\cos t}, P_{Q_{\text{manuf}}},$ *and* $I_{Q_{\text{manuf}},N_{\text{adoption}}}$ denote the fixed setup cost, per unit production cost and per unit inventory carrying cost, respectively. Additionally, let $P'_{Q_{\text{remanuf}}}, I_{Q,N_{\text{adoption}}}, I_{Q_{\text{remanuf}},N(t-\tau)}, C'_{\text{remanuf}}, C'_{\text{acq}},$ and C'_{disposal} denote the unit cost of producing Q_{remanuf} units, per unit inventory carrying cost for $\left(Q - N_{\text{adoption}}\right)$ units of inventory and inventory carrying cost for $\alpha N(t-\tau))$ units of inventory used in remanufacturing; unit cost of reverse logistics, refurbishing, and inspection; unit cost of acquiring the product back from the consumer; and unit cost of disposal or recycling of components declared unfit for remanufacturing, respectively.

The total cost comprising of fixed and the variable component at any time t and demand $N(t)$ without remanufacturing is then given as:

$$\begin{aligned} \psi'_{\text{prodcost}} &= F_{\cos t} + \left[P_{Q_{\text{manuf}}}.Q_{N,t} - P'_{Q_{\text{remanuf}}}.Q_{\text{remanuf}}\right] \\ &\quad + \left[I_{Q,N_{\text{adoption}}}.\left[Q_{N,t} - N(t)\right] - I_{Q,N_{\text{adoption}}}.\alpha.N\left(t-\tau\right)\right] \\ &\quad + C_{\text{remanuf}}.Q_{\text{remanuf}} + C_{\text{disposal}}.Q_{\text{disposal}} + C_{\text{acq}}.Q_{\text{acq}} \end{aligned} \tag{10}$$

where Q_{acq} is the quantity actually acquired from the consumer out of which a proportion would be finally declared to be fit for remanufacturing purpose and Q_{disposal} is quantity leftover for disposal or recycling from the remanufacturing process. It is important to note there that Q_{disposal} would be given by the product of return rate r% and leftover from the remanufacturing yield δ. Therefore, Q_{disposal} would be equal to $r(1-\delta)$ units denoting the unfit quantity of returned goods left for final disposal.

4.3 Determination of Optimal Time to Begin Remanufacturing

The optimal price r_S* and t* are those which maximize the overall profit function φ'_{profit}, i.e.,

$$\begin{aligned}\phi'_{\text{profit}} &= r_S.N(t)\text{-}\psi'_{\text{prodcost}}\\ &= r_S.N(t) - F_{\cos t} + \left[P_{Q_{\text{manuf}}}.Q_{N,t} - P'_{Q_{\text{remanuf}}}.Q_{\text{remanuf}}\right]\\ &\quad + \left[I_{Q,N_{\text{adoption}}}.\left[Q_{N,t} - N(t)\right] - I_{Q,N_{\text{adoption}}}.\alpha.N(t-\tau)\right]\\ &\quad + C_{\text{remanuf}}.Q_{\text{remanuf}} + C_{\text{disposal}}.Q_{\text{disposal}} + C_{\text{acq}}.Q_{\text{acq}}\end{aligned} \tag{11}$$

whereas, the overall profit earned by the manufacturing in the absence of remanufacturing system is:

$$\begin{aligned}\phi_{\text{profit}} &= r_S.N(t) - \psi_{\text{prodcost}}\\ &= r_S.N(t) - \left[F_{\cos t} + P_{Q_{\text{manuf}}}.Q_{N,t} + I_{Q,N_{\text{adoption}}}.\left[Q_{N,t} - N(t)\right]\right]\end{aligned} \tag{12}$$

In order to optimize φ, we kept a constraint on the minimum number of adopters along with the minimum level of production at any time point. We make use of nonlinear optimization technique with constraints in order to solve this problem. Although several software packages for numerical computation possess good optimization procedures, but we utilized Maple software with NLPSolve optimization function for our proposed problem.

5 Estimation of Parameter, Model Validation Data Set, and Data Analysis

We conduct a numerical analysis to examine the effect of the product diffusion dynamics on quantity reuse in manufacturing in general and then try and find out the optimal quantity to be reused by maximizing manufacturer's profit. In order to understand the diffusion dynamics on quantity reuse in manufacturing, we also run sensitivity analysis on the coefficient of innovation (p), coefficient of imitation (q), and profit maximization time based on ideal time to remanufacture the goods.

Table 2 Parameter estimates of the proposed model

Products	Average price	m (000)	p	Q	A
Electric refrigerators	300	40,001	0.002617	0.21566	0.803903/t
Home freezers	409.5	21,973	0.018119	0.1711	0.827605/t
Black-and-white TV	179.95	96,717	0.027877	0.25105	0.756595/t
Water softeners	35	5793	0.017703	0.29695	0.730042/t
Room air conditioners	395	16,895	0.010399	0.41861	0.651154/t
Clothes dryers	200	15,092	0.017206	0.35688	0.687918/t
Power lawnmowers	120	44,751	0.009184	0.3379	0.706746/t
Electric bed coverings	75	76,589	0.005876	0.24387	0.778999/t
Automatic coffee makers	98	58,838	0.017135	0.30145	0.727177/t
Steam irons	9.99	55,696	0.028632	0.32791	0.700093/t
Recover players	10	21,937	0.024796	0.6541	0.507177/t

5.1 Profit Maximization Model Estimates

In order to validate our model, we tested it on real data sets for 11 consumer durables (Bass [4] and used it for estimation of the model parameters and optimization purpose. The results of the bass innovation diffusion model parameter estimation are given in Table 2.

Now in order to solve the profit models proposed earlier, we have made use of the Maple software and the following numerical data. For the sake of illustration, we assume that the various per unit costs, i.e., fixed cost, production cost, inventory carrying cost, cost of remanufacturing, disposal cost, and acquisition cost, are a percentage of the average prices given in Table 2. Therefore, $F_{\cos t} = 5, P_{Q_{\mathrm{manuf}}} = 2, P'_{Q_{\mathrm{remanuf}}} = 2, I_{Q,N_{\mathrm{adoption}}} = 0.3, C_{\mathrm{remanuf}} = 0.2, C_{\mathrm{disposal}} = 0.05, C_{\mathrm{acq}} = 0.1.$

Further, we also assume that $\eta = 0.603$, $\theta = 4$, and $\delta = 0.9$. Using the results of parameter estimates, the time of adoption and number of adopters based on adopter categorization as discussed in Sect 2, we obtain the time and number of adopters estimates (Table 3).

We now apply the proposed model to the real data set of these 11 consumer durables to obtain optimal profit, time to optimal profit, and number of units adopted and remanufactured by optimal time (Table 4). It is important to note that the three optimal profits, time, and units adopted and remanufactured are based on the three adopter categorization time points, as discussed in Sect 2. Further, the behavior of profit functions for these 11 products is shown in Fig. 5.

We demonstrate the effect of different time to begin remanufacturing on overall units remanufactured and adopted by the optimal time T*. For electric refrigerators, the three time intervals based on adopter categorization of early adopter, early majority adopters, and later majority adopters are observed to be 3.22, 9.25, and 15.28 years, respectively. With each of these as the time to begin remanufacturing,

Table 3 Timing of adoption and number of adopters

			Timing of adoption			Number of adopters (000)		
Products	Peak time	Actual peak time	T1	T2	T3	T1	T2	T3
Electric refrigerators	20.21191	–	3.223433	6.0334	6.0334	8100.638138	11687.4	11687.4
Home freezers	11.866076	13	4.126922	6.96	6.96	2923.220508	7014.77	7014.77
Black-and-white TV	7.8796608	7	2.81423	4.7215	4.7215	12746.96146	31020.05	31020.05
Water softeners	8.9617123	9	2.362881	4.1854	4.1854	981.4358556	1771.991	1771.991
Room air conditioners	8.6134092	7	1.665309	3.0698	3.0698	3290.043381	4998.324	4998.324
Clothes dryers	8.1054662	7	1.962255	3.5205	3.5205	2690.420354	4566.731	4566.731
Power lawnmowers	10.387465	11	2.06409	3.7944	3.7944	8616.391821	13269.61	13269.61
Electric bed coverings	14.918193	14	2.8581	5.2732	5.2732	14859.65588	22642.06	22642.06
Automatic coffee makers	9.000678	10	2.326539	4.1338	4.1338	10087.27477	17950.53	17950.53
Steam irons	6.8385049	7	2.148535	3.6937	3.6937	8394.821242	17481.93	17481.93
Recover players	4.8204407	5	1.068572	1.9399	1.9399	4136.995775	6572.729	6572.729

Table 4 Optimal profit, time, and remanufactured units

Products	Profit earned			Optimal profit time			Units adopted			Units remanufactured		
	T1	T2	T3	T1	T2	T3	T1	T2	T3	T1	T2	T3
Electric refrigerators	2.79×10^7	2.69×10^7	2.58×10^7	37.59	42.61	48.68	39,110	39,699	39,920	817	714	625
Home freezers	1.58×10^7	1.52×10^7	1.46×10^7	31.74	37.41	44.45	21,420	21,781	21,922	542	453	382
Black-and-white TV	1.04×10^8	9.6×10^7	8.85×10^7	19.94	24.64	29.82	93,126	95,725	96,481	3382	2736	2282
Water softeners	1.13×10^6	1.05×10^6	9.78×10^5	19.77	23.81	28.32	5595	5736	5779	199	165	140
Room air conditioners	1.17×10^7	1.14×10^7	1.10×10^7	17.67	20.03	23.03	16,546	16,767	16,859	597	519	451
Clothes dryers	7.43×10^6	7.08×10^6	6.76×10^6	17.75	20.83	24.48	14,674	14,958	15,057	552	466	397
Power lawnmowers	2.93×10^7	2.73×10^7	2.56×10^7	20.45	24.02	28.05	43,397	44,350	44,651	1454	1232	1060
Electric bed coverings	5.66×10^7	5.24×10^7	4.86×10^7	28.76	33.82	39.45	74,204	75,900	76,419	1947	1650	1422
Automatic coffee makers	4.47×10^7	4.09×10^7	3.78×10^7	19.63	23.67	28.15	56,800	58,261	58,699	2026	1677	1421
Steam irons	3.86×10^7	3.48×10^7	3.16×10^7	16.07	19.86	23.94	53,520	55,117	55,560	2230	1810	1517
Recover players	8.71×10^6	7.95×10^6	7.3×10^6	9.78	11.7	13.81	21,178	21,726	21,886	1059	884	755

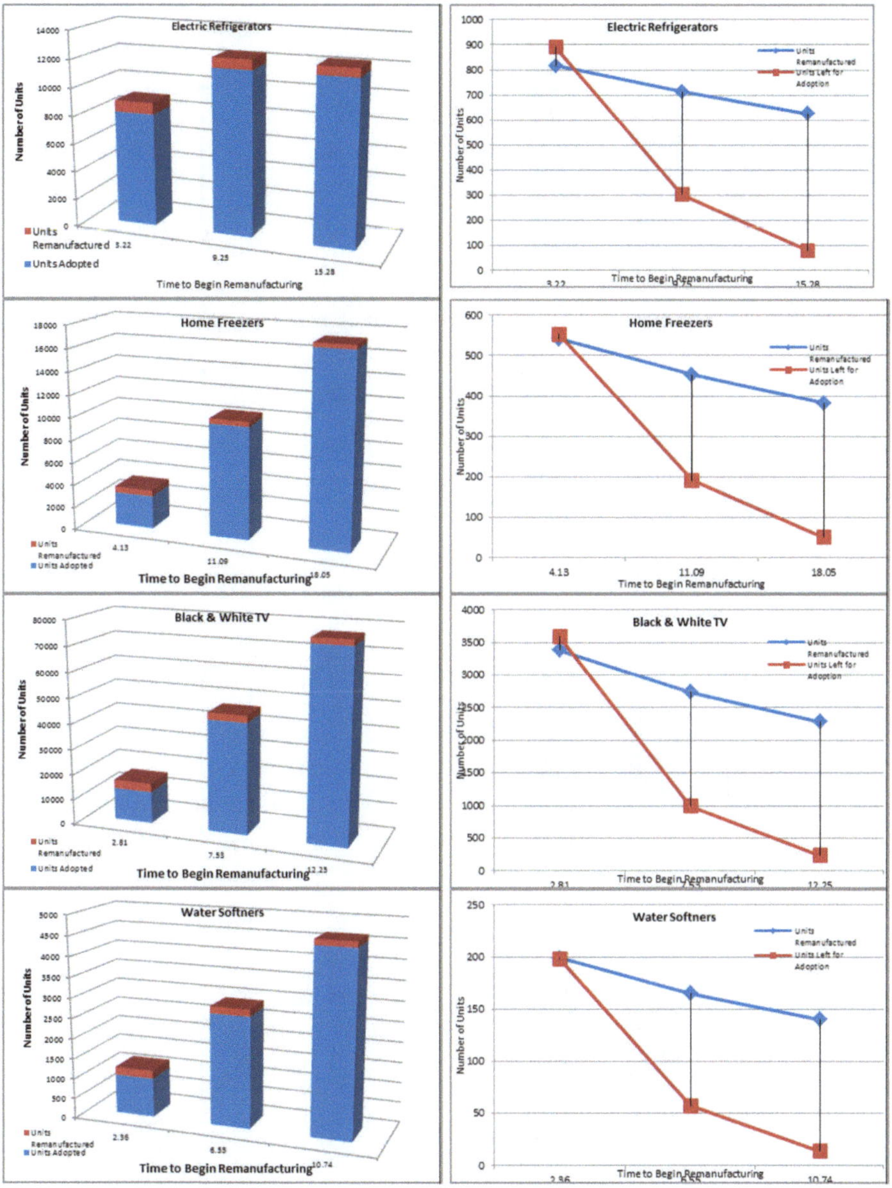

Fig. 5 Optimal time to begin remanufacturing and remanufactured units

we optimize the proposed profit function. We obtain the total units adopted till the optimal T* when remanufacturing begins at 3.22, 9.25, and 15.28 as 39,110, 39,699, and 39,920, respectively, with total quantity remanufactured as 817, 714, and 625, respectively.

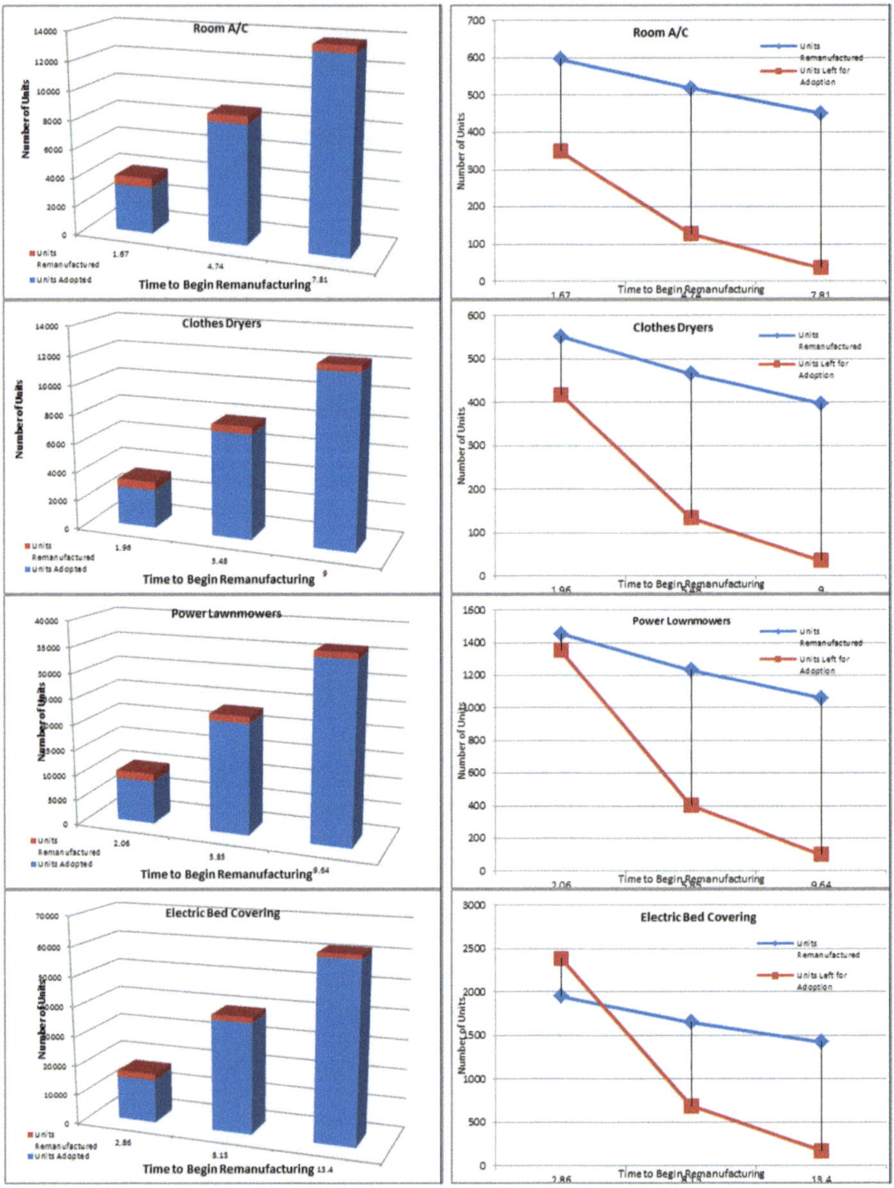

Fig. 5 (continued)

The optimal profit of 6.73×10^6 is obtained at a time when remanufacturing begins after four units of time. At the optimal time of 11.25 units, the total quantity available for reuse is approximately 814 units, and it increases to a level, wherein these reusable products are enough to meet the remaining market demand of the

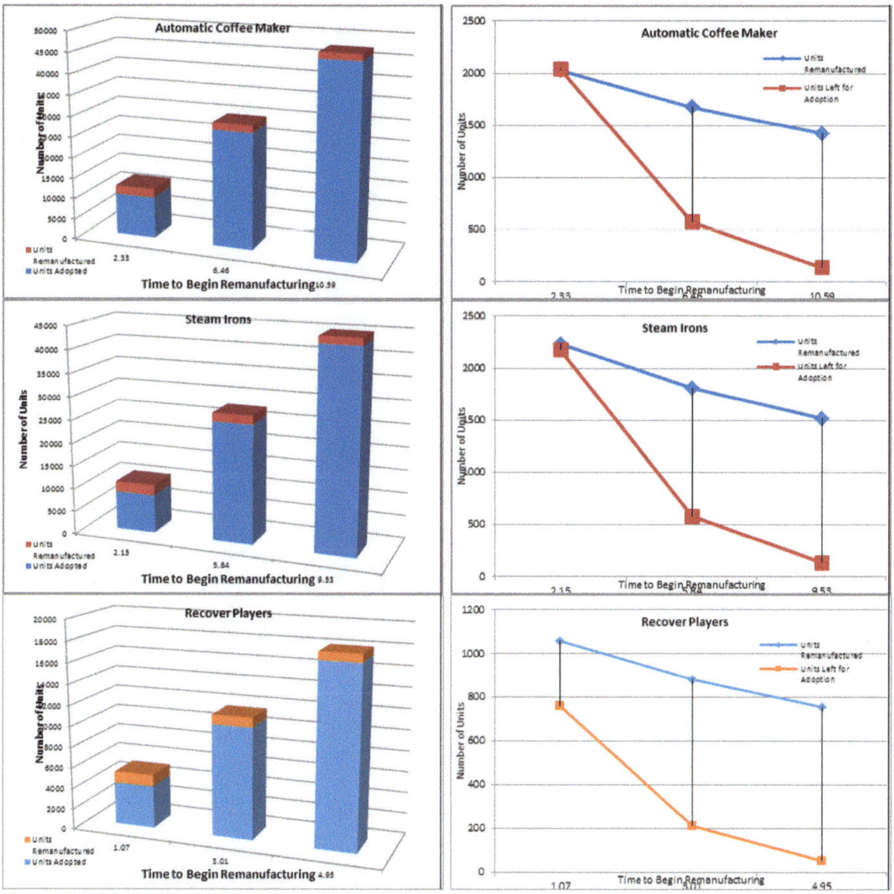

Fig. 5 (continued)

product. In the next subsection, we highlight this aspect of how the change in coefficient of imitation affects this quantity to be reused, in detail.

It is interesting to note here that when we compared the proposed model with a scenario where the manufacturer doesn't consider the option of remanufacturing, Eq. (12), the optimal profit of 2.21×10^5 is obtained at an optimal time of 10.76 units (Fig. 6). The huge profit difference is accounted for the fact that remanufacturing proves to be a really beneficial proposition for the manufacturers. Of course, in reality, getting the products back from the consumers would not be easy; however, if the consumer finds a good opportunity of trading off the used products with attractive deals from the producers, the idea seems to work for both the parties.

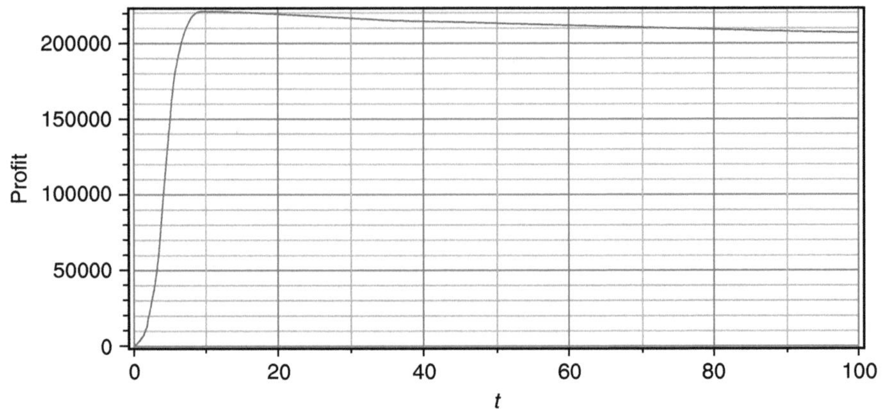

Fig. 6 Optimal profit without remanufacturing

Table 5 Products with smaller "q"

Product	Price	m	p	Q
Electric refrigerators	300	40,001	0.0026167	0.21566
Home freezers	409.5	21,973	0.018119	0.1711
Black-and-white TV	179.95	96,717	0.027877	0.25105
Water softeners	35	5793	0.017703	0.29695
Electric bed coverings	75	76,589	0.005876	0.24387
Automatic coffee makers	98	58,838	0.017135	0.30145

Table 6 Remanufactured units for products with smaller "q"

	Units left to be adopted when Remanufacturing begins at τ_i			Units remanufactured when Remanufacturing begins at τ_i		
Products	τ_1	τ_2	τ_3	τ_1	τ_2	τ_3
Electricrefrigerators	891	302	81	817	714	625
Home freezers	553	192	51	542	453	382
Black-and-white TV	3591	992	236	3382	2736	2282
Water softeners	198	57	14	199	165	140
Electric bed coverings	2385	689	170	1947	1650	1422
Automatic coffee makers	2038	577	139	2026	1677	1421

5.2 *Effect of Coefficient of Imitation on Time to Start Remanufacturing*

Considering the proposed optimization function, we estimate the optimal time to start remanufacturing using adopter categorization as suggested by Mahajan and Muller. The two Tables 5 and 6 demonstrate the effect of coefficients of innovation and imitation on optimal time to begin remanufacturing.

Table 7 Products with larger "*q*"

Product	Price	m	p	q
Room air conditioners	395	16,895	0.010399	0.41861
Clothes dryers	200	15,092	0.017206	0.35688
Power lawnmowers	120	44,751	0.0091837	0.3379
Steam irons	9.99	55,696	0.028632	0.32791
Recover players	10	21,937	0.024796	0.6541

Table 8 Remanufactured units for products with larger "*q*"

	Units left to be adopted when remanufacturing begins at τ_i			Units remanufactured when remanufacturing begins at τ_i		
Products	τ_1	τ_2	τ_3	τ_1	τ_2	τ_3
Room air conditioners	349	128	36	597	519	451
Clothes dryers	418	134	35	552	466	397
Power lawnmowers	1354	401	100	1454	1232	1060
Steam irons	2176	579	136	2230	1810	1517
Recover players	759	211	51	1059	884	755

We observe from Fig. 5 that the number of units remanufactured for products with relatively smaller q (Tables 5 and 6) is significantly different if remanufacturing begins at τ_1 than the products with relatively higher value of q. Moreover, it is more critical to note that for products with relatively smaller q, delaying the beginning of remanufacturing process considerably affects the overall profit (Table 4). In this case, the number of units produced by optimal time to profit is unable to meet the remaining number of potential units to be sold in the market. Hence, delaying the remanufacturing process from τ_1 to τ_2 helps to achieve the target of meeting future market potential at T^* (Table 5).

We also observe the difference in the number of units remanufactured for products with relatively larger q (Tables 7 and 8). Rather, it is critical to note that for such products, an early start to the remanufacturing process boosts the overall profit of the OEM at T* by remanufacturing enough units till T* so that the remaining market potential post optimal time to profit is met.

6 Conclusion, Discussion, and Further Research

We present a fundamentally new concept of profit maximization by considering estimated values of quantity of goods returned for remanufacturing and by using various costs of remanufacturing, refurbishing, acquiring, and disposal. The concept of quantity reused for remanufacturing is the major driver of our optimization modeling framework. We have also shown that the famous Bass model for innovation diffusion can serve as a foundation for reverse goods planning and can surely be utilized to estimate quantity returned from the market for optimizing overall manufacturer's profit. We also studied the effect of diffusion dynamics on quantity returned based on the existing Bass model. We are sure that using our proposed approach, product managers at manufacturing plants can quantify the total quantity of product that can be reused and time to begin remanufacturing and optimize overall profit.

As per the authors, proposed model is quite unique and hence provides for several avenues for future research in areas related to closed-loop supply chain, product diffusion in the presence of remanufacturing, etc. The proposed research can be easily extended to multigenerational products or product lines. Literature quite phenomenally supports the idea of maximizing overall profits in the presence of multigenerational products for the manufacturers, so an optimization model for maximizing total profit in the presence of multigenerational product and remanufacturing could provide some enriching insights into the economic benefits of remanufacturing.

References

1. Aras N, Aksen D (2008) Locating collection centers for distance and incentive dependent returns. Int J Prod Econ 111:316–333
2. Atasu A, Sarvary M, van Wassenhove LN (2008) Remanufacturing as a marketing strategy. Manag Sci 54(10):1731–1746
3. Ayres R, Ferrer G, van Leynsseele T (1997) Eco-efficiency, asset recovery and remanufacturing. Eur Manag J 15(5):557–574
4. Bass F (1969) A new product growth model for consumer durables. Manag Sci 15:215–227
5. Cobb CW, Douglas, P. H. (1928), A theory of production. Am Econ Rev 18:139–165(Supplement)
6. Das K, Chowdhury AH (2012) Designing a reverse logistics network for optimal collection, recovery and quality-based product-mix planning. Int J Prod Econ 135(1):209–221
7. Davis JB (1996) Product stewardship and the coming age of takeback, business and the environment. Cutter Information Corp., Arlington
8. Debo L, Toktay L, van Wassenhove L (2005) Market segmentation and product technology selection for remanufacturable products. Manag Sci 51(8):1193–1205
9. Dekker R, Fleischmann M, Inderfurth K, Van Wassenhove LN (eds) (2004) Reverse logistics quantitative models for closed loop supply chains. Springer, Berlin
10. Dobos (2003) Optimal production-inventory strategies for HMMS-type reverse logistics system. Int J Prod Econ 81:351–360
11. Flapper SDP, Van Nunen JAEE, Van Wassenhove LN (2005) Managing closed-loop supply chains. Springer, Berlin
12. Fourt LA, WoodLock JW (1960) Early prediction of market success for new grocery products. J Mark 25:31–38
13. Geyer R, van Wassenhove LN, Atasu A (2007) The economics of remanufacturing under limited component durability and finite product life cycles. Manag Sci 53(1):88–100
14. Govindan K, Soleimani H, Kannan D (2014) Reverse logistics and closed-loop supply chain: a comprehensive review to explore the future. Eur J Oper Res 240:603–626
15. Halldórsson A, Kotzab H, Skjøtt-Larsen T (2009) Supply chain management on the crossroad to sustainability: a blessing or a curse? Logistics Res 1(2):83–94
16. Kapur PK, Aggarwal AG, Garmabaki AHS, Singh G (2011) Modelling diffusion of successive generations of technology: a general framework. Int J Oper Res Accepted.

17. Kapur PK, Singh O, Chanda U, Basirzadeh M (2010) Determining adoption pattern with pricing using two dimensional innovation diffusion model. J High Technol Manag Res 21(2):136–146
18. Kenné JP, Dejax P, Gharbi A (2012) Production planning of a hybrid manufacturing-remanufacturing system under uncertainty within a closed-loop supply chain. Int J Prod Econ 135(1):81–93
19. Kerr W, Ryan C (2001) Eco-efficiency gains from remanufacturing a case study of photocopier remanufacturing at Fuji Xerox Australia. J Clean Prod 9(1):75–81
20. Lund RT (1983) Remanufacturing: United States experience and implications for developing nations. World Bank, Washington, DC
21. Mahajan V, Muller E, Srivastava RK (1990) Determination of adopter categories by using innovation diffusion models. J Market Res 27(1):37–50
22. Mazhar MI, Kara S, Kaebernick H (2004) Reuse potential of used parts in consumer products: assessment with Weibull analysis, proceedings of the 11th International CIRP life cycle engineering seminar on product life cycle – quality management, June 20–22, 2004, Belgrade, Serbia, pp 211–216
23. Mazhar MI, Kara S, Kaebernick H (2005) Reusability assessment of components in consumer products: a statistical and condition monitoring data analysis strategy. Proceedings of the 4th Australian LCA conference, Sydney, pp 1–8
24. Minner S, Kiesmuller G (2012) Dynamic product acquisition in closed loop supply chains. Int J Prod Res 50(11):2836–2851
25. Norton JA, Bass FM (1987) A diffusion theory model of adoption and substitution for successive generation of high-technology products. Manag Sci 33(9):1069–1086
26. OECD (2001) Extended producer responsibility: a guidance manual for governments. OECD, Paris
27. Ompal S, Kapur PK, Sachdeva N, Bibhu V (2014) Innovation diffusion models incorporating time lag between innovators and imitators adoption. Presented at ICRITO' 2014, IEEE Xplore. doi:331109/ICRITO.2014.7014710
28. Rahman S, Subramanian N (2012) Factors for implementing end-of-life computer recycling operations in reverse supply chains. Int J Prod Econ 140:239–248
29. Ray S, Boyaci T, Aras N (2005) Optimal prices and trade-in rebates for durable, remanufacturable products. Manuf Ser Oper Manag 7:208–288
30. Rogers EM (1962) Diffusion of innovations, 5th edn. The Press, New York
31. Sachdeva N, Kapur PK, Singh O (2016) An innovation diffusion model for consumer durables with three parameters. J Manag Analytics. doi:http://dx.doi.org/10.1080/23270012.2016.1197052
32. Savaskan RC, Bhattacharya S, van Wassenhove LN (2004) Closed-loop supply chain models with product remanufacturing. Manag Sci 50(2):239–252
33. Toffel MW (2003) Closing the loop: product take-back regulations and their strategic implications. Int J Corp Sustain 10(9):161–173
34. Wilson LO, Norton JA (1989) Optimal entry Timing for a product line extension. Marketing. Science 8(1):1–17